AF781208

Control Systems: Design, Applications and Implementation

Control Systems: Design, Applications and Implementation

Edited by
Anabelle Holmes

Willford Press
www.willfordpress.com

Published by Willford Press,
118-35 Queens Blvd., Suite 400,
Forest Hills, NY 11375, USA

Copyright © 2020 Willford Press

This book contains information obtained from authentic and highly regarded sources. Copyright for all individual chapters remain with the respective authors as indicated. All chapters are published with permission under the Creative Commons Attribution License or equivalent. A wide variety of references are listed. Permission and sources are indicated; for detailed attributions, please refer to the permissions page and list of contributors. Reasonable efforts have been made to publish reliable data and information, but the authors, editors and publisher cannot assume any responsibility for the validity of all materials or the consequences of their use.

Trademark Notice: Registered trademark of products or corporate names are used only for explanation and identification without intent to infringe.

ISBN: 978-1-68285-777-9

Cataloging-in-Publication Data

Control systems : design, applications and implementation / edited by Anabelle Holmes.
p. cm.
Includes bibliographical references and index.
ISBN 978-1-68285-777-9
1. Automatic control. 2. Control theory. 3. Automatic control--Design. 4. Digital control systems. I. Holmes, Anabelle.
TJ211 .C66 2020
629.892--dc23

For information on all Willford Press publications
visit our website at www.willfordpress.com

WILLFORD PRESS

Printed in China.

Contents

Preface

A control system directs, regulates, manages and commands the behavior of systems and devices using control loops. Control systems can range in size and functionality from the level of a home heating controller which uses a thermostat to control a domestic boiler, to massive industrial control systems which control entire processes or machines. Feedback control and automation are two important aspects of control systems. Some of the techniques used in the design of advanced control systems include adaptive control, dynamic-optimizing control, learning control and multivariable-noninteracting control. Computer technology is applied to industrial control problem for supervisory for optimizing control, hierarchy control and direct digital control. This book unfolds the innovative aspects in the design of control systems. Also included in this book is a detailed explanation of the various concepts, implementation and applications of control systems. It will help the readers in keeping pace with the rapid changes in control engineering.

Significant researches are present in this book. Intensive efforts have been employed by authors to make this book an outstanding discourse. This book contains the enlightening chapters which have been written on the basis of significant researches done by the experts.

Finally, I would also like to thank all the members involved in this book for being a team and meeting all the deadlines for the submission of their respective works. I would also like to thank my friends and family for being supportive in my efforts.

Editor

Modelling the Embedded Control System using iUML-B Pattern State Machine

Han Peng,[1] Chenglie Du,[1] Lei Rao,[2] and Zhouzhou Liu[1]

[1] *School of Computer Science, Northwestern Polytechnical University, Xi'an, China*
[2] *School of Software and Microelectronics, Northwestern Polytechnical University, Xi'an, China*

Correspondence should be addressed to Han Peng; hansbeng2016@gmail.com

Academic Editor: Carlos-Andrés García

Developing the formal model based on the Event-B design pattern is an excellent method to improve the development efficiency of the embedded control system and improve the reusability of the formal model. However, the instantiation of the Event-B design pattern requires the manual writing of a large number of model codes, which brings a great deal of learning cost and coding burden to the engineering staff. In this paper, we propose a modelling approach for formal development of control systems based on the application of iUML-B state machine patterns to model the four synchronization patterns of the typical control system. Then, we use the instantiation of iUML-B pattern state machine to establish a typical multilevel control system's Event-B model. The simulation results show that the event trace of the model obtained using our method is the same as that of the corresponding model obtained using the traditional Event-B design pattern. Compared with the traditional Event-B design pattern method, our method can greatly reduce the manual coding burden in the modelling process. The system model expressed using the iUML-B pattern state machine can be easily mapped to the labelled transition system so as to verify the behavioural properties of the model.

1. Introduction

The embedded control system has been widely used in aviation, aerospace, Internet of things, and cyber-physical system. Due to the complexity of the embedded control system, it is difficult to ensure its safety properties through test and simulation. Therefore, in a variety of safety standards, such as DO-178C [1] and IEC 61508 [2], it has been clearly stated that engineer must use formal methods to model and verify the embedded system.

Event-B [3] is a formal modelling language based on set theory and first order logic which is supported by a modern tool and built-in provers. There are a lot of successful cases of Event-B application including aircraft landing gear control system [4] and satellite communication system [5].

To improve the reusability of the Event-B model, the researchers proposed the concept of Event-B design patterns [6]. Unlike the software design pattern, the Event-B design pattern is a formal design pattern. The goal of Event-B design patterns is to instantiate some small formal models that have been proven to be correct into actual system models and use these small formal models to construct a larger formal model by model composition. This method not only avoids "reinvent the wheel" but also avoids repetitive refinement and proof.

However, according to our experience, one has to add a lot of guards and actions manually if he (she) wants to model the synchronization control flow of control-intensive system using Event-B. It is a heavy work for the modeller. Engineers need an intuitive, simple way to understand and model synchronous control flow patterns.

The main contribution of this paper is to show how to use iUML-B pattern state machine to model the synchronization patterns of the control-intensive embedded system and apply it to the Event-B model of the embedded control system. We proposed the concept of iUML-B pattern state machine and developed the pattern state machines of four synchronization patterns (strong synchronization, weak synchronization, strong-strong synchronization, and strong-weak synchronization) for the control-intensive embedded system. Then we establish a formal model of a typical embedded control system by instantiating these four pattern state machines. The experimental results show that, using our proposed method, the modeller can establish a model of the control-intensive embedded system easily according to some simple rules.

The remainder of this paper is organized as follows: Section 2 describes the related works of the Event-B design pattern. Section 3 introduces the basic knowledge of Event-B and its design patterns and iUML-B state machines. Section 4 uses the iUML-B pattern state machine to model the synchronization patterns of the embedded control system. In Section 5, we instantiated iUML-B pattern state machines into the Event-B model of the embedded control system. In Section 6, we present an evaluation of our approach. Section 7 summarizes the work of this paper and looks ahead to future work.

2. Related Work

Event-B design pattern has been widely used in many fields. Renato Silva [7, 8] proposed the design pattern and the "Generic Instantiation" approach and uses this technology to develop the Event-B model of a safety critical subway system. In the instantiation process of the design pattern, the "Generic Instantiation" method can use the rename plug-in to instantiate the design pattern, thus avoiding a lot of repetitive developments and proofs.

Abrial [3] proposed four synchronization patterns of embedded control system: strong synchronization, weak synchronization, strong-weak synchronization, and strong-strong synchronization pattern. Its main purpose is to model the "actuator-reactor" patterns of the reactive system. Abrial modelled a mechanical press controller model using these four synchronization patterns. We will introduce these patterns in detail in Section 4. Synchronization pattern is a template for control-intensive embedded systems. It makes modelling of multilayer control systems easy. The modeller needs to instantiate the patterns into the specific model and then composes them together to get a complex multilevel control system model.

Sanz Yeganefard [9] applied the monitored, controlled, mode, and commanded (MCMC) method to the control system and proposed four patterns, namely, the *monitor* pattern, *control* pattern, *mode* pattern, and *command* pattern, expressed by Event-B. Sanaz Yeganefard developed the Event-B model of the cruise control system [10], the automotive lane departure warning system [11], and the lane centering controller [12] using the four above-mentioned patterns. In the work of Sanaz Yeganefard, the composition of patterns is proposed to compose the simple patterns into a composition pattern. But the composition of patterns requires tool support. *MCMC* patterns can be considered as an extended "*actuator-reactor*" pattern because they add status monitoring to the "*actuator-reactor*" pattern to form a feedback loop. This is very useful for the development of the formal model of self-adaptive systems.

In addition to the typical Event-B design patterns described above, Ali Gondal [13] proposed some Event-B refinement patterns and decomposition/composition patterns to model the product line of feature-oriented control systems. Adisak Intana [14] proposed some Event-B refinement and composition patterns to model wireless sensor networks.

To the best of our knowledge, there is no literature on using iUML-B state machine to model the Event-B synchronous control flow patterns.

3. Preliminaries

3.1. Event-B Model. An Event-B model consists of two parts, machine and context. Context describes the static part of the system, including carrier sets, constants, axioms, and theorems. A machine uses variables and events to describe the changes of the system. The process of developing a system model with Event-B usually begins with an abstract model and then continues to refine the model until it approaches the implementation. In the process of machine refinement, new events and variables can be added step by step. The current development tool for Event-B is the mature Rodin platform [15].

In Event-B, an event is made up of the guards and actions parts. An event can usually be expressed as

e ≙ WHEN *guards* THEN *actions* END

When the guards of an event are all satisfied, the event can be triggered and the expression in actions parts describes the change of state variables when an event occurs.

3.2. iUML-B State Machine and Its Event Link Function. Snook and Butler invented a "UML-like" Event-B graphical front end, called UML-B [16]. UML-B uses the class diagrams and state diagrams familiar to the software engineers and system engineers to model the system requirements. System model expressed by UML-B can generate the corresponding Event-B machine directly on the Rodin platform. Recently, UML-B has evolved into iUML-B (integrated UML-B), which allows UML-B class diagrams and state diagrams to be embedded directly into an Event-B machine. iUML-B has been successfully applied to some large projects in Europe commission, such as [17–19].

The graphical symbol of the iUML-B state machine is similar to the state diagram of the UML, and we need not explain it. What we are interested in is a very powerful feature of the iUML-B state machine; that is, the transition edges in iUML-B state machine can be "linked" with the existing events in the Event-B model to control the order of these events. For example, suppose that we have already written four events: *INITIALISATION, a_on, a_off,* and *b_on.* If we want to control the order of these events like "after the *a_on* event has occurred, the *b_on* event must occur before the *a_off* event (*Req 1*)," then we can create an iUML-B state machine and "link" these four events to the transition edges of the state machine, as shown in Figure 1. The italic codes in Figure 1 are the codes automatically generated by the iUML-B state machine to control the event order.

We use the definition of "event trace" proposed by Butler [20] to describe the order of events in the Event-B model: "event trace represents a record of a possible execution trace

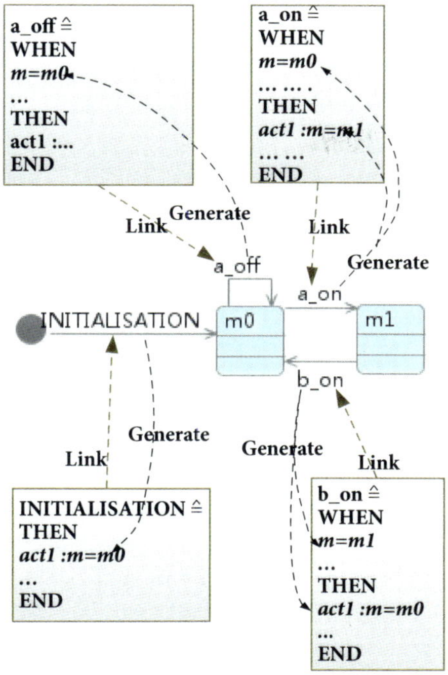

Figure 1: iUML-B state machine of Req 1.

of the model," and we use '$*$' to indicate an event that may occur 0 or more times. An event trace is usually represented by a set of events between "$\langle$" and "$\rangle$." For example, $\langle e1, e2 \rangle *$ indicates that the $e2$ event must be executed after the $e1$ event occurs, and this event sequence will be repeated indefinitely.

3.3. Event-B Design Pattern and Its Instantiation. The idea of the Event-B design pattern is to construct and prove the formal models of the relatively small problems in order to reuse these small formal models to construct the larger model. As with the design pattern in software engineering, the Event-B design pattern is an abstract model of a class of problems. For example, the classic "trigger-response" pattern can be expressed in a nonformalized language: "once the trigger event *a_on* occurs, then the response event *r_on* will be enabled." Its corresponding Event-B design pattern is shown in Figure 2(a) (we named it *P1*). Suppose that when the system is initialized, the values of a and r are $a0$ and $r0$, respectively. Then *r_on* event will be enabled after the *a_on* event occurs because the *a_on* event causes the value of variable a to change from $a0$ to $a1$, which makes all the guards of *r_on* become *TRUE*.

We can express the "trigger-response" relationship in the specific control system by instantiating the pattern P1. For example, in an actual mechanical control system, there is a relationship between a motor controller button and a motor indicator light: "when the motor controller button is pressed, the motor indicator light must be illuminated." This relationship is an instance of the *P1* pattern. To get this instance, we just need to rename the variables, constants, and event names of the pattern *P1* according to renaming rules in Figure 2(b). The resulting Event-B model is shown in Figure 2(c) (we named it *I1*).

In this way, the modeller does not have to prove the correctness of *I1* again if he has proven the correctness of *P1*. In other words, by using the Event-B design pattern, we can reuse not only the design strategy of the model but also the correctness of the model. Therefore, the direct benefit of the Event-B design pattern is that it can greatly reduce proof cost of the formal model.

4. Modelling Synchronous Patterns of the Embedded Control System

In this section, we use the iUML-B pattern state machine to model four typical synchronization patterns in the embedded control system, namely, strong synchronization pattern, weak synchronization pattern, strong-weak synchronization pattern, and strong-strong synchronization pattern. First, we explained in detail four synchronization patterns, which were proposed by Abrial in his book [3]. Then we modelled these four synchronization patterns with the iUML-B state machine.

We verify the correctness of our pattern state machine in three steps. First, we compare the Event-B code generated by our pattern state machine and that of the corresponding pattern proposed by Abrial (http://deploy-eprints.ecs.soton .ac.uk/113/3/ch3_pattern.zip). Then we use their corresponding labelled transitions system model to prove that they are equivalent in behaviour. Finally, we compared the event trace of our model with that of Abrial's Event-B model using the Rodin platform. Simulation results show that the event traces of them are identical.

4.1. Synchronization Requirements of Control System. In the embedded control system, the "trigger-response" problem is the simplest and most basic model. In this model, the actuator executes an "*action*" event and sends an instruction to the reactor. The reactor receives the instruction after the "*action*" event and executes the "*reaction*" event, as shown in Figure 3.

In this paper, we follow the following naming convention. The variables a and b represent the state changes of the actuators a and b, respectively. The variables r and s represent the state changes of the reactors r and s, respectively. The event *X_on* changes the value of the variable X from 0 to 1, where $X \in \{a, b, r, s\}$. The event *X_off* changes the value of the variable X from 1 to 0, where $X \in \{a, b, r, s\}$.

In addition, we refer to an "actuator-reactor" combination as a "subsystem." For instance, actuator a and reactor r compose a subsystem. We use "uninterruptible trace" to indicate an event trace that cannot be inserted in any other events. For example, if we require $\langle e1, e2, e3 \rangle$ to be an uninterruptible trace, then event trace $\langle e1, e2, e4, e3 \rangle$ is an illegal trace that deviated from our requirement.

(1) Strong Synchronization Requirement. A strong synchronization requirement means that once the actuator performs an action, the reactor must respond to it; otherwise, the actuator will not execute any further action. Figure 4 shows this strong synchronization relationship; that is, after *a_on* occurs, *r_on* must be executed; otherwise, the actuator will wait forever. Similarly, after *a_off* occurs, *r_off* must be

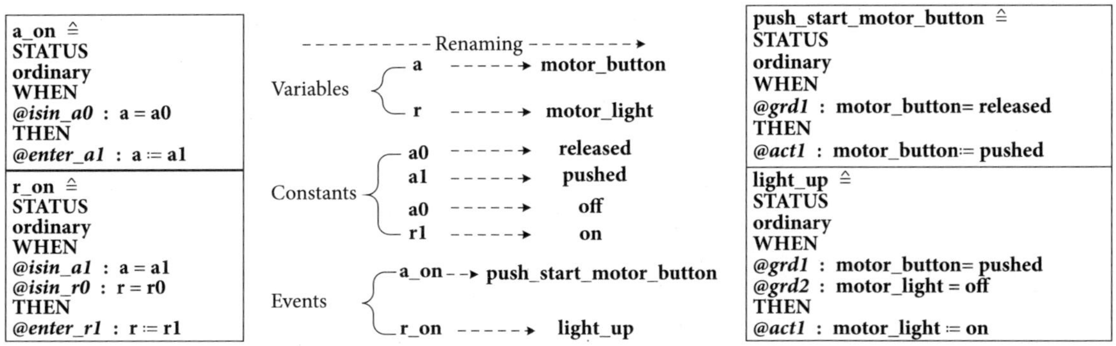

(a) Event-B pattern (b) Renaming rules (c) Instance of the pattern

FIGURE 2: Instantiation process of Event-B design pattern.

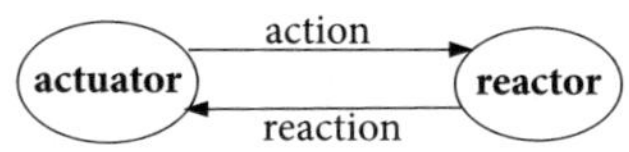

FIGURE 3: Actuator-reactor pattern.

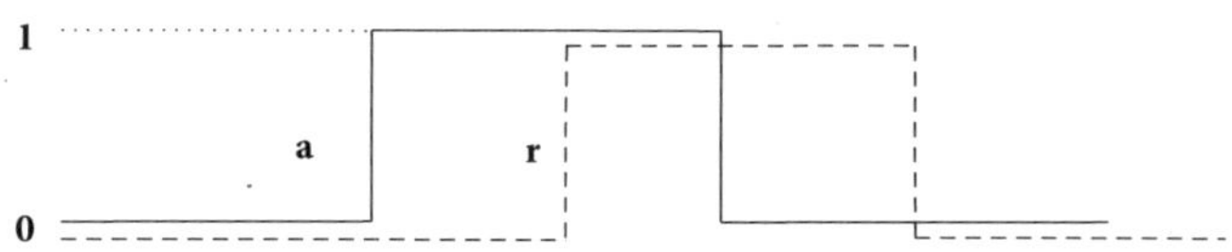

FIGURE 4: Strong synchronization requirement.

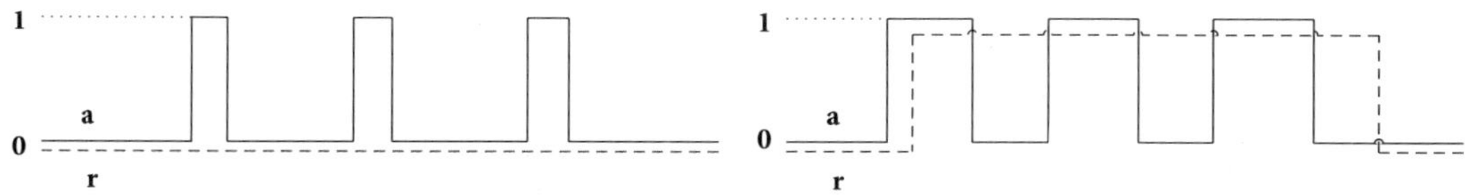

(a) (b)

FIGURE 5: Weak synchronization requirement.

executed. Therefore, the event trace of strong synchronization requirement is $\langle a\_on, r\_on, a\_off, r\_off \rangle *$.

(2) Weak Synchronization Requirement. The weak synchronization requirement means that after the actuator executed an action, the reactor may either respond to it or not respond. Thus, the behaviours of the actuator and the responder under the weak synchronization constraint will be the same as in the case of Figure 5. In Figure 5(a), the actuator performs *a_on* and *a_off* events multiple times and the reactor is always in the state $r = 0$. Figure 5(b) shows the case where the reactor is always in the $r = 1$ state.

(3) Strong-Strong Synchronization Requirement. Strong-strong synchronization requirement is the synchronization requirement between two subsystems. Assume that subsystem 1 comprises an actuator a and a reactor r and subsystem 2 comprises an actuator b and a reactor s.

The strong-strong synchronization requirement indicates that subsystem 1 and subsystem 2 are strongly synchronous; that is, the *b_on* event must occur after the *r_on* event of subsystem 1; otherwise subsystem 1 will wait forever. At the same time, after the *s_off* event has occurred in subsystem 2, subsystem 1 must execute *a_off* events. This is shown in Figure 6 (Abrial did not give the graphical presentation of strong-strong synchronization patterns in his book [3]. We obtain the graphical representation of the strong-strong synchronization shown in Figure 5 based on the Event-B code of strong-strong synchronous pattern given by Abrial and Figure 3.35 on page 147 in Abrial's book [3]). The event trace of strong-strong synchronization requirement is unique, that is, $\langle a\_on, r\_on, b\_on, s\_on, b\_off, s\_off, a\_off, r\_off \rangle *$.

(4) Strong-Weak Synchronization Requirement. Strong-weak synchronization requirement refers to weak synchronization relationship between subsystem 1 and subsystem 2. Specifically, there are two cases. The first case is that after the *r_on* event occurs in subsystem 1, the subsequent event may be either *b_on* event in subsystem 2 or *a_off* event in subsystem 1. The former indicates that subsystem 2 responds to the event of subsystem 1 and enters into an uninterruptible event trace $\langle b\_on, s\_on, b\_off, s\_off \rangle$; the latter

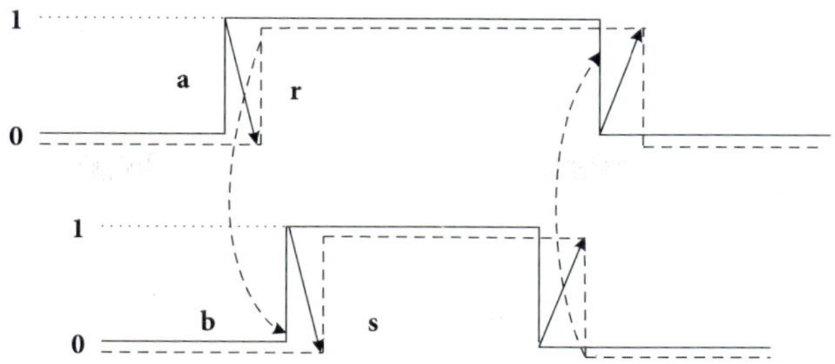

FIGURE 6: Strong-strong synchronization requirement.

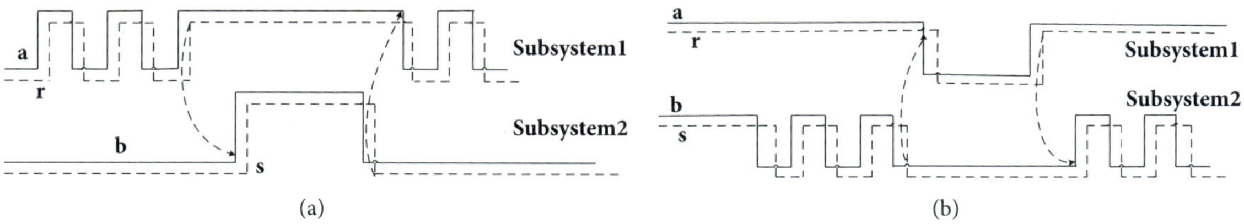

FIGURE 7: Strong-weak synchronization requirement.

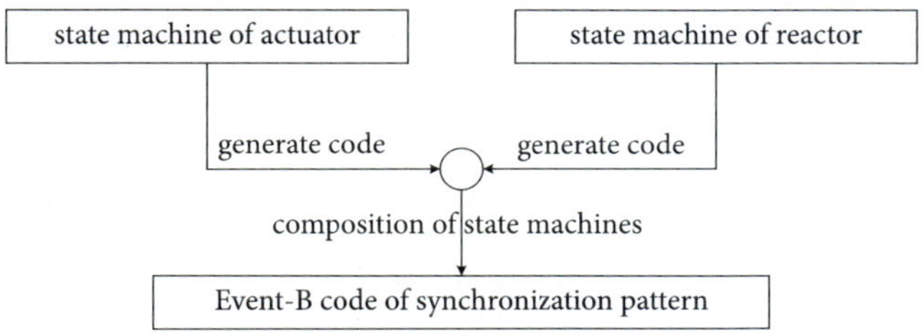

FIGURE 8: The principle of modelling.

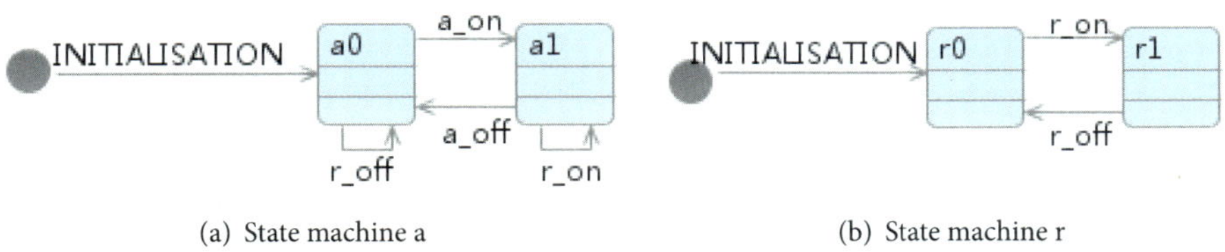

FIGURE 9: The iUML-B state machines of the weak synchronization pattern.

means that subsystem 2 does not respond to the event of subsystem 1, so subsystem 1 enters its uninterruptible event trace $\langle a\_off, r\_off, a\_on, r\_on\rangle$. The principle of strong-weak synchronization is shown in Figure 7(a) (based on the same reason as Section 4.1 (3), we obtain a graphical representation of the strong-weak synchronization pattern according to Figure 3.19 on page 129 in Abrial's book [3]). The second case is that after the *s_off* event in subsystem 2 occurs, subsystem 1 may either execute event *a_off* to respond to it or may not respond, which will cause subsystem 2 to execute the *b_on* event and return to its uninterruptible trace $\langle b\_on, s\_on, b\_off, s\_off\rangle$. This is shown in Figure 7(b).

4.2. Modelling the System Synchronization Pattern with the iUML-B State Machine. In this section, we use iUML-B state machine to model these synchronization patterns. The principle of our approach is to model the actuators and reactors in each design pattern (e.g., the actuators *a* and *b* and the reactors *r* and *s*) as an iUML-B state machine, respectively. Event-B code generated by these state machines is embedded in a single Event-B machine. The resulting Event-B machine is the model for various synchronous control flow patterns. The essence of this procedure is the combination of the state machine of the actuator and the state machine of the reactor, as shown in Figure 8.

(1) Modelling the Weak Synchronization Pattern. According to weak synchronization requirement, *r_on* must occur after *a_on*. Therefore, we first draw the state machines for *a* and *r* themselves; then we add a reflexive edge on its state *a1* and "link" this edge to event *r_on*. Then we can add a reflexive edge on its state *a0* and "link" this edge to event *r_off*, as shown in Figure 9.

The Event-B codes automatically generated according to iUML-B state machines in Figure 9 are

a_on ≙ **STATUS** **ordinary** **WHEN** ***@isin_a0*** **: a = a0** **THEN** ***@enter_a1*** **: a := a1**	**a_off ≙** **STATUS** **ordinary** **WHEN** ***@isin_a1*** **: a = a1** **THEN** ***@enter_a0*** **: a := a0**	**r_on ≙** **STATUS** **ordinary** **WHEN** ***@isin_a1*** **: a = a1** ***@isin_r0*** **: r = r0** **THEN** ***@enter_r1*** **: r := r1**	**r_off ≙** **STATUS** **ordinary** **WHEN** ***@isin_a0*** **: a = a0** ***@isin_r1*** **: r=r1** **THEN** ***@enter_r0*** **: r := r0**

We named it $Weak_{iUMLB}$. To prove that $Weak_{iUMLB}$ satisfies weak synchronization requirement, we give the Event-B code of weak synchronization proposed by Abrial:

a_on ≙ **STATUS** **Ordinary** **WHEN** ***@guard1*****:a = 0** **THEN** ***@action1*****:a := 1** **END**	**a_off ≙** **STATUS** **Ordinary** **WHEN** ***@guard1*****:a = 1** **THEN** ***@action1*****:a := 0** **END**	**r_on ≙** **STATUS** **Ordinary** **WHEN** ***@guard1*****:a=1** ***@guard2*****:r= 0** **THEN** ***@action1*****:r := 1** **END**	**r_off ≙** **STATUS** **Ordinary** **WHEN** ***@guard1*****:a=0** ***@guard2*****:r =1** **THEN** ***@action1*****:r := 0** **END**

We named Abrial's weak synchronization model as $Weak_{EventB}$. It is easy to see that the only difference between $Weak_{iUMLB}$ and $Weak_{EventB}$ is the value of variables a and r. According to the algorithm given in literature [21], we convert both models to labelled transitions system (LTS) and get two identical LTSs, as shown in Figure 10.

If we only concern the behaviour of the model, that is, the event trace, we can conclude that $Weak_{iUMLB}$ and $Weak_{EventB}$ are equivalent.

In order to confirm the above verification results, we simulated the event traces of $Weak_{iUMLB}$ and $Weak_{EventB}$ with the aid of the Rodin platform and found that they are identical. Due to space limitations, in the following discussion, we will no longer give the equivalence proof of the other three synchronization patterns.

(2) Modelling the Strong Synchronization Pattern. According to the requirement of strong synchronization pattern, we modify the state machine of the reactor on the basis of the weak synchronization pattern and add two reflexive edges *a_on* and *a_off* on the $r = r0$ state and $r = r1$ state, respectively, as shown in Figure 11.

(3) Modelling the Strong-Weak Synchronization Pattern. We model the strong-weak synchronization pattern using iUML-B state machines as shown in the four subgraphs (a), (b), (c), and (d) in Figure 12. It should be noted that the strong-weak synchronization pattern is the synchronization between the two subsystems, so we need to add more reflexive edge to limit the event order between the subsystems.

(4) Modelling the Strong-Strong Synchronization Pattern. The strong-strong synchronization pattern adds more constraints on the basis of strong-weak synchronization pattern. We added a new auxiliary state machine m to impose these constraints, as shown in Figure 13. We use dotted arrows to point to the Event-B code generated by each edge in the state machine m. It can be seen that the Event-B codes that generated by state machine m in Figure 13 constrained that event *b_on* must occur between *a_on* and *a_off* event.

As we have analyzed in Section 4.1 (4), in the strong-weak synchronization pattern, after the event sequence $\langle a\_on, r\_on \rangle$ occurs, both of *a_off* event and a *b_on* event are enabled. However, after adding a constraint variable m on the strong-weak synchronization pattern, *a_off* can be enabled only when $a = a1$ and $m = m0$. Therefore, in the Event-B model generated by the strong-strong synchronous pattern state machines, after the event sequence <*a_on, r_on*> occurs, the enabled event can only be *b_on*. Furthermore, as long as *b_on* occurs, the system will go into an uninterruptible event trace $\langle b\_on, s\_on, b\_off, s\_off \rangle$. Thus, the event trace of this Event-B model can only be $\langle a\_on, r\_on, b\_on, s\_on, b\_off, s\_off, a\_off, r\_off \rangle$. That is, the strong-weak synchronization pattern state machine becomes the strong-strong synchronization pattern state machine after adding a new state machine m, as shown in Figure 13.

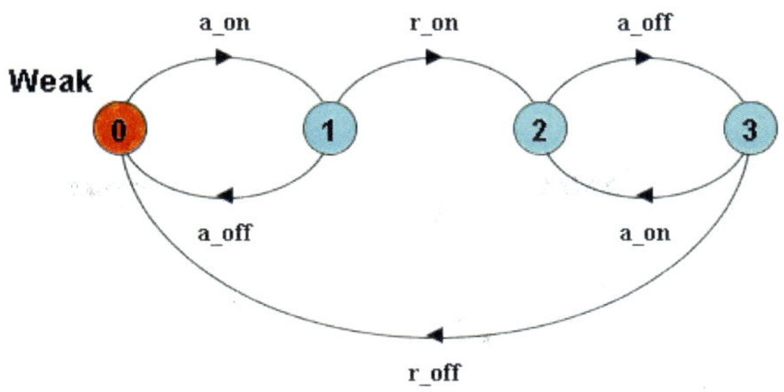

FIGURE 10: LTS model of weak synchronization pattern.

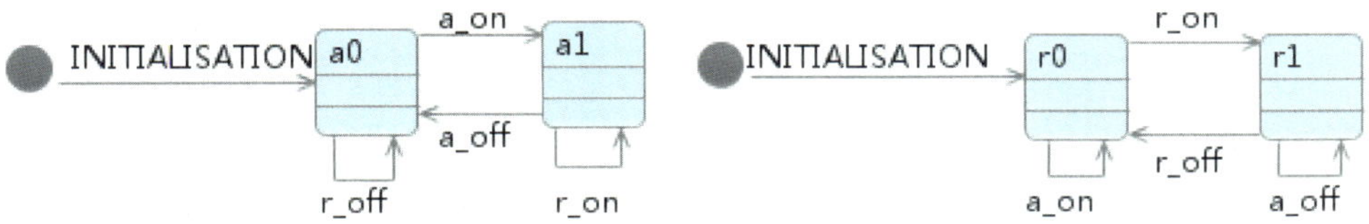

(a) State machine a (b) State machine r

FIGURE 11: The iUML-B state machines of the strong synchronization pattern.

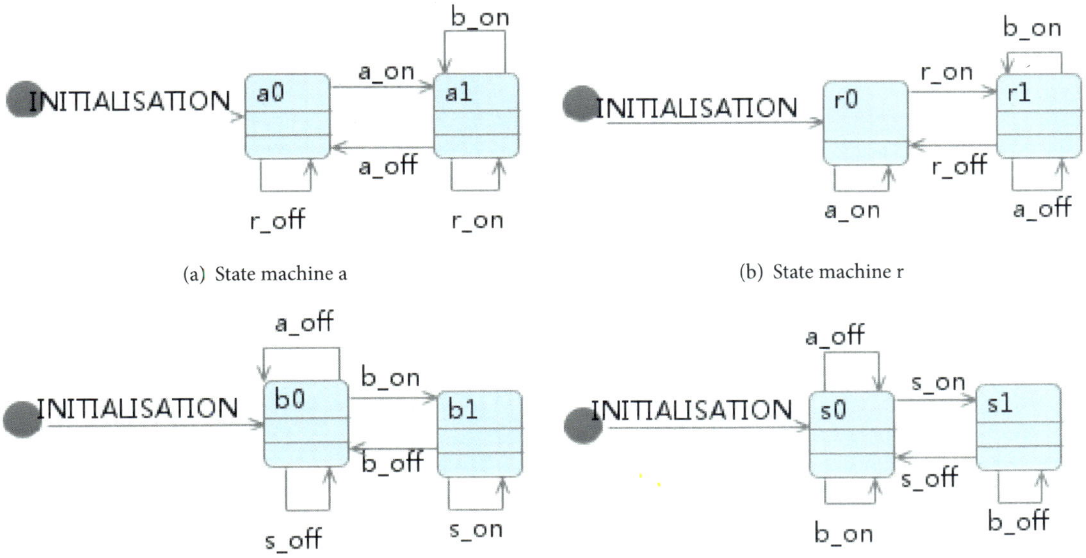

(a) State machine a (b) State machine r

(c) State machine b (d) State machine s

FIGURE 12: The iUML-B state machines of strong-weak synchronization pattern.

5. Modelling the Embedded Control System with Pattern State Machine Instantiation

In this section, we instantiate four synchronous pattern state machines and use these instances to build an Event-B model of an embedded control system. In order to prove the simplicity of our method and to compare it with the Event-B design pattern, we used the case of Abrial's mechanical press controller system in chapter 3 of literature [3]. In the remainder of this paper, we refer to this mechanical press controller system as the "**Press**" system.

5.1. System Overview. The principle of Press system is shown in Figure 14(a). The system consists of control parts and controlled parts. The controlled parts are *Motor*, *Clutch*, and *Door*. The control parts are four control buttons. The control buttons *B1* and *B2* control the start and stop of *motor*, respectively, and *B3* and *B4*, respectively, control the engage and disengage of the *Clutch*. When the *Motor* is working, as long as *Clutch* is engaged, the *Motor* will drive rod; thus slide is driven up and down. Then the tool below the slide will complete the processing of part. The *Door* is to ensure the safety of the staff; that is, when the *Motor* is working and

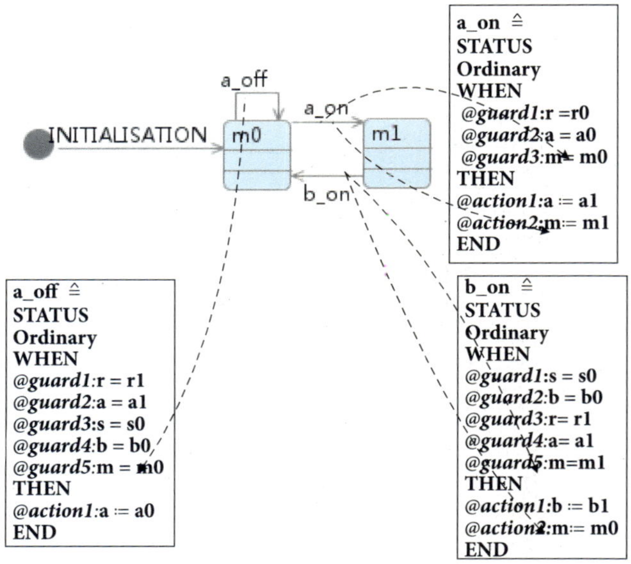

FIGURE 13: The iUML-B state machine *m* and the Event-B codes it generated.

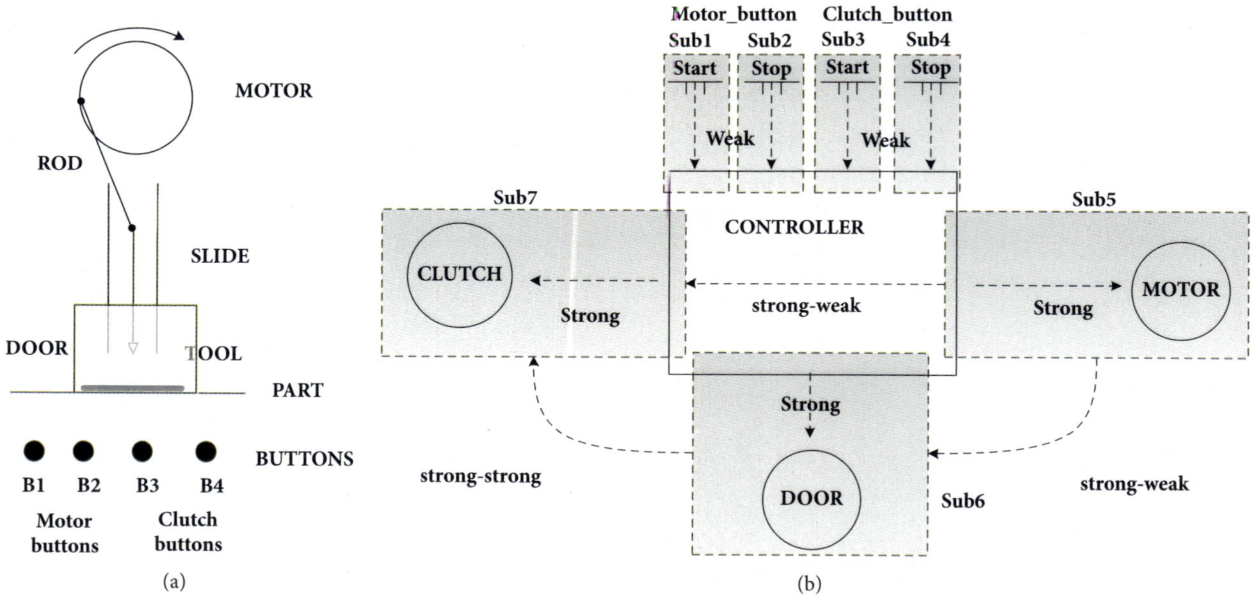

FIGURE 14: Mechanical press controller system.

the *Clutch* is engaged, the *Door* must be closed. Likewise, the *Door* can be opened only after the *Clutch* is disengaged. *Door* object is indirectly controlled by *B3* and *B4*.

The synchronization relationships of Press system are shown in Figure 14(b).The relationships between four buttons and the *controller* are weak synchronization. That is, pressing a control button (e.g., *B1*) does not guarantee that the corresponding controlled object will respond to it (e.g., the *Motor* is started). This is because a button may be pressed several times in an instant. But the controlled object can just respond to one of them. The relationship between the *controller* and the *Clutch* (the *Motor*, the *Door*) is a strong synchronization.

In addition, there are strong-weak synchronization or strong-strong synchronization relationships between subsystems, as shown in Figure 14(b).The strong-strong synchronization relationship between subsystems *Sub5* and *Sub7* means that, first, the *Clutch* can be engaged only when the *Motor* is working; at the same time, the *Motor* can be stopped only after the *Clutch* is disengaged; second, the *Motor* can be stopped and started more than once before the *Clutch* is engaged; similarly, the *Clutch* can be engaged and disengaged multiple times before the *Motor* is stopped. The strong-strong synchronization relationship between subsystems *Sub6* and *Sub7* is a safety requirement to ensure the safety of the operator. The strong-weak synchronization between subsystems

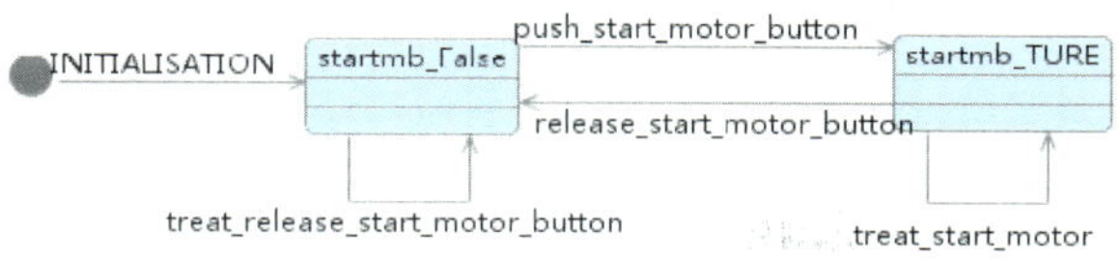

(a) State machine start_motor_button

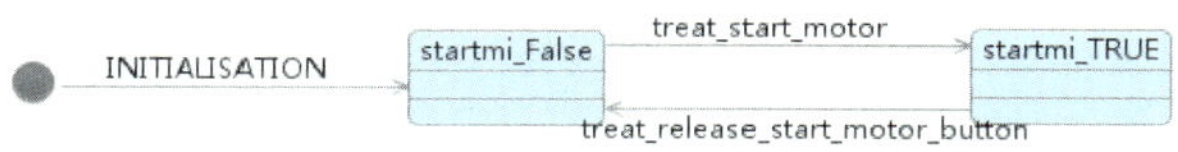

(b) State machine start_motor_impulse

FIGURE 15: Sub1's instance state machine.

Sub5 and *Sub6* is similar to the relationship between *Sub5* and *Sub7*.

5.2. System Modelling Process. Based on the analysis of the synchronization requirements in Section 4.1, we instantiate the four synchronous pattern state machines of iUML-B into the model of the mechanical press controller system. The instantiation of the Event-B design pattern is a renaming process for the variable names, constant names, and the event names of a pattern. We give only examples of instantiation of subsystem 1 (weak synchronization pattern) and subsystem 5-subsystem 7 (strong-weak synchronization pattern); the other subsystems can be instantiated in the same way as these two examples.

(1) Instantiation of the Weak Synchronization Pattern. According to the renaming relation, we get the two instance state machines *start_motor_button* and *start_motor_impulse* of *Sub1*, as shown in Figures 15(a) and 15(b), respectively.

(2) Instantiation of Strong-Weak Synchronization Patterns. Strong-weak synchronization pattern is the relationship between subsystems, which is more complex than the internal synchronization of subsystems because they contain the synchronization relationships inside each subsystem and the synchronization relationships between subsystems. Without the help of design pattern, this development will be a painful process. Here, we can get the iUML-B state machines of *Sub5* and *Sub7* just by instantiating the strong-weak synchronization pattern state machines into the instance state machines. We can get four instance state machines of *Sub5* and *Sub7*, namely, motor*_actuator, motor_sensor, clutch_actuator,* and *clutch_sensor*, as shown in Figure 16.

(3) Composition of Instance State Machines. To facilitate comparison with traditional design methods, we followed the same refinement approach that has been described by Abrial [3]. We get the iUML-B state machines of *Sub1* to *Sub7* through the instantiation of the pattern state machines and modelled the various synchronization relations between them. The final model of the system consists of 15 iUML-B state machines, of which 14 state machines are shown in Table 1. The 15th state machine is an auxiliary state machine named *m_s6_s7*, which is used to model the strong-strong

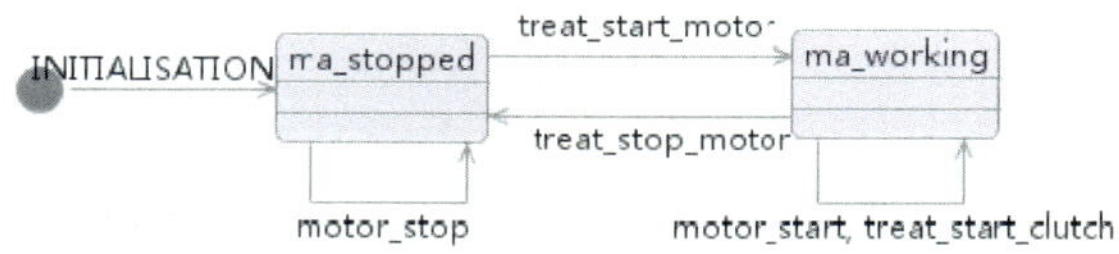

(a) State machine motor_actuator

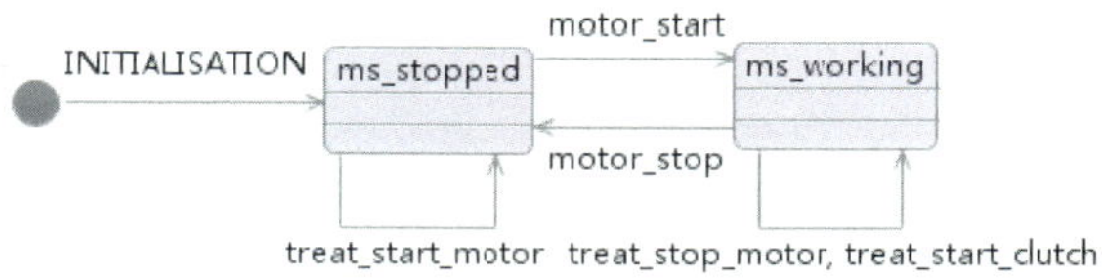

(b) State machine motor_sensor

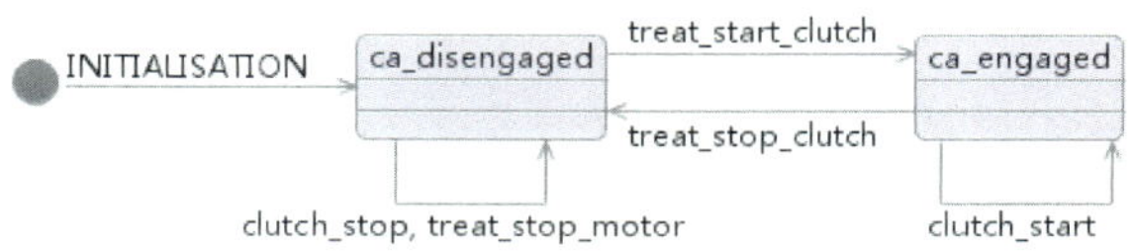

(c) State machine clutch_actuator

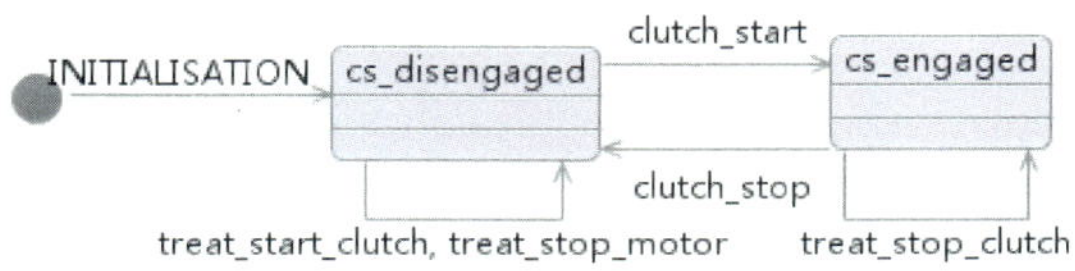

(d) State machine clutch_sensor

FIGURE 16: Instance state machines of Sub5 and Sub7.

synchronization relationship between *Sub6* and *Sub7*. It is an instance of the state machine *m* shown in Figure 13.

As we said at the beginning of Section 4.2, the state machine for each subsystem is actually a composed state machine for two state machines within the subsystem. For example, if we use "⊗" to express the composition of state machines, then we have

$$StateMachine_{sub1} = (start\_motor\_button \otimes start\_motor\_imulse) \quad (1)$$

We can get the state machine for the Press system by the composition of all 15 state machines, which is the composition of all 7 subsystems and *m_s6_s7*:

$$StateMachine_{Press} = \left(\left(\bigotimes_{i=1}^{7} Subi \right) \otimes m\_s6\_s7 \right) \quad (2)$$

The ultimate manifestation of $StateMachine_{Press}$ is the Press system's Event-B machine.

In fact, "composition of iUML-B state machines" here is not "composition of Event-B machines" but an operation similar to the composition of LTSs. In our previous work [22],we have proven that as long as an iUML-B state machine only describes a single variable's change (we refer to this kind of iUML-B state machine as an "atomic state machine"), we can easily convert it to its corresponding LTS model (we call it "atomic LTS"). We have also mapped the composition of iUML-B state machine to that of LTS. That is, if we

TABLE 1: Subsystems of Press system and its corresponding iUML-B state machines.

Subsystem	iUML-B state machine		Synchronization pattern
	Actuator	Reactor	
Sub1 (B1-controller)	start_motor_button	start_motor_impulse	Weak synchronization
Sub2 (B2-controller)	stop_motor_button	stop_motor_impulse	Weak synchronization
Sub3 (B3-controller)	start_clutch_button	start_clutch_impulse	Weak synchronization
Sub4 (B4-controller)	stop_clutch_button	stop_clutch_impulse	Weak synchronization
Sub5 (controller-Motor)	motor_actuator	motor_sensor	Strong synchronization
Sub6 (controller-Door)	door_actuator	door_sensor	Strong synchronization
Sub7 (controller-Clutch)	clutch_actuator	clutch_sensor	Strong synchronization

have multiple atomic state machines S_1, S_2... S_n, and their corresponding atomic LTS: $LTS(S_1)$, $LTS(S_2)$...$LTS(S_n)$, then we have

$$LTS\left(\bigotimes_{i=1}^{n} \mathrm{Si}\right) \sim \left(\|_{i=1}^{n} LTS(Si)\right) \quad (3)$$

where "‖" represents the composition of LTS and "~" represents the bisimulation equivalence relation between two LTSs.

Therefore, according to expression (3), for the iUML-B state machines of this paper, we have

$$\begin{aligned} & LTS\left(StateMachine_{sub1}\right) \\ & \quad \sim \left(LTS\left(start\_motor\_button\right) \parallel \right. \\ & \quad \left. LTS\left(start\_motor\_imulse\right)\right) \end{aligned} \quad (4)$$

and

$$\begin{aligned} & \mathrm{LTS}\left(StateMachine_{Press}\right) \\ & \quad \sim \left(\|_{i=1}^{7} \mathrm{LTS}\left(Sub_i\right)\right) \parallel \mathrm{LTS}\left(m\_s6\_s7\right) \end{aligned} \quad (5)$$

In this way, we can get the LTS model of an Event-B model, which is obtained by integrating of many iUML-B state machines. Further, it allows us to use a variety of model checking tools to verify the behaviour properties of the Event-B model.

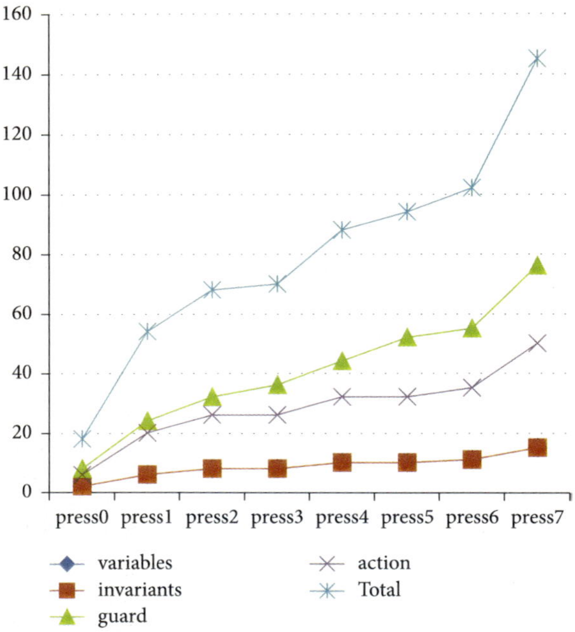

FIGURE 17: Statistics of Press model generated by iUML-B state machine.

6. Evaluation

In this section, we compared the development method based on iUML-B pattern state machine with the development method based on traditional Event-B design pattern. In general, using iUML-B state machines to model the synchronization control flow patterns has three benefits: (1) iUML-B is as easy as UML to learn and can reduce the manual coding costs; (2) iUML-B state machine can express the control flow explicitly; (3) iUML-B state machine can be easily converted to labelled transition system (LTS).

We used the iUML-B pattern state machine to establish the seven-level refinement model of the mechanical press controller system step by step, according to the refinement step that has been described by Abrial [3]. The simulation results on the Rodin platform show that, at each layer, the event trace of the model obtained using our method is the same as that of the corresponding model obtained using Abrial's traditional design pattern. The growth trend of these elements is shown in Figure 17.

We use Abrial's nongraphical model of the Press system (http://deploy-eprints.ecs.soton.ac.uk/113/2/ch3_press.zip), which is a direct hand-coded Event-B model, for comparison with our work. For comparison, we also present the statistical data for the mechanical press controller model developed by the traditional Event-B design pattern. The growth trend of its elements is shown in Figure 18 (the growth trend of variables and invariants in the figure is the same, so the two curves coincide).

From the statistical data, we can observe a phenomenon: no matter what method is used to model a control-intensive system, the number of model codes grows fast with the growth of refinement level. For the traditional Event-B design pattern method, the total number of variables, invariants, guards, and actions is 142 when refinement level is 7. For a method based on the iUML-B pattern state machine, this

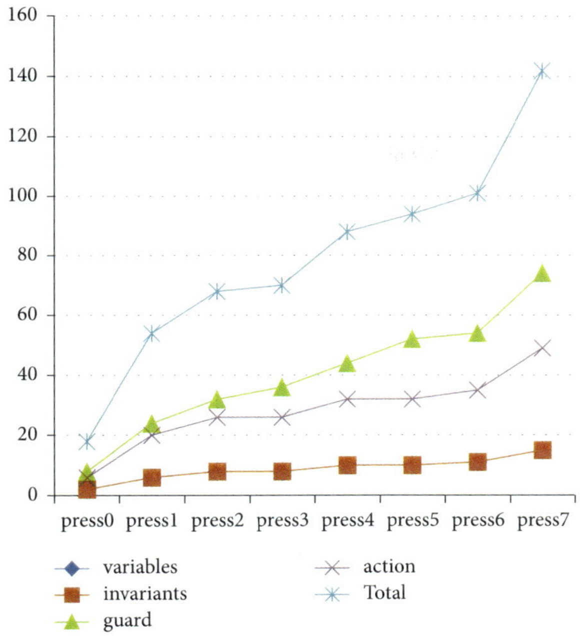

FIGURE 18: Statistics of Abrial's Press model.

number is 145. From the graphs in Figures 17 and 18, it can be seen that the growth rate of guards is the fastest, followed by actions. For more complex multilevel control system, the slope of these two curves will be greater.

Although the statistics in Figures 17 and 18 are similar, the guards and actions in Figure 17 are automatically generated from the iUML-B state machine and the guards and actions in Figure 18 must be manually coded by the modeller. In other words, using the iUML-B pattern state machine to model control-intensive systems, the amount of manual coding we can save at least is equal to (74 + 49)/142 = 86.6%.

Another benefit of our approach is that the control flow of the Event-B model becomes visible. Using the iUML-B state machine, we can express and analyze the event order of the system model easily. And the larger Event-B model is decomposed into smaller subsystems, each with its separated control flow. In this way, the complex multilevel control problem becomes a simple subsystem modelling problem. The final Event-B model is a composition of these subsystems. From the engineering point of view, at the time of decomposing the system, we also decompose the complexity of modelling problems and distribute this complexity to some smaller state machines.

Finally, the iUML-B state machine can be easily converted to LTS, so that the behaviour properties of the Event-B model can be analyzed and verified. From the point of view of behavioural semantic verification, this is a very desirable advantage.

7. Conclusion and Future Work

In this paper, we use the iUML-B pattern state machine to model the four synchronization design patterns in the embedded control system. Then we instantiate the four iUML-B pattern state machines to get the Event-B model of a complex, multilayer control system. The simulation results show that the event trace of the system model obtained using the iUML-B pattern state machine is the same as that of the model obtained using the traditional Event-B design pattern.

The advantage of our approach lies in the following points. First, the process of modelling synchronous control flow patterns is visible. The system's control flow synchronization patterns are expressed using the iUML-B state machines. This makes the system model be understood more easily. Second, the modelling process of the system becomes simpler. Using the traditional approach based on the Event-B design pattern, the modeller who wants to add a new control flow needs to find the locations in hundreds or even thousands of lines of code to insert the guards and actions. However, using the iUML-B pattern state machine, people only need to add some transition edges in a state machine to complete these tasks. Finally, the decomposition and composition of the subsystems are visible. The system's control flow model is decomposed into some iUML-B state machines, and each pair of iUML-B state machines forms a subsystem. The final system model is the composition of these iUML-B state machines.

Event-B is a data-oriented formal modelling language. Therefore, it cannot guarantee the consistency of behaviour between the refined model and the abstract model. In the future, we want to propose an integrated formal method that can guarantee the consistency of the refined model and abstract model in both behaviour aspect and data aspect.

Disclosure

Han Peng is a teacher at School of Computer Science, Xi'an Aeronautical University. Now he is studying for a doctoral degree at the School of Computer Science, Northwestern Polytechnical University.

Conflicts of Interest

The authors declare that there are no conflicts of interest regarding the publication of this paper

Acknowledgments

The authors are very grateful to Professor Colin Snook and Dr. Thai Son Hoang of the University of Southampton for their valuable constructive suggestions in this research work.

References

[1] G. Gigante and D. Pascarella, *Formal Methods in Avionic Software Certification: The DO-178C Perspective*, Springer, Berlin, Germany, 2012.

[2] F. De Rosa, R. Cesoni, S. Genta, and P. Maggiore, "Failure rate evaluation method for HW architecture derived from functional safety standards (ISO 19014, ISO 25119, IEC 61508)," *Reliability Engineering & System Safety*, vol. 165, pp. 124–133, 2017.

[3] J. R. Abrial, *Modeling in Event-B: System and Software Engineering*, Cambridge University Press, 2010.

[4] D. Méry and N. K. Singh, "Modeling an Aircraft Landing System in Event-B," *Communications in Computer and Information Science*, vol. 433, pp. 154–159, 2014.

[5] A. Iliasov, E. Troubitsyna, L. Laibinis et al., "Developing mode-rich satellite software by refinement in Event-B," *Science of Computer Programming*, vol. 78, no. 7, pp. 884–905, 2013.

[6] T. S. Hoang, A. Fürst, and J.-R. Abrial, "Event-B patterns and their tool support," *Software and Systems Modeling*, vol. 12, no. 2, pp. 229–244, 2013.

[7] R. Silva, "Application of Decomposition and Generic Instantiation," *Environmental Modelling Software*, vol. 47, pp. 138–147, 2011.

[8] R. Silva and M. Butler, "Supporting reuse of Event-B developments through generic instantiation," in *Proceedings of the International Conference on Formal Engineering Methods: Formal Methods and Software Engineering*, pp. 466–484, 2009.

[9] S. Yeganefard, M. Butler, and A. Rezazadeh, "Evaluation of a guideline by formal modelling of cruise control system in Event-B," in *Proceedings of the 2nd NASA Formal Methods Symposium*, pp. 182–191, 2010.

[10] S. Yeganefard and M. Butler, "Problem decomposition and sub-model reconciliation of control systems in Event-B," in *Proceedings of the IEEE 14th International Conference on Information Reuse and Integration*, pp. 528–535, August 2013.

[11] S. Yeganefard and M. Butler, "Structuring functional requirements of control systems to facilitate refinement-based formalisation," *Electronic Communications of the EASST*, vol. 46, 2011.

[12] S. Yeganefard and M. Butler, "Control systems: phenomena and structuring functional requirement documents," in *Proceedings of the IEEE 17th International Conference on Engineering of Complex Computer Systems*, pp. 39–48, July 2012.

[13] A. Gondal, M. Poppleton, and M. Butler, "Composing Event-B specifications—case-study experience," in *Software Composition*, vol. 6708 of *Lecture Notes in Computer Science*, pp. 100–115, Springer, Berlin, Germany, 2011.

[14] A. Intana, *Formal engineering methodologies for wireless sensor network development with simulation*, University of Southampton, 2015.

[15] J.-R. Abrial, M. Butler, S. Hallerstede, T. S. Hoang, F. Mehta, and L. Voisin, "Rodin: An open toolset for modelling and reasoning in Event-B," *International Journal on Software Tools for Technology Transfer*, vol. 12, no. 6, pp. 447–466, 2010.

[16] C. Snook and M. Butler, "UML-B: Formal modeling and design aided by UML," *ACM Transactions on Software Engineering and Methodology (TOSEM)*, vol. 15, no. 1, pp. 92–122, 2006.

[17] A. S. Fathabadi, C. Snook, and M. Butler, *Applying an integrated modelling process to run-time management of many-core systems*, 2014, International Conference on Integrated Formal Methods, Springer, Cham.

[18] T. S. Hoang, C. Snook, L. Ladenberger, and M. Butler, "Validating the requirements and design of a hemodialysis machine using iUML-B, BMotion studio, and co-simulation," in *Proceedings of the International Conference on Abstract State Machines, Alloy, B, TLA, VDM, and Z*, pp. 360–375, 2016.

[19] C. Snook, T. S. Hoang, and M. Butler, "Analysing security protocols using refinement in iUML-B," in *Proceedings of the NASA Formal Methods Symposium*, pp. 84–98, 2017.

[20] M. Butler, *Decomposition Structures for Event-B*, Springer, Berlin, Germany, 2009.

[21] D. L. Chaudhari and O. P. Damani, "Generating hierarchical state based representation from event-B models," *Electronic Notes in Theoretical Computer Science*, vol. 280, no. 1, pp. 35–46, 2011.

[22] H. Peng, C. Du, L. Rao, and F. Chen, "A LTS approach to control in event-B," *Scientific Programming*, vol. 2018, Article ID 8765186, 11 pages, 2018.

2

Dynamic Output Feedback Control of Discrete Markov Jump Systems based on Event-Triggered Mechanism

Zhen Zhao, Jinfeng Gao, and Tingting Zhang

Faculty of Mechanical Engineering and Automation, Zhejiang Sci-Tech University, Hangzhou, Zhejiang 310018, China

Correspondence should be addressed to Jinfeng Gao; gaojf163@163.com

Academic Editor: Cui-Qin Ma

This paper is devoted to the co-design strategy of event-triggered scheme and dynamic output feedback controller for a class of discrete-time networked control systems (NCSs) with random time delay. An event-triggered mechanism is given to ease the information transmission. Both the sensor and controller are set with mode-dependent quantizers in the system. A Markov process is used to model the time delay which is used to describe the quantization density. By employing the Lyapunov-Krasovskii functional and linear matrix inequality (LMI), sufficient conditions are obtained for the system. A specific example is given to demonstrate the proposed approach.

1. Introduction

In recent years, a boom of computer science and technology has led to the increasing role of network control in industry. Nevertheless, main problems in the network such as time delay, disorder, packet loss, and the occupation of network bandwidth cannot be ignored.

The issue of time delay is usually treated as the major cause of deterioration of system performance which has been dealt with by quite a few researchers, and they proposed a variety of approaches to these problems; see [1–5]. In [6], the authors propose sequential measurement fusion and state fusion estimation methods to deal with delayed data for clustered sensor networks, since they can handle the data that are available sequentially. In order to save network resources, some literatures have proposed time-triggered mechanism (such as [7–9]); this scheme transfers each sample data which reduces the utilization of network bandwidth. The event-triggering mechanism proposed in the 1990s is more advantageous than the time-triggered scheme (see [10–13]), which means that the data will not be transmitted unless condition set in advance is satisfied, and then it will reduce the time of accessing network and save the network resources. In [14], the event-triggered scheme is applied to the fault detection to guarantee the fault detection accuracy. Furthermore, an event-triggered scheme with an adaptive threshold would subserve the quality of the control; see [15, 16].

Quantization also has become a research hotspot these years, paper [17] notes that quantization process plays an important role in the networked control systems, an adaptive quantizer can help to alleviate network congestion and maintain system stability, paper [18] proposes a quantized state feedback strategy for global and asymptotical stabilization of the networked control systems, papers [19, 20] address the problem of output feedback control for networked control systems with limited information, it should be noted that the impact of quantification cannot be ignored, and, in [18, 19, 21], the quantization deviation is disposed as an uncertainty. Papers [20, 22, 23] design dynamic quantizer to reduce the impact of the quantization error.

Markov jump system is a random system with multimodes which has attracted much attention in the past few decades (see [24–26]); the jump transition of the system in each mode is determined by a Markov chain so it can be applied to the multimodes in many kinds of systems. In [4, 27, 28], the Markov process is applied to modeling the time delay, in [10, 20], the time delay is used to describe the quantization density in the form of function and the authors investigate the dynamic quantization output feedback controller; in [10], the author also takes into account an event-triggered scheme to

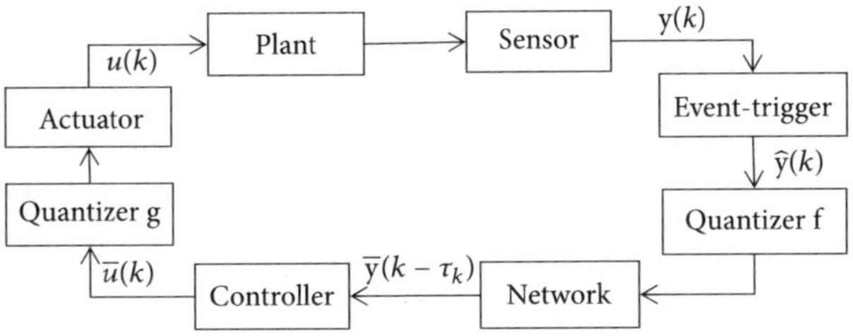

FIGURE 1: The structure of NCS with event-trigger and quantizers.

the aforementioned systems. Nevertheless, no article has considered a system with a dynamic output feedback controller and two quantizers based on event-triggered scheme.

In the view of the aforementioned analysis, in this paper, we aim to investigate the co-design strategy of the event-triggered scheme and dynamic output feedback controller. An adaptive event-triggered mechanism is presented to reduce the information transmission. A Markov chain is applied to modeling the time delay which is used to describe quantization density as a function to ease bandwidth usage. Methods utilizing Lyapunov-Krasovskii functional and linear matrix inequality (LMI) are presented to develop the sufficient conditions of stochastic stability. An event-triggered scheme based dynamic output feedback controller is adopted to guarantee the performance of the closed-loop system. Finally, an example demonstrates the proposed method in detail.

The remainder of this paper is organized as follows. The system description and formulation are provided in **Section 2**. The sufficient condition of stochastic stability and the co-design strategy are presented in **Section 3**. In **Section 4**, a simulation example is presented to demonstrate the feasibility of the proposed approach. The conclusion is given in **Section 5**.

2. Problem Formulation

The plant of NCS is considered as a discrete-time model

$$\begin{aligned} x(k+1) &= Ax(k) + Bu(k) \\ y(k) &= Cx(k) \end{aligned} \tag{1}$$

where $x(k) \in R^n$ is the system state, $u(k) \in R^m$ is the control input, and $y(k) \in R^n$ is the output, A, B, and C are constant matrices of appropriate dimensions, and the structure of our concerned NCS is shown as in Figure 1.

Throughout the paper, some assumptions as follows are needed for the considered NCS.

Assumption 1. Both the controller and actuator are event-driven, while the sensor is time-driven with a sampling period h.

Assumption 2. The data communication in the network is assumed to be single packet transmission without packet dropout.

Assumption 3. The network-induced delay is smaller than one sampling period and bounded by $0 \le \underline{\tau} \le \tau_k \le \overline{\tau}$.

The event-triggered scheme is shown as follows:

$$e^{\mathrm{T}}(k)\, e(k) \ge \sigma(k)\, y^{\mathrm{T}}(k)\, y(k) \tag{2}$$

where k is the sampling instant and $k = 0, 1, 2, \cdots$. $e(k)$ is the quantization error and defined as

$$e(k) = y(k) - \hat{y}(k-1) \tag{3}$$

where $\hat{y}(k-1)$ is the latest transmitted signal and $y(k)$ is the signal which is to be transmitted. Based on [10], $\sigma(k)$ is the adaptive threshold and defined as

$$\sigma(k) = \begin{cases} \sigma_1, & if\ e^{\mathrm{T}}(k-1)\, e(k-1) \ge \Delta \\ \sigma_2, & if\ e^{\mathrm{T}}(k-1)\, e(k-1) \le \Delta \end{cases} \tag{4}$$

and $\alpha(k)$ is defined as

$$\alpha(k) = \begin{cases} 1, & if\ (2)\ is\ satisfied \\ 0, & otherwise \end{cases} \tag{5}$$

and

$$\hat{y}(k) = \alpha(k)\, y(k) + (1 - \alpha(k))\, \hat{y}(k-1) \tag{6}$$

The quantization density in this paper is designed as a function of time delay $\tau_k = \tau(k)$, which is modeled as a discrete homogeneous Markov chain $\{r_k, k\}$, with a finite number of states $\omega = \{1, 2, \ldots, s\}$, and the transition probability of this Markov chain from mode i to mode j at time $k+1$ is

$$p_{ij} = Prob\{r_{k+1} = j \mid r_k = i\} \tag{7}$$

where $i, j \in \omega$ and $0 \le p_{ij} \le 1$.

The quantizer f is described as follows:

$$f(y_j, i) = \begin{cases} \rho^h(i), & if\ \dfrac{\rho^h(i)}{1+\delta(i)} < y_j < \dfrac{\rho^h(i)}{1-\delta(i)}, \\ & y_j > 0,\ h = 0, \pm 1, \pm 2, \cdots \\ 0, & if\ y_j = 0 \\ -f(-y_j, i), & if\ y_j < 0 \end{cases} \tag{8}$$

where $j = 1, 2, \ldots, s$, and the relation between $\delta_f(i)$ and $\rho_f(i)$ is

$$\delta_f(i) = \frac{1 - \rho_f(i)}{1 + \rho_f(i)} \tag{9}$$

and the quantized set is

$$\zeta = \{\pm\rho^h(i), h = 0, \pm 1, \pm 2, \cdots\} \tag{10}$$

Based on [21], $\Delta_f \in [-\delta_f(i), \delta_f(i)]$, and then

$$f\left(y_j, i\right) = \left(1 + \Delta_f\right) \bar{y}(k) \tag{11}$$

The quantized density and quantized set of the quantizer g is similar to the quantizer f, and it can be expressed as follows:

$$g\left(u_r, i\right) = \left(1 + \Delta_g\right) \bar{u}(k) \tag{12}$$

where $r = 1, 2, \cdots, m$ and $\Delta_g \in [-\delta_r(i), \delta_r(i)]$.

According to the above statement, the dynamic output feedback controller is designed as

$$\begin{aligned} \hat{x}(k+1) &= A_c(i)\hat{x}(k) + B_c(i)\bar{y}(k-\tau_k) \\ \bar{u}(k) &= C_c(i)\hat{x}(k) \end{aligned} \tag{13}$$

Combining (11) and (12), the closed-loop system of (1) can be represented as follows:

$$\begin{aligned} x(k+1) &= Ax(k) + B\left(1+\Delta_g\right)C_c(i)\hat{x}(k) \\ \hat{x}(k+1) &= A_c(i)\hat{x}(k) + B_c(i)\left(1+\Delta_f\right)Cx(k-\tau_k) \\ &\quad - B_c(i)\left(1+\Delta_f\right)(1-\alpha(k))e(k) \\ e(k+1) &= (CA-C)x(k) + CBC_c(i)\left(1+\Delta_g\right)\hat{x}(k) \\ &\quad + (1-\alpha(k))e(k) \end{aligned} \tag{14}$$

Define an augmented vector

$$\eta(k) = \left[x^{\mathrm{T}}(k) \;\; \hat{x}^{\mathrm{T}}(k) \;\; e^{\mathrm{T}}(k)\right]^{\mathrm{T}} \tag{15}$$

and then

$$\begin{aligned} &\eta(k+1) \\ &= \left(\Phi_{1\alpha(k)}(i) + \Phi_2(i) + \Phi_3(i)\left(1+\Delta_g\right)\right)\eta(k) \\ &\quad + \Phi_{4\alpha(k)}(i)\left(1+\Delta_f\right)\eta(k-\tau_k) \end{aligned} \tag{16}$$

where

$$\Phi_{1\alpha(k)}(i) = \begin{bmatrix} A & 0 & 0 \\ 0 & 0 & 0 \\ CA-C & 0 & (1-\alpha(k))I \end{bmatrix}$$

$$\Phi_2(i) = \begin{bmatrix} 0 & 0 & 0 \\ 0 & A_c(i) & 0 \\ 0 & 0 & 0 \end{bmatrix}$$

$$\Phi_3(i) = \begin{bmatrix} 0 & BC_c(i) & 0 \\ 0 & 0 & 0 \\ 0 & CBC_c(i) & 0 \end{bmatrix}$$

$$\Phi_{4\alpha(k)}(i) = \begin{bmatrix} 0 & 0 & 0 \\ B_c(i)C & 0 & -(1-\alpha(k))B_c(i) \\ 0 & 0 & 0 \end{bmatrix} \tag{17}$$

We define

$$J = \sum_{k=0}^{\infty} \begin{pmatrix} x(k) \\ u(k) \end{pmatrix}^{\mathrm{T}} Q' \begin{pmatrix} x(k) \\ u(k) \end{pmatrix} \tag{18}$$

as the cost function for system (16), where $Q' = \operatorname{diag}\{Q_1', Q_2'\} \in R^{2n\times 2n}$, $Q_1' \in R^{n\times n}$, and $Q_2' \in R^{n\times n}$.

3. Main Results

The stabilization of the closed-loop system (16) will be considered in this section, and, before going any further, a lemma and a definition are introduced, which will be helpful for deriving the following results.

Lemma 4 (see [10]). *Let $\bar{x}(k) = x(k+1) - x(k)$,*

$$\begin{aligned} &\tilde{\eta}(k) \\ &= \left[\eta^{\mathrm{T}}(k) \;\; \eta^{\mathrm{T}}(k)\Delta_g^{\mathrm{T}} \;\; \eta^{\mathrm{T}}(k-\tau_k) \;\; \eta^{\mathrm{T}}(k-\tau_k)\Delta_f^{\mathrm{T}}\right]^{\mathrm{T}}, \end{aligned} \tag{19}$$

and $\tilde{\eta}(k) \in R^l$, for any matrices $E \in R^{n\times n}$, $M \in R^{n\times l}$, and $Z \in R^{l\times l}$ satisfying

$$\begin{bmatrix} E & M \\ M^{\mathrm{T}} & Z \end{bmatrix} \geq 0 \tag{20}$$

Then the following inequality holds:

$$-\sum_{i=k-\tau}^{k-1} \bar{x}(i)^{\mathrm{T}} E\bar{x}(i) \leq \tilde{\eta}^{\mathrm{T}}(k)\left\{\gamma + \gamma^{T} + \bar{\tau}Z\right\}\tilde{\eta}(k) \tag{21}$$

where

$$\gamma = M^{\mathrm{T}}\left[\operatorname{diag}\{I,0,0\} \;\; 0 \;\; \operatorname{diag}\{-I,0,0\} \;\; 0\right] \tag{22}$$

Proof is given in the Appendix.

Definition 5 (see [20]). The closed-loop systems given in (23) is stochastically stable; i.e., there exists a constant $0 \leq \alpha \leq \infty$, such that $E\{\sum_{l=0}^{\infty}\eta^{\mathrm{T}}(l)\eta(l)\} < \alpha$ for any initial states $x(0)$ and r_0.

Theorem 6. *For given matrices $A_c(i)$, $B_c(i)$, and $C_c(i)$, positive constants $\bar{\tau}$, $\underline{\tau}$, and quantization density values $\rho_f(i)$ and $\rho_g(i)$, if there exists a set of positive definite matrices $E_{\alpha(k)}(i)$, $Z_{\alpha(k)}(i)$, Q', $P_0(i)$, $P_1(i)$, $W_0(i)$, $W_1(i)$, $Q_0(i)$, $Q_1(i)$, $T(i)$, $N(i)$, and $M(i)$ of appropriate dimensions the following inequalities hold:*

$$\min \quad tr\left(P_1\left(r_0\right)\right) \tag{23}$$

$$s.t. \quad \begin{bmatrix} E_{\alpha(i)}(i) & M(i) \\ * & Z_{\alpha(i)} \end{bmatrix} \geq 0 \tag{24}$$

$$\Lambda_0(i) + \Gamma_{10}^{\mathrm{T}}(i)\widetilde{P}_0(i)\Gamma_{10}(i) + \overline{\tau}Z_0(i) + \Gamma_2^{\mathrm{T}}(i)\overline{\tau}E_0(i)\Gamma_2(i) + \gamma(i) + \gamma^{\mathrm{T}}(i) + \sigma(i)I_1^{\mathrm{T}}C^{\mathrm{T}}CI_1 - I_2^{\mathrm{T}}I_2 < 0 \tag{25}$$

$$\Lambda_1(i) + \Gamma_{11}^{\mathrm{T}}(i)\widetilde{P}_1(i)\Gamma_{11}(i) + \overline{\tau}Z_1(i) + \Gamma_2^{\mathrm{T}}(i)\overline{\tau}E_1(i)\Gamma_2(i) + \gamma(i) + \gamma^{\mathrm{T}}(i) + I_2^{\mathrm{T}}I_2 - \sigma(i)I_1^{\mathrm{T}}C^{\mathrm{T}}CI_1 < 0 \tag{26}$$

$$\Gamma_{1\alpha(k)} = \left[\Phi_{1\alpha}(i) + \Phi_2(i) + \Phi_3(i) \quad \Phi_3(i) \quad \Phi_{4\alpha(k)}(i)\,\Phi_{4\alpha(k)}(i)\right]$$

$$\Gamma_2 = \left[\overline{A} \quad \overline{B} \quad 0 \quad 0\right]$$

$$\overline{A} = \left[A - I \quad BC_c(i) \quad 0\right]$$

$$I_1 = \left[\mathrm{diag}\{I, 0, 0\} \quad 0 \quad 0 \quad 0\right]$$

$$I_2 = \left[\mathrm{diag}\{0, 0, I\} \quad 0 \quad 0 \quad 0\right]$$

$$\gamma = M^{\mathrm{T}}\left[\mathrm{diag}\{I, 0, 0\} \quad 0 \quad \mathrm{diag}\{-I, 0, 0\} \quad 0\right]$$

$$\Lambda_0(i) = \mathrm{diag}\left\{(\overline{\tau} - \underline{\tau} + 1)Q - P_0(i) - T(i) - N(i) + \Psi^{\mathrm{T}}(i)Q'\Psi(i), -W_0(i), \delta^2(i)W_1(i) - Q_0, -W_1(i)\right\} \tag{27}$$

$$\Lambda_1(i) = \mathrm{diag}\left\{(\overline{\tau} - \underline{\tau} + 1)Q - P_1(i) - T(i) - N(i) + \Psi^{\mathrm{T}}(i)Q'\Psi(i), -W_2(i), \delta^2(i)W_3(i) - Q_1, -W_3(i)\right\}$$

$$\widetilde{P}_0(i) = \sum_{j=1}^{s} p_{ij}P_0(i)(j)$$

$$\widetilde{P}_1(i) = \sum_{j=1}^{s} p_{ij}P_1(i)(j)$$

$$\Psi(i) = \begin{bmatrix} I & 0 & 0 \\ 0 & C_c(i) & 0 \end{bmatrix}$$

Then the closed-loop system (16) is stochastically stable with the dynamic feedback controller.

Proof. Based on **Lemma 4**, (16) can be represented as follows:

$$\widetilde{\eta}(k+1) = \Gamma_{1\alpha(k)}\left(r_k\right)\widetilde{\eta}(k) \tag{28}$$

where

$$\Gamma_{1\alpha(k)} = \left[\Phi_{1\alpha}\left(r_k\right) + \Phi_2\left(r_k\right) + \Phi_3\left(r_k\right) \quad \Phi_3\left(r_k\right) \quad \Phi_{4\alpha(k)}\left(r_k\right)\Phi_{4\alpha(k)}\left(r_k\right)\right] \tag{29}$$

Then select Lyapunov-Krasovskii functional as

$$V\left(k, r_k\right) = V_1\left(k, r_k\right) + V_2\left(k, r_k\right) + V_3\left(k, r_k\right) + V_4\left(k, r_k\right) + V_5\left(k, r_k\right) \tag{30}$$

where

$$\begin{aligned} V_1\left(k, r_k\right) &= \eta_k^{\mathrm{T}}P\left(r_k\right)\eta_k, \\ V_2\left(k, r_k\right) &= \sum_{l=-\overline{\tau}}^{-1}\sum_{j=k+l}^{k-1}\overline{x}_j^{\mathrm{T}}E\left(r_k\right)\overline{x}_j, \\ V_3\left(k, r_k\right) &= \eta_k^{\mathrm{T}}T\left(r_k\right)\eta_k, \\ V_4\left(k, r_k\right) &= \sum_{l=k-\tau_k}^{k-1}\eta_k^{\mathrm{T}}Q\left(r_k\right)\eta_k \\ &\quad + \sum_{l=-\overline{\tau}+2}^{-\underline{\tau}+1}\sum_{j=k+l-1}^{k-1}\eta_j^{\mathrm{T}}Q\left(r_k\right)\eta_j, \\ V_5\left(k, r_k\right) &= \eta_k^{\mathrm{T}}T\left(r_k\right)\eta_k, \end{aligned} \tag{31}$$

Let

$$\begin{aligned} E_{\alpha(k+1)}\left(r_{k+1}\right) &= E_0\left(r_k\right) \\ Q_{\alpha(k+1)}\left(r_{k+1}\right) &= Q_0\left(r_k\right) \end{aligned} \tag{32}$$

Then we can obtain the following equalities and inequalities:

$$\begin{aligned}
\Delta V_1(k,r_k) &= \tilde{\eta}_k^{\mathrm{T}}\Gamma_{1\alpha(k)}^{\mathrm{T}}\tilde{P}(r_k)\Gamma_{1\alpha(k)}\tilde{\eta}_k - \eta_k^{\mathrm{T}}P(r_k)\eta_k,\\
\Delta V_2(k,r_k) &= \overline{x}_k^{\mathrm{T}}\overline{\tau}E_0(r_k)\overline{x}_k - \sum_{l=k-\overline{\tau}}^{k-1}\overline{x}_k^{\mathrm{T}}E_0(r_k)\overline{x}_k\\
&\le \tilde{\eta}(k)\left[\Gamma_2^{\mathrm{T}}(r_k)\overline{\tau}E_0(r_k)\Gamma_2(r_k)+\gamma(r_k)+\gamma^{\mathrm{T}}(r_k)\right.\\
&\quad \left.+\overline{\tau}Z(r_k)\right]\tilde{\eta},\\
\Delta V_3(k,r_k) &= \tilde{\eta}_k^{\mathrm{T}}\Gamma_{1\alpha(k)}^{\mathrm{T}}(r_k)T_0(r_k)\Gamma_{1\alpha(k)}\tilde{\eta}(k)\\
&\quad -\eta_k^{\mathrm{T}}T_0(r_k)\eta_k,\\
\Delta V_4(k,r_k) &\le (\overline{\tau}-\underline{\tau}+1)\eta_k^{\mathrm{T}}Q_0(r_k)\eta_k - \eta_{k-\tau_k}^{\mathrm{T}}Q_0(r_k)\\
&\quad \cdot\eta_{k-\tau_k},\\
\Delta V_5(k,r_k) &= \tilde{\eta}_k^{\mathrm{T}}\Gamma_{1\alpha(k)}^{\mathrm{T}}(r_k)N_0(r_k)\Gamma_{1\alpha(k)}\tilde{\eta}(k)\\
&\quad -\eta_k^{\mathrm{T}}N_0(r_k)\eta_k,\\
\Delta V&(k,r_k)\\
&\le \tilde{\eta}_k^{\mathrm{T}}\left[\Gamma_{1\alpha(k)}^{\mathrm{T}}\left(\tilde{P}(r_k)+T_0(r_k)+N_0(r_k)\right)\Gamma_{1\alpha(k)}\right.\\
&\quad +\Gamma_2^{\mathrm{T}}(r_k)\overline{\tau}E_0(r_k)\Gamma_2(r_k)+\gamma(r_k)+\gamma^{\mathrm{T}}(r_k)\\
&\quad \left.+\overline{\tau}Z(r_k)\right]\tilde{\eta}_k+\eta_k^{\mathrm{T}}\left((\overline{\tau}-\underline{\tau}+1)Q_0(r_k)-P(r_k)\right.\\
&\quad \left.-T_0(r_k)-N_0(r_k)\right)\eta_k-\eta_{k-\tau_k}^{\mathrm{T}}Q_0(r_k)\eta_{k-\tau_k}
\end{aligned}\tag{33}$$

Let

$$\begin{aligned}
J' &= \begin{pmatrix}x(k)\\u(k)\end{pmatrix}^{\mathrm{T}}Q'x\begin{pmatrix}x(k)\\u(k)\end{pmatrix}\\
&= \eta_k^{\mathrm{T}}\Psi^{\mathrm{T}}(r_k)Q'\Psi(r_k)\eta_k
\end{aligned}\tag{34}$$

according to (6), we know $\alpha(k)=0$ means that $e^{\mathrm{T}}(k)e(k) < \sigma(k)y^{\mathrm{T}}(k)y(k)$, then, adding $\eta_k^{\mathrm{T}}\Delta_g^{\mathrm{T}}W_0(r_k)\Delta_g\eta_k$ and subtracting $\eta_{k-\tau_k}^{\mathrm{T}}\Delta_f^{\mathrm{T}}W_1(r_k)\Delta_f\eta_{k-\tau_k}$ to (33),

$$\begin{aligned}
\Delta V(k,r_k)+J' &\le \tilde{\eta}_k^{\mathrm{T}}\left[\Lambda_0(i)\right.\\
&\quad +\Gamma_{10}^{\mathrm{T}}(i)\left(\tilde{P}(r_k)+T_0(r_k)+N_0(r_k)\right)\Gamma_{10}(i)\\
&\quad \left.+\Gamma_2^{\mathrm{T}}(i)\overline{\tau}E\Gamma_2(i)+\gamma(i)+\gamma^{\mathrm{T}}(i)+\overline{\tau}Z_0(i)\right]\tilde{\eta}_k
\end{aligned}\tag{35}$$

and, according to (25), $\Delta V(k,r_k)+J' \le 0$. When $\alpha(k)=1$, namely, (2) is satisfied, then, adding $\eta_k^{\mathrm{T}}\Delta_g^{\mathrm{T}}W_2(r_k)\Delta_g\eta_k$ and subtracting $\eta_{k-\tau_k}^{\mathrm{T}}\Delta_f^{\mathrm{T}}W_3(r_k)\Delta_f\eta_{k-\tau_k}$ to (35), in a similar way, $\Delta V(k,r_k)+J' \le 0$ is obtained, and then

$$\Delta V(k,r_k) \le -J' = -\begin{pmatrix}x(k)\\u(k)\end{pmatrix}^{\mathrm{T}}Q'x\begin{pmatrix}x(k)\\u(k)\end{pmatrix}\tag{36}$$

Then take expectation on both sides of it, and we obtain

$$\begin{aligned}
&E\{V(\infty,r_\infty)\}-E\{V(0,r_0)\} \le -\beta_0 E\left\{\sum_{k=0}^{\infty}\eta_k^{\mathrm{T}}\eta_k\right\}\\
&E\left\{\sum_{k=0}^{\infty}\eta_k^{\mathrm{T}}\eta_k\right\} < \frac{1}{\beta_0}E\{V(0,r_k)\}-\frac{1}{\beta_0}E\{V(\infty,r_\infty)\}\\
&< \beta
\end{aligned}\tag{37}$$

where $\beta_0 = \inf\{\lambda_{min}(-\Psi^T(r_k)Q'\Psi(r_k))\}$. It can be concluded from **Definition 5** that the system is stochastically stable, and the proof is completed. □

Theorem 7. *For given matrices $A_c(i)$, $B_c(i)$, and $C_c(i)$, positive constants $\overline{\tau}$, $\underline{\tau}$, and quantization density values $\rho_f(i)$ and $\rho_g(i)$, if there exists a set of positive definite matrices $E_{\alpha(k)}(i)$, $Z_{\alpha(k)}(i)$, $\overline{Q}$, $P_0(i)$, $P_1(i)$, $W_0(i)$, $W_1(i)$, $Q_0(i)$, $Q_1(i)$, $T(i)$, $N(i)$, and $M(i)$ of appropriate dimensions and constant $0<\mu<1$ the following inequalities hold:*

$$min \quad tr(P_1(r_0))\tag{38}$$

$$s.t. \quad \begin{bmatrix}E_{\alpha(i)}(i) & M(i)\\ * & Z_{\alpha(i)}\end{bmatrix} \ge 0\tag{39}$$

$$\Pi_0 = \begin{bmatrix}\Pi_{10} & \Gamma_{10}^{\mathrm{T}}(i) & \Gamma_2^{\mathrm{T}}(i) & \Psi^{\mathrm{T}}(i)\\ * & \overline{P_0} & 0 & 0\\ * & * & \overline{E_0} & 0\\ * & * & * & \tilde{Q}\end{bmatrix} < 0\tag{40}$$

$$\Pi_1 = \begin{bmatrix}\Pi_{11} & \Gamma_{11}^{\mathrm{T}}(i) & \Gamma_2^{\mathrm{T}}(i) & \Psi^{\mathrm{T}}(i)\\ * & \overline{P_1} & 0 & 0\\ * & * & \overline{E_1} & 0\\ * & * & * & \tilde{Q}\end{bmatrix} < 0\tag{41}$$

where

$$\begin{aligned}
\Pi_{10} &= \Theta_0(i)+\gamma(i)+\gamma^{\mathrm{T}}(i)+\overline{\tau}Z_0(i)+\mu I_1^{\mathrm{T}}C^{\mathrm{T}}CI_1\\
&\quad -I_2^{\mathrm{T}}I_2,\\
\Pi_{11} &= \Theta_1(i)+\gamma(i)+\gamma^{\mathrm{T}}(i)+\overline{\tau}Z_1(i)+I_2^{\mathrm{T}}I_2\\
&\quad -\mu I_1^{\mathrm{T}}C^{\mathrm{T}}CI_1,\\
\Theta_0(i) &= \mathrm{diag}\left\{(\overline{\tau}-\underline{\tau}+1)Q-P_0(i)-T(i)-N(i),\right.\\
&\quad \left.-W_0(i),\delta^2(i)W_1(i)-Q_0,-W_1(i)\right\}\\
\Theta_1(i) &= \mathrm{diag}\left\{(\overline{\tau}-\underline{\tau}+1)Q-P_1(i)-T(i)-N(i),\right.\\
&\quad \left.-W_2(i),\delta^2(i)W_3(i)-Q_1,-W_3(i)\right\}
\end{aligned}$$

$$\Gamma_{1\alpha(k)} = [\Phi_{1\alpha}(i) + \Phi_2(i) + \Phi_3(i)\ \ \Phi_3(i)\ \ \Phi_{4\alpha(k)}(i)\ \ \Phi_{4\alpha(k)}(i)]$$

$$\Gamma_2 = \left[\overline{A}\ \ \overline{B}\ \ 0\ \ 0\right]$$

$$\overline{A} = [A - I\ \ BC_c(i)\ \ 0]$$

$$I_1 = [\operatorname{diag}\{I, 0, 0\}\ \ 0\ \ 0\ \ 0]$$

$$I_2 = [\operatorname{diag}\{0, 0, I\}\ \ 0\ \ 0\ \ 0]$$

$$\gamma = M^{\mathrm{T}}[\operatorname{diag}\{I, 0, 0\}\ \ 0\ \ \operatorname{diag}\{-I, 0, 0\}\ \ 0]$$

$$\overline{P}_0(i) = -\left(\sum_{j=1}^{s} p_{ij}P_0(i)(j)\right)^{-1}$$

$$\overline{P}_1(i) = -\left(\sum_{j=1}^{s} p_{ij}P_1(i)(j)\right)^{-1}$$

$$\overline{E}_0(i) = -(\overline{\tau}E_0(i))^{-1}$$

$$\overline{E}_1(i) = -(\overline{\tau}E_1(i))^{-1}$$

$$\Psi(i) = \begin{bmatrix} I & 0 & 0 \\ 0 & C_c(i) & 0 \end{bmatrix}$$

$$\widetilde{Q} = (-Q')^{-1} \tag{42}$$

Proof. $\widetilde{P}_0(i) = \sum_{j=1}^{s} p_{ij}P_0(j)$ and $\widetilde{P}_1(i) = \sum_{j=1}^{s} p_{ij}P_1(j)$ have been proposed in **Theorem 6**, let $\mu = \sigma$, and then the Schur complement lemma to (25) and (40) is obtained. In a similar way, (41) can be obtained from (26), and then the proof is completed. □

4. Simulation

Consider a system as follows:

$$x(k+1) = \begin{bmatrix} 0 & 2 \\ -2 & -3 \end{bmatrix} x(k) + \begin{bmatrix} 0 \\ 0.1 \end{bmatrix} u(k)$$
$$y(k) = [0\ \ 1]\,x(k) \tag{43}$$

The time delay is bounded as $0.01 \le \tau_k \le 0.09$ and modeled by a Markov chain $\omega = \{1, 2\}$, which is corresponded to mode 1 and mode 2, assume that $\delta_f = \delta_g = 0.023$, $x(0) = [1\ \ 0]^T$, and the quantization density of each mode is defined as $\delta(1) = 0.02$ and $\delta(2) = 0.03$, $\mu = 0.0195$ is chosen as threshold, and the transition probability matrix is as follows:

$$p = \begin{bmatrix} 0.42 & 0.58 \\ 0.46 & 0.54 \end{bmatrix} \tag{44}$$

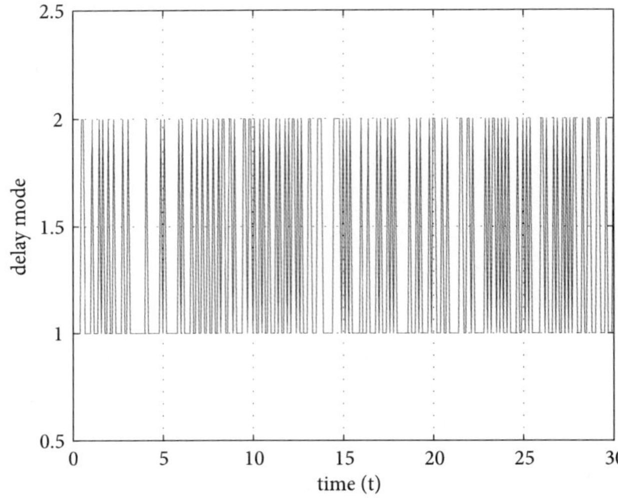

FIGURE 2: Group 1: the mode of time delay.

The linear matrix inequality is utilized and the matrices of controller are obtained as follows:

$$A_c(1) = \begin{bmatrix} -1.600 & 0.8703 \\ -0.6204 & -1.5032 \end{bmatrix}$$
$$A_c(2) = \begin{bmatrix} -1.4306 & 0.3010 \\ 1.3071 & 0 \end{bmatrix}$$
$$B_c(1) = [0.0012\ \ 0.1203]^{\mathrm{T}}$$
$$B_2(2) = [-0.0210\ \ -0.0013]^{\mathrm{T}}$$
$$C_c(1) = [-3.2122\ \ -0.0013]$$
$$C_c(2) = [-3.5421\ \ -0.0009] \tag{45}$$

and we take two groups parameters to make a comparison:

Group 1. $\sigma_1 = 0.01$; $\sigma_2 = 0.023$.

Group 2. $\sigma_1 = 0.01$; $\sigma_2 = 0.0195$.

Remark 8. Figures 2 and 5 show the switching of two modes of the time delay in a Markov chain, and the switching is determined by the transition probability. Figures 3 and 6 show the curves of the state response which eventually tend to be stable. The event-triggered release instants and intervals are shown in Figures 4 and 7, and the numbers of triggers in two figures are 79 and 73, which are much less than the number of the traditional sampling, which means that the network resources are greatly saved.

Remark 9. In literature [10], the structure of its system based on event-triggered scheme contains one quantizer, and the event-triggered release instants and interval are 74 and 84 while the sampling time is 10s. Comparing with [10], the method proposed in this article can save the network resources better.

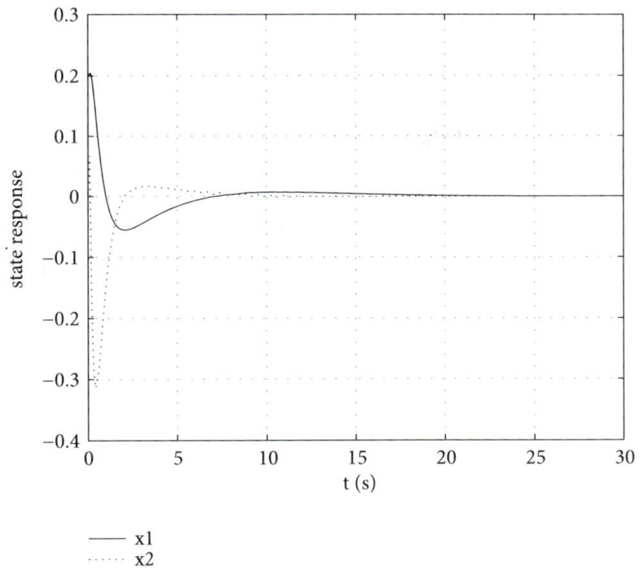

FIGURE 3: Group 1: the response of state.

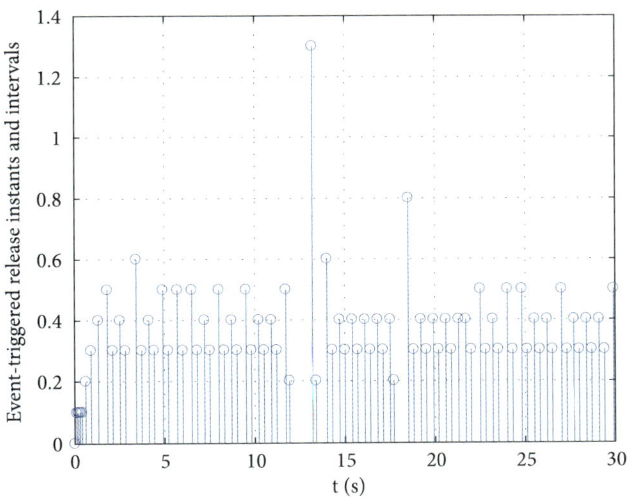

FIGURE 4: Group 1: the event-triggered release instants and intervals.

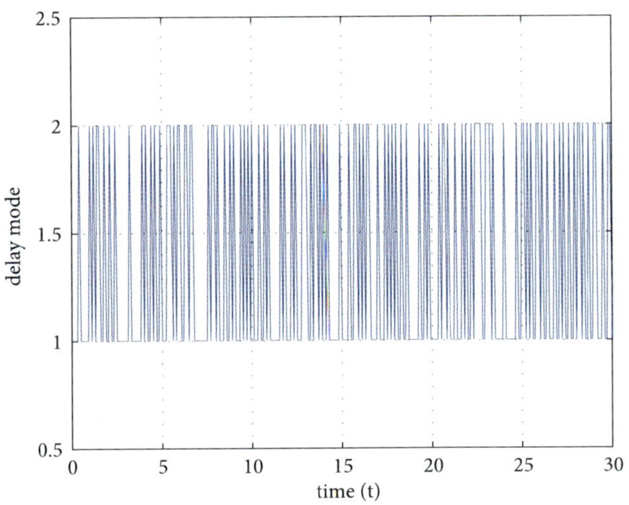

FIGURE 5: Group 2: the mode of time delay.

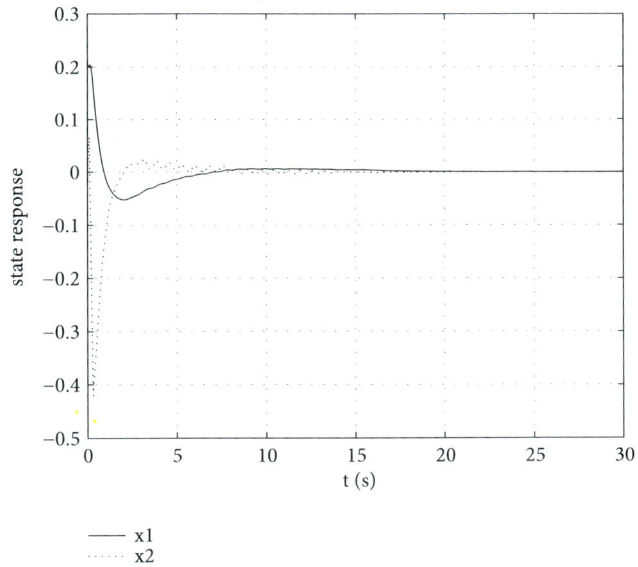

FIGURE 6: Group 2: the response of state.

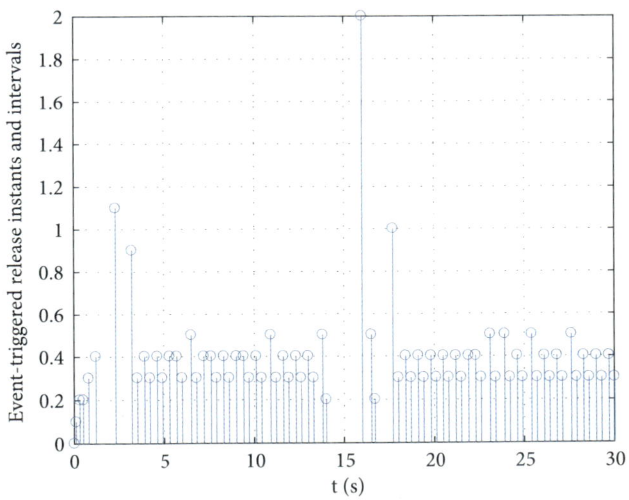

FIGURE 7: Group 2: the event-triggered release instants and intervals.

5. Conclusion

A co-design of event-triggered scheme and dynamic output feedback controller for a class of NCSs with random time delay is presented. Two mode-dependent quantizers are set to balance the network congestion and quantization error. A Markov process is used to model the time delay which is used to describe the quantization density as a function. Lyapunov-Krasovskii functional and LMI are used to derive the stability sufficient conditions. A numerical example is given to illustrate the proposed approach.

Appendix

Proof of Lemma 4 . From (20), we learn that

$$\sum_{i=k-\bar{\tau}}^{k-1} \begin{bmatrix} \bar{x}(i) \\ \bar{\eta}(i) \end{bmatrix}^{T} \begin{bmatrix} E & M \\ M^{T} & Z \end{bmatrix} \begin{bmatrix} \bar{x}(i) \\ \bar{\eta}(i) \end{bmatrix} \geq 0 \tag{A.1}$$

Then

$$\sum_{i=k-\overline{\tau}}^{k-1}\left[\overline{x}^{\mathrm{T}}(i)E\overline{x}(i)+\widetilde{\eta}^{\mathrm{T}}(k)M^{\mathrm{T}}\overline{x}(i)+\overline{x}^{\mathrm{T}}(i)M\widetilde{\eta}(k)+\widetilde{\eta}^{\mathrm{T}}(k)Z\widetilde{\eta}(k)\right]\geq 0,$$

$$\sum_{i=k-\overline{\tau}}^{k-1}\widetilde{\eta}^{\mathrm{T}}(k)M^{\mathrm{T}}\overline{x}(i)+\sum_{i=k-\overline{\tau}}^{k-1}\overline{x}^{\mathrm{T}}(i)M\widetilde{\eta}(k)+\sum_{i=k-\overline{\tau}}^{k-1}\widetilde{\eta}^{\mathrm{T}}(k)Z\widetilde{\eta}(k)\geq-\sum_{i=k-\overline{\tau}}^{k-1}\overline{x}^{\mathrm{T}}(i)E\overline{x}(i),\tag{A.2}$$

$$\widetilde{\eta}^{\mathrm{T}}(k)M^{\mathrm{T}}\{x(k)-x(k-\overline{\tau})\}+\{x(k)-x(k-\overline{\tau})\}^{\mathrm{T}}\cdot M\widetilde{\eta}(k)+\overline{\tau}\widetilde{\eta}^{\mathrm{T}}(k)Z\widetilde{\eta}(k)\geq-\sum_{i=k-\overline{\tau}}^{k-1}\overline{x}^{\mathrm{T}}(i)E\overline{x}(i),$$

Let

$$x(k)-x(k-\overline{\tau})=\left[diag\{I,0,0\}\ \ 0\ \ diag\{-I,0,0\}\ \ 0\right]\widetilde{\eta}(k)\tag{A.3}$$

$$\gamma=M^{\mathrm{T}}\left[diag\{I,0,0\}\ \ 0\ \ diag\{-I,0,0\}\ \ 0\right]$$

Then (21) is obtained. □

Conflicts of Interest

The authors declare that they have no conflicts of interest.

Acknowledgments

This work is supported by the National Natural Science Foundation of China under Grant (61374083) and the National Funds for Distinguished Young Scientists of China (61425009).

References

[1] W.-A. Zhang, A. Liu, and K. Xing, "Stability analysis and stabilization of aperiodic sampled-data systems based on a switched system approach," *Journal of The Franklin Institute*, vol. 353, no. 4, pp. 955–970, 2016.

[2] A. Liu, W. Zhang, L. Yu, S. Liu, and M. Z. Chen, "New results on stabilization of networked control systems with packet disordering," *Automatica*, vol. 52, pp. 255–259, 2015.

[3] P. Antsaklis and J. Baillieul, "Special issue on technology of networked control systems," *Proceedings of the IEEE*, vol. 95, no. 1, pp. 5–8, 2007.

[4] D. Huang and S. K. Nguang, "State feedback control of uncertain networked control systems with random time-delays," in *Proceedings of the IEEE Conference on Decision and Control*, pp. 3964–3969, 2007.

[5] L. Zhou and G. Lu, "Quantized feedback stabilization for networked control systems with nonlinear perturbation," *International Journal of Innovative Computing, Information and Control*, vol. 6, no. 6, pp. 2485–2495, 2010.

[6] W.-A. Zhang and L. Shi, "Sequential fusion estimation for clustered sensor networks," *Automatica*, vol. 89, pp. 358–363, 2018.

[7] R. Yang, P. Shi, G. P. Liu, and H. Gao, "Network-based feedback control for systems with mixed delays based on quantization and dropout compensation," *Automatica*, vol. 47, no. 12, pp. 2805–2809, 2011.

[8] Y. Xia, J. Yan, P. Shi, and M. Fu, "Stability analysis of discrete-time systems with quantized feedback and measurements," *IEEE Transactions on Industrial Informatics*, vol. 9, no. 1, pp. 313–324, 2013.

[9] Z. G. Wu, P. Shi, H. Y. Su, and J. Chu, "Network-based robust passive control for fuzzy systems with randomly occurring uncertainties," *IEEE Transactions on Fuzzy Systems*, vol. 21, no. 5, pp. 966–971, 2013.

[10] J. He, Q. Ma, J. Guo, and C. Zhou, "Event-triggered mechanism based dynamic quantized output feedback controller for networked control systems," in *Proceedings of the 29th Chinese Control and Decision Conference, CCDC 2017*, pp. 2569–2574, May 2017.

[11] P. Tabuada, "Event-triggered real-time scheduling of stabilizing control tasks," *IEEE Transactions on Automatic Control*, vol. 52, no. 9, pp. 1680–1685, 2007.

[12] S. Hu, Y. Zhang, and Z. Du, "Network-based H∞ tracking control with event-triggering sampling scheme," *Control Theory and Applications Iet*, vol. 6, no. 4, pp. 533–544, 2012.

[13] B. Wang, X. Meng, and T. Chen, "Event based pulse-modulated control of linear stochastic systems," *Institute of Electrical and Electronics Engineers Transactions on Automatic Control*, vol. 59, no. 8, pp. 2144–2150, 2014.

[14] K. Peng, M. Wang, and J. Dong, "Event-triggered fault detection framework based on subspace identification method for the networked control systems," *Neurocomputing*, vol. 239, pp. 257–267, 2017.

[15] H. Xu, A. Sahoo, and S. Jagannathan, "Stochastic adaptive event-triggered control and network scheduling protocol co-design for distributed networked systems," *IET Control Theory & Applications*, vol. 8, no. 18, pp. 2253–2265, 2014.

[16] J. Zhang, C. Peng, D. Du, and M. Zheng, "Adaptive event-triggered communication scheme for networked control systems with randomly occurring nonlinearities and uncertainties," *Neurocomputing*, vol. 174, pp. 475–482, 2016.

[17] N. Elia and S. K. Mitter, "Stabilization of linear systems with limited information," *IEEE Transactions on Automatic Control*, vol. 46, no. 9, pp. 1384–1400, 2002.

[18] C. Peng and Y. Tian, "Networked H_∞ control of linear systems with state quantization," *Information Sciences*, vol. 177, no. 24, pp. 5763–5774, 2007.

[19] E. Tian, D. Yue, and C. Peng, "Quantized output feedback control for networked control systems," *Information Sciences*, vol. 178, no. 12, pp. 2734–2749, 2008.

[20] F. Rasool, D. Huang, and S. K. Nguang, "Robust H∞ output feedback control of discrete-time networked systems with limited information," *Systems and Control Letters*, vol. 60, no. 10, pp. 845–853, 2011.

[21] M. Fu and L. Xie, "The sector bound approach to quantized feedback control," *IEEE Transactions on Automatic Control*, vol. 50, no. 11, pp. 1698–1711, 2005.

[22] K. Liu, E. Fridman, and K. H. Johansson, "Dynamic quantization of uncertain linear networked control systems," *Automatica*, vol. 59, pp. 248–255, 2015.

[23] G. Gu and L. Qiu, "Networked control systems for multi-input plants based on polar logarithmic quantization," *Systems & Control Letters*, vol. 69, pp. 16–22, 2014.

[24] Y. Kao, J. Xie, C. Wang, and H. R. Karimi, "A sliding mode approach to H∞ mathcontainer loading mathjax non-fragile observer-based control design for uncertain markovian neutral-type stochastic systems," *Automatica*, vol. 52, pp. 218–226, 2015.

[25] J. Cheng, J. H. Park, Y. Liu, Z. Liu, and L. Tang, "Finite-time H ∞ fuzzy control of nonlinear Markovian jump delayed systems with partly uncertain transition descriptions," *Fuzzy Sets and Systems*, vol. 314, pp. 99–115, 2016.

[26] L. Zha, J. A. Fang, X. Li, and X. Liu, "Event-triggered output feedback H∞ mathcontainer loading mathjax control for networked markovian jump systems with quantizations," *Nonlinear Analysis Hybrid Systems*, vol. 24, pp. 146–158, 2017.

[27] Y. Shi and B. Yu, "Output feedback stabilization of networked control systems with random delays modeled by Markov chains," *IEEE Transactions on Automatic Control*, vol. 54, no. 7, pp. 1668–1674, 2009.

[28] S. Chae, F. Rasool, S. K. Nguang, and A. Swain, "Robust mode delay-dependent H∞ control of discrete-time systems with random communication delays," *Control Theory and Applications Iet*, vol. 4, no. 6, pp. 936–944, 2010.

Wide-Area Damping Controller Design based on the Frequency Domain Self-Excitation Method

Lei Gao,[1,2] Yanchao Ma,[3] Yonghui Nie,[2] Shiyun Xu,[1] Xiaojie Chu,[1] and Long Peng[1]

[1]*China Electric Power Research Institute, Beijing 100192, China*
[2]*Northeast Electric Power University, Jilin 132012, Jilin Province, China*
[3]*State Grid Liaoning Power Co., Ltd., Yingkou Power Supply Company, Yingkou 115000, China*

Correspondence should be addressed to Shiyun Xu; xushiyun@epri.sgcc.com.cn

Academic Editor: Cui-Qin Ma

In this paper, a wide-area DC damping controller design method is proposed to damp the weak interarea oscillation modes. First, the low-order power system control model identification method is proposed based on the frequency domain self-excitation method to obtain the dominant modes and other related information. In what follows, the identified model is transformed into a low-order state space equation. By comparing the geometric metrics of different schemes, the various input signals and installation locations of the controller are selected. Furthermore, the damping controller parameter calculation is realized based on the self-excitation frequency domain identification method, which does not depend on the detailed model of the system. The design process is simple and it is easy to apply in the practical engineering of large-scale complex power grids. The applicability and effectiveness of the proposed controller design method are demonstrated through simulations of a two-area four-generator power system.

1. Introduction

The AC/DC interconnected power grid can transmit large capacity of power from the remote power basis to the load centers through communication networks. It plays an extremely important role in the energy allocation in China. Since the AC and DC transmission systems are generally coupled with each other, however, disturbances that occur in both the AC and the DC sides may trigger weakly damped low frequency oscillations. If the oscillations are of the interarea type, a large number of generators will participate in and the interconnected power grid will confront severe operational risk [1, 2].

However, due to the deficiency in observability, traditional power system stabilizers (PSS) which use local signals as inputs may fail to damp out the interarea oscillation and maintain system stability. To overcome the shortcoming of traditional PSS, wide-area damping controllers with remote signals obtained from wide-area measurement system (WAMS) have become a better choice in suppressing low frequency oscillation [3–5]. Since DC systems have fast response characteristics, it can effectively improve the stability of the system once faults have occurred in the AC system, and the DC damping control [6–9] that can suppress the interarea low frequency oscillation of AC/DC hybrid system has attracted the attention of researchers.

During the wide-area damping controller design, it is important to choose feedback control signal and installation location. The residue method [10, 11] is widely used for feedback signal selection, but it can only compare the same type of signals. For different types of signals, such as generator speed, line transmission power, and bus voltage, the amplitude of the above-mentioned residual value will not be accurate due to the ratio problem. In [12], a relative gain array (RGA) method is proposed. Although this method reduces the mutual influence among multiple controllers, it has the same disadvantages as the residual method. Other methods, such as the Hankel singular value (HSV) [13], the singular value decomposition (SVD) [14], and the minimum singular values (MSV) [15], have also been proposed to calculate the corresponding observability and controllability indices, but it becomes much more complicated to calculate for practical large-scale power systems. On the other hand, most of the existing damping control design methods require the precise

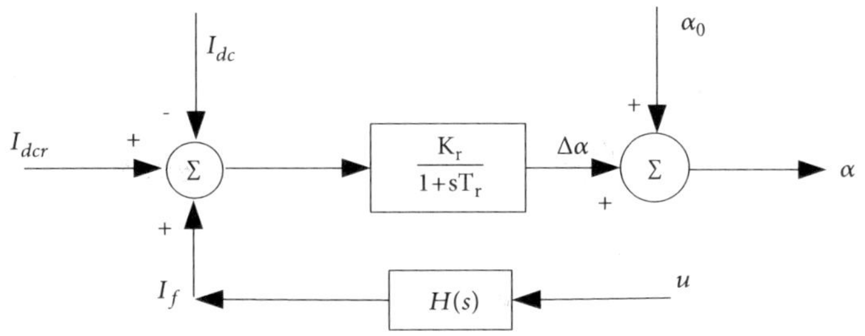

FIGURE 1: Rectifier current regulator.

system model through linearizing it around an equilibrium point. This may lead to the modeling error due to information incompleteness.

In this paper, the low frequency oscillation of an AC/DC hybrid system is damped by installing a wide-area damping controller in the DC system. Based on the above considerations, the improved geometric lateral index [11, 16] based on the frequency domain excitation identification theory is proposed to solve the wide-area DC supplementary damping control signal selection in AC/DC hybrid systems. This method can effectively solve the problem that different types of input signals have different amplitude ratios and can avoid solving the huge state matrix as well. Moreover, the method based on the self-excitation frequency domain identification is further utilized to configure the DC damping controller, which does not depend on the detailed model of the system. The design process is simple, and it is easy to use in the practical engineering of the complex large power grid. At the end of this study, the AC/DC hybrid system is taken as an example to demonstrate the effectiveness of the proposed controller design method.

2. Problem Formulation

2.1. DC Damping Controller. The DC damping control is a kind of small signal modulation, and the basic principle is to add a supplementary damping controller in the DC main control loop. The controller is usually installed at the rectifier side. In this regard, the power of the DC transmission line is adjusted by tuning the extinction angle α to damp the system oscillation.

The control block diagram is shown in Figure 1. By adding the DC additional control output I_f and the reference current input I_{dcr} to the current control of the inverter can improve the damping of the system. The dynamic equation of the linearized regulator is given as

$$\begin{aligned} \dot{\alpha} &= \frac{K_r}{T_r}\left(I_{dcr} - I_{dc} + I_f\right) - \frac{\alpha}{T_r} \\ \dot{I}_{dc} &= \frac{1}{T_{dc}}\left(-I_{dc} + I_{dcr} + I_f\right) \end{aligned} \qquad (1)$$

The supplementary wide-area damping controller transfer function is in the following form:

$$H(s) = \frac{K}{1+sT_1} \cdot \frac{sT_W}{1+sT_W} \cdot \left[\frac{1+sT_2}{1+sT_3}\right]^m \qquad (2)$$

where T_{dc}, T_r, T_1, T_2, T_3, T_W are the time constants, K_r and K are the controller proportional coefficient, I_{dc} is the DC current, and u is the feedback signal, which can be chosen as the generator speed deviation, the bus phase difference or any bus voltage, power, and so on.

In this study, the selected feedback signal is processed by the damping controller to the input of the current regulator in the DC main control system, based on the principle of DC damping control. The extinction angle α is maintained within the normal range ($\alpha_{min} < \alpha < \alpha_{max}$). In order to select the best alternative signal, the indicators defined by the improved geometric index can be calculated. In this regard, the designed damping controller can damp system oscillation.

2.2. System Model Analysis. By linearizing around the stable equilibrium point, the AC/DC hybrid power system equation can be given as

$$\begin{aligned} \Delta\dot{x} &= A\Delta x + B\Delta u \\ \Delta y &= C\Delta x \end{aligned} \qquad (3)$$

where A is an $n \times n$ state matrix, B is an $n \times p$ input matrix, and C is a $q \times n$ output matrix.

For power system with n generators, only part of the oscillation mode of the generator rotor is concerned, and the number of oscillating modes in which any one generator is significantly involved is generally 2~3. Based on this consideration, the state equation of the ith generator can be set as

$$\begin{bmatrix} \Delta\dot{\delta} \\ \Delta\dot{\omega} \end{bmatrix} = \begin{bmatrix} 0 & I \\ -M^{-1}K(s) & -M^{-1}D(s) \end{bmatrix} \cdot \begin{bmatrix} \Delta\delta \\ \Delta\omega \end{bmatrix} \qquad (4)$$

where $K(s)$ is the synchronization torque coefficient, $D(s)$ is the damping torque coefficient, and M is the diagonal matrix composed by the corresponding elements of the generator.

It can be seen from Figure 2 that if there is no external disturbance T_x, the differential equation of the corresponding equivalent system can be derived as

$$M(s)\Delta\ddot{\delta} + D(s)\Delta\dot{\delta} + K(s)\Delta\delta = 0 \qquad (5)$$

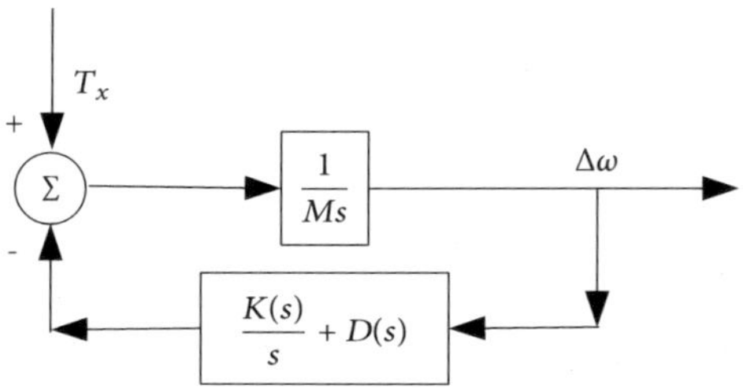

FIGURE 2: Equivalent system.

Assume that the eigenvalue of (5) is $\lambda_i = \sigma_i \pm j\omega_i$, then

$$\left(M\Delta\ddot{\delta} + D(s)\Delta\dot{\delta} + K(s)\Delta\delta\right)\Big|_{\lambda_i} \overset{\text{def}}{=} T_x \longrightarrow 0 \quad (6)$$

In order to calculate the electromechanical mode associated with system oscillation, the self-excited method [17] is utilized by deriving the eigenvalue of (6) through the Newton iteration method. However, the application of the traditional self-excited method is not only sensitive to the initial value but also easily leads to the loss of roots. In this regard, this paper proposes a frequency domain self-excited identification method, which is applied to choose controller inputs. This method avoids the problems of traditional self-excited method, which is described in detail in the following section.

3. Wide-Area DC Damping Controller Design

Generally speaking, the steps in designing a wide-area controller include wide-area signal selection, system reduction, controller design, and validation of the design by numerical simulations. Each of the above steps is briefly described in the following subsections.

3.1. Observable and Controllable Quantitative Index. In order to choose the feedback signal and the installation location of the wide-area damping controller, it is necessary to establish a unified index for the calculation of different types of signals or control points. The quantization index of the observability and controllability of the system based on the improving geometric lateral index can solve the defects of the different types of signals, which is described as follows.

For system (3), by denoting a new state variable $\Delta x = Wz$, the oscillation characteristics of the system can be decoupled and the system with a diagonal state matrix can be derived as

$$\begin{aligned} \dot{z} &= \Lambda z + B'\Delta u \\ \Delta y &= C'z + D\Delta u, \end{aligned} \quad (7)$$

where W is the modal matrix of A, $\Lambda = \operatorname{diag}\{\lambda_i\}$, $B' = W^{-1}B = N^T B$, $C' = CW$, W_i is the right eigenvector corresponding to λ_i, and N_i is the left eigenvector corresponding to λ_i.

The observability index of the ith modal of the system is given as

$$g_{oj}(i) = \cos\left(\alpha\left(W_i, c_j\right)\right) = \frac{\left|c_j W_i\right|}{\left\|W_i\right\| \left\|c_j\right\|} \quad (8)$$

while the controllability index is defined by

$$g_{ck}(i) = \cos\left(\theta\left(N_i, b_k\right)\right) = \frac{\left|b_k^T N_i\right|}{\left\|N_i\right\| \left\|b_k\right\|} \quad (9)$$

where $|\cdot|$ and $\|\cdot\|$, respectively, represent the modulo and Euclidean norm, c_j is the jth line of the matrix C, b_k is the kth column of the matrix B, $\alpha(W_i, c_j)$ is the angle between the output phasor c_j and the right eigenvector W_i, and $\theta(N_i, b_k)$ is the angle between the output phasor b_k and the left eigenvector N_i.

Thus, the comprehensive index corresponding to the ith modal is given by

$$\begin{aligned} g_{coi}(k, j) &= g_{oj}(i)\, g_{ck}(i) \\ &= \cos\left(\alpha\left(W_i, c_j\right)\right) \cos\left(\theta\left(N_i, b_k\right)\right) \end{aligned} \quad (10)$$

Quantitative analysis and comparison of the observability and controllability of different types of alternative signals can be realized by the calculation of the above-mentioned indices. However, in the practical power system, it is difficult to obtain the accurate full-order system model. Based on this consideration, this study carries out power system model identification according to the frequency domain self-excited method, and then the low-order system can be utilized to approximate the original system, which greatly reduces the calculation complicatedness and improves the calculation effectiveness.

3.2. Control Loop Selection. The feedback signals and installation locations selection are the basis for controller design of wide-area damping controller. In this subsection, a new method to select the feedback signal of wide-area damping controller is proposed by combining the basic principle of self-excitation and the method of frequency domain identification.

Based on the basic principle of the self-excited method, it can be derived from (6) that

$$T_x + \frac{dT_x}{ds}\left(\Delta\lambda_i\right) = 0 \quad (11)$$

By analyzing (11), the following necessary conditions are obtained:

$$\begin{aligned} T_x &\longrightarrow 0 \\ \frac{dT_x}{ds} &\longrightarrow 0 \end{aligned} \quad (12)$$

However, since the traditional self-excitation method iteratively solves the eigenvalue of (6) by Newton's method, it may lead to the incompleteness of electromechanical modes.

Therefore, the method proposed in this paper utilizes the frequency domain method to identify the transfer function of the system and then solve the relevant mode information to avoid the above-mentioned problems.

By carrying out the Laplace transform of (6) and (12), the system transfer function can be derived as

$$G(s) = \frac{\Delta\delta(s)}{\Delta T_x(s)} = \frac{s}{M(s)s^2 + D(s)s + K(s)} \quad (13)$$

If a more accurate model of the generator is considered, such as taking the excitation system model into account, (9) still holds, but the order will increase. In this case, $G(s)$ can be expressed as follows:

$$\begin{aligned} G(s) &= \frac{\Delta\delta_i(s)}{\Delta T_{xi}(s)} = K\frac{(s-z_{j1})(s-z_{j2})\cdots(s-z_{ji})}{(s-p_1)(s-p_2)\cdots(s-p_k)} \\ &= \frac{R_{j1}}{s-p_1} + \frac{R_{j2}}{s-p_2} + \cdots \frac{R_{jk}}{s-p_k} \end{aligned} \quad (14)$$

It can be analyzed from (6) and (12) that if the frequency of T_x is equal to the oscillation frequency of the system, only the torque with the amplitude close to zero can stimulate the large response of the oscillation frequency. Assume that T_x can increase the oscillation frequency to $\lambda\omega$, if different generators are selected as the "excitation point", then $G(\lambda_i\omega)$ can be expressed as

$$\begin{aligned} G(\lambda_i\omega) &= \frac{\Delta\delta_i(\lambda_i\omega)}{\Delta T_{xi}(\lambda_i\omega)} \\ &= \frac{R_{j1}}{\lambda_i\omega - p_1} + \frac{R_{j2}}{\lambda_i\omega - p_2} + \cdots \frac{R_{jk}}{\lambda_i\omega - p_k} \end{aligned} \quad (15)$$

Although the "excitation point" varies, the system inter-area oscillation mode keeps the same, where a pair of conjugate poles represents an oscillation mode. Let $\lambda_i = \sigma_i \pm j\omega_i$, then it can be used to calculate the system oscillation frequency and damping ratio.

Once the oscillation mode is determined, the index calculation of the candidate signal should be carried out, when it is necessary to transform the system into the state space form and $G(s)$ can be written as

$$G(s) = \frac{Y(s)}{U(s)} = \frac{b_m s^m + b_{m-1}s^{m-1} + \cdots + b_0}{s^n + a_{n-1}s^{n-1} + \cdots + a_0} \quad (16)$$

In this regard, the state space expression of the system can be derived as

$$\begin{aligned} \Delta\dot{x} &= E\Delta x + F\Delta u \\ \Delta y &= G\Delta x + H\Delta u \end{aligned} \quad (17)$$

where $E = \begin{bmatrix} 0 & I_{n-1} \\ A_0 & A_{n-1} \end{bmatrix}$ is the equivalent matrix of matrix A in (3), which contains the main mode of the original system. $F = [0 \ \cdots \ 1]^T$ and I are the corresponding control matrix and substantial matrix, respectively. $G = B_n - A_n b_n$, $H = b_n$. Equation (17) is the state space model derived by the frequency domain self-exited method.

The state space equation identified by the frequency domain excitation method preserves the main mode of operation of the system. Therefore, (17) can be used to substitute the original system to calculate the quantization index of the observability and controllability of the system.

The new method of selecting feedback signal of wide-area damping controller based on the self-excited frequency domain identification method proposed in this paper can be used in engineering application of large-scale power system without obtaining the full-state model of the system, and the calculation and operation are convenient, which is suitable for the engineering application of large power system.

3.3. Wide-Area Coordinated Controller Design. Once the state space equation is identified by the self-excited frequency domain identification, by carrying out the Laplace transform on (13), the transfer function between the input variable $\Delta\boldsymbol{U}_1$ and the output variable $\Delta\boldsymbol{Y}_1$ can be obtained as

$$G(s) = \frac{\Delta U_1(s)}{\Delta Y_1(s)} = G(sI - E)^{-1}F + H \quad (18)$$

Suppose dominant oscillation modes for (18) are $\lambda_{1,2} = \delta \pm j\omega$, and the desired damping ratio after adding damping controller is ξ. Substituting $s = j\omega_d = j\sqrt{1-\xi^2}\omega$ into (18) derives

$$G(j\omega_d) = \left.\frac{\Delta U_1(s)}{\Delta Y_1(s)}\right|_{s=j\omega_d} = G\angle\phi \quad (19)$$

where G and ϕ are the amplitude and phase of $G(s)$ at $s = j\omega_d$, respectively.

The lead lag link parameters of the wide-area DC damping controller (2) can be calculated by

$$\begin{aligned} \phi_x &= 180 - \phi \\ \alpha &= \frac{1+\sin\phi_x}{1-\sin\phi_x} \\ T_1 &= \frac{\sqrt{\alpha}}{\omega} \\ T_2 &= \frac{T_1}{\alpha} \end{aligned} \quad (20)$$

In order to prevent the damping controller from outputting a constant deviation voltage, the design principle is that, in the frequency range where damping controllers provide damping, no obvious phase displacement should be introduced in the isolating links. The typical value is chosen as 1~10s, and in this study, it is set to be 5s.

The advantage of the above wide-area DC damping controller parameter design based on the self-excited method is that it does not rely on the detailed model of the system, the design process is simple, the derived parameters are stable, and it is applicable to utilize in the practical complex large-scale power grid.

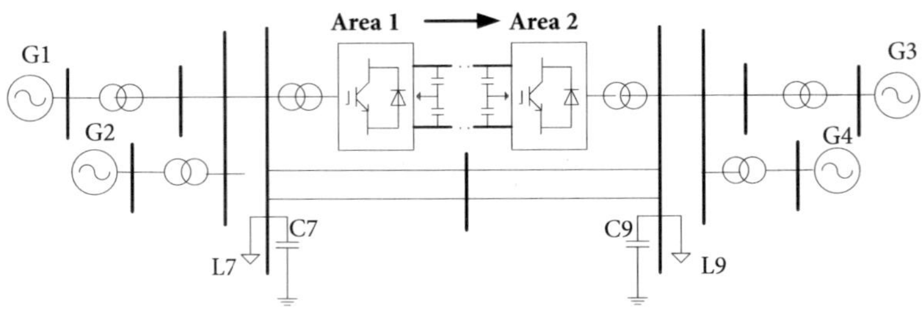

FIGURE 3: Two-area four-generator AC/DC hybrid system.

TABLE 1: Interarea oscillation frequency and damping ratio without control.

Operation mode	eigenvalue	frequency	Damping ratio
P=150MW	-0.1462+3.2683i	0.5202	0.0447
P=200MW	-0.1301+3.8321i	0.6099	0.0339
P=300MW	-0.2942+3.5092i	0.5585	0.0835

3.4. Wide-Area Damping Control Design Process. The basic steps of the selection of the feedback signal and the installation location of the wide-area damping controller based on the self-excited frequency domain identification method are as follows:

Step 1. Identify the low-order system based on the frequency domain self-excited method, and calculate the corresponding interarea oscillation modes $\lambda_{1,2} = \delta \pm j\omega$ and the oscillation frequency ω.

Step 2. Select different feedback signals and installation locations of the n groups of controllers, and then the system is able to be operated under the steady state with the open loop operation of the controller. The disturbance test for group i ($0<i<n$) is carried out, and for the disturbance $u_i(t)$ that does not destroy the stability of the original system, record the output response data $y_i(t)$.

Step 3. Extract the variation data $\Delta u_i(t)$ and $\Delta y_i(t)$ of $u_i(t)$ and $y_i(t)$ in the common cycle. Carry out the discrete Fourier transform, respectively, and obtain the varying phasors $\Delta u_i(j\omega)$ and $\Delta y_i(j\omega)$ corresponding to different frequency values, where $i = 1, 2, \ldots, n,\ 0 \leq j\omega \leq 2\pi$.

Step 4. The self-excited method is applied to identify the phasor data $\Delta u_i(j\omega)$ and $\Delta y_i(j\omega)$, and the corresponding transfer function $G_i(s)$ is fitted. Transform the transfer function into the state space form and obtain the corresponding system matrices A_i, B_i, C_i, D_i.

Step 5. Apply the geometric measure index method to calculate each index, which is used as the reference for the input signal of the comparison controller. Based on the above method, that best feedback signal and installation location are determined.

Step 6. Based on the state space equation and the dominant modes, identify the information such as the phase Φ and the oscillation mode ω that are required by controller design. Calculate the controller time constants T_i, $i = 1, 2, \ldots, 4$ by (20).

Step 7. Adjust the controller gain K_i to verify the damping effect, and the operation parameters of the wide-area DC damping controller are finally determined.

4. Case Study

Taking the two-area four-generator AC/DC hybrid system shown in Figure 3 as an example, the detailed model is built with MATLAB and verified by simulation. The specific parameters are taken from literature [18]. The wide-area damping controller is installed in the rectifier of the DC part. The optional input signals of the wide-area damping controller are the rotor speed difference of generator 1 (G1) and generator 3 (G3), the active power of the AC tie-line, the AC line voltage, AC line current amplitude, and difference of voltage phase between two sides of an AC contact line. By modifying the model parameters, the powers of transmission line between two regions are chosen as P = 150MW, P = 200MW, and P = 300MW, and thus three kinds of operation modes are derived (the positive direction of power flow is defined as from Area 1 to Area 2).

According to the procedure described in the previous section, the interarea oscillation modes are analyzed for the transmission power P = 150MW, P = 200MW, and P = 300MW, respectively, which are shown in Table 1.

By carrying out the wide-area damping controller design steps proposed in Section 3.4, the frequency domain self-excited method and the improved geometric lateral index are used to calculate the comprehensive and residue indices for different schemes. The results are shown in Tables 2–4.

As shown in Tables 2–4, the best candidate signals among the three operating modes are the rotor speed difference between G1 and G3 by comparing the comprehensive and residue indices.

In order to verify the validity of the analysis results, the wide-area damping control effect of different feedback signals

TABLE 2: System indices for P=150MW.

Alternative input signal	Residue index	geometric laterality index	frequency	Damping ratio	Compensation phase
the rotor speed difference of G1 and G3	0.2711	0.9409	0.5536	0.0870	93.7
the active power of the AC line	2.0689e-06	0.7241	0.5700	0.0533	147.4
AC line current amplitude	0.1376	0.8042	0.556	0.0773	16.5
difference of voltage phase between two sides of AC line	1.6235e-06	0.7171	0.567	0.0395	54.6

TABLE 3: System indices for P=200MW.

Alternative input signal	Residue index	geometric laterality index	frequency	Damping ratio	Compensation phase
the rotor speed difference of G1 and G3	0.3568	0.8951	0.6587	0.0473	97.2
the active power of the AC line	1.2302e-06i	0.7501	0.6331	0.0035	115.5
AC line current amplitude	0.0117	0.7276	0.6245	0.0172	31.4
difference of voltage phase between two sides of AC line	1.4371e-08	0.7330	0.5891	0.0063	29.8

TABLE 4: System indices for P=300MW.

Alternative input signal	Residue index	geometric laterality index	frequency	Damping ratio	Compensation phase
the rotor speed difference of G1 and G3	0.2967	0.9554	0.5572	0.0566	92.6
the active power of the AC line	6.3309e-06	0.8236	0.5401	0.0670	115.6
AC line current amplitude	0.1087	0.7183	0.5431	0.0070	55.1
difference of voltage phase between two sides of AC line	6.0653e-08	0.7496	0.5587	0.0636	165.8

is simulated by choosing the speed difference between G1 and G3 as the wide-area damping controller feedback signal, which is installed in the DC side of the main control loop. A three-phase short circuit occurred on the AC transmission line at t=10.4s, which is cleared at 10.9s. By carrying out the phase compensation method, parameters of the wide-area controller (2) can be calculated as

$$\begin{aligned} m &= 3, \\ K &= 30, \\ T_1 &= 0.02, \\ T_W &= 10, \\ \alpha &= 0.3218, \\ T_2 &= 0.5035, \\ T_3 &= 0.1620 \end{aligned} \tag{21}$$

where the operating mode corresponding to P=300MW is selected as the compensation phase of the controller parameter calculation, and let $\theta = 92.6^{\circ}$.

From Figures 4–6, it is observed that there are obvious differences in the control effect under the location of installation being selected. Considering the residue method, for the rotational speed differences of G1 and G3 as well as the

Table 5: Interarea oscillation frequency and damping ratio for system under control.

Operation mode	eigenvalue	frequency	Damping ratio
P=150MW	-0.6208+3.5014i	0.5573	0.1746
P=200MW	-0.4446+3.9169i	0.6234	0.1128
P=300MW	-0.4082+3.6602i	0.5825	0.1108

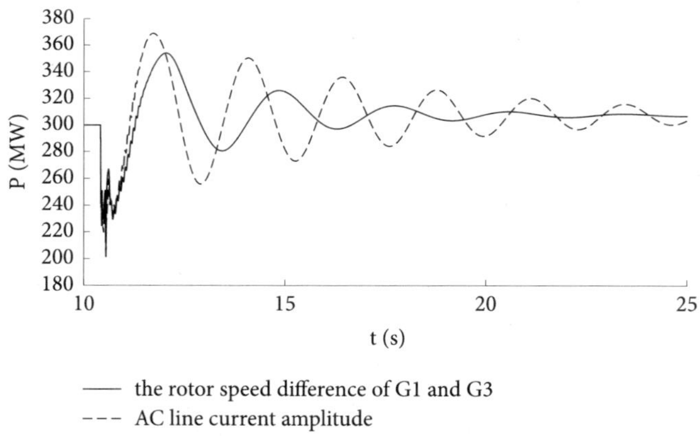

Figure 4: Comparison of power oscillation.

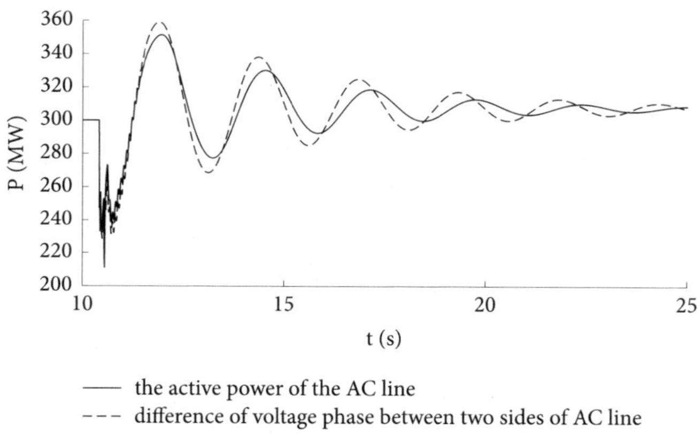

Figure 5: Comparison of power oscillation.

amplitude of the AC link current (Figure 4), the AC tie-line power P and the voltage phase difference of the tie-line (Figure 5), these two types of signals with different amplitudes are displayed as the larger the number of indicators, the better the control effect. However, since the magnitudes of the two types of signal indicators are significantly different, it is impossible to compare them at the same time. In this case, the improving geometric lateral index has a greater advantage, as it can compare different types of signals under the same order of magnitude. As shown in Figure 6, as the analysis results in Tables 2–4, the larger the index value the better the control effect, which verifies the effectiveness of the method in this paper.

For the system under wide-area damping control, the operating modes of the interarea oscillation frequency and damping ratio are calculated based on the frequency domain self-excited method as shown in Table 5.

It can be seen from Table 5 that, for system under control, the damping for each operating mode has been significantly increased, indicating that the system damping is improved under the designed controller. Simulation results of the AC tie-line transmission power with and without the designed damping controller are depicted in Figure 7. The above simulation results have verified the effectiveness and applicability of the proposed wide-area damping controller.

5. Conclusion

In this study, a wide-area damping controller design method for AC/DC hybrid power system is proposed. The input signal of the wide-area damping controller is selected by the improved geometric laterality index, which can be chosen as different types of signals to increase the range of the optional signal to improve the controller design effect. The system model can be identified through the frequency domain self-excitation method, based on which the phase compensation method is further utilized to design the DC

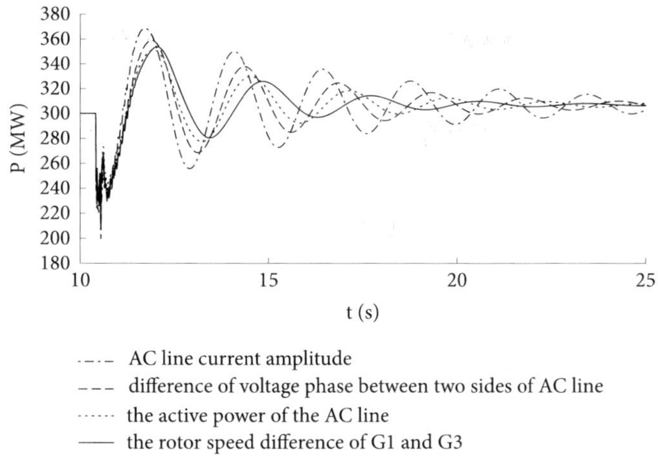

FIGURE 6: Active power oscillation of the tie-line.

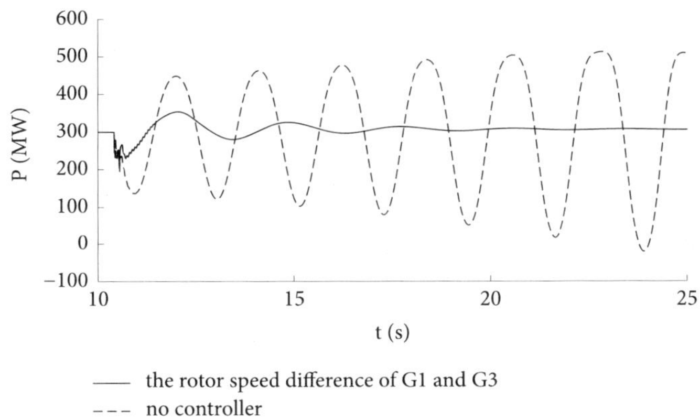

FIGURE 7: Transmission power comparison of the AC tie-line.

damping controller parameters. The proposed damping controller avoids the complex eigenvalue calculation in large-scale power systems and is not affected by the increase of system order. Moreover, the detailed system modeling is avoided, which may lead to inaccuracy and errors. The design process is simple, which is suitable for practical engineering applications.

Conflicts of Interest

The authors declare that they have no conflicts of interest.

Acknowledgments

This research has been supported by the Young Elite Scientists Sponsorship Program by CAST (YESS20160157) and CSEE (JLB-2017-305), State Grid Corporation of China, Fundamental and Forecast Projects (XT71-18-014), and the National Natural Science Foundation of China (U1766202, 51777195) as well as the Key Project of Smart Grid Technology and Equipment of National Key Research and Development Plan of China (2016YFB0900600).

References

[1] O. F. Ruiz, I. A. DIaz DIaz, and I. Cervantes, "On the implementation of advanced hybrid controllers of AC/DC converters," *IEEE Latin America Transactions*, vol. 15, no. 9, pp. 1677–1683, 2017.

[2] A. A. Eajal, M. A. Abdelwahed, E. F. El-Saadany, and K. Ponnambalam, "A unified approach to the power flow analysis of AC/DC hybrid microgrids," *IEEE Transactions on Sustainable Energy*, vol. 7, no. 3, pp. 1145–1158, 2016.

[3] R. Jalayer and B.-T. Ooi, "Co-Ordinated PSS tuning of large power systems by combining transfer function-eigenfunction analysis (TFEA), optimization, and eigenvalue sensitivity," *IEEE Transactions on Power Systems*, vol. 29, no. 6, pp. 2672–2680, 2014.

[4] N. Rashidirad, M. Hamzeh, K. Sheshyekani, and E. Afjei, "An effective method for low-frequency oscillations damping in multiBus DC microgrids," *IEEE Journal on Emerging and Selected Topics in Circuits and Systems*, vol. 7, no. 3, pp. 403–412, 2017.

[5] S. Debnath, J. Qin, and M. Saeedifard, "Control and stability analysis of modular multilevel converter under low-frequency operation," *IEEE Transactions on Industrial Electronics*, vol. 62, no. 9, pp. 5329–5339, 2015.

[6] R. Eriksson, "A new control structure for multiterminal DC grids to damp interarea oscillations," *IEEE Transactions on Power Delivery*, vol. 31, no. 3, pp. 990–998, 2016.

[7] Y. Liu, A. Raza, K. Rouzbehi, B. Li, D. Xu, and B. W. Williams, "Dynamic resonance analysis and oscillation damping of multiterminal DC grids," *IEEE Access*, vol. 5, pp. 16974–16984, 2017.

[8] L. Meng, T. Dragicevic, J. C. Vasquez, and J. M. Guerrero, "Tertiary and secondary control levels for efficiency optimization and system damping in droop controlled DC-DC converters," *IEEE Transactions on Smart Grid*, vol. 6, no. 6, pp. 2615–2626, 2015.

[9] A. Fuchs, M. Imhof, T. Demiray, and M. Morari, "Stabilization of large power systems using vsc-hvdc and model predictive control," *IEEE Transactions on Power Delivery*, vol. 29, no. 1, pp. 480–488, 2014.

[10] J. Zhang, C. Y. Chung, C. Lu, K. Men, and L. Tu, "A novel adaptive wide area PSS based on output-only modal analysis," *IEEE Transactions on Power Systems*, vol. 30, no. 5, pp. 2633–2642, 2015.

[11] W. Yao, L. Jiang, J. Wen, Q. H. Wu, and S. Cheng, "Wide-area damping controller of Facts devices for inter-area oscillations considering communication time delays," *IEEE Transactions on Power Systems*, vol. 29, no. 1, pp. 318–329, 2014.

[12] H. N. Villegas Pico, D. C. Aliprantis, J. D. McCalley, N. Elia, and N. J. Castrillon, "Analysis of hydro-coupled power plants and design of robust control to damp oscillatory modes," *IEEE Transactions on Power Systems*, vol. 30, no. 2, pp. 632–643, 2015.

[13] R. Shah, N. Mithulananthan, K. Y. Lee, and R. C. Bansal, "Wide-area measurement signal-based stabiliser for large-scale photovoltaic plants with high variability and uncertainty," *IET Renewable Power Generation*, vol. 7, no. 6, pp. 614–622, 2013.

[14] J. M. Lim and C. L. DeMarco, "SVD-based voltage stability assessment from phasor measurement unit data," *IEEE Transactions on Power Systems*, vol. 31, no. 4, pp. 2557–2565, 2016.

[15] E. T. González and J. G. C. Guizar, "The effect of induction motor modeling in thecontext of voltage sensitivity: Combined loads and reactive power generation, on electric networks steady state analysis," *IEEE Latin America Transactions*, vol. 10, no. 4, pp. 1924–1930, 2012.

[16] D. H. A. Lee, "Voltage stability assessment using equivalent nodal analysis," *IEEE Transactions on Power Systems*, vol. 31, no. 1, pp. 454–463, 2016.

[17] A. Thangali, H. Prasad, S. Kethamakka, D. Demirdjian, and N. Checka, "AESOP: Adaptive Event detection SOftware using Programming by example," in *Proceedings of the IEEE International Symposium on Technologies for Homeland Security, HST 2015*, Waltham, Mass, USA, April 2015.

[18] P. Kundur, *Power System Stability And Control [M.S. thesis]*, McGraw-Hill Education, 1994.

4

Consensus Control of Second-Order Multiagent Systems with Particle Swarm Optimization Algorithm

Xiongfeng Deng,[1] Xiuxia Sun,[1] Ri Liu,[2] and Shuguang Liu[1]

[1]Equipment Management and Unmanned Aerial Vehicle Engineering College, Air Force Engineering University, Xi'an 710051, China
[2]Theory Training Department, Air Force Harbin Flight Academy, Harbin 150001, China

Correspondence should be addressed to Xiuxia Sun; gcxysxx@126.com

Academic Editor: Carlos-Andrés García

This paper considers the consensus problem of second-order multiagent systems. Firstly, an improved consensus control protocol is proposed. Then, the convergence of the proposed control protocol is analyzed by applying Lyapunov stability theory. In order to improve the control effect of a given system, the particle swarm optimization (PSO) algorithm is introduced and an improved PSO algorithm is proposed. Additionally, a mapping relationship with agents and the individuals of PSO algorithm is designed. Finally, two simulation examples are provided to illustrate the effectiveness of proposed control protocol and the control effect of PSO algorithm.

1. Introduction

The cooperative control problem of multiagent systems has been attracting outstanding attention in the past few years. The main reason is its widespread application in various fields, such as robot systems [1], sensor networks [2], and unmanned aerial vehicle (UAV) systems [3]. The basic problem of the cooperative control of multiagent systems is consensus, which is to design a suitable control law such that the output of all agents can achieve synchronization.

In existing literature, the consensus problem of multiagent systems has been widely researched. In [4], an observer-based distributed output feedback was considered to solve the consensus tracking problem of multiagent systems. In [5, 6], the iterative learning control was applied to deal with the consensus problem of multiagent systems, while the same problem was studied by using linear quadratic regulator (LQR) in [7, 8]. Moreover, the cooperative tracking problem of nonlinear multiagent systems was analyzed in [9, 10], and the consensus control problem of a class of third-order nonlinear multiagent was studied in [11]. Furthermore, distributed control method [12], impulsive control method [13], and adaptive fuzzy output feedback control approach [14] were also applied to solve the consensus problem of multiagent systems.

To improve the control effect, some optimization algorithms, such as genetic algorithm [15], differential evolution algorithm [16], and PSO algorithm [17], have been considered by researchers. As an optimization method, PSO algorithm has the advantages of good search performance, simple implementation, and few adjustment parameters [18]. Due to these advantages, PSO algorithm has gained increasing interest since it was first presented. Currently, there are many results on the applications of the PSO algorithm. For example, to analyze how different algorithmic parameters in a distributed implementation affect the total evaluation and resulting fitness, a distributed PSO algorithm for a class of multirobot systems with the obstacle surroundings was researched in [19]. The nondeterministic navigation problem of UAVs was solved by using the PSO algorithm in [20], where a neighborhood control scheme was designed to eliminate the inherent weakness of PSO algorithm. In [21], the PSO algorithm was applied to deal with the three-dimensional path planning problem of UAVs.

Different from the above research results, a clustering routing algorithm with nonlinear dynamic adaptive PSO algorithm for wireless sensor networks was proposed in [22], where the whole sensor field was positioned via the improved nonlinear dynamic adaptive inertia weight and evaluation of fitness function. In [23], the PSO algorithm was applied to deal with the space trajectory optimization problem. Additionally, the result that the PSO algorithm was an efficient, reliable, and accurate method for determining optimal space trajectory was demonstrated. Furthermore, the relevant parameters of PSO algorithm are still prespecified with fixed value in some papers [19–21].

Motivated by these facts, we divert our attention to the consensus problem of a class of second-order multiagent systems in this work. The main contributions are outlined in two aspects. On the one hand, an improved control protocol for the given multiagent systems is proposed, and then the convergence is analyzed by the designed Lyapunov function. On the other hand, the PSO algorithm is introduced to improve control effect, where two adaptive laws are designed to adjust the updating laws of PSO algorithm. Also, a new mapping method is applied to solve the mapping problem of the agents' states and the individuals of PSO algorithm. Finally, some simulation examples are provided to illustrate the validity of our results.

The remainder of this paper is organized as follows. In Section 2, some preliminaries are briefly given. In Section 3, the consensus control problem of multiagent systems is analyzed. The fundamental theory of PSO algorithm and the mapping relationship with agents are introduced in Section 4, and simulation examples are provided in Section 5. Finally, some conclusions are drawn in Section 6.

2. Preliminaries

Let $\mathcal{G} = \{\mathcal{V}, \mathcal{E}, \mathcal{A}\}$ denote an undirected graph which consists of n nodes, where $\mathcal{V} = \{v_1, \cdots, v_N\}$ is the set of vertices and $\mathcal{E} = \{(v_i, v_j) : v_i, v_j \in \mathcal{V}\} \subseteq \mathcal{V} \times \mathcal{V}$ is the set of edges. The weighted adjacency matrix is denoted by $\mathcal{A} = [a_{ij}] \in R^{n\times n}$, where $a_{ij} > 0$ if and only if $i \neq j$ and $a_{ij} = 0$ otherwise. A path between v_i and v_j is a sequence of distinct vertices $v_i, v_{i_1}, \cdots, v_l$, where $(v_i, v_{i_1}) \in \mathcal{E}$. An undirected graph is connected if there exists a path from any node i to any other node j. Multiagent systems consist of n agents. Each agent can be described as a node and the exchange information among agents can be expressed as an edge with definite weight in a generalized graph. Therefore, the problem of multiagent systems can be solved through the graph theory.

For the sake of disscusion below, some difinitions and lemmas are given as follows.

Definition 1. The consensus of a multiagent systems is said to be achieved if, for any initial condition, there exist $\lim_{t\longrightarrow\infty} x_i(t) = x_j(t)$ and $\lim_{t\longrightarrow\infty} v_i(t) = v_j(t)$ for $i, j \in 1, \cdots, n$ and $i \neq j$.

Lemma 2 (see [24]). *Let* $\boldsymbol{\sigma} = [\sigma_1, \cdots, \sigma_m]^{\mathrm{T}} \in R^m$, $\boldsymbol{\rho} = [\rho_1, \cdots, \rho_m]^{\mathrm{T}} \in R^m$, *and* $\mathbf{C} = [c_{ij}] \in R^{m\times m}$; *if* $\mathbf{C}$ *is a symmetric matrix, that is,* $c_{ij} = c_{ji}$, *then we have*

$$\frac{1}{2}\sum_{i=1}^{m}\sum_{j=1}^{m} c_{ij}\left(\sigma_i - \sigma_j\right)\varphi\left(\rho_i - \rho_j\right) = \sum_{i=1}^{m}\sum_{j=1}^{m} c_{ij}\sigma_i\varphi\left(\rho_i - \rho_j\right) \quad (1)$$

where $\varphi(\cdot)$ *represents an odd function.*

3. Consensus Analysis of Multiagent Systems

Considering a general second-order multiagent systems with undirected graph, the dynamics of the ith agent are described as

$$\begin{aligned} \dot{x}_i(t) &= v_i(t) \\ \dot{v}_i(t) &= u_i(t) \end{aligned} \quad (2)$$

where $x_i(t) \in R$, $v_i(t) \in R$ and $u_i(t) \in R$ are the position, velocity, and control input of agent i, respectively; $i \in \{1, \cdots, n\}$ and n is the number of agents.

In this paper, a suitable control protocol $u_i(t)$ needs to be chosen for the multiagent systems (2) to achieve consensus. Inspired by the results of [25, 26], an improved control protocol is given as

$$\begin{aligned} u_i(t) = &-\alpha\sum_{j=1}^{n} a_{ij}\left(x_i(t) - x_j(t)\right)^3 \\ &- \beta\sum_{j=1}^{n} a_{ij}\left(v_i(t) - v_j(t)\right)^3 \end{aligned} \quad (3)$$

and the general form of (3) is described as

$$\begin{aligned} u_i(t) \\ = -\sum_{j=1}^{n} a_{ij}\left[\alpha\varphi\left(x_i(t) - x_j(t)\right) + \beta\varphi\left(v_i(t) - v_j(t)\right)\right] \end{aligned} \quad (4)$$

where α and β are positive constants; a_{ij} is the (i, j) element of $\mathcal{A}$ and $\varphi(\cdot)$ is an odd function.

For the convenience of analysis, the time variable t will be ignored from here on.

According to (4), (2) is rewritten as

$$\begin{aligned} \dot{x}_i &= v_i \\ \dot{v}_i &= -\sum_{j=1}^{n} a_{ij}\left[\alpha\varphi\left(x_i - x_j\right) + \beta\varphi\left(v_i - v_j\right)\right] \end{aligned} \quad (5)$$

Hence, we have the following Theorem.

Theorem 3. *Consider the multiagent systems (2) with the consensus protocol (4), and let the exchange information topology graph of agents be undirected and connected; then the consensus can be achieved, i.e.,* $x_i(t) = x_j(t)$ *and* $v_i(t) = v_j(t)$ *as* $t \longrightarrow \infty$ *for* $i, j \in 1, \cdots, n$ *and* $i \neq j$.

Proof. Let $x_{ij} = x_i - x_j$, and then have

$$\begin{aligned} \dot{x}_{ij} &= \dot{x}_i - \dot{x}_j = v_i - v_j \\ \dot{v}_i &= -\sum_{j=1}^{n} a_{ij}\left[\alpha\varphi\left(x_{ij}\right) + \beta\varphi\left(v_i - v_j\right)\right] \end{aligned} \quad (6)$$

Design the Lyapunov function candidate

$$V = V_1 + V_2 \quad (7)$$

with

$$V_1 = \frac{1}{4}\alpha\sum_{i=1}^{n}\sum_{j=1}^{n}a_{ij}\varphi^2\left(x_{ij}\right) \tag{8}$$

$$V_2 = \frac{1}{2}\sum_{i=1}^{n}v_i^2 \tag{9}$$

From (8) and (9), it is obvious that $V_1 \geq 0$ and $V_2 \geq 0$; then $V \geq 0$ can be obtained. Taking the derivative of function V_i $(i = 1, 2)$, we get

$$\dot{V}_1 = \frac{1}{2}\alpha\sum_{i=1}^{n}\sum_{j=1}^{n}a_{ij}\left(v_i - v_j\right)\varphi\left(x_{ij}\right) \tag{10}$$

$$\dot{V}_2 = -\alpha\sum_{i=1}^{n}\sum_{j=1}^{n}a_{ij}v_i\varphi\left(x_{ij}\right) - \beta\sum_{i=1}^{n}\sum_{j=1}^{n}a_{ij}v_i\varphi\left(v_i - v_j\right) \tag{11}$$

Considering Lemma 2, we have

$$\frac{1}{2}\alpha\sum_{i=1}^{n}\sum_{j=1}^{n}a_{ij}\left(v_i - v_j\right)\varphi\left(x_{ij}\right) = \alpha\sum_{i=1}^{n}\sum_{j=1}^{n}a_{ij}v_i\varphi\left(x_{ij}\right) \tag{12}$$

$$\begin{aligned}&\beta\sum_{i=1}^{n}\sum_{j=1}^{n}a_{ij}v_i\varphi\left(v_i - v_j\right)\\ &\quad = \frac{1}{2}\beta\sum_{i=1}^{n}\sum_{j=1}^{n}a_{ij}\left(v_i - v_j\right)\varphi\left(v_i - v_j\right)\end{aligned} \tag{13}$$

From (12) and (13), one gets

$$\dot{V} = \dot{V}_1 + \dot{V}_2 = -\frac{1}{2}\sum_{i=1}^{n}\sum_{j=1}^{n}a_{ij}\beta\left(v_i - v_j\right)\varphi\left(v_i - v_j\right) \tag{14}$$

Due to the assumption that $\mathcal{G}$ is undirected and connected and $\varphi(\cdot)$ is an odd function, it is obtained that $x\varphi(x) > 0$. Hence, the following result is held:

$$\begin{aligned}\dot{V} &= -\frac{1}{2}\beta\sum_{i=1}^{n}\sum_{j=1}^{n}a_{ij}\left[\left(v_i - v_j\right)^2\varphi^2\left(x_{ij}\right)\right]\\ &\quad - \frac{1}{2}\sum_{i=1}^{n}\sum_{j=1}^{n}a_{ij}\beta\left(v_i - v_j\right)\varphi\left(v_i - v_j\right) \leq 0\end{aligned} \tag{15}$$

Let $\Gamma = \{(x_{ij}, v_i) \mid \dot{V} = 0\}$. From (14), $\dot{V} = 0$ hints that $v_i = v_j$ $(i \neq j)$. Furthermore, $\dot{v}_i = \dot{v}_j$ $(i \neq j)$ can be deduced. Therefore, there exists $\xi \in R$ such that $v \in span(\mathbf{1} \times \xi)$, where $v = [v_1, \cdots, v_n]^{\mathrm{T}}$, $\mathbf{1} = [1, \cdots, 1]^{\mathrm{T}}$, and $span(\cdot)$ represents the vector space of v.

Combining $v_i = v_j$ $(i \neq j)$ and (7), then we have

$$\dot{v}_i = -\sum_{j=1}^{n}a_{ij}\alpha\varphi\left(x_{ij}\right) \tag{16}$$

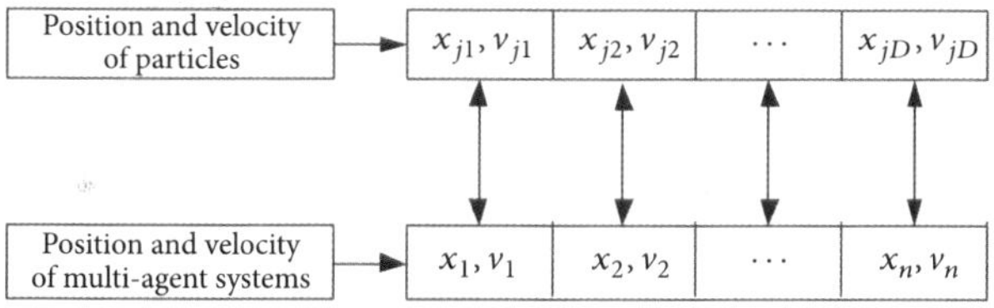

FIGURE 1: Mapping scheme.

In addition,

$$\begin{aligned}(\mathbf{1} \times \xi)^{\mathrm{T}}\dot{v} &= (\mathbf{1} \times \xi)^{\mathrm{T}}\left[-\sum_{j=1}^{n}a_{ij}\alpha\varphi\left(x_{ij}\right)\right]\\ &= -\xi\sum_{i=1}^{n}\sum_{j=1}^{n}a_{ij}\alpha\varphi\left(x_{ij}\right)\end{aligned} \tag{17}$$

In view of the exchange information topology graph $\mathcal{G}$ is undirected, and there exists $\varphi(x_{ij}) = -\varphi(x_{ji})$; we have

$$\begin{aligned}&\xi\sum_{i=1}^{n}\sum_{j=1}^{n}a_{ij}\alpha\varphi\left(x_{ji}\right) + \xi\sum_{i=1}^{n}\sum_{j=1}^{n}a_{ij}\alpha\varphi\left(x_{ij}\right)\\ &\quad = 2\xi\sum_{i=1}^{n}\sum_{j=1}^{n}a_{ij}\alpha\varphi\left(x_{ij}\right) = 0\end{aligned} \tag{18}$$

From (16) and (18), $(\mathbf{1} \times \xi)^{\mathrm{T}}\dot{v} = 0$ can be obtained, indicating that $\dot{v}$ is orthogonal to $\mathbf{1} \times \xi$. Hence, $\dot{v} = 0$ is deduced, and then we have

$$-\sum_{j=1}^{n}a_{ij}\alpha\varphi\left(x_{ij}\right) = 0 \tag{19}$$

and

$$-\sum_{i=1}^{n}x_i\sum_{j=1}^{n}a_{ij}\alpha\varphi\left(x_{ij}\right) = 0 \tag{20}$$

which hints that

$$-\frac{1}{2}\sum_{i=1}^{n}\sum_{j=1}^{n}a_{ij}\alpha\left(x_i - x_j\right)\varphi\left(x_{ij}\right) = 0 \tag{21}$$

From (21), we have $a_{ij}\alpha(x_i - x_j)\varphi(x_{ij}) = 0$, $\forall i \neq j$; then $x_i = x_j$, $\forall i \neq j$ is obtained. Thereby, the results that $x_i = x_j$ and $v_i = v_j$ for $i, j \in 1, \cdots, n$ and $i \neq j$ are obtained. The proof is completed. □

In order to improve the control effect, the PSO algorithm will be introduced in the following disscusion.

4. PSO Algorithm and the Mapping Relationship with Agents

4.1. PSO Algorithm. As an optimization method, PSO algorithm seeks the optimum solution of an optimization problem via selecting some particles. The characteristics of each

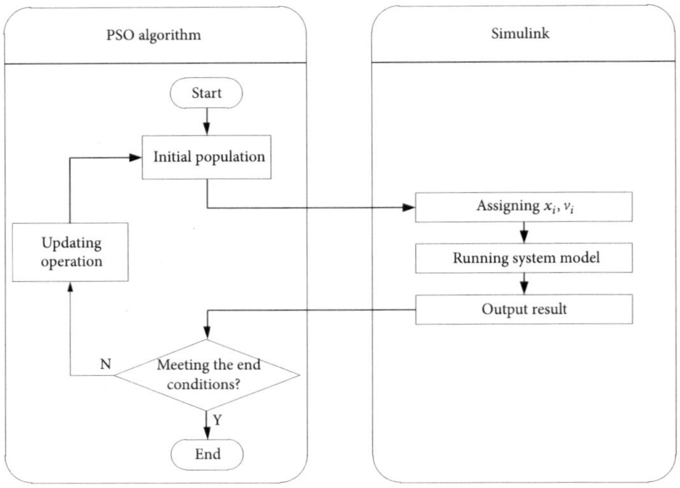

FIGURE 2: The flowchart of optimization.

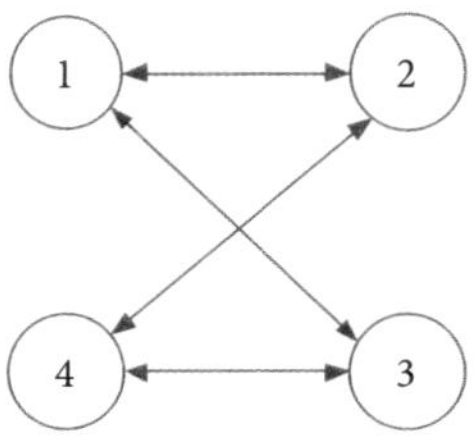

FIGURE 3: Exchange information topology graph.

particle include position, velocity, and fitness value. The fitness value is calculated by the fitness function, where the merit of the particle is described by the fitness valule. In the solution space, each paritcle's position and velocity are updated by the individual extremum P_{best} and the population extremum G_{best}, where P_{best} and G_{best} are updated by comparing the fitness value of the new particle.

Considering a D dimension search space, let $X = (X_1, \cdots, X_m)$ be a population with m particles. The position and velocity of the jth particle are defined as $X_j = (x_{j1}, \cdots, x_{jD})^{\mathrm{T}}$ and $V_j = (v_{j1}, \cdots, v_{jD})^{\mathrm{T}}$, respectively. The extremum of individual and population are defined as $P_{best,j} = (P_{best,j1}, \cdots, P_{best,jD})^{\mathrm{T}}$ and $G_{best} = (G_{best,1}, \cdots, G_{best,D})^{\mathrm{T}}$, respectively. The velocity and position updating laws of the jth particle at the $(k+1)$th iteration are given as

$$V_{jd}^{k+1} = \omega V_{jd}^{k} + \kappa_1 \gamma_1 \left(P_{best,jd}^{k} - X_{jd}^{k}\right) + \kappa_2 \gamma_2 \left(G_{best,d}^{k} - X_{jd}^{k}\right) \tag{22}$$

$$X_{jd}^{k+1} = X_{jd}^{k} + V_{jd}^{k+1} \tag{23}$$

where X_{jd}^k and V_{jd}^k are the dth position and velocity component of the jth individual, respectively; $\omega \in (0, 1]$ is inertia weight; κ_i $(i = 1, 2) \geq 0$ are acceleration factors; γ_i $(i = 1, 2) \in [0, 1]$ are random numbers between 0 and 1; $P_{best,jd}$ is the dth best position component of the jth individual and $G_{best,d}$ is the dth best position of the entire population.

In this paper, the updating laws (22) and (23) are improved and the results are given as

$$V_{jd}^{k+1} = \omega V_{jd}^{k} + \kappa_1 \gamma_1' \left(P_{best,jd}^{k} - X_{jd}^{k}\right) + \kappa_2 \gamma_2' \left(G_{best,d}^{k} - X_{jd}^{k}\right) \tag{24}$$

$$X_{jd}^{k+1} = X_{jd}^{k} - \Delta V_{jd}^{k+1} \operatorname{sgn}\left(\Delta V_{jd}^{k+1}\right) \tag{25}$$

where

$$\gamma_1' = \frac{\left\|P_{best,j}^{k}\right\|}{\sqrt{\left\|P_{best,j}^{k}\right\|^2 + \left\|G_{best}^{k}\right\|^2}} \tag{26}$$

$$\gamma_2' = \frac{\left\|G_{best}^{k}\right\|}{\sqrt{\left\|P_{best,j}^{k}\right\|^2 + \left\|G_{best}^{k}\right\|^2}} \tag{27}$$

$$\Delta V_{jd}^{k+1} = \left(V_{jd}^{k} - V_{jd}^{k+1}\right) \tag{28}$$

Remark 4. From (26) and (27), note that the attributes of γ_1' and γ_2' are the same as those of γ_1 and γ_2. However, compared with γ_1 and γ_2, γ_1' and γ_2' not only overcome the uncertainty of random variation but also have the capacity of adaptive adjustment. In addition, due to ΔV being introduced in the updating law (25), the smoothness of the convergence result can be guaranteed in the end of the algorithm.

As an important part of PSO algorithm, the fitness function in this paper is designed as

$$f_j^k = \sum_{d=1}^{D} \left(\left\|x_{jd}^{k} - x_{jd}^{k-1}\right\| + \left\|v_{jd}^{k} - v_{jd}^{k-1}\right\|\right) \tag{29}$$

where $D = n$ is the dimension of each particle; f_j^k is the fitness value of the jth particle in the kth iteration; x_{jd}^0 and v_{jd}^0 are the initial states of agents and $k \in (1, \cdots, K)$ represents the kth iteration.

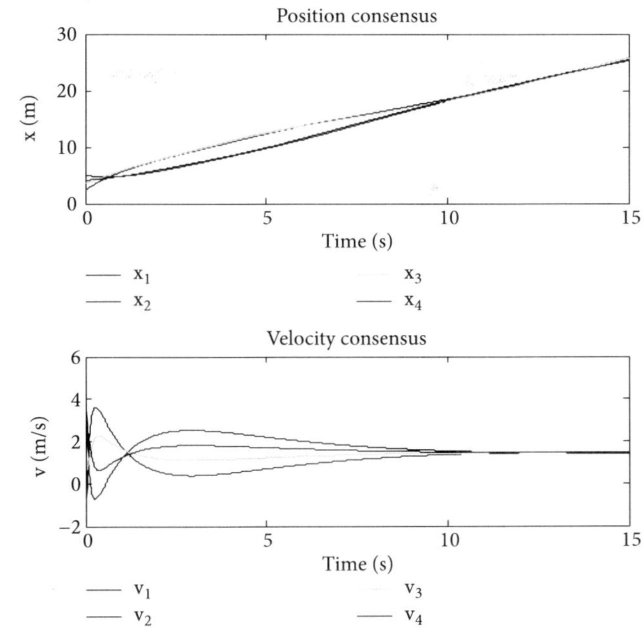

FIGURE 4: States consensus with control protocol (3).

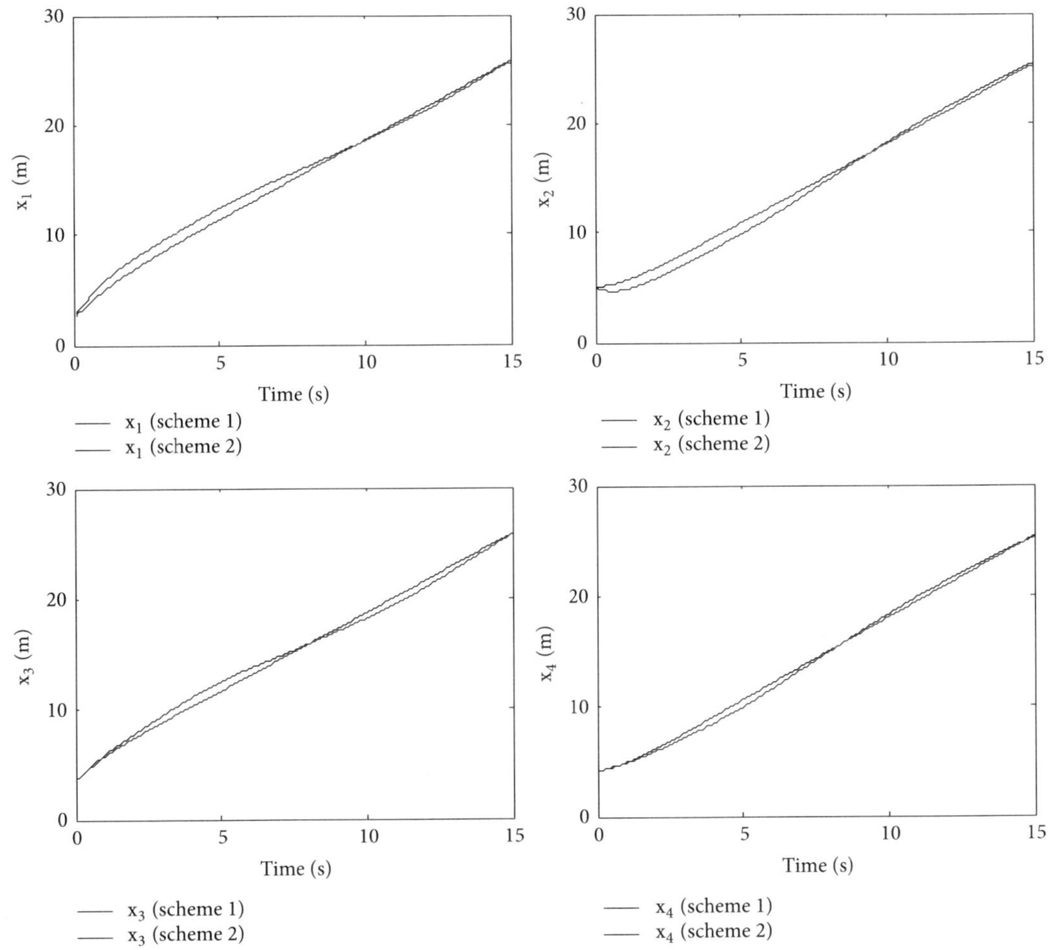

FIGURE 5: Position comparison results with standard PSO algorithm and control protocol (3).

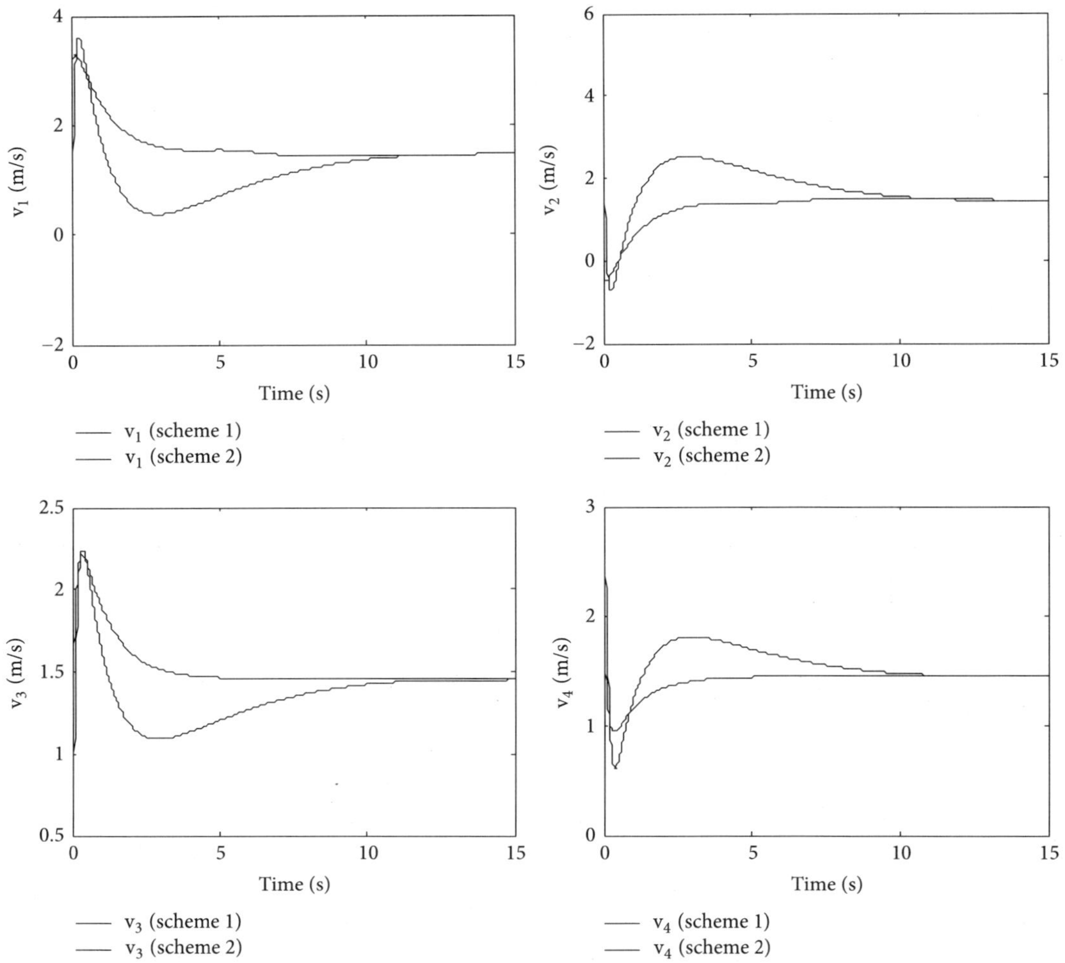

FIGURE 6: Velocity comparison results with PSO algorithm and control protocol (3).

Let $t_{\max}$ and Δt be the simulation time and simulation step, respectively. Hence, the maximum number of iterations K is defined as

$$K = \frac{t_{\max}}{\Delta t} \tag{30}$$

4.2. Mapping Relationship Design. For a multiagent systems with n agents, each agent is considered as a particle. Let x_i and v_i be the position and velocity of the ith agent, respectively. Then the sets of position and velocity are expressed as $X_{MAS} = (x_1, \cdots, x_n)$ and $V_{MAS} = (v_1, \cdots, v_n)$, respectively. In the PSO algorithm, let $X_{PSO,j} = (x_{j1}, \cdots, x_{jD})$ and $V_{PSO,j} = (v_{j1}, \cdots, v_{jD})$ be the position and velocity of the jth particle, respectively. Consequently, the mapping relationship between particles and agents is shown in Figure 1.

The operation process of PSO algorithm is given as follows:

Step 1. Initialize the population and the position and velocity of all agents.

Step 2. Assign the position and velocity x_i and v_i to each particle.

Step 3. Run system model and calculate the fitness function (29).

Step 4. Judge the termination codmitions and update X_{jd}^k and V_{jd}^k according to (24) and (25).

Step 5. Repeat Steps 2 and 4.

The flowchart is shown in Figure 2.

5. Simulation Analysis

Considering the second-order multiagent systems (2) with four agents, the exchange information topology graph $\mathcal{G}$ is shown in Figure 3.

The weighted adjacency matrix $\mathcal{A}$ is

$$\mathcal{A} = \begin{bmatrix} 0 & 1 & 1 & 0 \\ 1 & 0 & 0 & 1 \\ 1 & 0 & 0 & 1 \\ 0 & 1 & 1 & 0 \end{bmatrix} \tag{31}$$

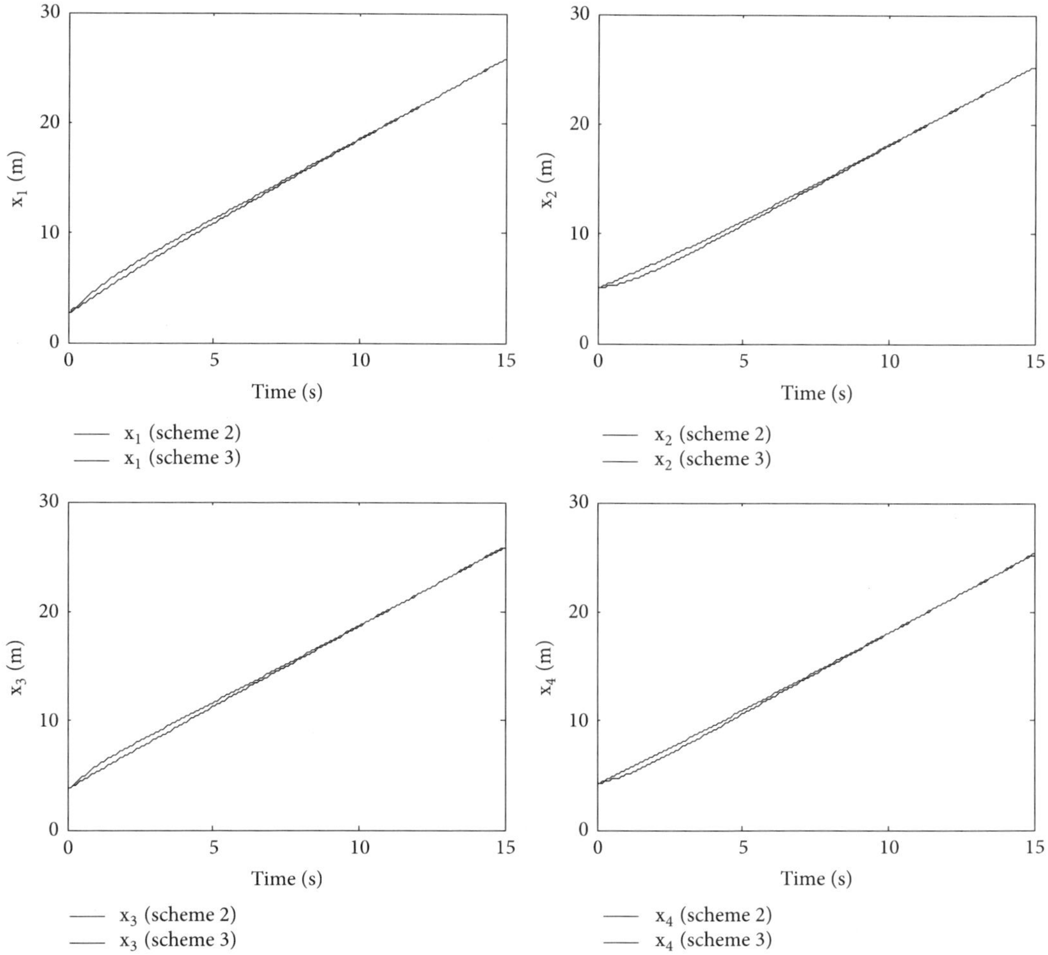

FIGURE 7: Position comparison results with standard PSO algorithm and improved PSO algorithm.

TABLE 1: List of parameters.

multi-agent systems	PSO algorithm
$n = 4$	$m = 100$
$\alpha = 2.5, \beta = 2.0$	$\omega = 0.2$
$t_{max} = 15s, \Delta t = 0.01s$	$\kappa_1 = 0.7, \kappa_2 = 1.5$
$x^0 = [2.5, 5, 3.6, 4]^T \ (m)$	$\lambda_1 = 0.4, \lambda_2 = 0.5$
$v^0 = [3.0, 3.7, 1.8, 1.4]^T \ (m/s)$	$K = 1500$

The parameter settings are shown in Table 1.

In addition, the updating laws (26), (27), and (28) are used for (24) and (25), respectively. According to the number of agents, it can be deducted that the dimension $D = 8$. To illustrate the validity of proposed control protocol (3) and the control effect of PSO algorithm, the following simulation examples are provided.

Example 5 (comparison the control protocol (3) and the standard PSO algorithm). In this example, we consider the consensus of multiagent systems (2) with control protocol (3) (scheme 1) and the standard PSO algorithm (22) and (23) (scheme 2); the simualtion results are shown in Figures 4–6.

The consensus of position and velocity with the control protocol (3) are shown in Figure 4. It can be found that the control effects are not good from Figure 4. Applying the standard PSO algorithm, the comparison results are drawn in Figures 5 and 6. Compared with control protocol (3), the consensus achieving time and the overshoot have been greatly improved by using the PSO algorithm.

Example 6 (comparison the standard PSO algorithm and the improved PSO algorithm). In this example, the consensus of the multiagent systems (2) with the standard PSO algorithm (22) and (23) (scheme 2) and the improved PSO algorithm (24) and (25) (scheme 3) is considered; the simualtion results are shown in Figures 7 and 8.

From Figures 7 and 8, it can be seen that the output of all agents have been further improved. Compared with the srtandard PSO alorithm, the improved PSO algorithm not only reduces the consensus achieving time and overshoot but also makes the curves smoother. In general, in the ability to improve the control effect of the multiagent systems, the

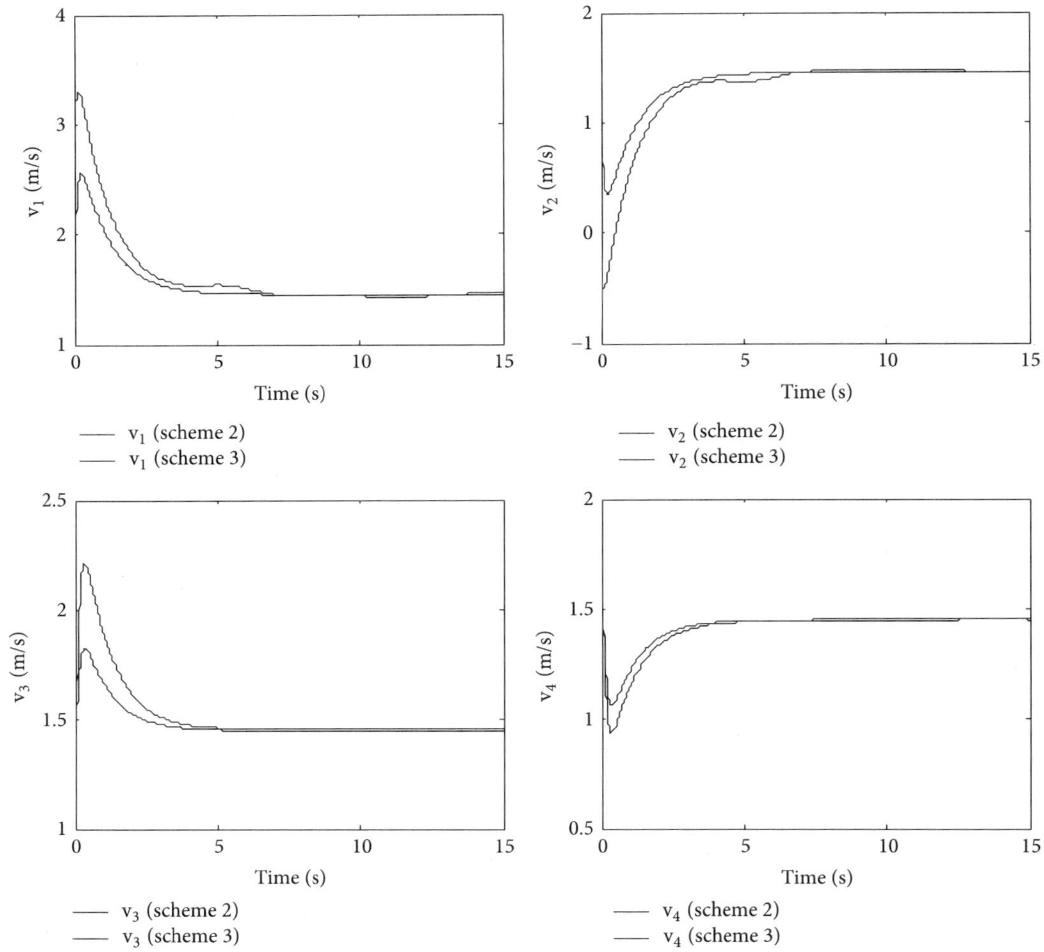

FIGURE 8: Velocity comparison results with standard PSO algorithm and improved PSO algorithm.

improved PSO algorithm proposed in this paper is superior to the other two proposed approaches.

6. Conclusion

In this paper, we discussed the consensus problem of a class of second-order multiagent systems with the PSO algorithm. An improved consensus protocol was designed, and then the convergence was analyzed by the Lyapunov stability theory. The PSO algorithm was introduced to improve the control effect of the multiagent systems. And then an improved PSO algorithm was proposed. In addition, we designed a mapping relationship with the agents and the individuals of PSO algorithm. Finally, the simulation results showed that the consensus with the proposed control protocol can be achieved. Also, the control effect can be improved by using PSO algorithm; and the better control effect can be obtained by utilizing the improved PSO algorithm.

Conflicts of Interest

The authors declare that there are no conflicts of interest regarding the publication of this paper.

Acknowledgments

This work is supported by the Aeronautical Science Foundation of China (Grant no. 20155896025).

References

[1] H. Chen, "Robust stabilization for a class of dynamic feedback uncertain nonholonomic mobile robots with input saturation," *International Journal of Control, Automation, and Systems*, vol. 12, no. 6, pp. 1216–1224, 2014.

[2] W. W. Yu, G. R. Chen, Z. D. Wang, and W. Yang, "Distributed consensus filtering in sensor networks," *IEEE Transactions on Systems, Man, and Cybernetics, Part B*, vol. 39, no. 6, pp. 1568–1577, 2009.

[3] W. Ren, R. W. Beard, and E. M. Atkins, "Information consensus in multivehicle cooperative control," *IEEE Control Systems Magazine*, vol. 27, no. 2, pp. 71–82, 2007.

[4] Y. L. Jiang, J. C. Liu, and S. Q. Wang, "Consensus tracking algorithm via observer-based distributed output feedback for multi-agent systems under switching topology," *Circuits, Systems and Signal Processing*, vol. 33, no. 10, pp. 3037–3052, 2014.

[5] X. F. Deng, X. X. Sun, and R. Liu, "Quantized consensus con-

trol for second-order nonlinear multi-agent systems with sliding mode iterative learning approach," *International Journal of Aeronautical and Space Sciences*, vol. 19, no. 2, pp. 518–533, 2018.

[6] X. F. Deng, X. X. Sun, R. Liu, and S. G. Liu, "Consensus control of time-varying delayed multiagent systems with high-order iterative learning control," *International Journal of Aerospace Engineering*, vol. 2018, pp. 1-12, 2018.

[7] D. H. Nguyen, "A sub-optimal consensus design for multi-agent systems based on hierarchical LQR," *Automatica*, vol. 55, pp. 88–94, 2015.

[8] J. Y. Ma, Y. S. Zheng, and L. Wang, "LQR-based optimal topology of leader-following consensus," *International Journal of Robust and Nonlinear Control*, vol. 25, no. 17, pp. 3404–3421, 2015.

[9] J. Q. Hu, J. D. Cao, K. Yuan, and T. Hayat, "Cooperative tracking for nonlinear multi-agent systems with hybrid time-delayed protocol," *Neurocomputing*, vol. 171, pp. 171–178, 2016.

[10] R. Abdulghafor, S. Turaev, A. Zeki, and I. Al-Shaikhli, "Reach a nonlinear consensus for MAS via doubly stochastic quadratic operators," *International Journal of Control*, vol. 91, no. 6, pp. 1431–1459, 2018.

[11] Y. M. Xin, Y. X. Li, X. Huang, and Z. S. Cheng, "Consensus of third-order nonlinear multi-agent systems," *Neurocomputing*, vol. 159, no. 1, pp. 84–89, 2015.

[12] G. Q. Hu, "Robust consensus tracking of a class of second-order multi-agent dynamic systems," *Systems Control Letters*, vol. 61, no. 1, pp. 134–142, 2012.

[13] D. D. Li, J. Ma, H. M. Zhu, and M. Sun, "The consensus of multi-agent systems with uncertainties and randomly occurring nonlinearities via impulsive control," *International Journal of Control, Automation, and Systems*, vol. 14, no. 4, pp. 1005–1011, 2016.

[14] W. Wang, D. Wang, Z. H. Peng, and H. Wang, "Cooperative adaptive fuzzy output feedback control for synchronization of nonlinear multi-agent systems in the presence of input saturation," *Asian Journal of Control*, vol. 18, no. 2, pp. 619–630, 2016.

[15] X. F. Zhang, Y. S. Ding, and Z. X. Liang, "Adaptive co-evolution model for multi-agent system," *Journal of Donghua University*, vol. 27, no. 2, pp. 285–288, 2010.

[16] D. K. He, G. Y. Gao, F. L. Wang, and Y. Liu, "Multi-agent differential evolution algorithm," *Control and Decision*, vol. 26, no. 7, pp. 961–966, 2011.

[17] D. S. Shu, Z. X. Huang, J. Y. Li, and X. Y. Zuo, "Application of multi-agent particle swarm algorithm in distribution network reconfiguration," *Journal of Electronics*, vol. 25, no. 6, pp. 1179–1185, 2016.

[18] R. C. Eberhart and J. Kennedy, "A new optimizer using particle swarm theory," in *Proceedings of the 6th International Symposium on Micro Machine and Human Science (MHS '95)*, pp. 39–43, Nagoya, Japan, October 1995.

[19] E. D. Mario and A. Martinoli, "Distributed particle swarm optimization for limited-time adaptation with real robots," *Robotica*, vol. 32, no. 2, pp. 193–208, 2014.

[20] A. Banks, J. Vincent, and K. Phalp, "Particle swarm guidance system for autonomous unmanned aerial vehicles in an air defence role," *Journal of Navigation*, vol. 61, no. 1, pp. 9–29, 2008.

[21] J. L. Foo, J. S. Knutzon, V. Kalivarapu, J. H. Oliver, and E. Wimer, "Path planning of unmanned aerial vehicles using B-splines and particle swarm optimization," *Journal of Aerospace Computing, Information and Communication*, vol. 6, no. 4, pp. 271–290, 2009.

[22] D. A. Li, H. L. Hao, G. L. Ji, and J. M. Zhao, "An adaptive clustering algorithm based on improved particle swarm optimisation in wireless sensor networks," *International Journal of High Performance Computing and Networking*, vol. 8, no. 4, pp. 370–380, 2015.

[23] M. Pontani and B. A. Conway, "Particle swarm optimization applied to space trajectories," *Journal of Guidance, Control, and Dynamics*, vol. 33, no. 5, pp. 1429–1441, 2010.

[24] W. Ren, "On consensus algorithms for double-integrator dynamics," *IEEE Transactions on Automatic Control*, vol. 53, no. 6, pp. 1503–1509, 2008.

[25] W. Ren and E. Atkins, "Distributed multi-vehicle coordinated control via local information exchange," *International Journal of Robust and Nonlinear Control*, vol. 17, no. 10-11, pp. 1002–1033, 2007.

[26] W. W. Yu, G. R. Chen, and M. Cao, "Some necessary and sufficient conditions for second-order consensus in multi-agent dynamical systems," *Automatica*, vol. 46, no. 6, pp. 1089–1095, 2010.

5

Solution of Some Types for Composition Fractional Order Differential Equations Corresponding to Optimal Control Problems

Sameer Qasim Hasan and Moataz Abbas Holel

Department of Mathematics, University of Mustansiriyah, Baghdad, Iraq

Correspondence should be addressed to Sameer Qasim Hasan; dr.sameer_kasim@yahoo.com

Academic Editor: Darong Huang

The approximate solution for solving a class of composition fractional order optimal control problems (FOCPs) is suggested and studied in detail. However, the properties of Caputo and Riemann-Liouville derivatives are also given with complete details on Chebyshev approximation function to approximate the solution of fractional differential equation with different approach. Also, the relation between Caputo and Riemann-Liouville of fractional derivative took a big role for simplifying the fractional differential equation that represents the constraints of optimal control problems. The approximate solutions are defined on interval [0, 1] and are compared with the exact solution of order one which is an important condition to support the working method. Finally, illustrative examples are included to confirm the efficiency and accuracy of the proposed method.

1. Introduction

The idea of fractional derivative dates back to a conversation between two mathematicians: Leibniz and L'Hopital. In 1695, they exchanged about the meaning of a derivative of order 1/2. Their correspondence has been well documented and is stated as the foundation of fractional calculus [1].

A fractional optimal control problem (FOCPs) is an optimal control problem focused on the performance index or the fractional differential equations governing the dynamics of the system or both contain at least one fractional order derivative term.

The formulation and solution of state and variables FOCPs were first established by Agrawal, where the applied fractional variational calculus (FVC) presented a general formulation and solution scheme for FOCPs in the Riemann-Liouville (RL) sense; it was based on variational virtual work coupled with the Lagrange multiplier technique. Since the Caputo fractional derivatives (CFDs) seems more natural and allows incorporating the usual initial conditions, it becomes a popular choice for researchers [2, 3].

The Chebyshev polynomials are used to solve composition FOCPs and Chebyshev polynomials$T_n(t)$ of degree n are important in approximation theory, since the roots of the Chebyshev polynomials of the first kind, which are also called Chebyshev nodes, are used as nodes in polynomial interpolation.

Agrawal and Baleanu [4] obtained necessary conditions for FOCPs with Riemann-Liouville derivative and then were able to solve the problem numerically.

Elbarbary introduced Chebyshev finite difference approximation for the boundary value problems of integer derivatives [5]. In [6], Khader and Hendy studied an efficient numerical scheme for solving fractional optimal control problems. In [7], Akbarian and Keyanpour studied a new approach to the numerical solution of fractional order optimal control Problems.

This paper presents a new approach for computing the approximate optimal control function for optimal control problems with some types of fractional order differential equation such as multi-fractional order differential equation, and composite fractional order differential equations are investigated with first kind Chebyshev approximate polynomial which works very efficiently with fractional orders and optimal control problems. All the discussion and details in addition to the transformations of the steps are given to compute all the useful of the new work. The obtained solutions of the method show that the technique of the approach is very convenient and efficient, and many calculations give high accuracy and may lead to near to exact solution by changing up the order of fractional positive integer numbers.

The multi-fractional differential equations corresponding to optimal control problems and basic theorems have been given with algorithm for multi-composite fractional order optimal control problems. Also, we give illustrative examples for the solution of the approximate systems.

The value of $x(t)$ is observed in detail to explain the activity of solution approximation with mixed boundary conditions. It is important to notice that the calculations are written by using the mathematical software MATHCAD by version (14.0.3.332).

This paper consists of the following sections: in Section 2, some basic definitions and properties of fractional order calculus are introduced (R-L and Caputo fractional derivatives). In Section 3, the shifted Chebyshev polynomials and numerical approximations of CFD and RLFD using Chebyshev polynomials are introduced. In Section 4, we derive the necessary optimality conditions of composition order fractional optimal control problems. In Section 5, we give numerical examples to solve composition FOCPs and show the accuracy of the presented method. Finally, conclusions are presented in Section 6.

2. Preliminaries

In this section, we give some definitions and properties of fractional derivatives that will be needed later on.

2.1. Fractional Order Calculus

Definition 1 (see [8]). The left (LRLFD) and right (RRLFD) Riemann-Liouville fractional derivatives of a function $f(t)$ are defined as

$$ {}_aD_t^\alpha f(t) = \frac{1}{\Gamma(n-\alpha)}\frac{d^n}{dt^n}\int_a^t (t-\tau)^{n-\alpha-1} f(\tau)\,d\tau, \quad t>a \tag{1}$$

$$ {}_tD_b^\alpha f(t) = \frac{1}{\Gamma(n-\alpha)}(-1)^n\frac{d^n}{dt^n}\int_t^b (\tau-t)^{n-\alpha-1} f(\tau)\,d\tau, \quad t<b. \tag{2}$$

where the order of the derivative α satisfies $n-1<\alpha\le n$, and $\Gamma(n)$ is the gamma function.

Definition 2 (see [8]). The left (LCFD) and right (RCFD) Caputo fractional derivatives of a function $f(t)$ are defined as

$$ {}_a^CD_t^\alpha f(t) = \frac{1}{\Gamma(n-\alpha)}\int_a^t (t-\tau)^{n-\alpha-1} f^{(n)}(\tau)\,d\tau. \quad (\alpha>0,\ t>0) \tag{3}$$

$$ {}_t^CD_b^\alpha f(t) = \frac{(-1)^n}{\Gamma(n-\alpha)}\int_t^b (\tau-t)^{n-\alpha-1} f^{(n)}(\tau)\,d\tau, \tag{4}$$

where $n-1<\alpha\le n,\ n\in N$.

2.2. Some Properties of the Fractional Calculus Are Presented in Detail Which Will Be Needed Later on. (i) The fractional operators are linear [9]:

$$P[\lambda f(t)+\mu g(t)] = \lambda Pf(t)+\mu Pg(t), \tag{5}$$

where P is one of ${}_aD_t^\alpha$, ${}_tD_b^\alpha$, ${}_a^CD_t^\alpha$, ${}_t^CD_b^\alpha$, ${}_aD_t^{-\alpha}$ or ${}_tD_b^{-\alpha}$ and λ, μ are real numbers.

(ii) Let $0<\alpha<1$, $n-1<\alpha+\beta<n$, $n-1<\beta<n$, n is a positive integer, $f\in C^n[a,t]$, and if $x^{(i)}(0)=0$, $i=0,1,\ldots\ldots n-1$.) [9], then

$$ {}_a^CD_t^\alpha\left({}_a^CD_t^\beta x(t)\right) = {}_a^CD_t^\beta\left({}_a^CD_t^\alpha x(t)\right) = {}_a^CD_t^{\alpha+\beta} x(t). \tag{6}$$

(iii) If $0<\alpha<1$, we obtain the following [2]:

$$\begin{aligned}\int_a^b g(t)\cdot {}_a^CD_t^\alpha f(t)\,dt &= \int_a^b f(t)\cdot {}_tD_b^\alpha g(t)\,dt + \left[{}_tI_b^{1-\alpha} g(t)\cdot f(t)\right]_a^b\\ \int_a^b g(t)\cdot {}_t^CD_b^\alpha f(t)\,dt &= \int_a^b f(t)\cdot {}_aD_t^\alpha g(t)\,dt + \left[{}_aI_t^{1-\alpha} g(t)\cdot f(t)\right]_a^b.\end{aligned} \tag{7}$$

Moreover, if f is a function such that $f(a)=f(b)=0$, we have simpler formulas:

$$\int_a^b g(t)\cdot {}_a^CD_t^\alpha f(t)\,dt = \int_a^b f(t)\cdot {}_tD_b^\alpha g(t)\,dt, \tag{8}$$

$$\int_a^b g(t)\cdot {}_t^CD_b^\alpha f(t)\,dt = \int_a^b f(t)\cdot {}_aD_t^\alpha g(t)\,dt. \tag{9}$$

(iv) The relation between the RLFD and the CFD is as follows [2].

For $\alpha>0$ and $n=[\alpha]+1$, the Riemann-Liouville and Caputo fractional derivatives are related by the following formulas:

$$ {}_aD_t^\alpha f(t) = {}_a^CD_t^\alpha f(t) + \sum_{k=0}^{n-1}\frac{f^{(k)}(a)}{\Gamma(k+1-\alpha)}(t-a)^{k-\alpha}, \tag{10}$$

$$ {}_tD_b^\alpha f(t) = {}_t^CD_b^\alpha f(t) + \sum_{k=0}^{n-1}\frac{f^{(k)}(b)}{\Gamma(k+1-\alpha)}(b-t)^{k-\alpha}. \tag{11}$$

(v) The constant function and power function of Caputo's derivative are as follows:

$$(1)\ {}_{a}^{C}D_{t}^{\alpha}C = 0, \quad \text{where } C \text{ is a constant.} \tag{12}$$

In particular, ${}_{a}^{C}D_{t}^{\alpha}1 = 0$, and, also from right integral, ${}_{t}^{C}D_{b}^{\alpha}1 = 0$.

$$(2)\ {}_{a}^{C}D_{t}^{\alpha}(t-a)^{\beta}$$

$$= \begin{cases} 0, & \text{for } \beta \in N_0 \text{ and } \beta < \lceil \alpha \rceil, \\ \dfrac{\Gamma(\beta+1)}{\Gamma(\beta-\alpha+1)}(t-a)^{\beta-\alpha}, & \text{for } \beta \in N_0 \text{ and } \beta \geq \lceil \alpha \rceil, \end{cases} \tag{13}$$

where $\lceil \alpha \rceil$ denotes the smallest integer greater than or equal to α and $N_0 = \{0, 1, 2, \ldots\}$. Recall that for $\alpha \in N$, [10].

3. The Shifted Chebyshev Polynomials

The well-known Chebyshev polynomials are defined on the interval $[-1, 1]$ and can be determined by the following recurrence formula [11]:

$$\begin{aligned} T_n(t) &= 2tT_{n-1}(t) - T_{n-2}(t), \\ T_0(t) &= 1, \\ T_1(t) &= t, \\ & \qquad n = 2, 3 \ldots. \end{aligned} \tag{14}$$

The Chebyshev polynomials can be expanded in power series as follows:

$$T_n(t) = \frac{n}{2}\sum_{m=0}^{\lfloor n/2 \rfloor}(-1)^m \frac{(n-m-1)!}{(m)!\,(n-2m)!}(2t)^{n-2m}, \tag{15}$$

$n = 0, 1, 2, \ldots$, $m = 0, 1, 2, \ldots, \lfloor n/2 \rfloor$ and $\lfloor n/2 \rfloor$ denotes the biggest integral less than or equal to $n/2$.

The Chebyshev polynomials $T_n(t)$ $(n = 0, 1, 2, \ldots)$ are orthogonal under integration over $[-1, 1]$ with the weighting function $w(t) = 1/\sqrt{1-t^2}$, with orthogonal condition:

$$\int_{-1}^{1}\frac{T_i(t)\,T_j(t)}{\sqrt{1-t^2}}dt = \begin{cases} 0 & \text{if } i \neq j, \\ \dfrac{\pi}{2} & \text{if } i = j \neq 0, \\ \pi & \text{if } i = j = 0. \end{cases} \tag{16}$$

where $i, j = 0, 1, 2 \ldots$.

In order to use these polynomials on the interval $[0, L]$ we define the so-called shifted pseudo-spectral Chebyshev polynomials by introducing the change of variable $z = 2t/L - 1$.

The shifted Chebyshev polynomials are defined as follows [12]:

$$T_n^p(t) = T_n\left(\frac{2t}{L} - 1\right), \quad \text{where } T_0^p(t) = 1,\ T_1^p(t) = \frac{2t}{L} - 1. \tag{17}$$

The analytic form of the shifted pseudo-spectral Chebyshev polynomial $T_n^p(t)$ of degree n is given by the following [10]:

$$T_n^p(t) = n\sum_{k=0}^{n}(-1)^{n-k}\frac{2^{2k}(n+k-1)!}{(2k)!\,(n-k)!L^k}t^k, \quad n = 1, 2 \ldots, \tag{18}$$

where $T_n^p(0) = (-1)^n$, and $T_n^p(L) = 1$.

The orthogonality relation is as follows:

$$\int_0^L T_n^p(t)\,T_m^p(t)\,w^p(t)\,dt = \delta m, \tag{19}$$

with the weight function $w^p(t) = 1/\sqrt{Lt - t^2}$, $\delta m = (b_k/2)\pi$, $b_0 = 2$, $b_k = 1$ for $k \geq 1$.

The function $x(t) \in L^2([0, L])$ can be expressed in terms of shifted Chebyshev polynomials as follows:

$$x(t) = \sum_{j=0}^{\infty}c_n T_n^p(t), \tag{20}$$

where the coefficients c_n are given by

$$c_n = \frac{1}{\delta n}\int_0^L x(t)\,T_n^p(t)\,w^p(t)\,dt, \quad n = 0, 1, \ldots. \tag{21}$$

3.1. The Chebyshev-Gauss-Lobatto Points. We choose the Chebyshev-Gauss-Lobatto points associated with the interval $[0, L]$, as follows:

$$t_r = \frac{L}{2} - \frac{L}{2}\cos\left(\frac{\pi r}{N}\right), \quad r = 0, 1, \ldots N. \tag{22}$$

These grids can be written as $L < x_N < x_{N-1} \cdots < x_1 < x_0 = 0$.

Clenshaw and Curtis [13] introduced an approximation of the function $x(t)$, as follows:

$$\begin{aligned} x_N(t) &= \sum_{n=0}^{N}{}'' a_n T_n^p(t) \\ a_n &= \frac{2}{N}\sum_{r=0}^{N}{}'' x(t_r)\,T_n^p(t_r). \end{aligned} \tag{23}$$

where $('')$ on the summation means that the first and last terms are to be taken with a factor $(1/2)$.

Theorem 3 (see [2]). *The fractional derivative of order α in the Caputo sense for the function $x(t)$ at the point t_s is given by*

$$ {}_0^C D_t^{\alpha} x_N(t_s) = \sum_{r=0}^{N} d_{s,r}^{(\alpha)} x(t_r), \quad \alpha > 0, \tag{24}$$

Such that

$$ d_{s,r}^{(\alpha)} = \frac{4\theta_r}{N} \sum_{n=\lceil\alpha\rceil}^{N} \sum_{j=0}^{N} \sum_{k=\lceil\alpha\rceil}^{n} \frac{n\theta_n}{b_j} \cdot \frac{(-1)^{n-k}(n+k-1)!\Gamma(k-\alpha+1/2)\, T_n^p(t_r)\, T_j^p(t_s)}{L^{\alpha}\Gamma(k+1/2)(n-k)!\Gamma(k-\alpha-j+1)\Gamma(k-\alpha+j+1)}, \tag{25}$$

where

$$ \begin{aligned} & s,r = 0,1,2,\dots,N \text{ and} \\ & \theta_0 = \theta_N = \frac{1}{2}, \\ & \theta_i = 1 \quad \forall i = 1,2,\dots,N-1. \end{aligned} \tag{26}$$

Proof. The fractional derivative of the approximate formula for the function $x(t)$ in (22) is given by

$$ \begin{aligned} {}_0^C D_t^{\alpha} x_N(t) &= \sum_{n=0}^{N}{}'' a_n\, {}_0^C D_t^{\alpha} T_n^p(t) \\ {}_0^C D_t^{\alpha} x_N(t) & \\ &= \sum_{n=0}^{N}{}'' a_n\, {}_0^C D_t^{\alpha} \left(n \sum_{k=0}^{n} (-1)^{n-k} \frac{2^{2k}(n+k-1)!}{(2k)!(n-k)!L^k} t^k \right) \end{aligned} \tag{27}$$

Using (12) and (13), in (18) we have

$$ {}_0^C D_t^{\alpha} T_n^p(t) = 0, \quad n = 0,1,\dots,\lceil\alpha\rceil - 1, \tag{28}$$

then

$$ {}_0^C D_t^{\alpha} x_N(t) = \sum_{n=\lceil\alpha\rceil}^{N}{}'' a_n\, {}_0^C D_t^{\alpha} T_n^p(t), \tag{29}$$

Therefore, for $n = \lceil\alpha\rceil, \lceil\alpha+1\rceil, \dots, N$ and by (12) and (13), in for formula (18), we get

$$ \begin{aligned} {}_0^C D_t^{\alpha} T_n^p(t) &= n \sum_{k=0}^{n} (-1)^{n-k} \frac{2^{2k}(n+k-1)!}{(2k)!(n-k)!L^k}\, {}_0^C D_t^{\alpha} t^k \\ &= n \sum_{k=\lceil\alpha\rceil}^{n} (-1)^{n-k} \frac{2^{2k}(n+k-1)!k!}{(2k)!(n-k)!L^k\Gamma(k-\alpha+1)} t^{k-\alpha}. \end{aligned} \tag{30}$$

Now, $t^{k-\alpha}$ can be expressed approximately in terms of shifted pseudo-spectral Chebyshev polynomial series, so we have

$$ t^{k-\alpha} \cong \sum_{j=0}^{N} a_{kj} T_j^p(t), \tag{31}$$

where a_{kj} is obtained from (21) with $x(t) = t^{k-\alpha}$. If only the first $(N+1)$-terms from pseudo-spectral shifted Chebyshev polynomials in (23) are considered, the approximate formula for the fractional derivative of the shifted pseudo-spectral Chebyshev polynomials is as follows:

$$ {}_0^C D_t^{\alpha} T_n^p(t) = \sum_{j=0}^{N} \sum_{k=\lceil\alpha\rceil}^{n} \frac{(-1)^{n-k} 2n(n+k-1)!\Gamma(k-\alpha+1/2)\, T_j^p(t)}{b_j L^{\alpha}\Gamma(k+1/2)(n-k)!\Gamma(k-\alpha-j+1)\Gamma(k-\alpha+j+1)}. \tag{32}$$

From (29) and (32), we have

$$ {}_0^C D_t^{\alpha} x_N(t) = \frac{4}{N} \sum_{n=\lceil\alpha\rceil}^{N}{}'' \sum_{r=0}^{N}{}'' \sum_{j=0}^{N} \sum_{k=\lceil\alpha\rceil}^{n} \frac{(-1)^{n-k} n(n+k-1)!\Gamma(k-\alpha+1/2)\, x(t_r)\, T_n^p(t_r)\, T_j^p(t)}{b_j L^{\alpha}\Gamma(k+1/2)(n-k)!\Gamma(k-\alpha-j+1)\Gamma(k-\alpha+j+1)}. \tag{33}$$

From (33), the fractional derivative of order α for the function $x(t)$ at the point t_s leads to the desired result. □

Theorem 4 (see [2]). *Let ${}_0^C D_t^{\alpha} x_N(t)$ be the approximation of the fractional derivative ${}_0^C D_t^{\alpha}$ of the function $x(t)$ as given by (24). Then it holds*

$$ \begin{aligned} & \left\| {}_0^C D_t^{\alpha} x(t) - {}_0^C D_t^{\alpha} x_N(t) \right\|_2 \\ & \le \sum_{n=0}^{N}{}'' a_n \Omega_n \left(\frac{F\left(t^{k-\alpha}; T_0^p, \dots, T_N^p\right)}{F\left(T_0^p, \dots, T_N^p\right)} \right)^{1/2}, \end{aligned} \tag{34}$$

where

$$\Omega_n = \sum_{k=\lceil\alpha\rceil}^{n} \frac{(-1)^{n-k}\, 2n\,(n+k-1)!\,\Gamma(k-\alpha+1/2)}{b_j L^{\alpha}\Gamma(k+1/2)\,(n-k)!\,\Gamma(k-\alpha-j+1)\,\Gamma(k-\alpha+j+1)}, \tag{35}$$

$$F(x; y_1, y_2, \dots y_n) = \begin{vmatrix} \langle x, x\rangle & \langle x, y_1\rangle & \cdots & \langle x, y_n\rangle \\ \langle y_1, x\rangle & \langle y_1, y_1\rangle & \cdots & \langle y_1, y_n\rangle \\ \vdots & \vdots & \ddots & \vdots \\ \langle y_n; x\rangle & \langle y_n; y_1\rangle & \cdots & \langle y_n; y_n\rangle \end{vmatrix}. \tag{36}$$

3.2. Approximation of the Right Riemann-Liouville Fractional Derivative. Using the right Caputo fractional derivative (RCFD) of a function $f(t)$ is defined in (4) when $0 < \alpha < 1$ and $f(t) \in AC\,[a,b]$,

$$ {}_{s}^{c}D_b^{\alpha} f(s) = \frac{-1}{\Gamma(1-\alpha)} \int_s^b (t-s)^{-\alpha} f'(t)\,dt, \tag{37}$$

From relation between the RLFD and the CFD (11), when $0 < \alpha < 1$, (that is $n = 1$), we have

$$ {}_{s}D_b^{\alpha} f(s) = \frac{f(b)}{\Gamma(1-\alpha)} (b-s)^{-\alpha} + {}_{s}^{c}D_b^{\alpha} f(s), \tag{38}$$

Use (37) in (38) to obtain

$$ {}_{s}D_b^{\alpha} f(s) = \frac{f(b)}{\Gamma(1-\alpha)} (b-s)^{-\alpha} + \frac{1}{\Gamma(1-\alpha)} \int_s^b (t-s)^{-\alpha} f'(t)\,dt \tag{39}$$

Let f be a sufficiently smooth function in $[0,b]$ and let $J(s; f)$ be defined as follows:

$$ J(s; f) = \int_s^b (t-s)^{-\alpha} f'(t)\,dt, \quad 0 < s < b. \tag{40}$$

Substituting (40) in (39), we deduce

$$ {}_{s}D_b^{\alpha} f(s) = \frac{f(b)}{\Gamma(1-\alpha)} (b-s)^{-\alpha} + \frac{J(s; f)}{\Gamma(1-\alpha)}. \tag{41}$$

Now, we approximate $f(t)$, $0 \le t \le b$, by a sum of shifted Chebyshev polynomials $T_k(2t/b - 1)$ according to

$$ f(t) \approx p_N(t) = \sum_{k=0}^{N}{}'' a_k T_k\left(\frac{2t}{b} - 1\right), \qquad a_k = \frac{2}{N} \sum_{j=0}^{N}{}'' f(t_j)\, T_k\left(\frac{2t_j}{b} - 1\right), \tag{42}$$

where

$$ t_j = \frac{b}{2} - \frac{b}{2}\cos\left(\frac{\pi j}{N}\right), \quad j = 0, \dots, N, \text{ and obtain} \tag{43}$$

$$ J(s; f) \approx J(s; p_N) = \int_s^b p_N{}'(t)\,(t-s)^{-\alpha}\,dt.$$

Lemma 5 (see [7]). *Let p_N be the polynomial of degree N as given by (42); then there exists a polynomial F_{N-1} of degree $N-1$ such that*

$$ \int_s^x \left[p_N{}'(t) - p_N{}'(s)\right](t-s)^{-\alpha}\,dt = \left[F_{N-1}(x) - F_{N-1}(s)\right](x-s)^{1-\alpha}. \tag{44}$$

Proof. Let $p_N{}'(t) - p_N{}'(s)$ be expanded in a Taylor series at $t = s$

$$ p_N{}'(t) - p_N{}'(s) = \sum_{k=1}^{N-1} A_k(s)\,(t-s)^k. \tag{45}$$

Then,

$$ \begin{aligned} \int_s^x \left[p_N{}'(t) - p_N{}'(s)\right](t-s)^{-\alpha}\,dt &= \sum_{k=1}^{N-1} A_k(s) \int_s^x (t-s)^{k-\alpha}\,dt \\ &= \left[(t-s)^{1-\alpha} \sum_{k=1}^{N-1} \frac{A_k(s)\,(t-s)^k}{k-\alpha+1}\right]_s^x. \end{aligned} \tag{46}$$

The assertion follows, if we choose $F_{N-1}(x) = \sum_{k=0}^{N-1}(A_k(s)(x-s)^k/(k-\alpha+1))$, with an arbitrary constant $A_0(s)$. From (44), we have

$$ \begin{aligned} J(s; p_N) &= \int_s^b p_N{}'(t)\,(t-s)^{-\alpha}\,dt \\ &= \left[\frac{p_N{}'(s)}{1-\alpha} + \left[F_{N-1}(b) - F_{N-1}(s)\right]\right](b-s)^{1-\alpha} \end{aligned} \tag{47}$$

Moreover, ${}_sD_b^\alpha f(s)$ can be approximated by means of

$$ {}_sD_b^\alpha f(s) \approx \frac{f(b)}{\Gamma(1-\alpha)}(b-s)^{-\alpha} + \frac{J(s;p_N)}{\Gamma(1-\alpha)}. \quad (48)$$

We express $F_{N-1}(t)$ in (47) by a sum of Chebyshev polynomials and provide the recurrence relation satisfied by the Chebyshev coefficients. Differentiating both sides of (44) with respect to x yields

$$\begin{aligned}&\left[p_N'(x) - p_N'(s)\right](x-s)^{-\alpha}\\ &= F_{N-1}'(x)(x-s)^{1-\alpha}\\ &\quad + \left[F_{N-1}(x) - F_{N-1}(s)\right](1-\alpha)(x-s)^{-\alpha}.\end{aligned} \quad (49)$$

$$\begin{aligned}&p_N'(x) - p_N'(s)\\ &= F_{N-1}'(x)(x-s)\\ &\quad + \left[F_{N-1}(x) - F_{N-1}(s)\right](1-\alpha).\end{aligned} \quad (50)$$

To evaluate $F_{N-1}(s)$ in (47), we expand $F_{N-1}'(x)$ in terms of the shifted Chebyshev polynomials:

$$F_{N-1}'(x) = {\sum_{k=0}^{N-2}}' b_k T_k\left(\frac{2x}{b}-1\right), \quad 0 \le x \le b, \quad (51)$$

where (′) on the summation means that the first term is to be taken with a factor (1/2).

$$\begin{aligned}&F_{N-1}(x) - F_{N-1}(s)\\ &= \frac{b}{4}\sum_{k=1}^{N-1}\frac{b_{k-1}-b_{k+1}}{k}\left(T_k\left(\frac{2x}{b}-1\right) - T_k\left(\frac{2s}{b}-1\right)\right),\end{aligned} \quad (52)$$

where $b_{N-1} = b_N = 0$. On the other hand, we have

$$\begin{aligned}&(x-s)F_{N-1}'(x)\\ &= \frac{b}{2}F_{N-1}'(x)\left[\left(\frac{2x}{b}-1\right) - \left(\frac{2s}{b}-1\right)\right].\end{aligned} \quad (53)$$

By using the relation $T_{k+1}(v) + T_{k-1}(v) = 2vT_k(v)$ and from (51), it follows that

$$\begin{aligned}&(x-s)F_{N-1}'(x) = \frac{b}{4}\\ &\cdot {\sum_{k=1}^{N-1}}'\left(b_{k+1} - 2\left(\frac{2s}{b}-1\right)b_k + b_{k-1}\right)T_k\left(\frac{2x}{b}-1\right),\end{aligned} \quad (54)$$

Such that $b_{-1} = b_1$.

Let

$$p_N'(x) = {\sum_{k=0}^{N-1}}' c_k T_k\left(\frac{2x}{b}-1\right). \quad (55)$$

Inserting $F_{N-1}(x) - F_{N-1}(s)$ and $(x-s)F_{N-1}'(x)$ as given (52) and (54) into (50) and taking (55) into account, we get

$$\begin{aligned}&\left(1-\frac{1-\alpha}{k}\right)b_{k+1} - 2\left(\frac{2s}{b}-1\right)b_k + \left(1+\frac{1-\alpha}{k}\right)b_{k-1}\\ &= \frac{4}{b}c_k, \quad 1 \le k.\end{aligned} \quad (56)$$

The Chebyshev coefficients c_k of $p_N'(x)$ as given by (55) can be evaluated by integrating and comparing it with (42):

$$c_{k-1} = c_{k+1} + \frac{4k}{b}a_k, \quad k = N, N-1, \ldots, 1, \quad (57)$$

with starting values $c_N = c_{N+1} = 0$, where a_k are the Chebyshev coefficients of $p_N(x)$. □

4. The Necessary Optimality Conditions of Composition Order Fractional Optimal Control Problems $M\,{}_a^CD_t^\alpha(t) + N\,{}_a^CD_t^\beta({}_a^CD_t^\mu x(t)) = f(t, x(t), u(t))$

In this problem, the constraint has multiorder and composition for the three fractional derivatives α, β, and μ, where α, β and $\mu > 0$. The multi-order fractional optimal control problems refers to the minimization of an objective functional subject to dynamical constraints on the state and the control which have three fractional derivatives order models.

The necessary optimality conditions of this type are introduced as follows.

Let α, β and μ be real numbers and α, β and $\mu \in (0,1)$, and let f_0 and f be two differentiable functions with domain $[a,\infty[\times\mathbb{R}^2$. We consider a general form of multicomposition fractional optimal control problem:

$$J(x,u,T) = \int_a^T f_0(t, x(t), u(t))\,dt, \quad (58)$$

which subject to the multi derivative-composite fractional dynamic control system,

$$M\,{}_a^CD_t^\alpha(t) + N\,{}_a^CD_t^\beta\left({}_a^CD_t^\mu x(t)\right) = f(t, x(t), u(t)), \quad (59)$$

and the boundary conditions,

$$\begin{aligned}x(a) &= x_a,\\ x(T) &= x_T.\end{aligned} \quad (60)$$

where $M, N \neq 0, T,$ x_a *and* x_T are fixed real numbers.

Theorem 6. *If (x,u,T) is a minimizer of (58), (59), and (60), then there exists a function $\lambda(t)$ for which (x,u,T) satisfies the multi-composite fractional optimality conditions:*

(i) The Hamiltonian system is defined as

$$\begin{aligned} &M\ {}_tD_T^{\alpha}\lambda(t) + N\ {}_tD_T^{\beta}\left({}_tD_T^{\mu}\lambda(t)\right) \\ &\quad = \frac{\partial H}{\partial x}(t, x(t), u(t), \lambda(t)) \\ &M\ {}_a^CD_t^{\alpha}x(t) + N\ {}_a^CD_t^{\beta}\left({}_a^CD_t^{\mu}x(t)\right) \\ &\quad = \frac{\partial H}{\partial \lambda}(t, x(t), u(t), \lambda(t)). \end{aligned} \tag{61}$$

(ii) The stationary condition is

$$\frac{\partial H}{\partial u}(t, x(t), u(t), \lambda(t)) = 0 \quad \text{For all } t \in [a, T]. \tag{62}$$

Proof. We consider the following multi-composite fractional optimal control problem:

$$\begin{aligned} &\text{minimize J}(x, u, T) = \int_a^T f_0(t, x(t), u(t))\,dt, \\ &M\ {}_a^CD_t^{\alpha}(t) + N\ {}_a^CD_t^{\beta}\left({}_a^CD_t^{\mu}x(t)\right) = f(t, x(t), u(t)), \end{aligned} \tag{63}$$

Thus,

$$\begin{aligned} J^*(x, u, T, \lambda) = &\int_a^T \Big\{ f_0(t, x(t), u(t)) + \lambda(t) \\ &\cdot \Big[f(t, x(t), u(t)) \\ &- \left(M\ {}_a^CD_t^{\alpha}(t)\,x(t) + N\ {}_a^CD_t^{\beta}\left({}_a^CD_t^{\mu}x(t)\right)\right)\Big]\Big\}\,dt \end{aligned} \tag{64}$$

Suppose that

$$\begin{aligned} F = &f_0(t, x(t), u(t)) + \lambda(t)\, f(t, x(t), u(t)) \\ &- M\lambda(t)\ {}_a^CD_t^{\alpha}(t)\,x(t) - N\lambda(t)\ {}_a^CD_t^{\beta}\left({}_a^CD_t^{\mu}x(t)\right) \end{aligned} \tag{65}$$

Using (8) in (65), we get

$$\begin{aligned} F = &f_0(t, x(t), u(t)) + \lambda(t)\, f(t, x(t), u(t)) \\ &- Mx(t)\ {}_tD_T^{\alpha}\lambda(t) - Nx(t)\ {}_tD_T^{\beta}\left({}_tD_T^{\mu}\lambda(t)\right), \end{aligned} \tag{66}$$

Evaluating the Euler equation for $x(t)$ of (66),

$$\frac{\partial F}{\partial x(t)} - \frac{d}{dt}\left(\frac{\partial F}{\partial x^{\cdot}(t)}\right) = 0, \tag{67}$$

$$\begin{aligned} &\frac{\partial f_0}{\partial x(t)} + \lambda(t)\frac{\partial f}{\partial x(t)} - M\ {}_tD_T^{\alpha}\lambda(t) \\ &- N\ {}_tD_T^{\beta}\left({}_tD_T^{\mu}\lambda(t)\right) - \frac{d}{dt}\left(\frac{\partial F}{\partial x^{\cdot}(t)}\right) = 0 \end{aligned} \tag{68}$$

From (67), we get $(d/dt)(\partial F/\partial x^{\cdot}(t)) = 0$, as follows:

$$\begin{aligned} &\frac{\partial f_0}{\partial x(t)} + \lambda(t)\frac{\partial f}{\partial x(t)} - M\ {}_tD_T^{\alpha}\lambda(t) \\ &- N\ {}_tD_T^{\beta}\left({}_tD_T^{\mu}\lambda(t)\right) = 0 \end{aligned} \tag{69}$$

Now, evaluating Euler's equation for $\lambda(t)$ of (65):

$$\frac{\partial F}{\partial \lambda(t)} - \frac{d}{dt}\left(\frac{\partial F}{\partial \lambda^{\cdot}(t)}\right) = 0, \tag{70}$$

$$\begin{aligned} &\frac{\partial f_0}{\partial \lambda(t)} + \lambda(t)\frac{\partial f}{\partial \lambda(t)} - M\ {}_a^CD_t^{\alpha}(t)\,x(t) \\ &- N\ {}_a^CD_t^{\beta}\left({}_a^CD_t^{\mu}x(t)\right) = 0, \end{aligned} \tag{71}$$

Finally, the Euler's equation for $u(t)$ of (65) is

$$\frac{\partial F}{\partial u(t)} - \frac{d}{dt}\left(\frac{\partial F}{\partial u^{\cdot}(t)}\right) = 0. \tag{72}$$

From (65) it is clear that $(d/dt)(\partial F/\partial u^{\cdot}(t)) = 0$, then we get

$$\frac{\partial F}{\partial u(t)} = \frac{\partial f_0}{\partial u(t)} + \lambda(t)\frac{\partial f}{\partial u(t)} = 0 \tag{73}$$

From (69), (71), and (73) we have that

$$\begin{aligned} &M\ {}_tD_T^{\alpha}\lambda(t) + N\ {}_tD_T^{\beta}\left({}_tD_T^{\mu}\lambda(t)\right) \\ &\quad = \frac{\partial H}{\partial x}(t, x(t), u(t), \lambda(t)) \\ &M\ {}_a^CD_t^{\alpha}(t)\,x(t) + N\ {}_a^CD_t^{\beta}\left({}_a^CD_t^{\mu}x(t)\right) \\ &\quad = \frac{\partial H}{\partial \lambda}(t, x(t), u(t), \lambda(t)) \\ &0 = \frac{\partial H}{\partial u}(t, x(t), u(t), \lambda(t)) \quad \text{For all } t \in [a, T]. \end{aligned} \tag{74}$$

□

5. Illustrative Example

In this section, we consider the following linear–quadratic of multi-composite fractional optimal control problem:

$$\min \text{J}(x, u) = \int_0^1 (u(t) - x(t))^2\,dt, \tag{75}$$

which is subject to the multi-composite fractional dynamical system,

$$\begin{aligned} &{}_0^CD_t^{\alpha}x(t) + {}_0^CD_t^{\beta}\left({}_0^CD_t^{\mu}x(t)\right) \\ &\quad = u(t) - x(t) + \frac{6t^{\alpha+1}}{\Gamma(\alpha+2)} + \frac{6t^{\beta+\mu+1}}{\Gamma(\beta+\mu+2)}, \end{aligned} \tag{76}$$

and the boundary conditions,

$$x(0) = 0,$$
$$x(1) = \frac{6}{\Gamma(\alpha+\beta+\mu+2)}. \tag{77}$$

The exact solution for $\alpha = \beta = \mu = 1$ is given by the following:

$$\bar{x}(t) = \bar{u}(t) = \frac{6t^{\alpha+\beta+\mu+1}}{\Gamma(\alpha+\beta+\mu+2)}. \tag{78}$$

Now, we develop algorithm for solution (75), (76), and (77). It is based on the necessary optimality conditions from Theorem 6 and implements the following steps.

Step 1. Compute the Hamiltonian function:

$$H = (u(t) - x(t))^2 + \lambda(t) \cdot \left(u(t) - x(t) + \frac{6t^{\alpha+1}}{\Gamma(\alpha+2)} + \frac{6t^{\beta+\mu+1}}{\Gamma(\beta+\mu+2)}\right). \tag{79}$$

Step 2. Derive the necessary optimality conditions from Theorem 6:

$$ {}_tD_1^{\alpha}\lambda(t) + {}_tD_1^{\beta}\left({}_tD_1^{\mu}\lambda(t)\right) = \frac{\partial H}{\partial x} = -\left(2(u(t) - x(t)) + \lambda(t)\right), \tag{80}$$

$$ {}_0^CD_t^{\alpha}x(t) + {}_0^CD_t^{\beta}\left({}_0^CD_t^{\mu}x(t)\right) = \frac{\partial H}{\partial \lambda} = u(t) - x(t) + \frac{6t^{\alpha+1}}{\Gamma(\alpha+2)} + \frac{6t^{\beta+\mu+1}}{\Gamma(\beta+\mu+2)}, \tag{81}$$

$$2(u(t) - x(t)) + \lambda(t) = 0. \tag{82}$$

Step 3. Find the coupled system.

(i) From substitution (82) in (80), we get

$$ {}_tD_1^{\alpha}\lambda(t) + {}_tD_1^{\beta}\left({}_tD_1^{\mu}\lambda(t)\right) = 0. \tag{83}$$

(ii) Find control variable $u(t)$ from (82) and substitution in (80) to obtain

$$ {}_0^CD_t^{\alpha}x(t) + {}_0^CD_t^{\beta}\left({}_0^CD_t^{\mu}x(t)\right) = -\frac{1}{2}\lambda(t) + \frac{6t^{\alpha+1}}{\Gamma(\alpha+2)} + \frac{6t^{\beta+\mu+1}}{\Gamma(\beta+\mu+2)}, \tag{84}$$

Using property (6) for (83) and (84), we get the coupled system:

$$ {}_tD_1^{\alpha}\lambda(t) + {}_tD_1^{\beta+\mu}\lambda(t) = 0, \tag{85a}$$

$$ {}_0^CD_t^{\alpha}x(t) + {}_0^CD_t^{\beta+\mu}x(t) = -\frac{1}{2}\lambda(t) + \frac{6t^{\alpha+1}}{\Gamma(\alpha+2)} + \frac{6t^{\beta+\mu+1}}{\Gamma(\beta+\mu+2)}. \tag{85b}$$

Step 4a. Solve (85a) by Chebyshev expansion method, and use (48) to approximate$\lambda(t)$.

$$ {}_tD_1^{\alpha}\lambda(t) \approx \frac{\lambda(1)}{\Gamma(1-\alpha)}\left(1-t_s\right)^{-\alpha} + \frac{J(t_s;p_N)}{\Gamma(1-\alpha)}, \tag{86}$$

$$ {}_tD_1^{\beta+\mu}\lambda(t) \approx \frac{\lambda(1)}{\Gamma(1-(\beta+\mu))}\left(1-t_s\right)^{-(\beta+\mu)} + \frac{J(t_s;p_N)}{\Gamma(1-(\beta+\mu))}, \tag{87}$$

$s = 1, 2, \ldots, N-1$, where $J(t_s;p_{N1})$ and $J(t_s;p_{N2})$ is defined in (47).
Sum both (86) and (87) to get

$$\lambda(1)\left[\frac{\left(1-t_s\right)^{-\alpha}}{\Gamma(1-\alpha)} + \frac{\left(1-t_s\right)^{-(\beta+\mu)}}{\Gamma(1-(\beta+\mu))}\right] + \frac{J(t_s;p_{N1})}{\Gamma(1-\alpha)} + \frac{J(t_s;p_{N2})}{\Gamma(1-(\beta+\mu))} = 0. \tag{88}$$

$s = 1, 2, \ldots, N-1$, where $J(t_s;p_{N1})$ and $J(t_s;p_{N2})$ is defined in (47).

Step 4b. Solve (85b) by Chebyshev expansion method, and use (24) to approximate $x(t)$.

$$ {}_0^CD_t^{\alpha}x_N(t_s) = \sum_{r=0}^{N} d_{s,r}^{(\alpha)}x(t_r), \tag{89}$$

$$ {}_0^CD_t^{\beta+\mu}x_N(t_s) = \sum_{r=0}^{N} d_{s,r}^{(\beta+\mu)}x(t_r), \quad \alpha, (\beta+\mu) > 0. \tag{90}$$

$s, r = 0, 1, 2, \ldots, N$, where $d_{s,r}^{(\alpha)}$ is defined in (25).
Using (89) and (90) in (85a) to get

$$\sum_{r=0}^{N} d_{s,r}^{(\alpha)}x(t_r) + \sum_{r=0}^{N} d_{s,r}^{(\beta+\mu)}x(t_r) = -\frac{1}{2}\lambda(t_s) + \frac{6t_s^{\alpha+1}}{\Gamma(\alpha+2)} + \frac{6t_s^{\beta+\mu+1}}{\Gamma(\beta+\mu+2)}. \tag{91}$$

where $d_{s,r}^{(\alpha)}$ and $d_{s,r}^{(\beta+\mu)}$ are defined in (25).

Step 5. Find the results at the shifted Gauss-Lobatto nodes t_s, from (22) when $s = 1, 2, \ldots, N-1$, and α, β and $\mu \in (0, 1)$, in two cases.

Case 1. Let $N = 2$, $\alpha = 0.4$, $\beta = 0.2$ and $\mu = 0.1 \longrightarrow \beta + \mu = 0.3$.

$s = 1, 2, \ldots, N-1$, and since $N = 2$, then $s = 1$. $t_1 = 1/2 - (1/2)\cos(\pi/2) = 0.5$.

Table 1: The values of Caputo fractional derivative when ($\alpha = 0.4,\ \beta = 0.2$ and $\mu = 0.1$ in Case 1) and ($\alpha = 0.7,\ \beta = 0.4$ *and* $\mu = 0.3$ in Case 2 needed later on) are shown.

n	j	k	$d_{1,1}^{(0.4)}$	$d_{1,1}^{(0.3)}$	$d_{1,2}^{(0.4)}$	$d_{1,2}^{(0.3)}$	$d_{1,1}^{(0.7)}$	$d_{1,2}^{(0.7)}$
1	0	1	0	0	0.6722	0.6274	0	0.8154
1	1	1	0	0	0	0	0	0
1	2	1	0	0	0.0387	0.0287	0	0.0572
2	0	1	0.6722	0.6274	-0.6722	-0.6274	0.8154	-0.8154
2	1	1	0	0	0	0	0	0
2	2	1	0.0387	0.0287	-0.0387	-0.0287	0.0572	-0.0572
2	0	2	-0.5777	-0.5210	0.5777	0.5210	-0.7720	0.7720
2	1	2	0	0	0	0	0	0
2	2	2	0.0592	0.0620	-0.0592	-0.0620	0.0396	-0.0396
	sum		0.1924	0.1971	0.5185	0.459	0.1402	0.7324
	$d_{s,r}^{(\alpha)} = 4\theta_r/N.sum$		0.3848	0.3942	0.5185	0.459	0.2804	0.7324

Now, substituting the value of $\alpha,\ \beta,\ \mu$ and t_1, in (88), we get

$$\lambda(1)[1.8346] + (0.6715) J(t_s; p_{N1}) + (0.7704) J(t_s; p_{N2}) = 0 \tag{92}$$

where $s = 1, 2, \ldots, N-1$, where $J(t_s; p_{N1})$ and $J(t_s; p_{N2})$ are defined in (47).

After calculating (92) at the shifted Gauss-Lobatto nodes $t_s = t_1 = 0.5$ and substitution in (88) we get $\lambda(t_1) = 0$.

Now, we calculate (91) and, using the boundary conditions (78), we have

$$d_{1,0}^{(0.4)} x(t_0) + d_{1,1}^{(0.4)} x(t_1) + d_{1,2}^{(0.4)} x(t_2) + d_{1,0}^{(0.3)} x(t_0) + d_{1,1}^{(0.3)} x(t_1) + d_{1,2}^{(0.3)} x(t_2) = -\frac{1}{2}\lambda(t_1) + \frac{6{t_1}^{1.4}}{\Gamma(2.4)} + \frac{6{t_s}^{1.3}}{\Gamma(2.3)}. \tag{93}$$

Use (25) to evaluate the computation of $d_{1,1}^{(0.4)}$, $d_{1,2}^{(0.4)}$, $d_{1,1}^{(0.3)}$ and $d_{1,2}^{(0.3)}$, which is shown in Table 1.

when $r = 1 \longrightarrow 4\theta_1/N = 4(1)/2 = 2$ and $r = 2 \longrightarrow 4\theta_2/N = 4(0.5)/2 = 1$.

The results of Table 1 and boundary conditions from (78) can be substituted in (93); we get

$$d_{1,0}^{(0.4)} x(0) + (0.3848) x(t_1) + (0.5185) x(1) + d_{1,0}^{(0.3)} x(0) + (0.3942) x(t_1) + (0.459) x(1) = \frac{6{t_1}^{1.4}}{\Gamma(2.4)} + \frac{6{t_1}^{1.3}}{\Gamma(2.3)}, \tag{94}$$

$$d_{1,0}^{(0.4)}(0) + (0.3848) x(t_1) + (0.5185)\left(\frac{6}{\Gamma(2.7)}\right) + d_{1,0}^{(0.3)}(0) + (0.3942) x(t_1) + (0.459)\left(\frac{6}{\Gamma(2.7)}\right) = \frac{6(0.5)^{1.4}}{\Gamma(2.4)} + \frac{6(0.5)^{1.3}}{\Gamma(2.3)}, \tag{95}$$

$$x(t_1) = 0.1566. \tag{96}$$

Step 6. Compute approximation solution of $x(t)$ and $u(t)$ at $N = 2$.

$$x_2(t) = \frac{2}{N}\sum_{n=0}^{N}{}'' \sum_{r=0}^{N}{}'' x(t_r) T_n^p(t_r) T_n^p(t), \tag{97}$$

$$x_2(t) = \frac{1}{2} a_0 T_0^p(t) + a_1 T_1^p(t) + \frac{1}{2} a_2 T_2^p(t). \tag{98}$$

To find $a_0,\ a_1$ and a_2 from (23), we get

If $n = 0 \longrightarrow$

$$a_0 = \sum_{r=0}^{2}{}'' x(t_r) T_0^p(t_r), \quad i.e.\ T_0^p(t) = 1.$$

$$= 2.0204.$$

If $n = 1 \longrightarrow$

$$a_1 = \sum_{r=0}^{2}{}'' x(t_r) T_1^p(t_r), \quad i.e.\ T_1^p(t) = \frac{2t}{b} - 1 \tag{99}$$

$$= 1.9421.$$

If $n = 2 \longrightarrow$

$$a_2 = \sum_{r=0}^{2}{}'' x(t_r) T_2^p(t_r),$$

$$= 1.8638.$$

Now, we substitute $a_0,\ a_1$, and a_2, in (98), to get

$$x_2(t) = 7.4552t^2 - 3.571t. \tag{100}$$

To compute the approximation solution of the control $u(t)$ from (76), we get

$$u(t) = {}_0^C D_t^{\alpha} x(t) + {}_0^C D_t^{\beta}\left({}_0^C D_t^{\mu} x(t)\right) + x(t) - \frac{6t^{\alpha+1}}{\Gamma(\alpha+2)} - \frac{6t^{\beta+\mu+1}}{\Gamma(\beta+\mu+2)}, \tag{101}$$

TABLE 2: The numerical values of the state $x(t)$ and control $u(t)$ with $N = 2$ which converged to exact values are shown.

t	$\alpha = 0.4$ $\beta = 0.2$ $\mu = 0.1$	$\alpha = 0.7$ $\beta = 0.4$ $\mu = 0.3$	$\alpha = 1$ $\beta = 1$ $\mu = 1$	$\alpha = 0.4$ $\beta = 0.2$ $\mu = 0.1$	$\alpha = 0.7$ $\beta = 0.4$ $\mu = 0.3$	$\alpha = 1$ $\beta = 1$ $\mu = 1$
0	0	0.0001	0	0	0.0001	0
0.1	-0.283	0.1966	0.078	-2.0660	0.5152	0.78
0.2	-0.416	0.3942	0.0252	-2.9336	0.1920	0.252
0.3	-0.4	0.5928	0.502	-2.9336	0.2078	0.502
0.4	-0.236	0.7925	0.818	-3.0727	0.6427	0.818
0.5	0.078	0.9932	1.196	-2.4852	1.0983	1.196
0.6	0.541	1.1950	1.63	-1.5024	1.5673	1.63
0.7	1.153	1.3978	2.118	-0.1432	2.0459	2.118
0.8	1.915	1.6017	2.658	-0.1432	2.5314	2.658
0.9	2.825	1.8067	3.247	-0.1432	3.0221	3.247
1	3.884	2.0127	3.884	6.0665	3.5170	3.884

$$ {}_{0}^{C}D_t^{0.4} x_2(t) = 10.4290t^{1.6} - 3.9963t^{0.6}. \tag{102}$$

$$ {}_{0}^{C}D_t^{0.2} \left({}_{0}^{C}D_t^{0.1} x_2(t) \right) = 9.6527t^{1.7} - 3.9301t^{0.7}. \tag{103}$$

Using (102), (103), and (100) in (101), we have:

$$\begin{aligned} u(t) = {} & 10.4290t^{1.6} - 3.9963t^{0.6} + 9.6527t^{1.7} \\ & - 3.9301t^{0.7} + 7.4552t^2 - 3.571t \\ & - 4.8303t^{1.4} - 5.1427t^{1.3}. \end{aligned} \tag{104}$$

Case 2. Let $N = 2$, $\alpha = 0.7$, $\beta = 0.4$ and $\mu = 0.3 \longrightarrow \beta + \mu = 0.7$.

$s = 1, 2, \ldots, N-1$, and since $N = 2$, then $s = 1$. $t_1 = 1/2 - (1/2)\cos(\pi/2) = 0.5$.

Now, we calculate (91) and using the boundary conditions (78), we have

$$\begin{aligned} & d_{1,0}^{(0.7)} x(t_0) + d_{1,1}^{(0.7)} x(t_1) + d_{1,2}^{(0.7)} x(t_2) + d_{1,0}^{(0.7)} x(t_0) \\ & + d_{1,1}^{(0.7)} x(t_1) + d_{1,2}^{(0.7)} x(t_2) \\ & = -\frac{1}{2}\lambda(t_1) + \frac{6t_1^{\ 1.7}}{\Gamma(2.7)} + \frac{6t_s^{\ 1.7}}{\Gamma(2.7)}. \end{aligned} \tag{105}$$

To find $d_{1,1}^{(0.7)}$ and $d_{1,2}^{(0.7)}$ from Table 1 and we get $\lambda(t_1) = 0$, in the same steps of Case 1.

$$\begin{aligned} & d_{1,0}^{(0.7)}(0) + (0.2804)\,x(t_1) + (0.7324)(2.0126) \\ & + d_{1,0}^{(0.7)}(0) + (0.2804)\,x(t_1) \\ & + (0.7324)(2.0126) = 2.3911,, \\ & x(t_1) = 0.9932. \end{aligned} \tag{106}$$

Now, compute approximation solution of $x(t)$ and $u(t)$ at $N = 2$, in Case 2:

$$ x_2(t) = \frac{2}{N} \sum_{n=0}^{N}{}'' \sum_{r=0}^{N}{}'' x(t_r)\, T_n^p(t_r)\, T_n^p(t), \tag{107}$$

$$ x_2(t) = \frac{1}{2} a_0 T_0^p(t) + a_1 T_1^p(t) + \frac{1}{2} a_2 T_2^p(t). \tag{108}$$

To find a_0, a_1 and a_2, from (23), we get

$$\begin{aligned} a_0 &= 1.9995, \\ a_1 &= 1.0063 \text{ and} \\ a_2 &= 0.0131. \end{aligned} \tag{109}$$

We substitute a_0, a_1, and a_2, in (108), to get

$$ x_2(t) = 0.0528t^2 + 1.9598t + 0.0001. \tag{110}$$

To compute the approximation solution of the control $u(t)$, we get

$$\begin{aligned} u(t) = {} & {}_{0}^{C}D_t^{\alpha} x(t) + {}_{0}^{C}D_t^{\beta} \left({}_{0}^{C}D_t^{\mu} x(t) \right) + x(t) \\ & - \frac{6t^{\alpha+1}}{\Gamma(\alpha+2)} - \frac{6t^{\beta+\mu+1}}{\Gamma(\beta+\mu+2)}, \end{aligned} \tag{111}$$

$$ {}_{0}^{C}D_t^{0.7} x_2(t) = 0.0905t^{1.3} + 2.1837t^{0.3}. \tag{112}$$

Now, using (110) and (112) in (111), we get

$$\begin{aligned} u(t) = {} & 0.0905t^{1.3} + 2.1837t^{0.3} + 0.0528t^2 + 1.9598t \\ & + 0.0001 - 3.8843t^{1.7}. \end{aligned} \tag{113}$$

In Table 2, we give the computation of $x(t)$ and $u(t)$ when values t = 0...1 with different values of orders α, β *and* μ, for fractional derivative ($\alpha = 0.4$, $\beta = 0.2$, *and* $\mu = 0.1$.); see Figures 1(b) and 2(b) and ($\alpha = 0.7$, $\beta = 0.4$, *and* $\mu = 0.3$); see Figures 1(c) and 2(c) for $N = 2$, which converge to exact values with ($\alpha = \beta = \mu = 1$); see Figures 1(a) and 2(a).

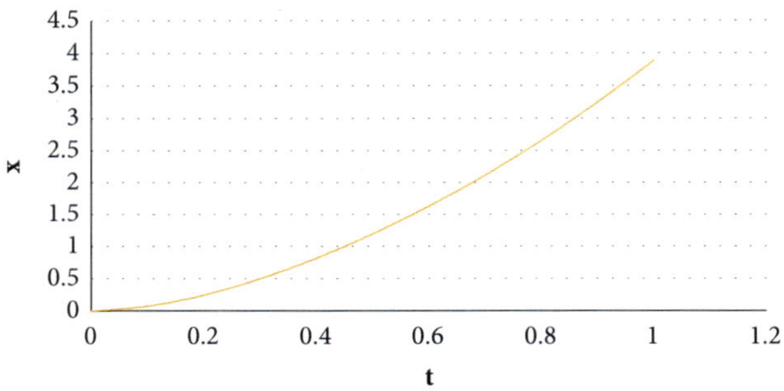

(a) Exact solution of $x(t)$ for $(N = 2)$ and $\mu = \alpha = \beta = 1$.

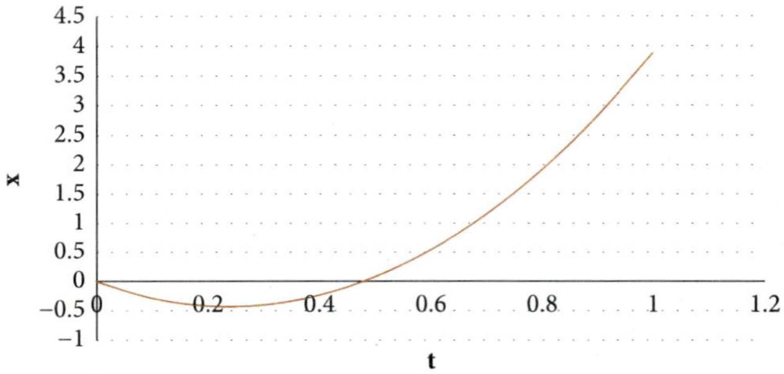

(b) Approximate solutions of $x(t)$ with $(N = 2)$ and $\alpha = 0.4$, $\beta = 0.2$ *and* $\mu = 0.1$

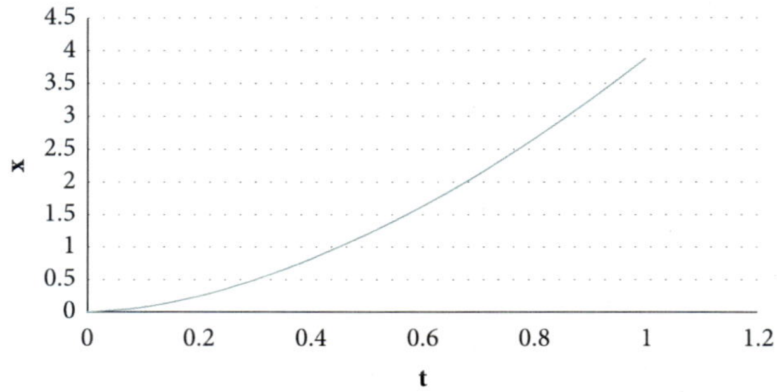

(c) Approximate solutions of $x(t)$ with $(N = 2)$ $\alpha = 0.7$, $\beta = 0.4$ *and* $\mu = 0.3$.

FIGURE 1: Display the exact and approximate values of state x(t) with different types of α, β, μ.

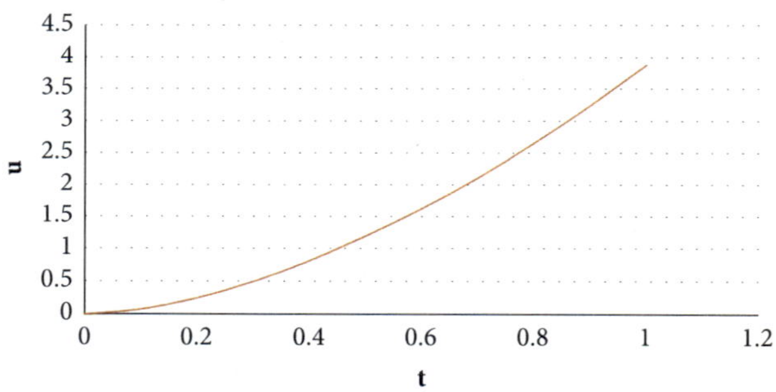

(a) Exact solution of $u(t)$ with $(N = 2)$ and $\alpha = \beta = \mu = 1$.

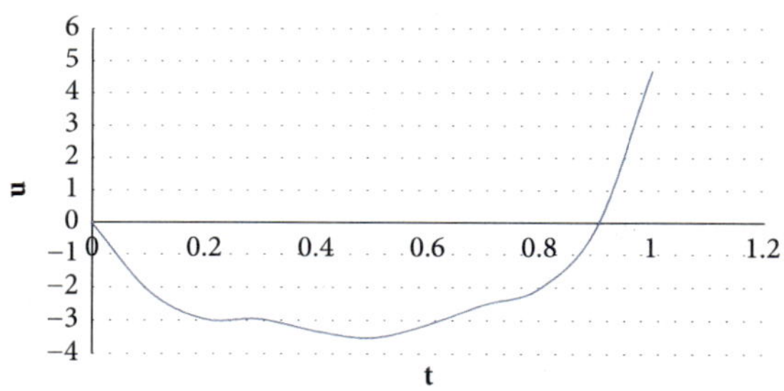

(b) Approximate solutions of $u(t)$ for $(N = 2)$ and $\alpha = 0.4$, $\beta = 0.2$ *and* $\mu = 0.1$

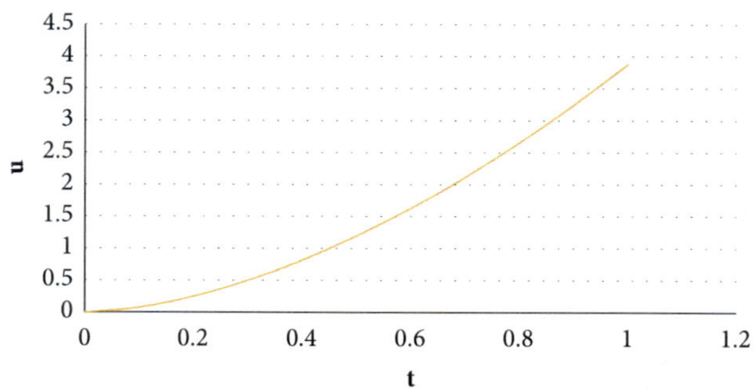

(c) Approximate solutions of $u(t)$ with $(N = 2)$ $\alpha = 0.7$, $\beta = 0.4$ *and* $\mu = 0.3$.

FIGURE 2: Display the exact and approximate values of control input u(.) with different types of α, β, μ.

6. Conclusions

In this paper, we have presented algorithm for solution a class of multi-composite order fractional optimal control problems, in two cases one when $\alpha = 0.4$, $\beta = 0.2$ and $\mu = 0.1$, and the other one when $\alpha = 0.7$, $\beta = 0.4$ *and* $\mu = 0.3$ at $(N = 2)$. In both cases, the solution is approximated by Chebyshev series. Numerical results for illustrative example show that the algorithm converges from the exact solution when $\alpha = \beta = \mu = 1$, and we note that the convergent to the exact solution is dependent on increasing of the fractional order of derivative.

Conflicts of Interest

The authors declare that they have no conflicts of interest.

References

[1] K. Oldham and J. Spanier, *The fractional calculus theory and applications of differentiation and integration to arbitrary order*, vol. 111, Elsevier, 1974.

[2] R. Almeida, S. Pooseh, and D. F. Torres, *Computational methods in the fractional calculus of variations*, World Scientific Publishing Co Inc, 2015.

[3] C. Tricaud and Y. Q. Chen, "An approximate method for numerically solving fractional order optimal control problems of general form," *Computers & Mathematics with Applications*, vol. 59, no. 5, pp. 1644–1655, 2010.

[4] O. P. Agrawal and D. Baleanu, "A hamiltonian formulation and a direct numerical scheme for fractional optimal control problems," *Journal of Vibration and Control*, vol. 13, no. 9-10, pp. 1269–1281, 2007.

[5] E. M. Elbarbary and M. El-Kady, "Chebyshev finite difference approximation for the boundary value problems," *Applied Mathematics and Computation*, vol. 139, no. 2-3, pp. 513–523, 2003.

[6] M. M. Khader and A. Hendy, "An efficient numerical scheme for solving fractional optimal control problems," *International Journal of Nonlinear Science*, vol. 14, pp. 287–296.

[7] T. Akbarian and M. Keyanpour, "A new approach to the numerical solution of fractional order optimal control problems," *Applications and Applied Mathematics. An International Journal*, vol. 8, no. 2, pp. 523–534, 2013.

[8] Y. Zhou, *Basic Theory of Fractional Differential Equations*, World Scientific, Singapore, 2014.

[9] A. B. Malinowska and D. F. Torres, *Introduction to the fractional calculus of variations*, World Scientific Publishing Co Inc, 2012.

[10] Q. M. Al-Mdallal, M. I. Syam, and M. N. Anwar, "A collocation-shooting method for solving fractional boundary value problems," *Communications in Nonlinear Science and Numerical Simulation*, vol. 15, no. 12, pp. 3814–3822, 2010.

[11] N. H. Sweilam and M. M. Khader, "A Chebyshev pseudo-spectral method for solving fractional-order integro-differential equations," *Anziam Journal*, vol. 51, no. 4, pp. 464–475, 2010.

[12] B. Obsieger, *Numerical Methods III-Approximation of Functions*, University-Books, 2013.

[13] C. W. Clenshaw and A. R. Curtis, "A method for numerical integration of an automatic computer," *Numerische Mathematik*, vol. 2, no. 1, pp. 197–205, 1960.

6

A Modified RBF Neuro-Sliding Mode Control Technique for a Grid Connected PMSG based Variable Speed Wind Energy Conversion System

Rostand Marc Douanla,[1,2] Godpromesse Kenné,[1] François Béceau Pelap,[3] and Armel Simo Fotso[1,2]

[1] *Unité de Recherche d'Automatique et d'Informatique Appliquée (LAIA), Département de Génie Électrique, IUT FOTSO Victor Bandjoun, Université de Dschang, BP 134, Bandjoun, Cameroon*
[2] *Unité de Recherche de Matière Condensée, d'Electronique et de Traitement du Signal (LAMACETS), Département de Physique, Université de Dschang, BP 69, Dschang, Cameroon*
[3] *Unité de Recherche de Mécanique et de Modélisation des Systèmes Physiques (L2MSP), Département de Physique, Université de Dschang, BP 69, Dschang, Cameroon*

Correspondence should be addressed to Godpromesse Kenné; godpromesse@gmail.com

Academic Editor: Daniela Proto

A modified control scheme based on the combination of online trained neural network and sliding mode techniques is proposed to enhance maximum power extraction for a grid connected permanent magnet synchronous generator (PMSG) wind turbine system. The proposed control method does not need the knowledge of the uncertainty bounds nor the exact model of the nonlinear system. Since the neural network is trained online, the time to estimate good weights can affect the dynamic performance of the process during the startup phase. Therefore an appropriate way to smoothly and explicitly accelerate the neural network rate of convergence during the startup phase is proposed. Furthermore, a flexible grid side voltage source converter control structure which can handle both grid connected and standalone modes based on conventional proportional integral (PI) control method is presented. Simulations are done in Matlab/Simulink environment to verify the effectiveness and assess the performance of the proposed controller. The results analysis shows the superiority of the proposed RBF neuro-sliding mode controller compared to a nonlinear controller based on sliding mode control method when the system undergoes parameter uncertainties.

1. Introduction

In recent years, a new challenge that seeks to reduce the effects of climate change has emerged in the world energy production domain. The exploitation of renewable energy sources has been intensified so as to tackle the growing energy demand with reduced environmental impact. Wind energy source, as a sustainable and renewable resource, has gained more interest for electrical power generation [1]. Wind energy conversion systems (WECSs) can be divided into two categories: fixed speed and variable speed. Fixed speed wind generators are directly connected to the electrical network but have limited controllability and range of operation [2]. Variable speed wind turbines (WTs) although equipped with additional components bring more advantages such as maximum power point tracking (MPPT) control strategy, increased power capture, and improved power quality [3]. The captured electrical energy can be injected into the utility grid or used to power up standalone loads. The electrical generator involved in the variable wind conversion process can be interfaced to the load by means of different topologies of controlled power electronic converters [4, 5].

Various investigations on wind turbine generators have shown that permanent magnet synchronous generator (PMSG) has many advantages compared to other kinds of electrical machines [6, 7] and therefore becomes an attractive choice for variable speed wind turbines. Meanwhile, the PMSG is a nonlinear dynamic and highly coupled system. In addition, the PMSG is subjected to many disturbances including change in wind speed and parameter uncertainties such as variations in stator resistance, stator inductance, rotor flux, and rotor inertia during operation. Hence, in order to tap efficiently power from PMSG based WT, it is important to take into consideration the above disturbances in the control design.

Linear controllers [8–14] have been studied and implemented. But the reduced range of operating points and poor robustness of these controllers with respect to machine uncertainties are their main limitations. Optimal tuning of control parameters for PMSG WECS using democratic joint operation algorithm has been reported in [15] to boost the performance of a linear controller. Meanwhile an alternative way to cope efficiently with the nonlinear nature and varying operation conditions of PMSG WECS is to exploit nonlinear control designs. Accordingly feedback [16, 17] and sliding mode [18–25] techniques have been successfully processed. An association of passivity theory with sliding mode control was proposed in [26] to remove the PMSG WECS nonlinearities and cope with various uncertainties. However, the implementation of the above control algorithms required the knowledge of the mathematical model of the plant which is not accurate due to various assumptions considered. In addition, the knowledge of the upper limit of the system uncertainties, including parametric variations and external disturbances, is assumed. But the difficulty to obtain the exact system uncertainties bound in practical application may lead to unsatisfactory global control performance.

In searching appropriate solutions to the aforementioned shortcomings, artificial neural network (NN) control approaches have been explored. NNs are considered to be powerful tools for identification/nonlinear control by imitating cognitive expertise of the human [27–30], thanks to their self-learning abilities and capabilities to approximate nonlinear functions. Some results from the application of NN based control schemes for PMSG WECSs can be found in [31–34]. In [31], an online learning modified recurrent Elman NN control method is designed for a standalone PMSG WECS with practical implementation. An estimation algorithm is developed to observe the upper limit of the network approximation error. This estimation helped to improve the dynamic performance of the controller. In [32], online trained wavelet recurrent NN control method to improve the transient performance of a standalone PMSG WT has been reported and some experimental results were provided. The approaches in [31, 32] require extra memory to store previous information and high number of neurones which in turn induces large computation time. Another fact is that the total efficiency of the controlled PMSG WECS is limited since the MPPT algorithm, which is essential for variable speed WTs, was not addressed. In [33], an adaptive RBF NN control strategy is adopted to achieve tip speed ratio MPPT control of a PMSG WT with experimental results. The NN is trained offline. The parameters of the RBF NN are adjusted using a gradient method coupled to a modified particle swarm optimization algorithm. But the difficulty to obtain offline a training pattern which matches the exact behaviour of the closed loop system can hinder the performance of this scheme. In [34], sliding mode RBF NN control is applied to extract maximum power from PMSG WT. Simulation results demonstrated the robustness of the proposed method against wind speed variations. Nevertheless, the last two control methods assumed that the uncertainties of the system lie within a known boundaries. The main contributions of this paper can be summarized as follows:

(i) A new control design of NN controller to tap maximum power from a grid connected PMSG wind turbine is investigated.

(ii) Online trained (as opposed to offline trained methods, e.g., [33]) RBF neuro-sliding mode controller is selected to benefit from the simple structure and the properties of approximation capability of RBF NN and the robustness feature of sliding mode.

(iii) An estimation of the upper limit of the system uncertainty bound which is not obtainable in practice is provided (as opposed to some existing works [33, 34]).

(iv) A solution to reduce the control effort and accelerate the rate of convergence of the network during the startup phase (as opposed to the method in [30]) is proposed.

(v) A PI control scheme for the grid side which is able to handle both grid connected and stand modes of operation is proposed.

The remainder of this paper is organized as follows. Section 2 focuses on the system modeling. In Section 3, the proposed hybrid control development is presented and the GSC control design is provided in Section 4. Computer simulations using a nonlinear controller based on sliding mode control method and the proposed sliding neural control approach are reported in Section 5. The conclusions of the paper are drawn up in Section 6.

2. PMSG WECS Modeling

Figure 1 presents an overview of the grid connected PMSG WECS under analysis. A wind turbine rotor is coupled to a PMSG shaft via a gearbox. The PMSG is controlled through a PWM machine side converter (MSC) to enhance MPPT. An electronic load controller (ELC), composed of a single switch PWM dc-to-dc converter feeding an electronic load, is used to provide a fixed dc voltage needed for PWM grid side converter (GSC) operation. The electronic load, working in a constant voltage mode, is designed such that when the duty cycle of the dc-to-dc converter is unity (during fault or over generation), it should consume the maximum output power of the generator. A LCL filter is introduced to improve the quality of the power injected into the local AC load or the grid by lessening the harmonics produced by the GSC switching effects. A circuit breaker is used in case of grid connected/standalone transition.

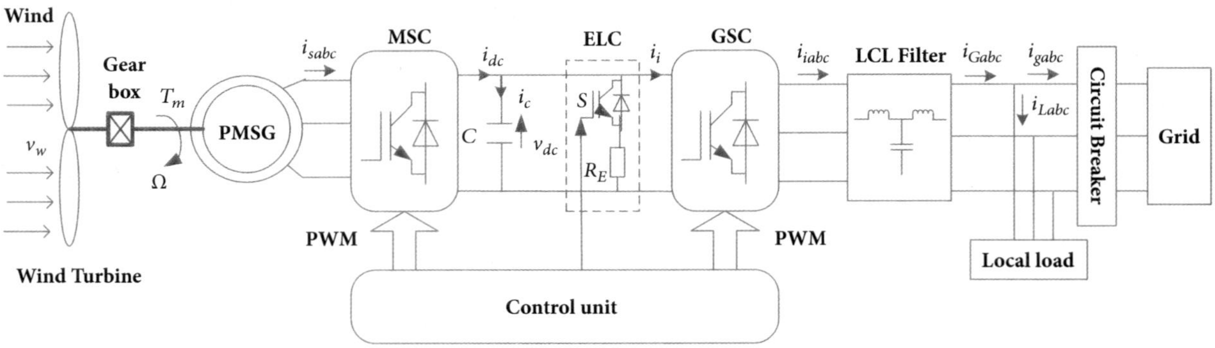

FIGURE 1: Control system diagram.

2.1. Turbine Modeling. The mechanical power P_m extracted by the WT from the kinetic energy of the wind is expressed in terms of the air density ρ, the area swept by the blades πr^2 (r is the blade radius), the power coefficient $C_p(\zeta,\beta)$, and the wind speed v_w as follows [35]:

$$P_m = 0.5\rho\pi r^2 C_p(\zeta,\beta)\, v_w^3. \quad (1)$$

The power coefficient is a function of the tip speed ratio ζ and the blade pitch angle β and can be stated as

$$C_p(\zeta,\beta) = 0.5176\left(\frac{116}{\zeta_i} - 0.4\beta - 5\right)e^{-21/\zeta_i} + 0.0068\zeta$$
$$\text{with } \frac{1}{\zeta_i} = \frac{1}{\zeta + 0.08\beta} - \frac{0.035}{\beta^3+1}. \quad (2)$$

The tip speed ratio ζ is expressed as

$$\zeta = \frac{r\Omega_t}{v_w} = \frac{r\Omega}{G v_w},$$
$$G = \frac{\Omega}{\Omega_t} = \frac{T_t}{T_m}. \quad (3)$$

where G represents the gearbox ratio. Ω_t and T_t are rotational speed and torque developed by the turbine at the blade side of the gearbox while Ω and T_m are the machine side shaft speed and mechanical torque, respectively. T_m is computed by combining (1) and (3) as

$$T_m = \frac{P_m}{\Omega} = 0.5\rho\pi r^5 \frac{C_p(\zeta,\beta)\,\Omega^2}{(G\zeta)^3}. \quad (4)$$

The turbine equation of motion referred to as the generator side of the gearbox is given by

$$\dot{\Omega} = \frac{T_m}{J} - \frac{f\Omega}{J} - \frac{T_e}{J} \quad (5)$$

with J and f being the total rotating parts inertia and viscous factor, respectively. T_e is the electromagnetic torque of the generator and is defined later.

When the wind speeds belong to region 2 [36, 37], i.e., between the cut-in and the rated values, the MPPT algorithm should be enabled to increase the efficiency of the turbine captured power. Therefore the power coefficient should be kept at its maximum value C_{pmax} for each wind speed in this region. This is possible if the turbine is working at the optimum tip speed ratio ζ_{opt} which in turn leads to generator optimum speed as follows:

$$\Omega_{opt} = \frac{\zeta_{opt} G v_w}{r}. \quad (6)$$

Hence the optimum power caption is achieved if the generator speed is steered to the optimum speed $\Omega^* = \Omega_{opt}$. This MPPT scheme, known as tip speed ratio MPPT, is adopted in the present work. Thus the target optimum power of the generator is computed as

$$P_{mopt} = K_{opt}\Omega_{opt}^3 \quad (7)$$

where $K_{opt} = 0.5\rho\pi r^5 (C_{pmax}/(G\zeta_{opt})^3)$ is the optimal torque constant.

2.2. Modeling of PMSG Associated with MSC and dc-Link. The nonlinear dynamics of a surface mounted PMSG coupled with PWM voltage source rectifier and PWM dc-to-dc converter (see Figure 1) in the synchronously rotating (d,q) coordinates system is described as follows [38–40]:

$$\dot{i}_{sd} = -\frac{R_s}{L_s} i_{sd} + p\Omega i_{sq} - \frac{v_{sd}}{L_s} \quad (8)$$

$$\dot{i}_{sq} = -\frac{R_s}{L_s} i_{sq} - p\Omega i_{sd} + \frac{p\Omega\phi_r}{L_s} - \frac{v_{sq}}{L_s} \quad (9)$$

$$T_e = \frac{3}{2} p\phi_r i_{sq},$$
$$P_e = \frac{3p\Omega\phi_r i_{sq}}{2} \quad (10)$$

$$\phi_{sd} = L_s i_{sd} + \phi_r,$$
$$\phi_{sq} = L_s i_{sq} \quad (11)$$

$$v_{sd} = S_d v_{dc},$$
$$v_{sq} = S_q v_{dc} \tag{12}$$

$$\dot{v}_{dc} = \frac{1}{C}\left(S_d i_{sd} + S_q i_{sq}\right) - \frac{S v_{dc}}{C R_E} - \frac{i_i}{C} \tag{13}$$

$$\dot{v}_{dc}^2 = \frac{3p\Omega\phi_r i_{sq}}{C} - \frac{2S v_{dc}^2}{C R_E} - \frac{2P_i}{C} \tag{14}$$

where i_{sdq} denote the stator current components and R_s and L_s are the stator windings resistance and inductance. p is the number of pole pairs. ϕ_{sdq} are the stator flux components and ϕ_r is the rotor flux. P_e is the electrical power delivered by the generator. v_{sdq} are the stator voltage components while v_{dc} is the voltage across the dc-link capacitor C. R_E is the electronic load resistance. S is the duty cycle of the dc-to-dc converter switch. i_i and P_i are the GSC current and active power, respectively. S_{dq} state the (a,b,c) to (d,q) reference frame transformation of the PWM duty ratios S_{abc} of the MSC switching devices.

2.3. GSC and LCL Filter Dynamics. The mathematical model of the GSC and LCL filter in the (d,q) coordinates system is expressed as follows [41]:

$$L_1 \dot{i}_{id} = -R_1 i_{id} + L_1\omega_g i_{iq} + v_{id} - v_d \tag{15}$$

$$L_1 \dot{i}_{iq} = -R_1 i_{iq} - L_1\omega_g i_{id} + v_{iq} - v_q \tag{16}$$

$$C_f \dot{v}_d = i_{id} - i_{Gd} \tag{17}$$

$$C_f \dot{v}_q = i_{iq} - i_{Gq} \tag{18}$$

$$L_2 \dot{i}_{Gd} = -R_2 i_{Gd} + L_2\omega_g i_{Gq} + v_d - v_{gd} \tag{19}$$

$$L_2 \dot{i}_{Gq} = -R_2 i_{Gq} - L_2\omega_g i_{Gd} + v_q - v_{gq} \tag{20}$$

where i_{idq}, v_{dq}, and $i_{Gdq} = i_{gdq} + i_{Ldq}$ are inverter current, filter capacitor voltage, and grid side current components. i_{Ldq} and i_{gdq} represent, respectively, the local AC load current and injected grid current components. $v_{id} = S_{id} v_{dc}$ and $v_{iq} = S_{iq} v_{dc}$ denote the inverter output voltage components. S_{id} and S_{iq} are the duty ratios of the GSC. ω_g is the angular frequency of the grid voltage. L_1, L_2, and C_f denote, respectively, the LCL filter inverter side inductance, capacitance, and grid side inductance. The filter inductors have internal resistances R_1 and R_2.

3. Proposed RBF Neuro-Sliding Mode Control Design

The MSC, acting as voltage source rectifier, is controlled to efficiently extract the maximum power from the wind turbine while the dc-to-dc converter control task is to track the voltage across the dc-link capacitor to the desired level. It is important to regulate the dc voltage for the efficient working of the GSC. The d-axis current reference is set to zero in order to obtain unity power factor of the PMSG while the generator speed reference is equal to the optimal rotor speed defined earlier for MPPT.

In [30] authors proposed a combined online trained RBF neural network and sliding mode control algorithm for a class of SISO nonlinear uncertain systems. In the present study the same approach is modified and exploited. Unless otherwise specified, the proposed control development is conducted assuming the following:

(A1) The state variables as well as the system disturbances are bounded; i.e., they belong to a compact subset $\Gamma \subset \Re$.

(A2) The rate of change of the system parameters is slow compared to the fast dynamics of the system state variables.

(A3) The upper bound of the system uncertainties is unknown.

Before processing the proposed RBF neuro-sliding mode control scheme, let us derive each state variable to be control (i.e., i_{sd}, Ω and v_{dc}) until its control input explicitly appears and thus rearrange the d-axis current dynamic in (8) as

$$\dot{i}_{sd} = f_{i_{sd}} + g_{i_{sd}} v_{sd} + d_{i_{sd}}(t) \tag{21}$$

$$\text{with } f_{i_{sd}} = -\frac{R_s}{L_s} i_{sd} + p\Omega i_{sq}$$
$$\text{and } g_{i_{sd}} = -\frac{1}{L_s}. \tag{22}$$

The term $d_{i_{sd}}(t)$ has been added to represent the unknown uncertainties of the system. Since the mechanical torque is unmeasurable, replace it by its optimal expression $K_{opt}\Omega^2$. Then consider the electromagnetic torque expression in (10) and derive (5) with respect to time as follows:

$$\ddot{\Omega} = f_\Omega + g_\Omega v_{sq} + d_\Omega(t) \tag{23}$$

$$\text{with } f_\Omega = \frac{2K_{opt}\Omega - f}{J^2}\left(K_{opt}\Omega^2 - f\Omega - \frac{3p\phi_r i_{sq}}{2}\right)$$
$$+ \frac{3p\phi_r}{2JL_s}\left(R_s i_{sq} + p\Omega L_s i_{sd} - p\Omega\phi_r\right) \tag{24}$$
$$\text{and } g_\Omega = \frac{3p\phi_r}{2JL_s}.$$

$d_\Omega(t)$ represents the unknown uncertainties of the system.

From (14), rewrite the dc-link voltage dynamic as ($u = v_{dc}^2$ and $d_{v_{dc}}(t)$ state, respectively, the square dc voltage, and the system uncertainties):

$$\dot{u} = f_u + g_u v_u + d_{v_{dc}}(t) \text{ with} \tag{25}$$

$$f_u = \frac{3p\Omega\phi_r}{C} i_{sq} - \frac{2P_i}{C},$$
$$g_u = -\frac{2}{CR_E}, \tag{26}$$
$$v_u = Su.$$

Now define the d-axis current, the rotor speed, and square dc voltage tracking errors as $e_{i_{sd}} = i_{sd} - i_{sd}^*$, $e_\Omega = \Omega - \Omega^*$, and $e_u = u - u^*$, respectively (the reference signals are marked with $*$). Let also the associated sliding variables be $S_{i_{sd}}$, S_Ω, and S_u such that ($h_1 > 0$, h_2 and $h_3 > 0$)

$$S_{i_{sd}} = e_{i_{sd}} + h_1 \int e_{i_{sd}} dt,$$
$$S_\Omega = \dot{e}_\Omega + h_2 e_\Omega + h_3 \int e_\Omega dt, \tag{27}$$
$$S_u = e_u.$$

Derive the sliding surfaces with respect to time and obtain

$$\dot{S}_{i_{sd}} = f_{i_{sd}} + g_{i_{sd}} v_{sd} + d_{i_{sd}}(t) + \mu_{i_{sd}}, \tag{28}$$
$$\dot{S}_\Omega = f_\Omega + g_\Omega v_{sq} + d_\Omega(t) + \mu_\Omega, \tag{29}$$
$$\dot{S}_\Omega = f_u + g_u v_u + d_u(t) + \mu_u. \tag{30}$$

with $\mu_{i_{sd}} = h_1 e_{i_{sd}} - \dot{i}_{sd}^*$, $\mu_\Omega = h_2 \dot{e}_\Omega + h_3 e_\Omega - \ddot{\Omega}^*$ and $\mu_u = -\dot{u}^*$.

Consider systems (21), (23), and (25) with proposed respective controllers in (31), (32), and (33)

$$v_{sd}^* = -\frac{1}{g_{i_{sd}}}\left(f_{i_{sd}} + \mu_{i_{sd}} + d_{i_{sd}}(t)\right) - \frac{S_{i_{sd}}}{\epsilon_{i_{sd}}}, \tag{31}$$
$$v_{sq}^* = -\frac{1}{g_\Omega}\left(f_\Omega + \mu_\Omega + d_\Omega(t)\right) - \frac{S_\Omega}{\epsilon_\Omega} \tag{32}$$
$$v_u^* = -\frac{1}{g_u}\left(f_u + \mu_u + d_u(t)\right) - \frac{S_u}{\epsilon_u} \tag{33}$$

where $0 < \epsilon_{i_{sd}} < 1$, $0 < \epsilon_\Omega < 1$, and $0 < \epsilon_u < 1$ are design parameters. Thus (28), (29), and (30) become

$$\dot{S}_{i_{sd}} = -\frac{S_{i_{sd}}}{\epsilon_{i_{sd}}}, \tag{34}$$
$$\dot{S}_\Omega = -\frac{S_\Omega}{\epsilon_\Omega}, \tag{35}$$
$$\dot{S}_u = -\frac{S_u}{\epsilon_u}. \tag{36}$$

From (34), (35), and (36) one can conclude that $S_{i_{sd}}$, S_Ω, and S_u converge exponentially to 0.

Remark 1. Considering the assumption that the functions $f_{i_{sd}}$, $g_{i_{sd}}$, f_Ω, g_Ω, f_u, and g_u along with the terms $\mu_{i_{sd}}$, $d_{i_{sd}}(t)$, μ_Ω, $d_\Omega(t)$, μ_u, and $d_{v_{dc}}(t)$ are unknown, the controllers in (31), (32), and (33) can not be implemented in real operation condition. Consequently three sets of RBF neuro-sliding mode controllers are employed to estimate the control signals v_{sd}^*, v_{sq}^*, and v_u^*.

Remark 2 (the conditions $g_{i_{sd}} < 0$, $g_\Omega > 0$, and $g_u < 0$ can be verified). Compute the proposed approximations of the desired control actions v_{sd}^*, v_{sq}^*, and v_u^* as follows:

$$\hat{v}_{sd}^*\left(\chi_{i_{sd}}, t\right) = \Psi\left(\chi_{i_{sd}}, \hat{\omega}_{i_{sd}}\right) + b_{i_{sd}}(t), \tag{37}$$
$$\hat{v}_{sq}^*\left(\chi_\Omega, t\right) = \Psi\left(\chi_\Omega, \hat{\omega}_\Omega\right) + b_\Omega(t), \tag{38}$$
$$\hat{v}_u^*\left(\chi_u, t\right) = \Psi\left(\chi_u, \hat{\omega}_u\right) + b_u(t), \tag{39}$$

where $\chi_{i_{sd}}^T = (i_{sd}, S_{i_{sd}}, S_{i_{sd}}/\epsilon_{i_{sd}})$, $\chi_\Omega^T = (\Omega, S_\Omega, S_\Omega/\epsilon_\Omega)$, and $\chi_u^T = (u, S_u, S_u/\epsilon_u)$ are the NNs input vectors. $\Psi(\chi_{i_{sd}}, \hat{\omega}_{i_{sd}})$, $\Psi(\chi_\Omega, \hat{\omega}_\Omega)$, and $\Psi(\chi_u, \hat{\omega}_u)$ denote the RBF NNs. $\hat{\omega}_{i_{sd}}$, $\hat{\omega}_\Omega$, and $\hat{\omega}_u$ are estimations of optimal weight vectors $\omega_{i_{sd}}^*$, ω_Ω^*, and ω_u^*. Bearing in mind Remark 2, the terms $b_{i_{sd}}(t)$, $b_\Omega(t)$, and $b_u(t)$, which help to improve the neural networks rate of convergence in the presence of uncertainties, are described as follows:

$$b_{i_{sd}}(t) = \hat{\lambda}_{i_{sd}} \operatorname{sign}\left(S_{i_{sd}}\right) \tag{40}$$
$$b_\Omega(t) = -\hat{\lambda}_\Omega \operatorname{sign}\left(S_\Omega\right) \tag{41}$$
$$b_u(t) = \hat{\lambda}_u \operatorname{sign}\left(S_u\right). \tag{42}$$

$\hat{\lambda}_{i_{sd}}$, $\hat{\lambda}_\Omega$, and $\hat{\lambda}_u$ are estimates of the positive unknown limits of the NNs uncertainties and are computed as

$$\dot{\hat{\lambda}}_{i_{sd}} = \begin{cases} \alpha_{i_{sd}}, & \text{if } S_{i_{sd}} \neq 0 \\ 0, & \text{if } S_{i_{sd}} = 0 \end{cases} \tag{43}$$
$$\dot{\hat{\lambda}}_\Omega = \begin{cases} \alpha_\Omega, & \text{if } S_\Omega \neq 0 \\ 0, & \text{if } S_\Omega = 0 \end{cases} \tag{44}$$
$$\dot{\hat{\lambda}}_u = \begin{cases} \alpha_u, & \text{if } S_u \neq 0 \\ 0, & \text{if } S_u = 0 \end{cases} \tag{45}$$

where $\alpha_{i_{sd}}$, α_Ω, $\alpha_u > 0$ and $\hat{\lambda}_{i_{sd}}(0)$, $\hat{\lambda}_\Omega(0)$, $\hat{\lambda}_u(0) = 0$. For N number of nodes in the hidden layer, the functions Ψ and $\hat{\omega}$ are described as follows:

$$\Psi\left(\chi_x, \hat{\omega}_x\right) = \sum_{j=1}^{N} \hat{\omega}_{xj} \psi\left(\left\|\chi_x - c_{xj}\right\|, v_{xj}\right); \tag{46}$$

$$\dot{\hat{\omega}}_{xj} = \operatorname{Proj}\left[-S_x \left.\frac{\partial \Psi}{\partial \omega_{xj}}\right|_{\omega_{xj}=\hat{\omega}_{xj}}\right]$$
$$= \begin{cases} -S_x \left.\dfrac{\partial \Psi}{\partial \omega_{xj}}\right|_{\omega_{xj}=\hat{\omega}_{xj}}, & \text{if } \left|\hat{\omega}_{xj}\right| < \Re_\omega \\ 0, & \text{otherwise} \end{cases} \tag{47}$$
$$x = i_{sd}, \Omega, u;\ j = 1, \ldots, N,$$

where $\psi = \exp(-\|\chi_x - c_{xj}\|^2 / 2v_{xj}^2)$ is a Gaussian activation function. c_{xj} and v_{xj} are, respectively, the center and the width of the jth node. Proj is the projection function on the compact $\Gamma_{\omega_x} = \{\omega_x : \|\omega_x\| \leq \Re_{\omega_x}\}$. The weight vectors $\widehat{\omega}_{xj}$ are initialized in a uniformly random manner while the centers and the widths are selected according to the clustering method as [42]

$$c_{xj} = \chi_{xmin} + \frac{2j-1}{2}\nu_{xj}$$
$$\nu_{xj} = \frac{\chi_{xmax} - \chi_{xmin}}{N}, \tag{48}$$

where χ_{xmin} and χ_{xmax} state, respectively, the lower and the upper bounds of the input vector χ_x^T.

The proof of the convergence of the above NNs controllers is found in [30].

Remark 3. During the startup phase, given that the weights vector is randomly initialized, the network may take a relatively long time to converge. This implies that the network uncertainties are more significant during this phase. Unfortunately the estimations of the network uncertainties bounds given in (43), (44), and (45) start from 0 before increasing to sufficient values. An obvious solution is to take a too high estimation rate α_x, $x = i_{sd}, \Omega, u$. But a too large rate α_x may excite high frequency unmodeled dynamics, maintain higher control effort, and render the system unstable.

To address this issue, a transformation for the estimation of uncertainties bound is proposed. This transformation seeks to steer the tracking error to the origin and also allow the tracking process to be explicitly controllable so that the rate of convergence during the startup phase can be smoothly pre-specified and faster. In concrete terms, define

$$\widehat{\lambda}_x^* = \widehat{\lambda}_x \kappa_x(t), \quad x = i_{sd}, \Omega, u \tag{49}$$

where $\widehat{\lambda}_x^*$ is the transformed $\widehat{\lambda}_x$ and $\kappa_x(t)$, a smooth decreasing function that satisfies

$$\kappa_x(t) > 0,$$
$$\lim_{t\longrightarrow\infty} \kappa_x(t) = 1, \tag{50}$$
$$\kappa_x(0) = \kappa_{xmax},$$

κ_{xmax} is the maximum value of $\kappa_x(t)$. For example, the function

$$\kappa_x(t) = 1 + \gamma_x e^{-\sigma_x t}, \quad \gamma_x, \sigma_x > 0, \tag{51}$$

is a good candidate and may be used in (49).

Considering the aforementioned transformation, the bias terms in (40), (41), and (42) become

$$b_{i_{sd}}^*(t) = \widehat{\lambda}_{i_{sd}} \kappa_{i_{sd}}(t) \operatorname{sign}(S_{i_{sd}}) \tag{52}$$

$$b_{\Omega}^*(t) = -\widehat{\lambda}_{\Omega} \kappa_{\Omega}(t) \operatorname{sign}(S_{\Omega}) \tag{53}$$

$$b_u^*(t) = \widehat{\lambda}_u \kappa_u(t) \operatorname{sign}(S_u) \tag{54}$$

$$\kappa_{i_{sd}}(t) = 1 + \gamma_{i_{sd}} e^{-\sigma_{i_{sd}} t},$$
$$\kappa_{\Omega}(t) = 1 + \gamma_{\Omega} e^{-\sigma_{\Omega} t}, \tag{55}$$

$$\kappa_u(t) = 1 + \gamma_u e^{-\sigma_u t} \tag{56}$$

where $\gamma_x > 0$ and $\sigma_x > 0$, $x = i_{sd}, \Omega$, are specified by the designer.

It is well known that online identification and adaptation of a time varying parameter in a controller contribute to relax the requirement of the system parameter knowledge and help to improve the control performance [43]. Such is the purpose of the following rotor flux identifier. Substitute T_m by $K_{opt}\Omega^2$ in (5) and rearrange it as

$$\dot{\Omega} = \frac{K_{opt}\Omega^2 - f\Omega}{J} - \frac{3p}{2J}\phi_r i_{sq} \tag{57}$$

Based on (57), a generator speed observer can be written as ($K_1 > 0$ is a constant gain)

$$\dot{\widehat{\Omega}} = \frac{K_{opt}\Omega^2 - f\Omega}{J} - \frac{3p}{2J}\widehat{\phi}_r i_{sq} - \nu, \tag{58}$$

$$\nu = K_1 \operatorname{sign}(\widetilde{\Omega}). \tag{59}$$

Computing the dynamic of observation error yields

$$\dot{\widetilde{\Omega}} = -\frac{3p}{2J}\widetilde{\phi}_r i_{sq} - \nu, \tag{60}$$

where $\widehat{\Omega}$ and $\widehat{\phi}_r$ denote the estimated quantities while $\widetilde{\Omega} = \widehat{\Omega} - \Omega$ and $\widetilde{\phi}_r = \widehat{\phi}_r - \phi_r$ are the error quantities. Assuming that the parameter identifiability condition $-3pi_{sq}/2J \neq 0\ \forall t \geq 0$ is verified, the rotor flux estimation law can be processed as ($K_2, \tau > 0$ and s is Laplace variable)

$$\dot{\widehat{\Omega}} = \frac{K_{opt}\Omega^2 - f\Omega}{J} - \frac{3p}{2J}\widehat{\phi}_r i_{sq} - \nu \tag{61}$$

$$\dot{\widetilde{\phi}}_r = \dot{\widehat{\phi}}_r = -K_2\widetilde{\phi}_r,$$
$$\widetilde{\phi}_r = \frac{\nu_{1eq}}{-(3p/2J)\, i_{sq}} \tag{62}$$

$$\nu = K_1 \operatorname{sign}(\widetilde{\Omega}),$$
$$\nu_{eq} = \frac{\nu}{1 + \tau s}. \tag{63}$$

The proof of convergence of the identification algorithm in ((61)–(63)) can be found in [24].

The schematic description of the proposed control algorithm is presented in Figure 2. In this figure, graphs (a), (b), and (c) show the equivalent control systems while graph (d) exhibits the interface of the control inputs with PWM units. θ is the generator shaft position.

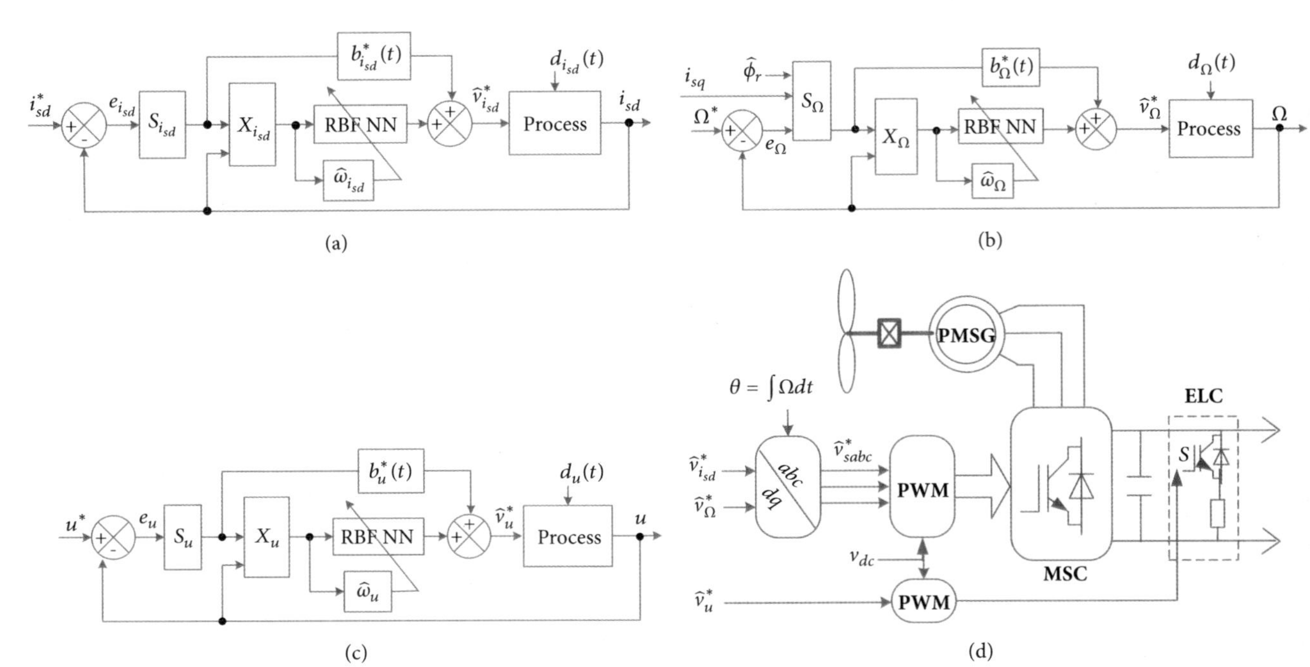

FIGURE 2: Block diagram of the proposed RBF neuro-sliding controller.

4. Control Design for GSC

The GSC, acting as a voltage source inverter, is controlled to track the active and reactive power of the grid by using grid current regulator. Conventional PI control method is exploited in the present study to achieve the control objective. Details concerning the selection of PI gains are described in Appendix. Figure 3 illustrates the proposed GSC control structure. θ_g is the grid voltage angular position estimated by a PLL for grid synchronization.

The references of the grid current components are computed from the reference active power P^* and reference reactive power Q^* as follows:

$$i_{Gd}^* = \frac{2}{3}\frac{P^* v_{gd} + Q^* v_{gq}}{v_{gd}^2 + v_{gq}^2},$$
$$i_{Gq}^* = \frac{2}{3}\frac{P^* v_{gq} - Q^* v_{gq}}{v_{gd}^2 + v_{gq}^2}. \quad (64)$$

In order to allow the GSC to operate at unity power factor, Q^* is set to zero. The reference active power is derived directly from the optimum power captured from the WT as shown in (65).

$$P^* = K_{opt}\Omega_{opt}^3 - \frac{Su}{CR_E} \quad (65)$$

Remark 4. In standalone mode, the GSC is controlled to regulate the amplitude and frequency of the AC load voltage. From the grid current control structure in Figure 3, it is possible to switch to standalone mode by disconnecting the grid and setting manually the reference components (v_d^* and v_q^*) of the voltage across the filter capacitor and the value of the load voltage angular frequency for (d, q)-(a, b, c) reference frame transformation. The switches in Figure 3 together with the circuit breaker are dedicated to this end. That is an interesting feature of the proposed inverter control structure.

5. Numerical Simulation Results

Numerical simulations of the studied WECS (see Figure 1 with control structures in Figures 2 and 3) are carried out in Matlab/Simulink software to assess the performance of the proposed RBF neuro-sliding mode controller (RBF-NSMC). Simulations based on a sliding mode control (SMC) method [24] are conducted and the results are presented to compare the capabilities of the proposed RBF-NSMC. The WECS parameters [44] and the numerical values of the control parameters are listed in Tables 1 and 2, respectively. It is worth mentioning that the PMSG can directly be connected to the WT. However, a gearbox ratio $G = 1.2$ is selected to comply with realistic transmission condition. An inductive RL load ($R = 250\Omega$ and $L = 50mH$) is used to materialize a general local AC load. More details on the selection criteria of the LCL filter parameters are given in Appendix. Each neural network is built up with four neurons in the hidden layer.

The analysis of the system response is conducted considering three operation conditions. Firstly the effectiveness of the proposed RBF-NSMC is investigated under nominal WECS parameters (see Figures 4–9). Secondly the robustness of the controller is examined under WECS parameters uncertainties (see Figures 10–12). The wind speed is time varying in both cases and its behaviour is shown in Figure 4. Finally the performance of the control system is evaluated under transition from grid connected to standalone mode during low wind speeds (see Figures 13 and 14).

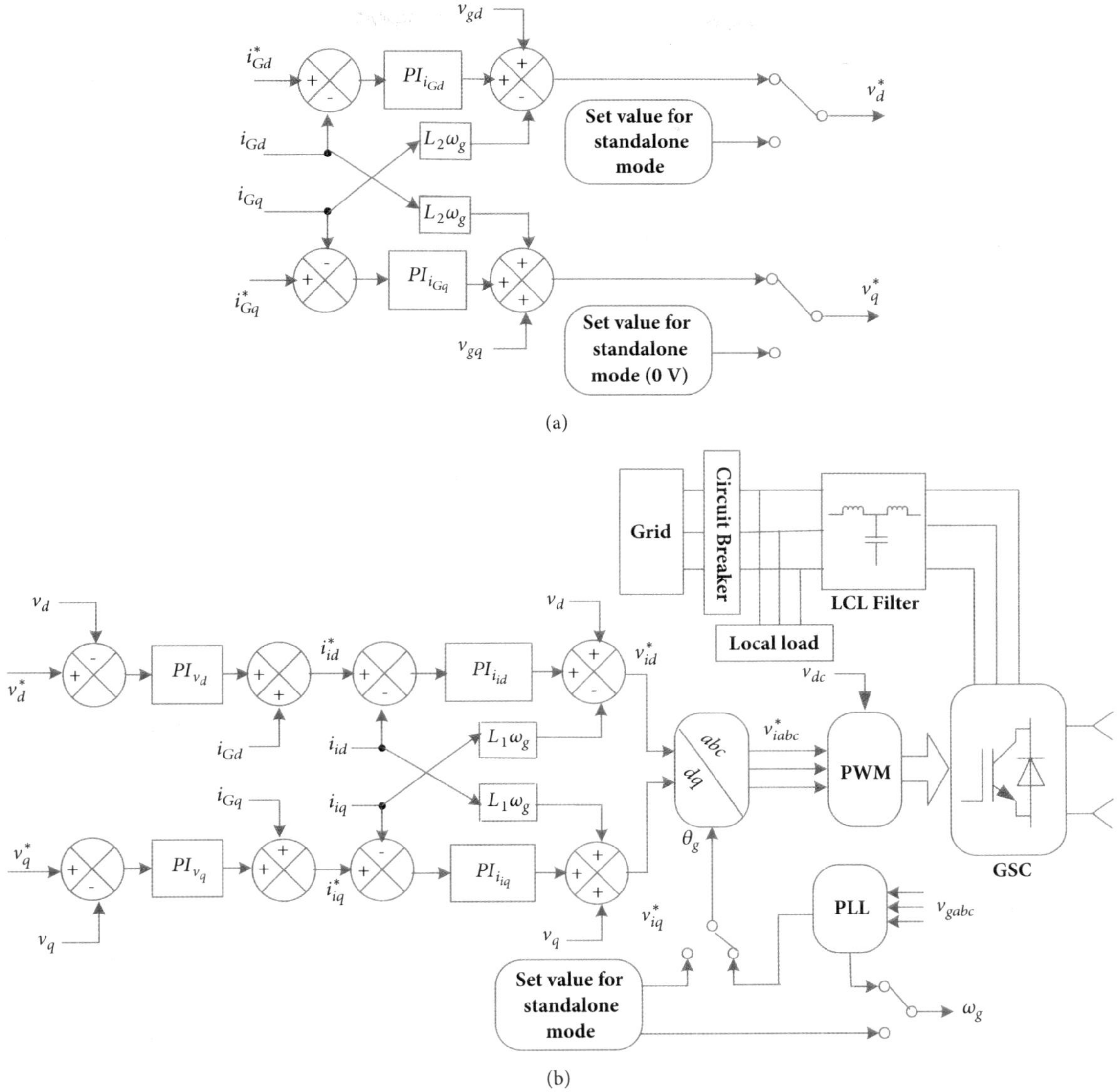

FIGURE 3: GSC control diagram. (a) Grid current control loop. (b) Control loop for LCL filter capacitor voltage.

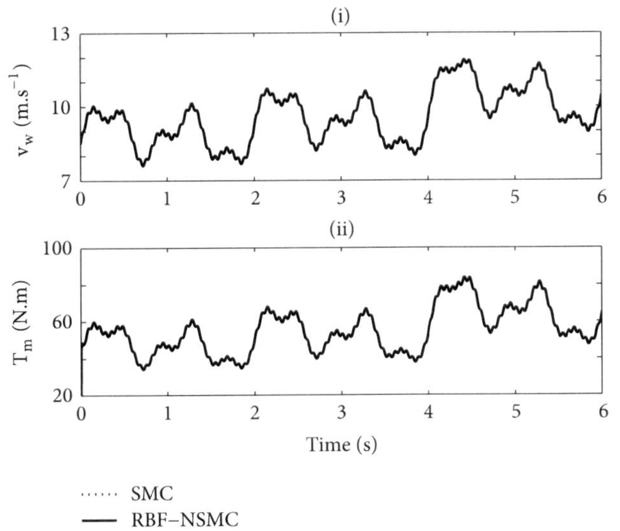

FIGURE 4: Wind speed profile and consequent mechanical torque.

Case 1 (test under varying wind speed with WECS nominal parameters). It can be observed on Figures 4–7 that the control objective of the generator side converter and dc-link is achieved since the generator d–axis current is rapidly set to zero (see Figure 5), the dc voltage is regulated to the desired reference value, and the maximum power is accurately harnessed from the wind (see Figure 6). For this test, one can remark clearly on Figure 7 that the results provided by both proposed RBF-NSMC and SMC methods are practically similar. Figures 8 and 9 exhibit the behaviour of the grid side signals when the proposed control approach is maintained on the machine side and dc-link. It is noticeable from Figure 8(a) that the grid dq current components are steered to their estimated references with small response time. Figure 8(b) shows that the regulation of the voltage across the LCL filter capacitor is satisfactory. This voltage regulation is necessary for local AC load feeding and grid synchronization. The sinusoidal waveforms of the filter capacitor voltage and the total grid plus local load current are depicted in Figure 9. From

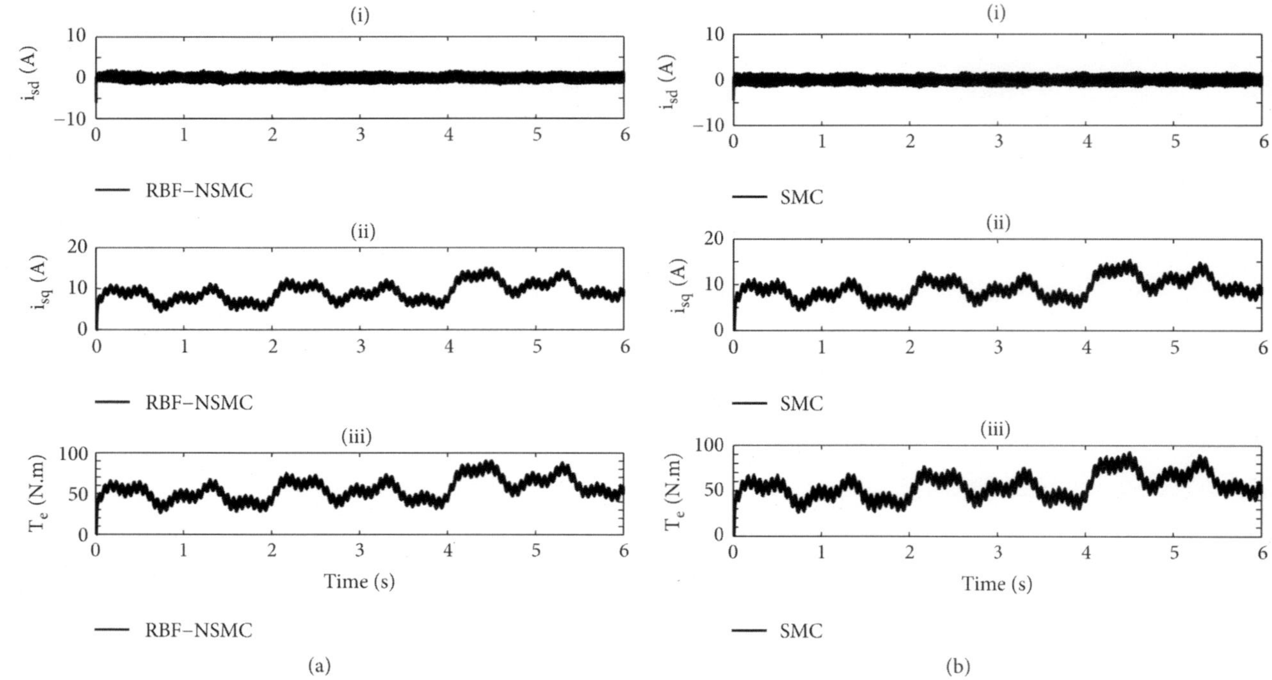

FIGURE 5: Control system response to wind speed fluctuations with nominal values of parameters. (a) Proposed controller. (b) Sliding mode controller. (i) Generator d–axis current. (ii) Generator q–axis current. (iii) Electromagnetic torque.

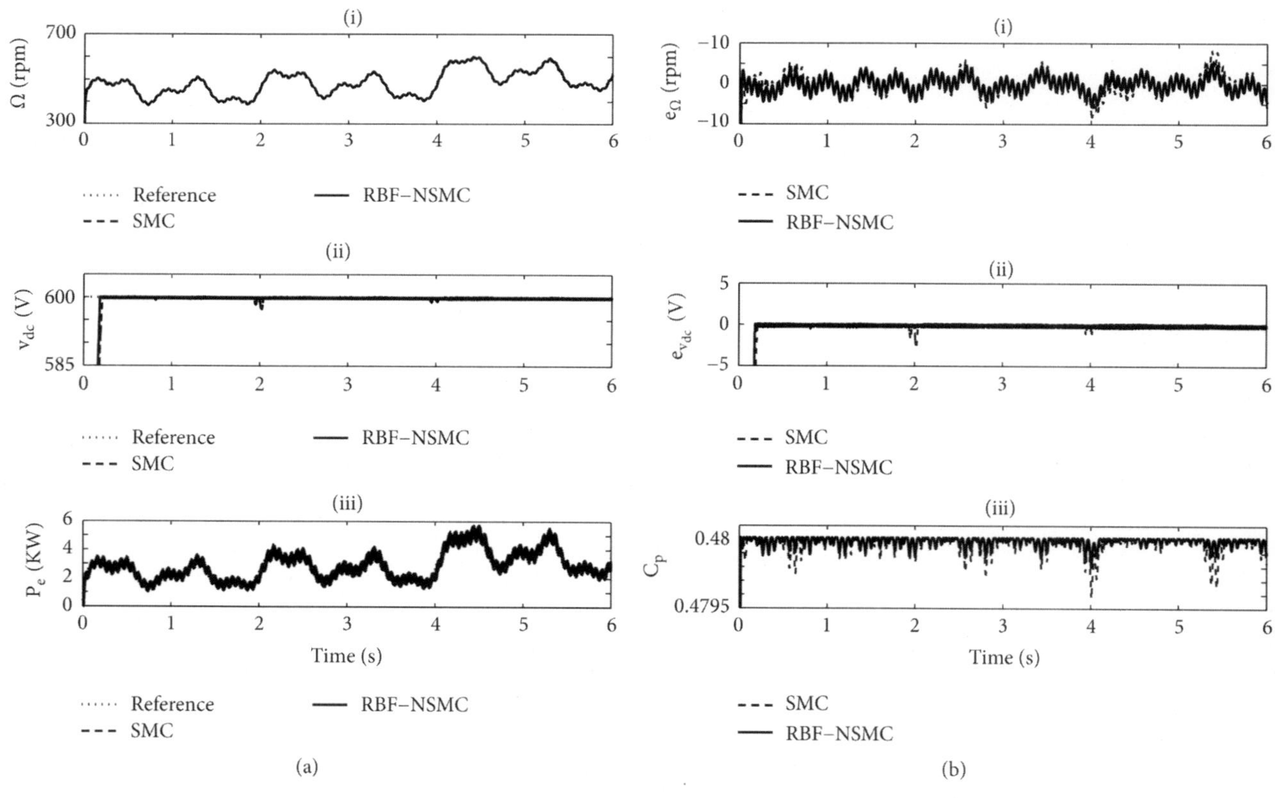

FIGURE 6: Control system response to wind speed fluctuations with nominal values of parameters. (a) (i) Generator speed, (ii) dc-link voltage, and (iii) generator electrical power. (b) (i) Generator speed error, (ii) dc-link voltage error, and (iii) power coefficient.

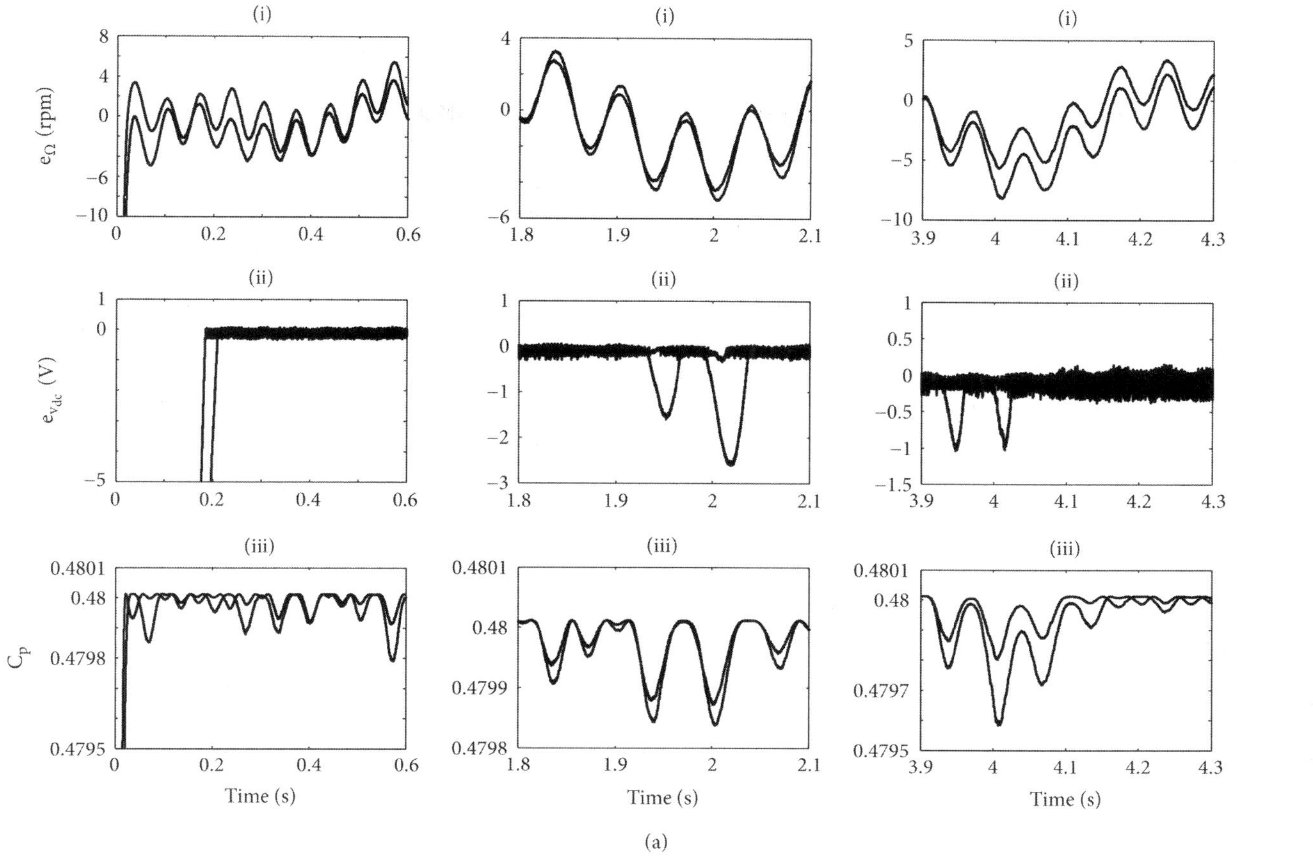

FIGURE 7: Control system response to wind speed fluctuations with nominal values of parameters. Zoom on (blue line: RBF-NSMC, black line: SMC): (i) Generator speed error, (ii) dc-link voltage error, and (iii) power coefficient.

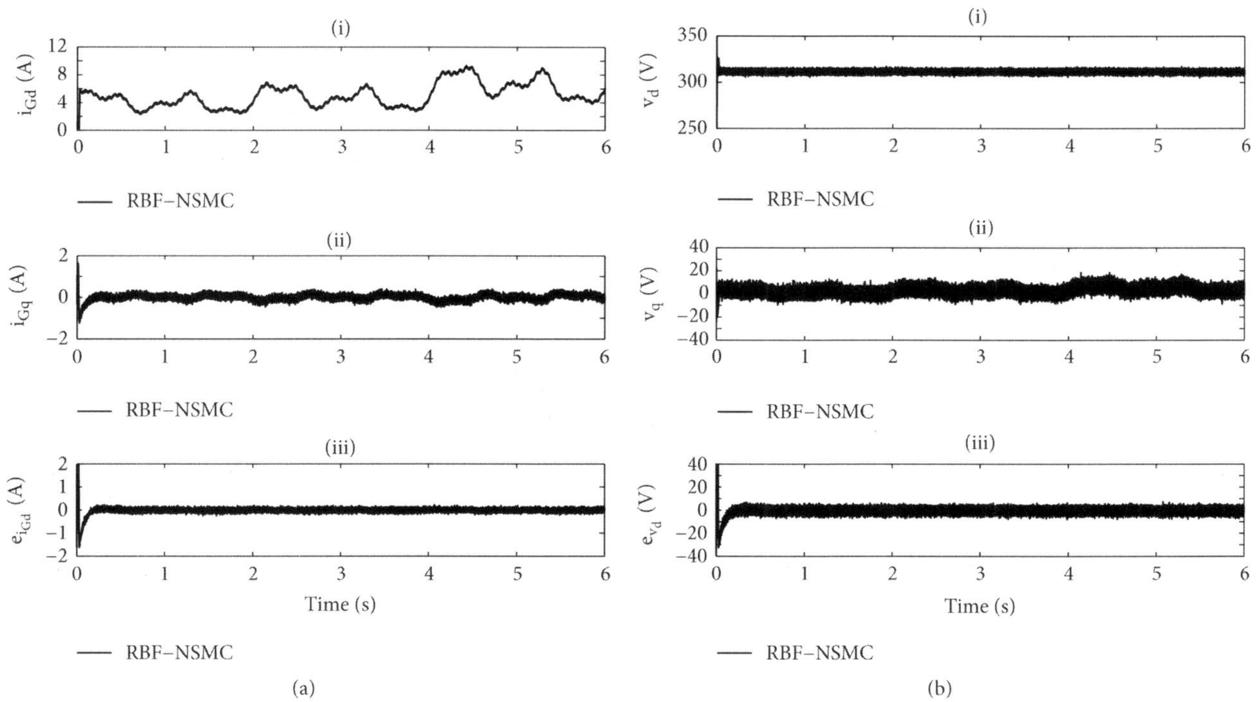

FIGURE 8: Control system response to wind speed fluctuations with nominal values of parameters. (a) (i) Grid d–axis current, (ii) grid q-axis current, and (iii) grid d–axis current error. (b) (i) d–axis voltage across LCL filter capacitor, (ii) d–axis voltage across LCL filter capacitor, and (iii) error signal of d–axis voltage across LCL filter capacitor.

Table 1: Nominal values of PMSG WECS parameters.

Wind turbine	$\beta = 0$ $G = 1.2$	$\zeta_{opt} = 8.1$ $r = 1.84m$	$C_{pmax} = 0.48$ $J = 7.856Kg{\cdot}m^2$	$\rho = 1.225Kg{\cdot}m^{-3}$
PMSG	$R_s = 0.3676\Omega$	$L_s = 3.55mH$	$\phi_r = 0.2867Wb$	$2p = 28$
dc-link	$C = 2200\mu F$	$v_{dc} = 600V$		

Table 2: Numerical values of control parameters.

MSC control						
$d-$axis current controller	$h_1 = 10$	$\alpha_{i_{sd}} = 20000$	$\gamma_{i_{sd}} = 10$	$\sigma_{i_{sd}} = 5$	$\epsilon_{i_{sd}} = 0.1$	
Rotor speed regulator	$h_2 = 190$	$h_3 = 1200$	$\alpha_\Omega = 20000$	$\gamma_\Omega = 15$	$\sigma_\Omega = 5$	$\epsilon_\Omega = 0.01$
dc voltage control	$\alpha_u = 5000$	$\gamma_u = 0$	$\sigma_u = 0$	$\epsilon_u = 0.1$		
Rotor flux estimation	$K_1 = 10$	$K_2 = 8000$				
GSC Control						
Grid current regulation	$K_{P_{iGd}} = 18$	$K_{I_{iGd}} = 45.5$	$K_{P_{iGq}} = 18$	$K_{I_{iGq}} = 45.5$		
LCL capacitor voltage control	$K_{P_{v_d}} = 0.07$	$K_{I_{v_d}} = 1$	$K_{P_{v_q}} = 0.07$	$K_{I_{v_q}} = 1$		
GSC current control	$K_{P_{i_{id}}} = 75$	$K_{I_{i_{id}}} = 482$	$K_{P_{i_{iq}}} = 75$	$K_{I_{i_{iq}}} = 482$		

the above observations one can undoubtedly conclude that the proposed control scheme is effective and the extracted optimum power is transferred to the load with unity power factor.

Case 2 (test under varying wind speed with WECS parameters uncertainties). The graphs on Figures 10–12 show the results obtained when uncertainties are introduced on the process parameters at time $t = 1s$. These perturbations consist of +100%, +50%, −20%, and +30% sudden changes on the nominal values of stator resistance, stator inductance, rotor flux, and rotor inertia, respectively. From Figure 10, one can note that the proposed controller presents fast dynamic response compared to that of the SMC one. The absolute values of tracking errors representing maximum deviations from the optimal curve, for different time periods, are picked up from Figure 10(b) and presented in Table 3. It comes out from this table that the deviations under the proposed RBF-NSMC method are always smaller than those provided by the SMC. This demonstrates that the proposed control scheme features higher robustness with respect to varying wind speed and parameter disturbances. It is shown in Figure 11(a) that the transformed estimation of the unknown uncertainties bound starts with a high value to accelerate the process and converges to a low value to reduce the energy consumed by the controller and also prevent from unwanted high frequency unmodeled dynamics. The resulting smooth control actions for both MSC and GSC are shown in Figure 11(b). These control actions are very significant since they are not saturated and therefore are favorable for interfacing with PWM units. This remark supports the fact that the proposed RBF-NSMC is feasible for real time implementation. The total harmonic distortions (THDs) of the LCL filter capacitor voltage and injected grid current are both less than 5% which is the upper limit as seen in Figure 12. Although this study is focused on the grid connected mode, one may say that, for wind speeds lower than those depicted in the used wind profile, one can switch to standalone mode as stated in Remark 4. This helps not to exceed the 5% THD maximum value on the injected grid current.

Case 3 (performance evaluation under grid connected to standalone mode transition). In this test, the system starts with the same conditions as in Case 2. But at time $t = 1.84s$, the mean value of the wind profile changes to low value (from $8.5m{\cdot}s^{-1}$ to $5.5m{\cdot}s^{-1}$, see Figure 13). Simultaneously the system is switched to standalone mode (see Remark 4). It is observed on Figure 14(a) that the global stability of the system is preserved and better transient performance is kept by the proposed RBF-NSMC as compared to the SMC. It is seen on Figure 14(b) that the filter capacitor voltage is still well regulated by the proposed grid side PI controller despite the grid disconnection. It is also noted that, after this grid disconnection, the injected grid current vanishes and only the portion absorbed by the local load remains. These observations demonstrate the flexible feature of the proposed grid side control structure.

6. Conclusions

A control scheme of a grid connected variable speed PMSG WECS is investigated. The control approach based on online learning RBF-sliding network has been proposed to extract efficiently maximum power without the knowledge of neither the exact model of the nonlinear system nor the requirement of the upper limit of the system disturbances. A transformation of the estimated system uncertainties bound for explicitly accelerating the rate of convergence of the NN during startup phase has been proposed. Simulation results have shown the effectiveness and the feasibility of the proposed control scheme. The proposed RBF-NSMC exhibited higher robustness with fast dynamic response face to wind speed variations and WECS parameters changes. The proposed grid side inverter control structure is flexible so as to switch between grid connected mode and standalone mode in order to cover both high and low wind speeds.

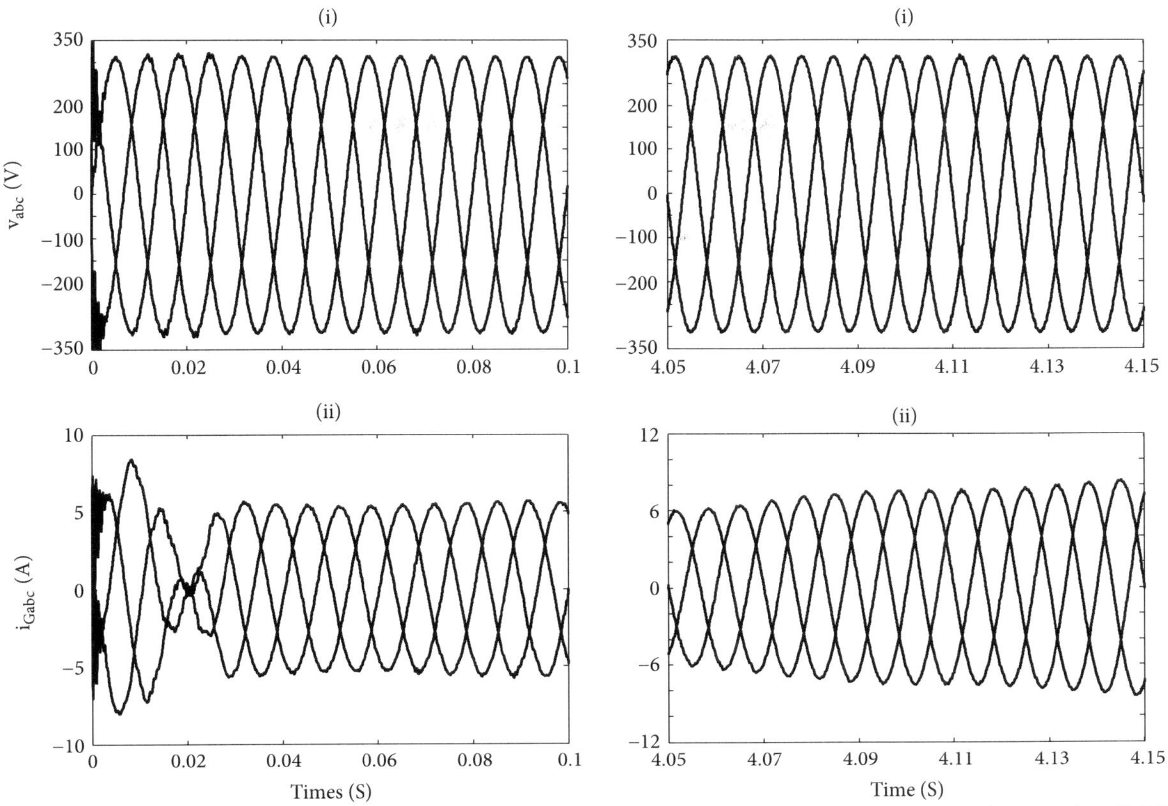

FIGURE 9: Control system response to wind speed fluctuations with nominal values of parameters. Zoom on (proposed RBF-NSMC): (i) 3-phase voltage across the LCL filter capacitor and (ii) 3-phase total grid plus local load current.

TABLE 3: Comparison of maximum deviations from the optimal curve under varying wind speed with parametric perturbations.

Time period (s)	$t \in [0.9\ 1.3]$		$t \in [1.8\ 2.2]$		$t \in [3.9\ 4.2]$	
Controller	RBF-NSMC	SMC	RBF-NSMC	SMC	RBF-NSMC	SMC
$\lvert e_{\Omega}\rvert_{max}(rpm)$	8.5336	9.7521	6.0294	9.8529	6.3361	8.4202
$\lvert e_{v_{dc}}\rvert_{max}(V)$	0.8992	1.6050	3.5966	7.8319	3.9286	12.5000
$\lvert C_p - C_{pmax}\rvert_{max}(-)$	0.0005	0.0007	0.0003	0.0007	0.0002	0.0004

Appendix

A. Selection of GSC PI Control Gains

In this reasoning K_{Px} and K_{Ix} represent proportional and integral gains, respectively, and $e_x = x^* - x$ quotes the error signal, while x is a signal with x^* being the corresponding reference. Rearrange (19) and (20) in the following form:

$$L_2 \dot{i}_{Gd} = -R_2 i_{Gd} + v'_d, \quad \text{(A.1)}$$

$$v'_d = L_2 \omega_g i_{Gq} - v_{gd} \quad \text{(A.2)}$$

$$L_2 \dot{i}_{Gq} = -R_2 i_{Gq} + v'_q, \quad \text{(A.3)}$$

$$v'_q = -L_2 \omega_g i_{Gd} - v_{gq} \quad \text{(A.4)}$$

v'_d and v'_q are the transformed control inputs. And compute the open loop transfer functions of (A.1) and (A.3)

$$H^{ol}_{i_{Gd}}(s) = H^{ol}_{i_{Gq}}(s) = \frac{1/R_2}{\tau^{ol}_{i_G} s + 1} \quad \text{(A.5)}$$

where $\tau^{ol}_{i_G} = L_2/R_2$ is the open loop time constant. Now select the correctors in the form of

$$C_{i_{Gd}}(s) = C_{i_{Gq}}(s) = K_{Pi_G} + \frac{K_{Ii_G}}{s} = \frac{K_{Pi_G}\left(a_{i_G} s + 1\right)}{a_{i_G} s}, \quad \text{(A.6)}$$

$$a_{i_G} = \frac{K_{Pi_G}}{K_{Ii_G}} \quad \text{(A.7)}$$

Proceed to the compensation of the open loop time constant by choosing a_{i_G} as

$$a_{i_G} = \tau^{ol}_{i_G} \quad \text{(A.8)}$$

and calculate the corrected closed loop transfer functions in the form of first-order systems as

$$H^{cl}_{i_{Gd}}(s) = H^{cl}_{i_{Gq}}(s) = \frac{1}{\tau^{cl}_{i_G} s + 1} \quad \text{(A.9)}$$

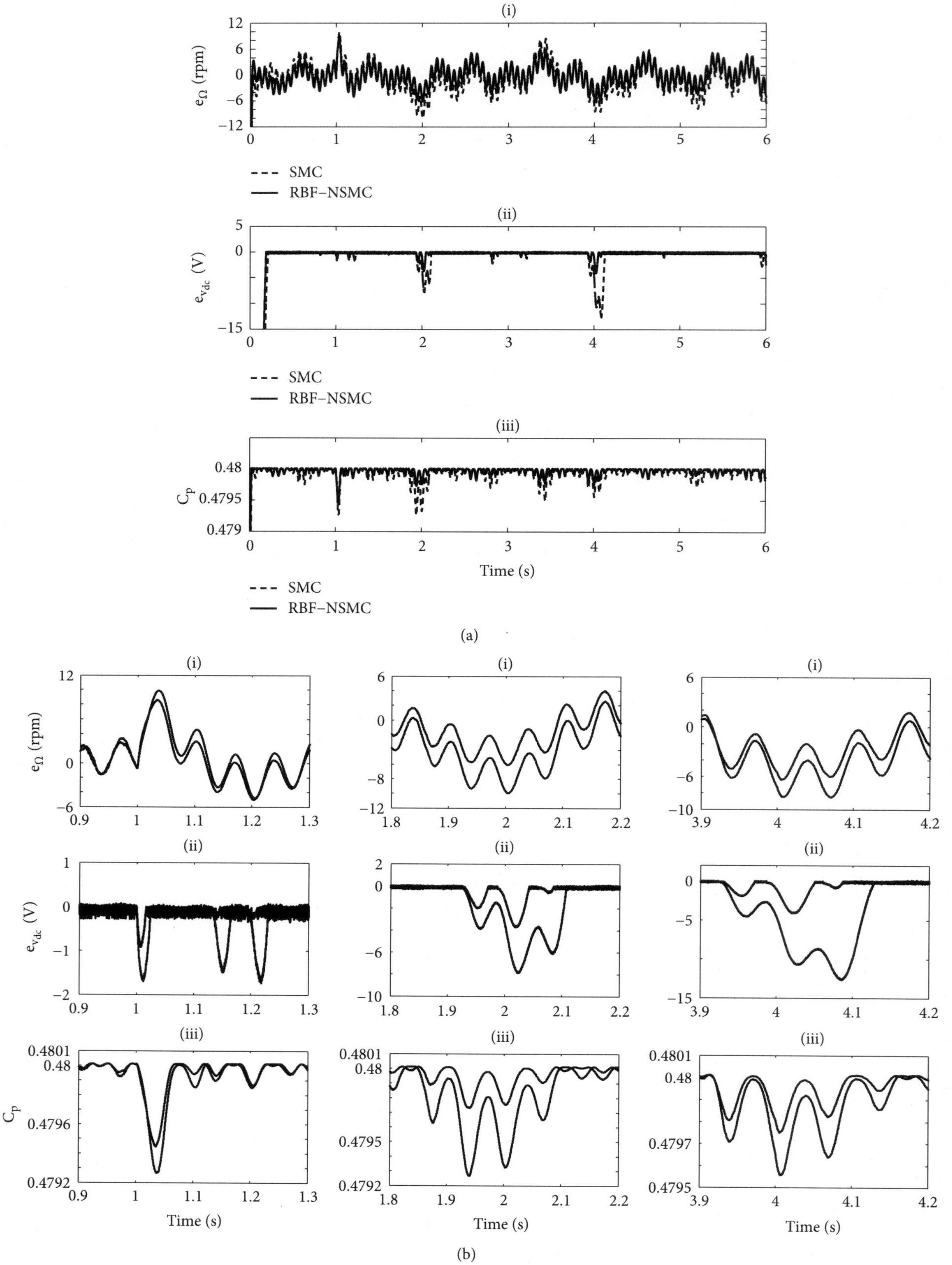

FIGURE 10: Control system response to wind speed fluctuations with parameters uncertainties (at time $t = 1s$, variations of +100% for R_s, +50% for L_s, −20% for ϕ_r, and +30% for J are introduced). (a) Signals trajectories and (b) corresponding zoom (blue line: RBF-NSMC, black line: SMC). (i) Generator speed error, (ii) dc-link voltage error, and (iii) power coefficient.

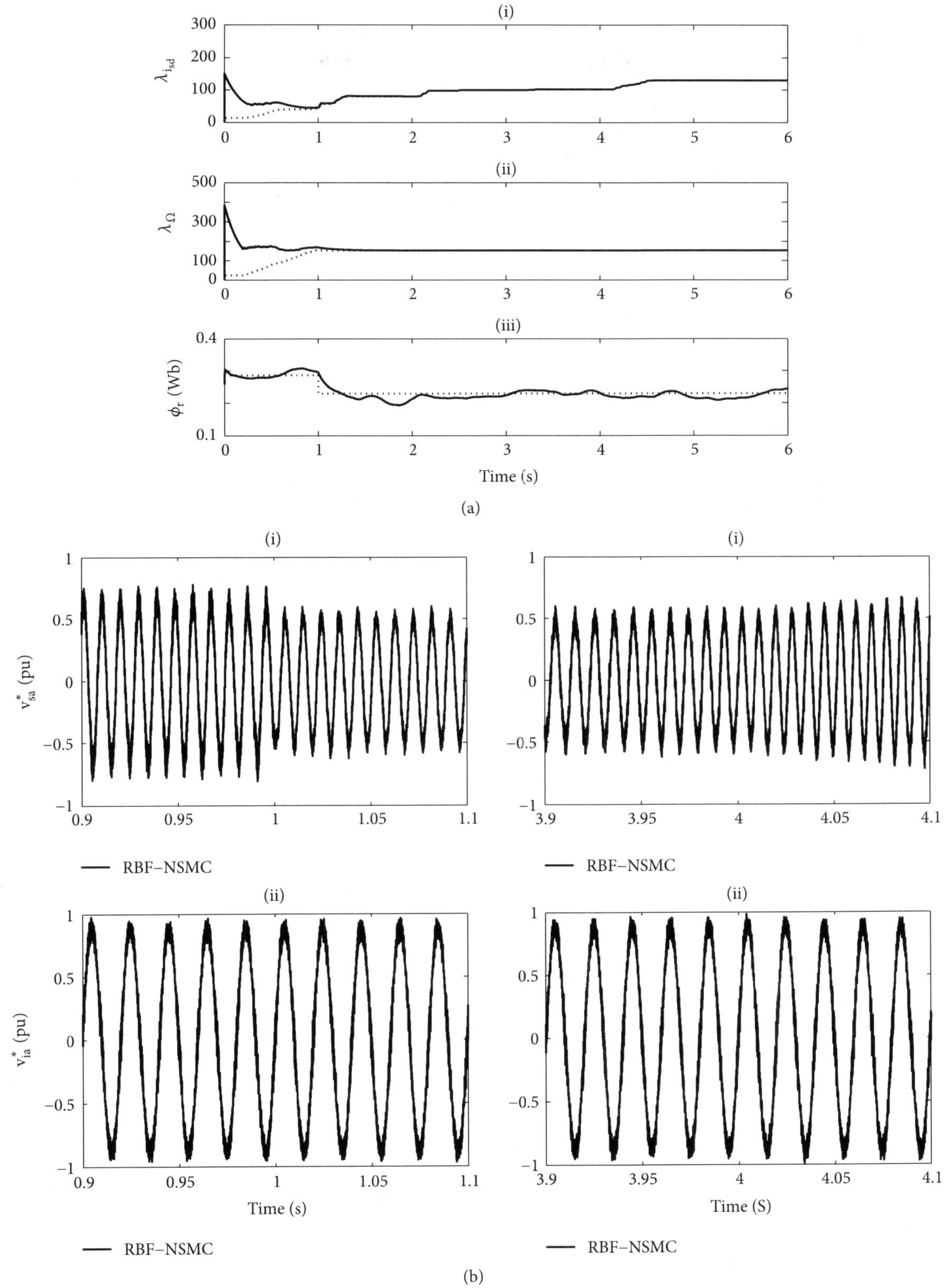

FIGURE 11: Control system response to wind speed fluctuations with parameters uncertainties (at time $t = 1s$, variations of $+100\%$ for R_s, $+50\%$ for L_s, -20% for ϕ_r, and $+30\%$ for J are introduced). (a) Estimated parameters. (i) $\lambda_{i_{sd}}$: (dotted line) estimate $(\widehat{\lambda}_{i_{sd}})$ and (solid line) its transformed estimate $(\widehat{\lambda}^*_{i_{sd}})$, (ii) λ_Ω: (dotted line) estimate $(\widehat{\lambda}_\Omega)$ and (solid line) its transformed estimate $(\widehat{\lambda}^*_\Omega)$, and (iii) ϕ_r: (dotted line) reference and (solid line) its estimate. (b) Zoom on control signals: (i) approximation of phase$-a$ voltage of the MSC $\widehat{v}^*_{sa}$; (ii) phase$-a$ voltage of the GSC v^*_{ia}.

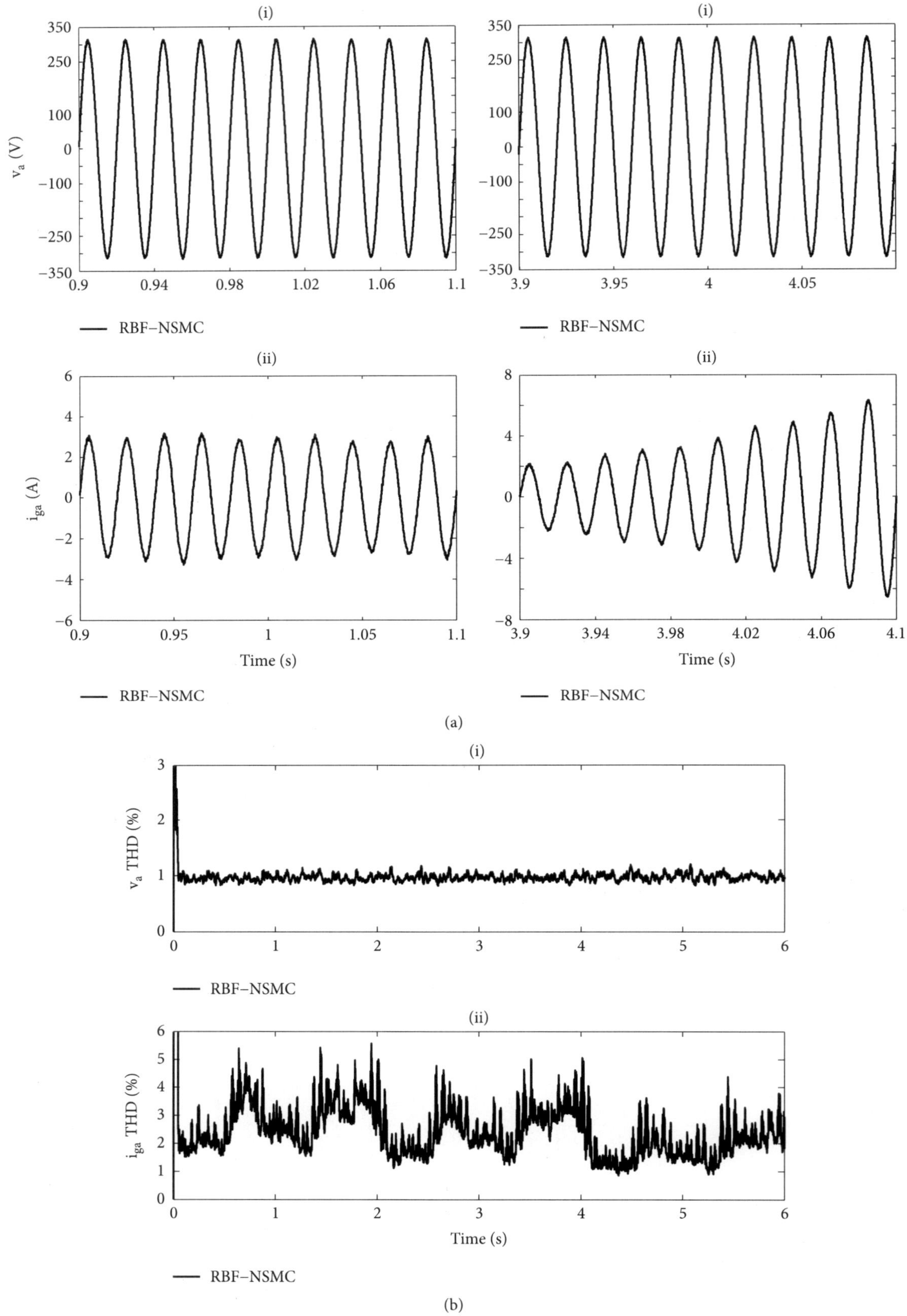

FIGURE 12: Control system response to wind speed fluctuations with parameters uncertainties (at time $t = 1s$, variations of +100% for R_s, +50% for L_s, −20% for ϕ_r, and +30% for J are introduced). (a) Zoom on LCL filter capacitor voltage and injected grid current trajectories with (b) corresponding THD behaviours. (i) Filter capacitor phase–a voltage; (ii) injected grid phase–a current.

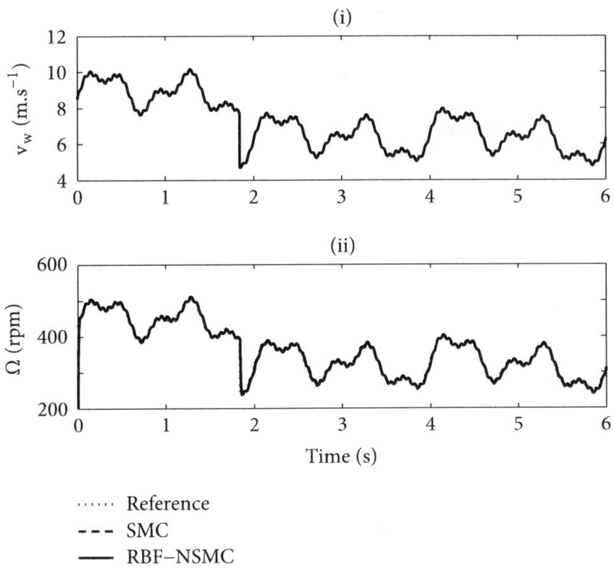

FIGURE 13: Control system response to wind speed fluctuations with parameters uncertainties and grid connected to standalone mode transition. The transition occurs at time $t = 1.84s$. (i) Wind profile with step change of $-3m{\cdot}s^{-1}$ at time $t = 1.84s$ and consequent (ii) rotor speed tracking response.

where $\tau_{i_G}^{cl} = L_2/K_{Pi_G}$ is the closed loop time constant. The practical settling time with 5% criteria ($t_{s5\%}$) of a first-order system is equal to three times its time constant. On the other hand the settling time of the closed loop system should be at least equal to the open loop settling time; that is,

$$t_{s5\%}^{cl} \le t_{s5\%}^{ol} \Longrightarrow \quad K_{Pi_G} \ge R_2 \tag{A.10}$$

Hence the outputs of the closed loop systems (A.1) and (A.3) will exponentially converge to their reference values if conditions (A.8) and (A.10) are satisfied. Thus the reference values of dq components for the voltage across the LCL filter capacitor are expressed as

$$v_d^* = K_{Pi_{Gd}} e_{i_{Gd}} + K_{Ii_{Gd}} \int e_{i_{Gd}} dt - L_2 \omega_g i_{Gq} + v_{gd} \tag{A.11}$$

$$v_q^* = K_{Pi_{Gq}} e_{i_{Gq}} + K_{Ii_{Gq}} \int e_{i_{Gq}} dt + L_2 \omega_g i_{Gd} + v_{gq}. \tag{A.12}$$

Adopt the same methodology for the GSC currents control loops and obtain

$$v_{id}^* = K_{Pi_{id}} e_{i_{id}} + K_{Ii_{id}} \int e_{i_{id}} dt - L_1 \omega_g i_{iq} + v_d \tag{A.13}$$

$$v_{iq}^* = K_{Pi_{iq}} e_{i_{iq}} + K_{Ii_{iq}} \int e_{i_{iq}} dt + L_1 \omega_g i_{id} + v_d \tag{A.14}$$

For the control design of the LCL filter capacitor voltage, rewrite (17) and (18) in the form of

$$C_f \dot{v}_d = i_d', \quad i_d' = i_{id} - i_{Gd} \tag{A.15}$$

$$C_f \dot{v}_q = i_q', \quad i_q' = i_{iq} - i_{Gq}. \tag{A.16}$$

The corresponding transfer functions are written as

$$H_{v_d}^{ol}(s) = H_{v_q}^{ol}(s) = \frac{1}{C_f s} \tag{A.17}$$

Let the correctors be

$$C_{v_d}(s) = C_{v_q}(s) = K_{Pv_c} + \frac{K_{Iv_c}}{s} = \frac{K_{Pv_c}\left(a_{v_c} s + 1\right)}{a_{v_c} s}, \tag{A.18}$$

$$a_{v_c} = \frac{K_{Pv_c}}{K_{Iv_c}} \tag{A.19}$$

Then calculate the closed loop transfer functions and arrange them in the form of second-order systems as

$$H_{v_d}^{cl}(s) = H_{v_q}^{cl}(s) = \frac{a_{v_c} s + 1}{s^2/\omega_0^2 + \left(2\xi/\omega_0\right) s + 1} \tag{A.20}$$

$$\text{with } a_{v_c} = \frac{2\xi}{\omega_0}, \quad K_{Pv_c} = a_{v_c} C_f \omega_0^2. \tag{A.21}$$

ξ and ω_0 are, respectively, the system oscillation damping factor and undamped natural angular frequency. Thus the reference currents of the GSC are computed as

$$i_{id}^* = K_{Pv_d} e_{v_d} + K_{Iv_d} \int e_{v_d} dt + i_{iGd} \tag{A.22}$$

$$i_{iq}^* = K_{Pi_q} e_{v_q} + K_{Iv_q} \int e_{v_q} dt + i_{iGq} \tag{A.23}$$

B. LCL Filter Design

Grid side inverter rated power $P_n = 5KW$, nominal grid line to line voltage $U_n = 220\sqrt{3}V(rms)$, switching frequency $f_{sw} = 5KHz$, and grid frequency $f_g = 50Hz$. In this study the calculation of LCL filter parameters is based on the methodology presented in [45].

Base Values Definition. The base impedance Z_b, inductance L_b, capacitance C_b, and nominal (rms) converter current are computed, respectively, as follows:

$$Z_b = \frac{U_n^2}{P_n}, \quad L_b = \frac{Z_b}{\omega_n}, \quad C_b = \frac{1}{\omega_n Z_b} \quad \text{and } I_n = \frac{P_n}{\sqrt{3}U_n} \tag{B.1}$$

where $\omega_n = \omega_g = 2\pi f_g$ is the grid angular frequency.

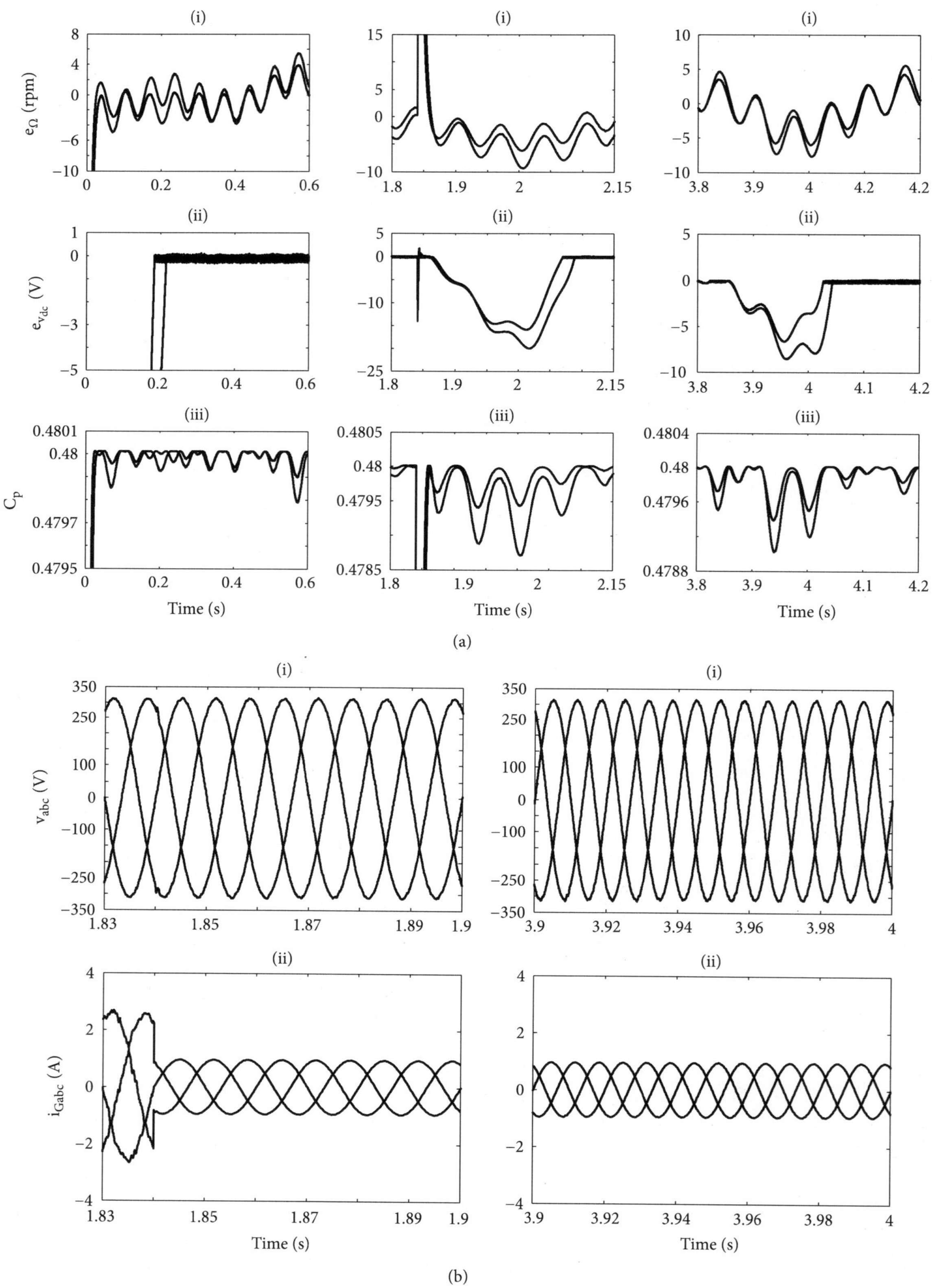

FIGURE 14: Control system response to wind speed fluctuations with parameters uncertainties and grid connected to standalone mode transition. The transition occurs at time $t = 1.84s$. (a) Zoom on (blue line: RBF-NSMC, black line: SMC): (i) generator speed error, (ii) dc-link voltage error, and (iii) power coefficient. (b) Zoom on (proposed RBF-NSMC): (i) 3-phase voltage across the LCL filter capacitor; (ii) 3-phase total grid plus local load current.

Converter Side Inductance L_1 Selection. The converter side inductance is chosen according to the following expression:

$$L_1 \geq \frac{v_{dc}^*}{6 f_{sw} I_{maxrip}}. \quad (B.2)$$

where I_{maxrip} is the maximum ripple current which must lie between 10% and 25% of the nominal converter current.

Filter Capacitance C_f Selection. The filter capacitance should be less than 5% of the base value; i.e., $C_f \leq 0.05 C_b$.

Grid Side Inductance L_2 Selection. The grid inductance is selected as

$$L_2 = r_a L_1 \quad \text{with } r_a = \left| \frac{1 - HA}{HA\left(1 - L_1 C_f W_{sw}^2\right)} \right|. \quad (B.3)$$

r_a is the ratio between converter side and grid side inductances and HA is the harmonics attenuation of the phase current. In this study the parameters have been chosen by considering 12% attenuation of maximum converter ripple current, 4.5% base capacitance, and $HA = 0.05$. The resulting LCL filter parameters are $L_1 = 15.6mH$, $C_f = 5\mu F$, and $L_2 = 4mH$. Thus the filter resonant frequency $f_{res} = (1/2\pi)\sqrt{(L_1 + L_2)/L_1 L_2} = 1276\text{Hz}$ is well between the desired range $10 f_g < f_{res} < 0.5 f_{sw}$. The inductors internal resistances have been taken as $R_1 = 0.1\Omega$ and $R_2 = 0.01\Omega$.

Conflicts of Interest

There are no conflicts of interest.

References

[1] H. Sun, Y. Han, and L. Zhang, "Maximum Wind Power Tracking of Doubly Fed Wind Turbine System Based on Adaptive Gain Second-Order Sliding Mode," *Journal of Control Science and Engineering*, vol. 2018, Article ID 5342971, 11 pages, 2018.

[2] M. G. Zanchettin and R. Reginatto, "Grid integration limits for fixed-speed wind turbines with induction generators," *Journal of Control, Automation and Electrical Systems*, vol. 24, no. 6, pp. 873–884, 2013.

[3] F. Blaabjerg, Z. Chen, and S. B. Kjaer, "Power electronics as efficient interface in dispersed power generation systems," *IEEE Transactions on Power Electronics*, vol. 19, no. 5, pp. 1184–1194, 2004.

[4] J. A. Baroudi, V. Dinavahi, and A. M. Knight, "A review of power converter topologies for wind generators," *Journal of Renewable Energy*, vol. 32, no. 14, pp. 2369–2385, 2007.

[5] Z. Chen, J. M. Guerrero, and F. Blaabjerg, "A review of the state of the art of power electronics for wind turbines," *IEEE Transactions on Power Electronics*, vol. 24, no. 8, pp. 1859–1875, 2009.

[6] N. Goudarzi and W. D. Zhu, "A review on the development of wind turbine generators across the world," *International Journal of Dynamics and Control*, vol. 1, no. 2, pp. 192–202, 2013.

[7] M. M. Hossain and M. H. Ali, "Future research directions for the wind turbine generator system," *Renewable & Sustainable Energy Reviews*, vol. 49, pp. 481–489, 2015.

[8] M. Chinchilla, S. Arnaltes, and J. C. Burgos, "Control of permanent-magnet generators applied to variable-speed wind-energy systems connected to the grid," *IEEE Transactions on Energy Conversion*, vol. 21, no. 1, pp. 130–135, 2006.

[9] H. Geng and D. Xu, "Stability analysis and improvements for variable-speed multipole permanent magnet synchronous generator-based wind energy conversion system," *IEEE Transactions on Sustainable Energy*, vol. 2, no. 4, pp. 459–467, 2011.

[10] S. Li, T. A. Haskew, R. P. Swatloski, and W. Gathings, "Optimal and direct-current vector control of direct-driven PMSG wind turbines," *IEEE Transactions on Power Electronics*, vol. 27, no. 5, pp. 2335–2337, 2012.

[11] C. N. Wang, W. C. Lin, and X. K. Le, "Modelling of a pmsg wind turbine with autonomous control," *Mathematical Problems in Engineering*, Article ID 856173, p. 9, 2014.

[12] L. Fateh, O. Ahmed, O. Amar, D. Abdelhak, and B. Lakhdar, "Modeling and control of a permanent magnet synchronous generator dedicated to standalone wind energy conversion system," *Frontiers in Energy*, vol. 10, no. 2, pp. 155–163, 2016.

[13] M. F. M. Arani and Y. A.-R. I. Mohamed, "Assessment and Enhancement of a Full-Scale PMSG-Based Wind Power Generator Performance under Faults," *IEEE Transactions on Energy Conversion*, vol. 31, no. 2, pp. 728–739, 2016.

[14] E. Hamatwi, I. E. Davidson, and M. N. Gitau, "Rotor Speed Control of a Direct-Driven Permanent Magnet Synchronous Generator-Based Wind Turbine Using Phase-Lag Compensators to Optimize Wind Power Extraction," *Journal of Control Science and Engineering*, vol. 2017, Article ID 6375680, p. 17, 2017.

[15] B. Yang, T. Yu, H. Shu, X. Zhang, K. Qu, and L. Jiang, "Democratic joint operations algorithm for optimal power extraction of PMSG based wind energy conversion system," *Energy Conversion and Management*, vol. 159, pp. 312–326, 2018.

[16] M. S. Merzoug, H. Benalla, and L. Louze, "Nonlinear control of permanent magnet synchronous generators PMSG using feedback linearization," *Revue des Energies Renouvelables*, vol. 14, pp. 357–367, 2011.

[17] A. El Magri, F. Giri, G. Besançon, A. El Fadili, L. Dugard, and F. Z. Chaoui, "Sensorless adaptive output feedback control of wind energy systems with PMS generators," *Control Engineering Practice*, vol. 21, no. 4, pp. 530–543, 2013.

[18] S. Ciampichetti, M. L. Corradini, G. Ippoliti, and G. Orlando, "Sliding mode control of permanent Magnet Synchronous Generators for wind turbines," in *Proceedings of the 37th Annual Conference of the IEEE Industrial Electronics Society, IECON '11*, pp. 740–745, Australia, November 2011.

[19] S. Benelghali, M. E. Benbouzid, J. F. Charpentier, T. Ahmed-Ali, and I. Munteanu, "Experimental Validation of a Marine Current Turbine Simulator: Application to a Permanent Magnet Synchronous Generator-Based System Second-Order Sliding Mode Control," *IEEE Transactions on Industrial Electronics*, vol. 58, no. 1, pp. 118–126, 2011.

[20] M. S. Merzoug, H. Benalla, and L. Louze, "Sliding mode control (SMC) of permanent magnet synchronous generators (PMSG)," in *Proceedings of the Terragreen: Clean Energy Solutions for Sustainable Environment, CESSE '12*, pp. 43–52, Lebanon, February 2012.

[21] S. Li-Xia, W. Zhe, H. Feng-Ling, and Y. Feng, "A chattering-free terminal sliding mode control of direct-drive PWM forwind generation system," in *Proceedings of the IEEE IECON, 4th Annual IEEE International Conference on Cyber Technology in Automation*, pp. 296–301, 2014.

[22] S. Shoja-Majidabad, H. Toosian-Shandiz, and A. Hajizadeh, "Robust fractional-order control of PMSG-based WECS," *International Journal of Automation and Control Engineering*, vol. 9, no. 2, pp. 107–129, 2015.

[23] Y. Errami, M. Ouassaid, and M. Maaroufi, "Design of variable structure control for wind energy system-based permanent magnet synchronous generator and operating under different grid conditions," *International Journal of Systems, Control and Communications*, vol. 7, no. 2, pp. 164–185, 2016.

[24] G. Kenne, R. M. Douanla, and F. B. Pelap, "An adaptive nonlinear control strategy for a stand-alone permanent magnet synchronous generator driven by a variable speed wind turbine," *International Journal of Dynamics and Control*, vol. 5, no. 4, pp. 1103–1113, 2017.

[25] Y. Errami, A. Obbadi, S. Sahnoun, M. Ouassaid, and M. Maaroufi, "Power Extraction Control of Variable Speed Wind Turbine Systems Based on Direct Drive Synchronous Generator in All Operating Regimes," *Journal of Electrical and Computer Engineering*, vol. 2018, Article ID 3837959, 17 pages, 2018.

[26] B. Yang, T. Yu, H. Shu et al., "Passivity-based sliding-mode control design for optimal power extraction of a PMSG based variable speed wind turbine," *Journal of Renewable Energy*, vol. 119, pp. 577–589, 2018.

[27] M. A. Mayosky and G. I. E. Cancelo, "Direct adaptive control of wind energy conversion systems using Gaussian networks," *IEEE Transactions on Neural Networks and Learning Systems*, vol. 10, no. 4, pp. 898–906, 1999.

[28] L. Jinkun, *Radial Basis Function Neural Network Control for Mechanical Systems*, Tsinghua University Press, Beijing, China, 2013.

[29] C. C. Hua, C. X. Yu, and X. P. Guan, "Neural network observer-based networked control for a class of nonlinear systems," *Neurocomputing*, vol. 133, pp. 103–110, 2014.

[30] G. Kenné, A. S. Fotso, and F. Lamnabhi-Lagarrigue, "A new adaptive control strategy for a class of nonlinear system using rbf neuro-sliding-mode technique: application to SEIG wind turbine control system," *International Journal of Control*, vol. 90, p. 18, 2017.

[31] C.-H. Lin, "Recurrent modified Elman neural network control of permanent magnet synchronous generator system based on wind turbine emulator," *Journal of Renewable and Sustainable Energy*, vol. 5, no. 5, 2013.

[32] C.-H. Lin, "Dynamic control for permanent magnet synchronous generator system using novel modified recurrent wavelet neural network," *Nonlinear Dynamics*, vol. 77, no. 4, pp. 1261–1284, 2014.

[33] C.-M. Hong, C.-H. Chen, and C.-S. Tu, "Maximum power point tracking-based control algorithm for PMSG wind generation system without mechanical sensors," *Energy Conversion and Management*, vol. 69, pp. 58–67, 2013.

[34] S. Boulouma and H. Belmili, "RBF neural network sliding mode control of a PMSG based wind energy conversion system," in *Proceedings of the International Renewable and Sustainable Energy Conference, IRSEC '16*, pp. 438–443, Morocco, November 2016.

[35] A. Barra, H. Ouadi, F. Giri, and R. Chakib, "Sensorless Nonlinear Control of Wind Energy Systems with Doubly Fed Induction Generator," *Journal of Control, Automation and Electrical Systems*, vol. 27, no. 5, pp. 562–578, 2016.

[36] S. Ghoudelbourk, D. Dib, A. Omeiri, and A. T. Azar, "MPPT control in wind energy conversion systems and the application of fractional control (PIα) in pitch wind turbine," *International Journal of Modelling, Identification and Control*, vol. 26, no. 2, pp. 140–151, 2016.

[37] F. Jaramillo-Lopez, G. Kenne, and F. Lamnabhi-Lagarrigue, "Maximum power extraction on wind turbine systems using block-backstepping with gradient dynamics control," *International Journal of Adaptive Control and Signal Processing*, vol. 31, no. 6, pp. 835–858, 2017.

[38] J. S. Thongam, P. Bouchard, R. Beguenane, A. F. Okou, and A. Merabet, "Control of variable speed wind energy conversion system using a wind speed sensorless optimum speed MPPT control method," in *Proceedings of the 37th Annual Conference of the IEEE Industrial Electronics Society, IECON '11*, pp. 855–860, Australia, November 2011.

[39] S. A. Deraz and F. E. Abdel Kader, "A new control strategy for a stand-alone self-excited induction generator driven by a variable speed wind turbine," *Journal of Renewable Energy*, vol. 51, pp. 263–273, 2013.

[40] F. Jaramillo-Lopez, G. Kenne, and F. Lamnabhi-Lagarrigue, "A novel online training neural network-based algorithm for wind speed estimation and adaptive control of PMSG wind turbine system for maximum power extraction," *Journal of Renewable Energy*, vol. 86, pp. 38–48, 2016.

[41] N. Pogaku, M. Prodanović, and T. C. Green, "Modeling, analysis and testing of autonomous operation of an inverter-based microgrid," *IEEE Transactions on Power Electronics*, vol. 22, no. 2, pp. 613–625, 2007.

[42] T. Ahmed-Ali, G. Kenné, and F. Lamnabhi-Lagarrigue, "Identification of nonlinear systems with time-varying parameters using a sliding-neural network observer," *Neurocomputing*, vol. 72, no. 7-9, pp. 1611–1620, 2009.

[43] V. I. Utkin, "Sliding mode control: mathematical tools, design and applications," in *Nonlinear and optimal control theory*, vol. 1932 of *Lecture Notes in Math.*, pp. 289–347, Springer, Berlin, 2008.

[44] F. Mancilla-David and R. Ortega, "Adaptive passivity-based control for maximum power extraction of stand-alone windmill systems," *Control Engineering Practice*, vol. 20, no. 2, pp. 173–181, 2012.

[45] T. Orlowska-Kowalska, F. Blaabjerg, and J. Rodriguez, *Advanced and intelligent control in power electronics and drives*, Springer, Heidelberg, 2014.

7

Event-Based Nonfragile $H\infty$ Filter Design for Networked Control Systems with Interval Time-Varying Delay

Zhongda Lu,[1,2] Guangtao Ran,[1] Guoliang Zhang,[1] and Fengxia Xu[1]

[1] *School of Computer and Control Engineering, Qiqihar University, Qiqihar 161006, China*
[2] *College of Computer Science and Technology, Harbin University of Science and Technology, Harbin 150080, China*

Correspondence should be addressed to Zhongda Lu; luzhongda@163.com

Academic Editor: Yun-Bo Zhao

This paper first investigates the event-triggered nonfragile H_∞ filter design for a class of nonlinear NCSs with interval time-varying delay. An event-triggered scheme is addressed to determine sampled data to be transmitted so that network communication resource can be saved significantly. The nonfragile filter design is assumed to include multiplicative gain variations according to the filter's implement. Under the event-triggered scheme, the filtering error system is modeled as a system with interval time-varying delay. By constructing a new Lyapunov-Krasovskii functional and employing Wirtinger inequality, a sufficient condition is derived, which guarantees that the filtering error system is asymptotically stable with the prescribed H_∞ performance. The nonfragile filter parameters are obtained by solving a set of linear matrix inequalities. Two numerical examples are given to show the usefulness and the effectiveness of the proposed method.

1. Introduction

During the past several years, Takagi-Sugeno (T-S) fuzzy model approach has got considerable attention due to its great merits on modeling for nonlinear networked control systems (NCSs) [1]. Many efforts have been proposed based on T-S fuzzy model for nonlinear systems [2–5]. On one hand, some state variables cannot be directly measured in practical system; therefore, the filtering problem attracted the attention of many researchers to estimate the unmeasured states. In comparison with traditional Kalman filtering, the H_∞ filter does not require statistical assumptions on the exogenous signals. Thus, the H_∞ filtering theory has got considerable development. The phenomenon of H_∞ filter design with time-varying delays, sensor faults, and packet dropouts was studied for nonlinear systems [6]. The authors in [7] proposed new results on a delay-derivative-dependent fuzzy H_∞ filter.

It should be mentioned that all the above works are based on accuracy assumptions that filter can be implemented exactly. However, inaccuracies or uncertainties do occur in filter implementation, which will reduce the performance of systems and lead the filter system to be fragile. Thus, a nonfragile filter must be designed to handle uncertainties and maintain the performance of systems. Up to now, a few results on the nonfragile filter have been proposed. Recently, the nonfragile H_∞ filtering problem for a class of discrete-time Takagi-Sugeno fuzzy systems with both randomly occurring gain variations and channel fading was investigated in [8]. Based on vertex theory and probabilistic algorithm, deterministic and randomised filtering algorithms are proposed in [9]. Aiming at the fuzzy stochastic systems, the authors in [10] proposed a nonfragile robust H_∞ filter design and a desired H_∞ performance level. It should be pointed out that the nonfragile H_∞ filtering problem has not been fully studied, which is one of motivations of this paper.

On the other hand, one topic that has been increasingly important is how to mitigate the network bandwidth while guaranteeing the stability and other desired performance of systems [11–15]. In traditional time-triggered control systems, the sample data will be transmitted into controller via network communication whatever the data are desired or not, which leads to network resource waste and low efficiency. Therefore, an event-triggered scheme was proposed to reduce data transmission and improve the effective utilities of network resource. Recently, a novel discrete model for networked control systems was introduced in [16], which contained trigger parameters, dynamic quantization, and fault detection. In [17], event-triggering load frequency control was employed in multiarea power systems. It should be noted that although the works on event-triggered NCSs are rich, the limitations still remain and there still exist some problems to be handled [18–20]. To the best of our knowledge, there are no works that investigate how to use the event-triggered scheme in nonfragile H_∞ filtering systems, which is another motivation of this paper.

According to the above discussion, we first consider the problem of the event-triggered nonfragile H_∞ filter design for a class of nonlinear networked control systems. Our main contributions of this paper are summarized as follows:

(1) An event-triggered communication scheme is proposed to save network resource significantly. By considering the event-triggered scheme, filter's multiplicative gain variations, and interval time-varying delays, a novel filtering error system is established.

(2) Different from some existing works, the Wirtinger inequality is used to tackle the integral items of the derivative of Lyapunov-Krasovskii; a more relaxed H_∞ performance stability criterion is derived.

The rest of this paper is organized as follows. The problem formulation is given in Section 2; under the event-triggered scheme, the filtering error system is modeled as a system with interval time-varying delay. Stability analysis for filtering error system is presented in Section 3. The nonfragile H_∞ filter design method is first addressed in Section 4. Numerical examples are provided in Section 5 to demonstrate the effectiveness of the proposed method.

Notations. Throughout this paper, R^n denote the n-dimensional Euclidean space and $R^{n\times m}$ is the set of $n \times m$ real matrices. Superscript $(\bullet)^T$ stands for the matrix transposition, I represents the identity matrix, and diag$\{\cdots\}$ denotes the block-diagonal matrix. The notation $P > 0$ means that the matrix P is a real symmetric positive define matrix. In symmetric block matrices, "$*$" is used as ellipsis for terms induced by symmetry.

2. Problem Formulation

2.1. T-S Fuzzy System. Consider the following nonlinear system, which can be described by T-S fuzzy model with i plant rules:

Plant Rule i

$$\text{IF } \varsigma_1(t) \text{ is } M_{i1}, \varsigma_2(t) \text{ is } M_{i2}, \ldots, \text{and } \varsigma_r(t) \text{ is } M_{ip}, \quad \text{THEN} \tag{1}$$

$$\begin{aligned} \dot{x}(t) &= A_i x(t) + B_i \omega(t) \\ z(t) &= C_i x(t) \\ y(t) &= D_i x(t), \end{aligned} \tag{2}$$

where M_{ij} are the fuzzy sets, $i = 1, 2, \ldots, r$, and $j = 1, 2, \ldots, p$ is the number of IF-THEN rules. $\varsigma_1(t), \varsigma_2(t) \cdots \varsigma_r(t)$ are the premise variables. $x(t) \in R^n$ and $y(t) \in R^m$ are the system state vector and the measured output, respectively. $\omega(t) \in R^q$ is the external disturbance and $\omega(t) \in l_2[0,\infty)$. $z(t) \in R^p$ is the signal vector to be estimated; A_i, B_i, C_i, and D_i are known parameter matrices with appropriate dimensions.

By using center-average defuzzifier, product inference, and a singleton fuzzifier, the T-S fuzzy system (2) can be rewritten as follows:

$$\begin{aligned} \dot{x}(t) &= \sum_{i=1}^{r} \mu_i(\varsigma(t)) \left[A_i x(t) + B_i \omega(t)\right] \\ z(t) &= \sum_{i=1}^{r} \mu_i(\varsigma(t)) C_i x(t) \\ y(t) &= \sum_{i=1}^{r} \mu_i(\varsigma(t)) D_i x(t) \end{aligned} \tag{3}$$

where $\varsigma = [\varsigma_1, \ldots, \varsigma_r]^T$; the fuzzy basis functions are given by

$$\begin{aligned} \mu_i(\varsigma(t)) &= \frac{h_i(\varsigma(t))}{\sum_{i=1}^{r} h_i(\varsigma(t))}, \\ h_i(\varsigma(t)) &= \prod_{j=1}^{p} M_{ij}\left(\varsigma_j(t)\right). \end{aligned} \tag{4}$$

$M_{ij}(\varsigma_j(t))$ represents the grade of membership for $\varsigma_j(t)$ in M_{ij}; $0 \le \mu_i(\varsigma(t)) \le 1$, $(i = 1, 2, \ldots, r)$, $\sum_{i=1}^{r} \mu_i(\varsigma(t)) > 0$.

Therefore, we have

$$\sum_{i=1}^{r} \mu_i(\varsigma(t)) = 1, \quad \mu_i(\varsigma(t)) \ge 0 \quad (i = 1, 2, \ldots, r). \tag{5}$$

2.2. Event-Triggered Scheme. In this section, we consider the communication networks with limited bandwidth; an event generator under the event-triggered scheme is employed. Inspired by the method in [12], the event-triggered scheme is adopted as follows:

$$t_{k+1}h = t_k h + \min_{j\ge1} \{\Delta(t_k h + jh)\} \tag{6}$$

for $\Delta(t_k h + jh) = e_k(t_k h + jh)^T \psi e_k(t_k h + jh) \ge \lambda y(t_k h)^T \psi y(t_k h)$, where $e_k(t_k h + jh) = y(t_k h + jh) - y(t_k h)$ is the threshold error among the current sampling data and the last transmitted data. ψ and λ are the triggering parameters.

Remark 1. It should be mentioned that the current sampling data will be transmitted only when the condition proposed in (6) is satisfied. Therefore, in comparison with periodic transmission communication scheme, the event-triggered mechanism can reduce the transmission rate and utilize the limited bandwidth effectively.

2.3. Nonfragile Fuzzy Filter. In some previous results of filter design, the implicit assumption is made that there are no multiplicative gain uncertainties. However, in fact, there inevitably exist filter parameter uncertainties in filter implementation. Therefore, in this section, we consider a full-order nonfragile fuzzy filter with gain variations as follows:

Plant Rule j

$$\text{IF } \varsigma_1(t_kh) \text{ is } W_{i1}, \quad \varsigma_2(t_kh) \text{ is } W_{i2},\dots,\text{and } \varsigma_r(t_kh) \text{ is } W_{ip}, \text{THEN} \tag{7}$$

$$\begin{aligned}\dot{x}_f(t) &= \left(A_{fj} + \Delta A_{fj}(t)\right)x_f(t) + \left(B_{fj} + \Delta B_{fj}(t)\right) \cdot y_f(t) \\ z_f(t) &= \left(C_{fj} + \Delta C_{fj}(t)\right)x_f(t),\end{aligned} \tag{8}$$

where $x_f(t) \in R^n$ is the filter state vector, $y_f(t) \in R^n$ is the real input of the filter, $z_f(t)$ is the estimated signal vector of $z(t)$, and A_{fj}, B_{fj}, and C_{fj} are the filter parameters to be designed.

Similarly, we represent the nonfragile fuzzy filter as

$$\begin{aligned}\dot{x}_f(t) &= \sum_{j=1}^{r}\mu_j(\varsigma(t_kh))\left[\left(A_{fj} + \Delta A_{fj}(t)\right)x_f(t) + \left(B_{fj} + \Delta B_{fj}(t)\right)y_f(t)\right] \\ z_f(t) &= \sum_{j=1}^{r}\mu_j(\varsigma(t_kh))\left(C_{fj} + \Delta C_{fj}(t)\right)x_f(t),\end{aligned} \tag{9}$$

where ΔA_{fj}, ΔB_{fj}, and ΔC_{fj} are the multiplicative gain uncertainties, which can be defined as

$$\begin{aligned}\Delta A_{fj}(t) &= A_{fj}M_{1j}K_A(t)N_{1j} \\ \Delta B_{fj}(t) &= B_{fj}M_{2j}K_B(t)N_{2j} \\ \Delta C_{fj}(t) &= C_{fj}M_{3j}K_C(t)N_{3j},\end{aligned} \tag{10}$$

where M_{ij} and N_{ij} $(i = 1,2,3)$ are constant matrices with appropriate dimension. $K_A(t), K_B(t), K_C(t)$ are uncertain bounded matrices:

$$\begin{aligned}K_A^T(t)K_A(t) &\le I, \\ K_B^T(t)K_B(t) &\le I, \\ K_C^T(t)K_C(t) &\le I.\end{aligned} \tag{11}$$

Next, consider the effect of the logic ZOH; the last transmission data instant is maintained with the holding interval as follows:

$$y_z = y(t_kh), \quad t \in \left[t_kh + \tau_{t_k}, t_{k+1}h + \tau_{t_{k+1}}\right) \tag{12}$$

In order to facilitate analysis, the holding interval can be represented by the following subsets:

$$\left[t_kh + \tau_{t_k}, t_{k+1}h + \tau_{t_{k+1}}\right) = \bigcup_{l=0}^{m}\Omega_l, \tag{13}$$

where

$$\Omega_i = \left[t_kh + lh + \tau_{t_{k+i}}, t_kh + lh + h + \tau_{t_{k+i+1}}\right), \quad l = 0,1,\dots,m. \tag{14}$$

Then, we define the function of the interval time-varying delays as follows:

$$\tau(t) = \begin{cases}(1-\Delta t_0)t & t \in \Omega_0 \\ (1-\Delta t_1)t & t \in \Omega_1 \\ \vdots & \vdots \\ (1-\Delta t_m)t & t \in \Omega_m,\end{cases} \tag{15}$$

where $\Delta t_m = \alpha t = t_kh + lh$, $0 \le \alpha < 1$. Therefore, we can easily derive

$$\begin{aligned}0 < \tau_1 \le \tau(t) \le \tau_2, \\ \dot{\tau}(t) \le 1 - \alpha = d \le 1,\end{aligned} \tag{16}$$

where τ_1 and τ_2 are lower bound and upper bound of the delays, respectively. d is the bound of the delay variation.

To summarize, the inputs of the filter are described as

$$y_f(t) = y_z(t) = y(t-\tau(t)) - e_k(t-\tau(t)) \tag{17}$$

2.4. Fuzzy Filtering Error System. For simplicity, we let μ_i represent $\mu_i(\varsigma(t))$ and let μ_j represent $\mu_j(\varsigma(t_kh))$, and

$$\begin{aligned}\xi(t) &= \operatorname{col}\left\{x(t), x_f(t)\right\}, \\ e_f(t) &= z(t) - z_f(t).\end{aligned} \tag{18}$$

By combining (3), (9), and (17), we can obtain the following filtering error system:

$$\begin{aligned}\dot{\xi}(t) &= \sum_{i=1}^{r}\sum_{j=1}^{r}\left\{\overline{A}_{ij}(t)\xi(t) + \overline{B}_{ij1}(t)\xi(t-\tau(t)) + \overline{B}_{ij2}(t)e_k(t-\tau(t)) + \overline{B}_{wij}(t)\omega(t)\right\} \\ e_f(t) &= \sum_{i=1}^{r}\sum_{j=1}^{r}\overline{C}_{ij}(t)\xi(t),\end{aligned} \tag{19}$$

where

$$\begin{aligned}\overline{A}_{ij}(t) &= \widetilde{A}_{ij} + \Delta\widehat{A}_{ij}(t), \\ \overline{B}_{ij1}(t) &= \widetilde{B}_{ij1} + \Delta\widehat{B}_{ij1}(t), \\ \overline{B}_{ij2}(t) &= \widetilde{B}_{ij2} + \Delta\widehat{B}_{ij2}(t), \\ \overline{C}_{ij}(t) &= \widetilde{C}_{ij} + \Delta\widehat{C}_{ij}(t), \\ \widetilde{A}_{ij} &= \sum_{i=1}^{r}\sum_{j=1}^{r}\mu_i\mu_j\begin{bmatrix}A_i & 0 \\ 0 & A_{fj}\end{bmatrix},\end{aligned}$$

$$\Delta\widehat{A}_{ij}(t) = \sum_{i=1}^{r}\sum_{j=1}^{r}\mu_i\mu_j\begin{bmatrix}0 & 0\\ 0 & \Delta A_{fj}(t)\end{bmatrix},$$

$$\widetilde{B}_{ij1} = \sum_{i=1}^{r}\sum_{j=1}^{r}\mu_i\mu_j\begin{bmatrix}0 & 0\\ 0 & B_{fj}D_i\end{bmatrix},$$

$$\Delta\widehat{B}_{ij1} = \sum_{i=1}^{r}\sum_{j=1}^{r}\mu_i\mu_j\begin{bmatrix}0 & 0\\ 0 & \Delta B_{fj}(t)D_i\end{bmatrix},$$

$$\widetilde{B}_{ij2} = \sum_{i=1}^{r}\sum_{j=1}^{r}\mu_i\mu_j\begin{bmatrix}0\\ -B_{fj}\end{bmatrix},$$

$$\Delta\widehat{B}_{ij2} = \sum_{i=1}^{r}\sum_{j=1}^{r}\mu_i\mu_j\begin{bmatrix}0\\ -\Delta B_{fj}(t)\end{bmatrix},$$

$$\widetilde{C}_{ij} = \sum_{i=1}^{r}\sum_{j=1}^{r}\mu_i\mu_j\begin{bmatrix}C_i & -C_{fj}\end{bmatrix},$$

$$\Delta\widehat{C}_{ij} = \sum_{i=1}^{r}\sum_{j=1}^{r}\mu_i\mu_j\begin{bmatrix}0 & -\Delta C_{fj}(t)\end{bmatrix},$$

$$\widetilde{B}_{wij} = \sum_{i=1}^{r}\sum_{j=1}^{r}\mu_i\mu_j\begin{bmatrix}B_i\\ 0\end{bmatrix}. \tag{20}$$

Before the end of this section, the following definitions and lemmas are needed for the fuzzy filtering error system.

Definition 2. The fuzzy filtering error system is asymptotically stable with an H_∞ performance, if the following holds:

(1) The filtering error system (19) is asymptotically stable when $\omega(t) = 0$.

(2) The filtering error system (19) has a prescribed $H\infty$ performance γ; under zero initial condition,

$$\int_0^L \|e(t)\|^2\,dt \le \gamma^2\int_0^L \|\omega(t)\|^2\,dt \tag{21}$$

is satisfied for any nonzero $\omega(t) \in L_2[0,\infty)$.

Lemma 3 (see [21]). *Let X, Y, and G be real matrices with appropriate dimensions and $G_i^T(t)G_i(t) \le I$. Then, for any constant $\varepsilon > 0$, one has*

$$XG(t)Y + Y^TG(t)X^T \le \varepsilon XX^T + \varepsilon^{-1}Y^TY \tag{22}$$

Lemma 4 (see [22]). *For any vectors $x, y \in R^n$ and any scalar $\varepsilon > 0$, matrices D, F, and E are real matrices of appropriate dimensions with $F^T(t)F(t) \le I$; then the following inequalities hold:*

$$2xDF(t)Ey \le \varepsilon x^TDD^Tx + \varepsilon^{-1}y^TEE^Ty \tag{23}$$

Lemma 5 (see [23]). *Let V, H, E, Q, and F be real matrices of appropriate dimensions such that $Q > 0$ and $F^TF \le I$. Then, for any scalar $\varepsilon > 0$ such that $Q - \varepsilon DD^T > 0$, one has*

$$\begin{aligned}&(V+HFE)^TQ(V+HFE)\\ &\le V^T\left(Q^{-1} - \varepsilon HH^T\right)V + \varepsilon^{-1}E^TE\end{aligned} \tag{24}$$

Lemma 6 (see [24]). *Let $P > 0$ and let A_i and A_l be any real matrices of appropriate dimensions. Then*

$$A_i^TPA_l + A_l^TPA_i \le A_i^TPA_i + A_l^TPA_l \tag{25}$$

Lemma 7 (Seuret and Gouaisbaut [25]). *For a given symmetric and positive matrix $R > 0$ of appropriate dimensions and different signal x in $[a,b] \to R^n$, the following inequality holds:*

$$\begin{aligned}&-\int_a^b \dot{\xi}^T(s)R\dot{\xi}(s)\,ds\\ &\le -\frac{1}{b-a}\begin{bmatrix}\xi(b)\\ \xi(a)\\ v\end{bmatrix}^T\begin{bmatrix}a_1R & a_2R & a_3R\\ * & a_1R & a_3R\\ * & * & a_4R\end{bmatrix}\begin{bmatrix}\xi(b)\\ \xi(a)\\ v\end{bmatrix},\end{aligned} \tag{26}$$

where

$$\begin{aligned}v &= \frac{1}{b-a}\int_a^b \xi(s)\,ds\\ a_1 &= \frac{\pi^2}{4} + 1,\\ a_2 &= \frac{\pi^2}{4} - 1,\\ a_3 &= -\frac{\pi^2}{2},\\ a_4 &= \pi^2.\end{aligned} \tag{27}$$

3. Stability Analysis

In this section, we first present the H_∞ performance analysis for the filtering error system (19) under event-triggered scheme. The following results are established to guarantee that the filtering error system (19) is asymptotically stable.

Theorem 8. *For given positive parameters τ_1, τ_2, $d \le 1$, $\gamma > 0$, $\varepsilon_i > 0$, $(i = 1, 2, \ldots, 10)$, and $0 < \lambda < 1$, the filtering error system in (19) is asymptotically stable with H_∞ performance γ under the event-triggered scheme (16), if there exist matrices $Q_i > 0$, $(i = 1, 2, 3)$, $\psi > 0$, $R_i > 0$ $(i = 1, 2)$, $a_i > 0$ $(i = 1, 2, 3, 4)$, and $P > 0$ and A_{fj}, B_{fj}, and C_{fj} with appropriate dimensions such that the following matrix inequalities hold:*

$$\begin{bmatrix}I & \widetilde{M}_3\\ * & \varepsilon_{10}^{-1}\end{bmatrix} > 0 \tag{28}$$

$$\begin{aligned}&\begin{bmatrix}R_1^{-1} & \widetilde{M}_1\\ * & \varepsilon_4^{-1}\end{bmatrix} > 0,\\ &\begin{bmatrix}R_2^{-1} & \widetilde{M}_1\\ * & \varepsilon_7^{-1}\end{bmatrix} > 0\end{aligned} \tag{29}$$

$$\begin{bmatrix} R_1^{-1} & \widetilde{M}_{21} \\ * & \varepsilon_5^{-1} \end{bmatrix} > 0, \quad \begin{bmatrix} R_2^{-1} & \widetilde{M}_{21} \\ * & \varepsilon_8^{-1} \end{bmatrix} > 0 \tag{30}$$

$$\begin{bmatrix} R_1^{-1} & \widetilde{M}_{22} \\ * & \varepsilon_6^{-1} \end{bmatrix} > 0, \quad \begin{bmatrix} R_2^{-1} & \widetilde{M}_{22} \\ * & \varepsilon_9^{-1} \end{bmatrix} > 0 \tag{31}$$

$$\Pi = \begin{bmatrix} \Lambda_{11} & \Lambda_{12} \\ * & \Lambda_{22} \end{bmatrix} < 0 \tag{32}$$

for

$$\Lambda_{11} = \begin{bmatrix} \Pi_{11} & \Pi_{12} & \Pi_{13} & 0 & \Pi_{15} \\ * & \Pi_{22} & \Pi_{23} & 0 & \Pi_{25} \\ * & * & \Pi_{33} & \Pi_{34} & 0 \\ * & * & * & \Pi_{44} & 0 \\ * & * & * & * & \Pi_{55} \end{bmatrix},$$

$$\Lambda_{12} = \begin{bmatrix} 0 & 0 & \Pi_{18} & \Pi_{19} \\ \Pi_{26} & 0 & 0 & 0 \\ \Pi_{36} & \Pi_{37} & 0 & 0 \\ 0 & \Pi_{47} & 0 & 0 \\ 0 & 0 & 0 & 0 \end{bmatrix}, \tag{33}$$

$$\Lambda_{22} = \begin{bmatrix} \Pi_{66} & 0 & 0 & 0 \\ * & \Pi_{77} & 0 & 0 \\ * & * & \Pi_{88} & 0 \\ * & * & * & \Pi_{99} \end{bmatrix},$$

where

$$\begin{aligned}
\Pi_{11} = {} & P\widetilde{A}_{ij} + \widetilde{A}_{ij}^T P + Q_1 + Q_2 + Q_3 - a_1 R_1 \\
& + \varepsilon_1 P\widetilde{M}_1 \widetilde{M}_1^T P + \varepsilon_1^{-1} \widetilde{N}_1^T \widetilde{N}_1 \\
& + \varepsilon_2 P\widetilde{M}_{21} \widetilde{M}_{21}^T P + \varepsilon_3 P\widetilde{M}_{22} \widetilde{M}_{22}^T P \\
& + \varepsilon_{10}^{-1} \widetilde{N}_3^T \widetilde{N}_3 \\
& + \alpha_1^2 \widetilde{A}_{ij}^T \left[R_1^{-1} - \varepsilon_4 \widetilde{M}_1 \widetilde{M}_1^T\right]^{-1} \widetilde{A}_{ij} \\
& + \alpha_1^2 \varepsilon_4^{-1} \widetilde{N}_1^T \widetilde{N}_1 \\
& + \alpha_2^2 \widetilde{A}_{ij}^T \left[R_2^{-1} - \varepsilon_7 \widetilde{M}_1 \widetilde{M}_1^T\right]^{-1} \widetilde{A}_{ij} \\
& + \alpha_2^2 \varepsilon_7^{-1} \widetilde{N}_1^T \widetilde{N}_1 + \widetilde{C}_{ij}^T \left[I - \varepsilon_{10} \widetilde{M}_3 \widetilde{M}_3^T\right]^{-1} \widetilde{C}_{ij},
\end{aligned}$$

$$\Pi_{12} = -a_2 R_1,$$

$$\Pi_{13} = P\widetilde{B}_{ij1},$$

$$\Pi_{15} = -a_3 R_1,$$

$$\Pi_{18} = P\widetilde{B}_{ij2},$$

$$\Pi_{19} = P\widetilde{B}_{wij},$$

$$\Pi_{22} = -Q_2 - a_1 R_1 - a_1 R_2,$$

$$\Pi_{23} = -a_2 R_2,$$

$$\Pi_{25} = -a_3 R_1,$$

$$\Pi_{26} = -a_3 R_2,$$

$$\begin{aligned}
\Pi_{33} = {} & -(1-d)\, Q_1 + \varepsilon_2^{-1} \widetilde{N}_{21}^T \widetilde{N}_{21} + \lambda D_i^T \psi D_i - 2a_1 R_2 \\
& + \alpha_1^2 \varepsilon_5^{-1} \widetilde{N}_{21}^T \widetilde{N}_{21} + \alpha_2^2 \varepsilon_8^{-1} \widetilde{N}_{21}^T \widetilde{N}_{21} \\
& + \alpha_1^2 B_{ij1}^T \left[R_1^{-1} - \varepsilon_5 \widetilde{M}_{21} \widetilde{M}_{21}^T\right]^{-1} B_{ij1} \\
& + \alpha_2^2 B_{ij1}^T \left[R_2^{-1} - \varepsilon_8 \widetilde{M}_{21} \widetilde{M}_{21}^T\right]^{-1} B_{ij1},
\end{aligned}$$

$$\Pi_{34} = -a_2 R_2,$$

$$\Pi_{36} = -a_3 R_2,$$

$$\Pi_{37} = -a_3 R_2,$$

$$\Pi_{44} = -Q_3 - a_1 R_2,$$

$$\Pi_{47} = -a_3 R_2,$$

$$\Pi_{55} = -a_4 R_1,$$

$$\Pi_{66} = -a_4 R_2,$$

$$\Pi_{77} = -a_4 R_2,$$

$$\begin{aligned}
\Pi_{88} = {} & \varepsilon_3^{-1} \widetilde{N}_{22}^T \widetilde{N}_{22} + (\lambda - 1)\, \psi + \alpha_1^2 \varepsilon_6^{-1} \widetilde{N}_{22}^T \widetilde{N}_{22} \\
& + \alpha_2^2 \varepsilon_9^{-1} \widetilde{N}_{22}^T \widetilde{N}_{22} \\
& + \alpha_1^2 B_{ij2}^T \left[R_1^{-1} - \varepsilon_6 \widetilde{M}_{22} \widetilde{M}_{22}^T\right]^{-1} B_{ij2} \\
& + \alpha_2^2 B_{ij2}^T \left[R_2^{-1} - \varepsilon_9 \widetilde{M}_{22} \widetilde{M}_{22}^T\right]^{-1} B_{ij2},
\end{aligned}$$

$$\Pi_{99} = \alpha_1^2 \widetilde{B}_{wij}^T R_1 \widetilde{B}_{wij} + \alpha_2^2 \widetilde{B}_{wij}^T R_2 \widetilde{B}_{wij} - \gamma^2 I_n,$$

$$\widetilde{M}_1 = \begin{bmatrix} 0 \\ A_{fj} M_{1j} \end{bmatrix},$$

$$\widetilde{M}_{21} = \begin{bmatrix} 0 \\ B_{fj} M_{2j} \end{bmatrix},$$

$$\widetilde{M}_{22} = \begin{bmatrix} 0 \\ -B_{fj} M_{2j} \end{bmatrix},$$

$$\widetilde{M}_3 = C_{fj} M_{3j},$$

$$\begin{aligned}\widetilde{N}_1 &= [0 \;\; N_{1j}],\\ \widetilde{N}_{21} &= [0 \;\; N_{2j}D_i],\\ \widetilde{N}_{22} &= [0 \;\; N_{2j}],\\ \widetilde{N}_3 &= [0 \;\; N_{3j}],\\ a_1 &= \frac{\pi^2}{4} + 1,\\ a_2 &= \frac{\pi^2}{4} - 1,\\ a_3 &= -\frac{\pi^2}{2},\\ a_4 &= \pi^2,\\ \alpha_1 &= 2\tau_1,\\ \alpha_2 &= 2(\tau_2 - \tau_1).\end{aligned} \tag{34}$$

Proof. We construct a candidate of Lyapunov-Krasovskii function as follows:

$$V(t,\xi_t) = \sum_{j=1}^{3} V_j(t,\xi_t), \tag{35}$$

where

$$\begin{aligned}V_1(t,\xi_t) &= \xi^T(t)P\xi(t),\\ V_2(t,\xi_t) &= \int_{t-\tau(t)}^{t} \xi^T(s)Q_1\xi(s)\,ds\\ &\quad + \int_{t-\tau_1}^{t} \xi^T(s)Q_2\xi(s)\,ds\\ &\quad + \int_{t-\tau_2}^{t} \xi^T(s)Q_3\xi(s)\,ds,\\ V_3(t,\xi_t) &= \tau_1\int_{-\tau_1}^{0}\int_{t+\theta}^{t} \dot{\xi}^T(s)R_1\dot{\xi}(s)\,ds\,d\theta\\ &\quad + (\tau_2-\tau_1)\int_{-\tau_2}^{-\tau_1}\int_{t+\theta}^{t} \dot{\xi}^T(s)R_2\dot{\xi}(s)\,ds\,d\theta.\end{aligned} \tag{36}$$

Taking the time derivation of $V(t)$ for t, we have

$$\dot{V}(t,\xi_t) = \dot{V}_1(t,\xi_t) + \dot{V}_2(t,\xi_t) + \dot{V}_3(t,\xi_t), \tag{37}$$

where

$$\begin{aligned}\dot{V}_1(t,\xi_t) &= 2\xi^T(t)P\overline{A}_{ij}(t)\xi(t)\\ &\quad + 2\xi^T(t)P\overline{B}_{ij1}(t)\xi(t-\tau(t))\\ &\quad + 2\xi^T(t)P\overline{B}_{ij2}(t)e_k(t-\tau(t))\\ &\quad + 2\xi^T(t)P\overline{B}_{wij}(t)\omega(t)\\ &= 2\xi^T(t)P\widetilde{A}_{ij}\xi(t) + 2\xi^T(t)P\Delta\widehat{A}_{ij}(t)\xi(t)\\ &\quad + 2\xi^T(t)P\widetilde{B}_{ij1}\xi(t-\tau(t))\\ &\quad + 2\xi^T(t)P\Delta\widehat{B}_{ij1}(t)\xi(t-\tau(t))\\ &\quad + 2\xi^T(t)P\widetilde{B}_{ij2}e_k(t-\tau(t))\\ &\quad + 2\xi^T(t)P\Delta\widehat{B}_{ij2}(t)e_k(t-\tau(t))\\ &\quad + 2\xi^T(t)P\overline{B}_{wij}(t)\omega(t)\end{aligned} \tag{38}$$

$$\begin{aligned}\dot{V}_2(t,\xi_t) &= \xi^T(t)(Q_1+Q_2+Q_3)\xi(t)\\ &\quad - (1-\dot{\tau}(t))\xi^T(t-\tau(t))Q_1\xi(t-\tau(t))\\ &\quad - \xi^T(t-\tau_1)Q_2\xi(t-\tau_1)\\ &\quad - \xi^T(t-\tau_2)Q_3\xi(t-\tau_2)\\ &\le \xi^T(t)(Q_1+Q_2+Q_3)\xi(t)\\ &\quad - (1-d)\xi^T(t-\tau(t))Q_1\xi(t-\tau(t))\\ &\quad - \xi^T(t-\tau_1)Q_2\xi(t-\tau_1)\\ &\quad - \xi^T(t-\tau_2)Q_3\xi(t-\tau_2)\end{aligned} \tag{39}$$

$$\begin{aligned}\dot{V}_3(t,\xi_t) &= \tau_1^2\dot{\xi}^T(t)R_1\dot{\xi}(t) + (\tau_2-\tau_1)^2\dot{\xi}^T(t)R_2\dot{\xi}(t)\\ &\quad - \tau_1\int_{t-\tau_1}^{t}\dot{\xi}^T(s)R_1\dot{\xi}(s)\,ds\\ &\quad - (\tau_2-\tau_1)\int_{t-\tau_2}^{t-\tau_1}\dot{\xi}^T(s)R_2\dot{\xi}(s)\,ds\end{aligned} \tag{40}$$

By Lemma 3, for $\forall\varepsilon_1 > 0$, we can obtain

$$\begin{aligned}2\xi^T(t)P\Delta\widehat{A}_{ij}(t)\xi(t) &= 2\xi^T(t)P\left[\widetilde{M}_1K_A(t)\widetilde{N}_1\right]\xi(t)\\ &\le \xi^T(t)\left(\varepsilon_1P\widetilde{M}_1\widetilde{M}_1^TP + \varepsilon_1^{-1}P\widetilde{N}_1^T\widetilde{N}_1P\right)\xi(t)\end{aligned} \tag{41}$$

By Lemma 4, if there exist $\forall\varepsilon_2 > 0$ and $\forall\varepsilon_3 > 0$, we have

$$\begin{aligned}&2\xi^T(t)P\Delta\widehat{B}_{ij1}(t)\xi(t-\tau(t))\\ &\quad = 2\xi^T(t)P\left[\widetilde{M}_{21}K_B(t)\widetilde{N}_{21}\right]\xi(t-\tau(t))\\ &\quad \le \xi^T(t)\left[\varepsilon_2P\widetilde{M}_{21}\widetilde{M}_{21}^TP\right]\xi(t)\\ &\qquad + \xi^T(t-\tau(t))\left[\varepsilon_2^{-1}P\widetilde{N}_{21}^T\widetilde{N}_{21}P\right]\xi(t-\tau(t))\end{aligned} \tag{42}$$

$$\begin{aligned}&2\xi^T(t)P\Delta\widehat{B}_{ij2}(t)e_k(t-\tau(t))\\ &\quad = 2\xi^T(t)P\left[\widetilde{M}_{22}K_B(t)\widetilde{N}_{22}\right]e_k(t-\tau(t))\\ &\quad \le \xi^T(t)\left[\varepsilon_3P\widetilde{M}_{22}\widetilde{M}_{22}^TP\right]\xi(t)\\ &\qquad + e_k^T(t-\tau(t))\left[\varepsilon_3^{-1}P\widetilde{N}_{22}^T\widetilde{N}_{22}P\right]e_k(t-\tau(t))\end{aligned} \tag{43}$$

From (41) to (43), we derive

$$\begin{aligned} &2\xi^T(t)P\Delta\widehat{A}_{ij}(t)\xi(t)+2\xi^T(t)P\Delta\widehat{B}_{ij1}(t)\xi(t-\tau(t)) \\ &+2\xi^T(t)P\Delta\widehat{B}_{ij2}(t)e_k(t-\tau(t))\le\xi^T(t) \\ &\cdot\left\{\varepsilon_1P\widetilde{M}_1\widetilde{M}_1^TP+\varepsilon_2P\widetilde{M}_{21}\widetilde{M}_{21}^TP+\varepsilon_1^{-1}P\widetilde{N}_1^T\widetilde{N}_1P\right. \\ &\left.+\varepsilon_3P\widetilde{M}_{22}\widetilde{M}_{22}^TP\right\}\xi(t)+e_k^T(t-\tau(t)) \\ &\cdot\left[\varepsilon_3^{-1}P\widetilde{N}_{22}^T\widetilde{N}_{22}P\right]e_k(t-\tau(t))+\xi^T(t-\tau(t)) \\ &\cdot\left[\varepsilon_2^{-1}P\widetilde{N}_{21}^T\widetilde{N}_{21}P\right]\xi(t-\tau(t)) \end{aligned} \tag{44}$$

By using Lemmas 5 and 6, if there exists $\forall\varepsilon_i>0 (i=4,5,6,7,8)$, we derive

$$\begin{aligned} &\tau_1^2\dot{\xi}^T(t)R_1\dot{\xi}(t)=\tau_1{}^2\left[\overline{A}_{ij}(t)\xi(t)\right. \\ &+\overline{B}_{ij1}(t)\xi(t-\tau(t))+\overline{B}_{ij2}(t)e_k(t-\tau(t)) \\ &\left.+\overline{B}_{wij}(t)w(t)\right]^TR_1\left[\overline{A}_{ij}(t)\xi(t)\right. \\ &+\overline{B}_{ij1}(t)\xi(t-\tau(t))+\overline{B}_{ij2}(t)e_k(t-\tau(t)) \\ &\left.+\overline{B}_{wij}(t)w(t)\right]\le4\tau_1^2\xi^T(t) \\ &\cdot\left\{\widetilde{A}_{ij}^T\left[R_1^{-1}-\varepsilon_4\widetilde{M}_1\widetilde{M}_1^T\right]^{-1}\widetilde{A}_{ij}+\varepsilon_4^{-1}\widetilde{N}_1^T\widetilde{N}_1\right\}\xi(t) \\ &+4\tau_1^2\xi^T(t-\tau(t))\left\{\widetilde{B}_{ij1}^T\left[R_1^{-1}-\varepsilon_5\widetilde{M}_{21}\widetilde{M}_{21}^T\right]^{-1}\widetilde{B}_{ij1}\right. \\ &\left.+\varepsilon_5^{-1}\widetilde{N}_{21}^T\widetilde{N}_{21}\right\}\xi(t-\tau(t))+4\tau_1^2e_k^T(t-\tau(t)) \\ &\cdot\left\{\widetilde{B}_{ij2}^T\left[R_1^{-1}-\varepsilon_6\widetilde{M}_{22}\widetilde{M}_{22}^T\right]^{-1}\widetilde{B}_{ij2}+\varepsilon_6^{-1}\widetilde{N}_{22}^T\widetilde{N}_{22}\right\} \\ &\cdot e_k(t-\tau(t))+4\tau_1^2\omega^T(t)\overline{B}_{wij}^T(t)R_1\overline{B}_{wij}(t)\omega(t) \end{aligned} \tag{45}$$

$$\begin{aligned} &(\tau_2-\tau_1)^2\dot{\xi}^T(t)R_2\dot{\xi}(t)=(\tau_2-\tau_1)^2\left[\overline{A}_{ij}(t)\xi(t)\right. \\ &+\overline{B}_{ij1}(t)\xi(t-\tau(t))+\overline{B}_{ij2}(t)e_k(t-\tau(t)) \\ &\left.+\overline{B}_{wij}(t)w(t)\right]^TR_1\left[\overline{A}_{ij}(t)\xi(t)\right. \\ &+\overline{B}_{ij1}(t)\xi(t-\tau(t))+\overline{B}_{ij2}(t)e_k(t-\tau(t)) \\ &\left.+\overline{B}_{wij}(t)w(t)\right]\le4(\tau_2-\tau_1)^2\xi^T(t) \\ &\cdot\left\{\widetilde{A}_{ij}^T\left[R_2^{-1}-\varepsilon_7\widetilde{M}_1\widetilde{M}_1^T\right]^{-1}\widetilde{A}_{ij}+\varepsilon_7^{-1}\widetilde{N}_1^T\widetilde{N}_1\right\}\xi(t) \\ &+4(\tau_2-\tau_1)^2\xi^T(t-\tau(t)) \\ &\cdot\left\{\widetilde{B}_{ij1}^T\left[R_2^{-1}-\varepsilon_8\widetilde{M}_{21}\widetilde{M}_{21}^T\right]^{-1}\widetilde{B}_{ij1}+\varepsilon_8^{-1}\widetilde{N}_{21}^T\widetilde{N}_{21}\right\}\xi(t \\ &-\tau(t))+4(\tau_2-\tau_1)^2e_k^T(t-\tau(t)) \\ &\cdot\left\{\widetilde{B}_{ij2}^T\left[R_2^{-1}-\varepsilon_9\widetilde{M}_{22}\widetilde{M}_{22}^T\right]^{-1}\widetilde{B}_{ij2}+\varepsilon_9^{-1}\widetilde{N}_{22}^T\widetilde{N}_{22}\right\} \\ &\cdot e_k(t-\tau(t))+4(\tau_2-\tau_1)^2\omega^T(t)\overline{B}_{wij}^T(t) \\ &\cdot R_2\overline{B}_{wij}(t)\omega(t) \end{aligned} \tag{46}$$

By applying Lemma 5, if there exists $\forall\varepsilon_{10}>0$, we can obtain

$$\begin{aligned} e_f^T(t)e_f(t)&=\xi^T(t)\overline{C}_{ij}^T(t)\overline{C}_{ij}\xi(t) \\ &=\xi^T(t)\left(\widetilde{C}_{ij}+\Delta\widehat{C}_{ij}(t)\right)^T\left(\widetilde{C}_{ij}+\Delta\widehat{C}_{ij}(t)\right)\xi(t) \\ &\le\xi^T(t)\widetilde{C}_{ij}^T\left(I-\varepsilon_{10}\widetilde{M}_3\widetilde{M}_3^T\right)^{-1}\widetilde{C}_{ij}\xi^T(t) \\ &+\varepsilon_{10}^{-1}\xi^T(t)\widetilde{N}_3^T\widetilde{N}_3\xi(t) \end{aligned} \tag{47}$$

Furthermore, we use Lemma 7 to deal with the integral items in (40); it follows that

$$\begin{aligned} &-\tau_1\int_{t-\tau_1}^{t}\dot{\xi}^T(s)R_1\dot{\xi}(s)\,ds\le\begin{bmatrix}\xi(t)\\ \xi(t-\tau_1)\\ \upsilon_1\end{bmatrix}^T \\ &\cdot\begin{bmatrix}a_1R_1 & a_2R_1 & a_3R_1\\ * & a_1R_1 & a_3R_1\\ * & * & a_4R_1\end{bmatrix}\begin{bmatrix}\xi(t)\\ \xi(t-\tau_1)\\ \upsilon_1\end{bmatrix} \end{aligned} \tag{48}$$

$$\begin{aligned} &-(\tau_2-\tau_1)\int_{t-\tau_2}^{t-\tau_1}\dot{\xi}^T(s)R_2\dot{\xi}(s)\,ds=-(\tau_2-\tau_1) \\ &\cdot\int_{t-\tau(t)}^{t-\tau_1}\dot{\xi}^T(s)R_2\dot{\xi}(s)\,ds-(\tau_2-\tau_1) \\ &\cdot\int_{t-\tau_2}^{t-\tau(t)}\dot{\xi}^T(s)R_2\dot{\xi}(s)\,ds\le-\begin{bmatrix}\xi(t-\tau_1)\\ \xi(t-\tau(t))\\ \upsilon_2\end{bmatrix}^T \\ &\cdot\begin{bmatrix}a_1R_2 & a_2R_2 & a_3R_2\\ * & a_1R_2 & a_3R_2\\ * & * & a_4R_2\end{bmatrix}\begin{bmatrix}\xi(t-\tau_1)\\ \xi(t-\tau(t))\\ \upsilon_2\end{bmatrix} \\ &-\begin{bmatrix}\xi(t-\tau(t))\\ \xi(t-\tau_2)\\ \upsilon_3\end{bmatrix}^T \\ &\cdot\begin{bmatrix}a_1R_2 & a_2R_2 & a_3R_2\\ * & a_1R_2 & a_3R_2\\ * & * & a_4R_2\end{bmatrix}\begin{bmatrix}\xi(t-\tau(t))\\ \xi(t-\tau_2)\\ \upsilon_3\end{bmatrix}, \end{aligned} \tag{49}$$

where

$$\begin{aligned} \upsilon_1&=\frac{1}{\tau_1}\int_{t-\tau_1}^{t}\xi(s)\,ds, \\ \upsilon_2&=\frac{1}{\tau(t)-\tau_1}\int_{t-\tau(t)}^{t-\tau_1}\xi(s)\,ds, \\ \upsilon_3&=\frac{1}{\tau_2-\tau(t)}\int_{t-\tau_2}^{t-\tau(t)}\xi(s)\,ds. \end{aligned} \tag{50}$$

By combining (44) to (49), we obtain

$$\dot{V}(t,\xi_t)+e_f^T(t)e_f(t)-\gamma^2\omega^T(t)\omega(t)\le\widetilde{\eta}^T(t)\Pi\widetilde{\eta}(t) \tag{51}$$

where

$$\widetilde{\eta}(t) = \operatorname{col}\{\xi(t), \xi(t-\tau_1), \xi(t-\tau(t)), \xi(t-\tau_2), v_1, v_2, v_3, e_k(t-\tau(t)), \omega(t)\}. \tag{52}$$

Then, from (28) to (32), under the zero initial condition, we integrate the right and left sides of (51) from 0 to L for all $L > 0$; we can easily derive

$$\int_0^L e_f^T(t) e_f(t)\,dt \le \gamma^2 \int_0^L \omega^T(t)\,\omega(t)\,dt \tag{53}$$

This completes the proof. □

4. Fuzzy $H\infty$ Filter Design

It is worth mentioning that the condition in (32) cannot be directly used for filter design. Therefore, in this section, we provide a sufficient condition for the fuzzy H_∞ filter design, and a suitable filter parameter matrix is obtained.

Theorem 9. *For given positive parameters τ_1, τ_2, $d \le 1$, $\gamma > 0$, $\delta > 0$, $\varepsilon_i > 0$, $(i = 1, 2, \ldots, 10)$, and $0 < \lambda < 1$, if there exist symmetric positive definite matrices $P > 0$, $Q_1 > 0$, $Q_2 > 0$, $Q_3 > 0$, $\psi > 0$, $R_1 > 0$, and $R_2 > 0$ and $\widetilde{A}_{fj}$, $\widetilde{B}_{fj}$, and $\widetilde{C}_{fj}$ with appropriate dimensions, the following hold:*

$$\begin{bmatrix} I & \widetilde{M}_3 \\ * & \varepsilon_{10}^{-1} \end{bmatrix} > 0 \tag{54}$$

$$\begin{bmatrix} R_1^{-1} & \widetilde{M}_1 \\ * & \varepsilon_4^{-1} \end{bmatrix} > 0, \qquad \begin{bmatrix} R_2^{-1} & \widetilde{M}_1 \\ * & \varepsilon_7^{-1} \end{bmatrix} > 0 \tag{55}$$

$$\begin{bmatrix} R_1^{-1} & \widetilde{M}_{21} \\ * & \varepsilon_5^{-1} \end{bmatrix} > 0, \qquad \begin{bmatrix} R_2^{-1} & \widetilde{M}_{21} \\ * & \varepsilon_8^{-1} \end{bmatrix} > 0 \tag{56}$$

$$\begin{bmatrix} R_1^{-1} & \widetilde{M}_{22} \\ * & \varepsilon_6^{-1} \end{bmatrix} > 0, \qquad \begin{bmatrix} R_2^{-1} & \widetilde{M}_{22} \\ * & \varepsilon_9^{-1} \end{bmatrix} > 0 \tag{57}$$

$$\widetilde{\Pi} = \begin{bmatrix} \widetilde{\Lambda}_{11} & \widetilde{\Lambda}_{12} \\ * & \widetilde{\Lambda}_{22} \end{bmatrix} < 0 \tag{58}$$

for

$$\widetilde{\Lambda}_{11} = \begin{bmatrix} \widetilde{\Pi}_{11} & \widetilde{\Pi}_{12} & \widetilde{\Pi}_{13} & 0 & \widetilde{\Pi}_{15} \\ * & \widetilde{\Pi}_{22} & \widetilde{\Pi}_{23} & 0 & \widetilde{\Pi}_{25} \\ * & * & \widetilde{\Pi}_{33} & \Pi_{34} & 0 \\ * & * & * & \Pi_{44} & 0 \\ * & * & * & * & \Pi_{55} \end{bmatrix},$$

$$\widetilde{\Lambda}_{12} = \begin{bmatrix} 0 & 0 & \widetilde{\Pi}_{18} & \widetilde{\Pi}_{19} \\ \widetilde{\Pi}_{26} & 0 & 0 & 0 \\ \Pi_{36} & \Pi_{37} & 0 & 0 \\ 0 & \Pi_{47} & 0 & 0 \\ 0 & 0 & 0 & 0 \end{bmatrix}, \tag{59}$$

$$\widetilde{\Lambda}_{22} = \begin{bmatrix} \Pi_{66} & 0 & 0 & 0 \\ * & \Pi_{77} & 0 & 0 \\ * & * & \widetilde{\Pi}_{88} & 0 \\ * & * & * & \widetilde{\Pi}_{99} \end{bmatrix},$$

where

$$\widetilde{\Pi}_{11} = \begin{bmatrix} \widetilde{\varphi}_{11} & \widetilde{\varphi}_{12} & \widetilde{N}_1^T & \widetilde{\varphi}_{14} & \widetilde{\varphi}_{15} & \widetilde{N}_3^T & \widetilde{\varphi}_{17} & \widetilde{\varphi}_{18} & \widetilde{C}_{ij}^T \\ * & -\varepsilon_1^{-1} & 0 & 0 & 0 & 0 & 0 & 0 & 0 \\ * & * & \widetilde{\varphi}_{33} & 0 & 0 & 0 & 0 & 0 & 0 \\ * & * & * & -\varepsilon_2^{-1} & 0 & 0 & 0 & 0 & 0 \\ * & * & * & * & -\varepsilon_3^{-1} & 0 & 0 & 0 & 0 \\ * & * & * & * & * & -\varepsilon_{10} & 0 & 0 & 0 \\ * & * & * & * & * & * & \widetilde{\varphi}_{77} & 0 & 0 \\ * & * & * & * & * & * & * & \widetilde{\varphi}_{88} & 0 \\ * & * & * & * & * & * & * & * & \widetilde{\varphi}_{99} \end{bmatrix},$$

$$\widetilde{\varphi}_{11} = \begin{bmatrix} P_{11}A_i & \widetilde{A}_{fj} \\ P_{12}^T A_i & \widetilde{A}_{fj} \end{bmatrix} + \begin{bmatrix} P_{11}A_i & \widetilde{A}_{fj} \\ P_{12}^T A_i & \widetilde{A}_{fj} \end{bmatrix}^T + Q_1 + Q_2 + Q_3 - a_1 R_1,$$

$$\widetilde{\varphi}_{12} = \begin{bmatrix} \widetilde{A}_{fj} M_{1j} \\ \widetilde{A}_{fj} M_{1j} \end{bmatrix},$$

$$\widetilde{\varphi}_{14} = \begin{bmatrix} \widetilde{B}_{fj} M_{2j} \\ \widetilde{B}_{fj} M_{2j} \end{bmatrix},$$

$$\widetilde{\varphi}_{15} = -\widetilde{\varphi}_{14},$$

$$\widetilde{\varphi}_{17} = \alpha_1 \begin{bmatrix} P_{11}A_i & \widetilde{A}_{fj} \\ P_{12}^T A_i & \widetilde{A}_{fj} \end{bmatrix}^T,$$

$$\widetilde{\varphi}_{18} = \alpha_2 \begin{bmatrix} P_{11}A_i & \widetilde{A}_{fj} \\ P_{12}^T A_i & \widetilde{A}_{fj} \end{bmatrix}^T,$$

$$\widetilde{\varphi}_{33} = -\left(\varepsilon_1^{-1} + \alpha_1^2 \varepsilon_4^{-1} + \alpha_2^2 \varepsilon_7^{-1}\right)^{-1},$$

$$\widetilde{\varphi}_{77} = \begin{bmatrix} -2\delta P + \delta^2 R_1 & \widetilde{\varphi}_{12} \\ * & -\varepsilon_4^{-1} \end{bmatrix},$$

$$\widetilde{\varphi}_{88} = \begin{bmatrix} -2\delta P + \delta^2 R_2 & \widetilde{\varphi}_{12} \\ * & -\varepsilon_7^{-1} \end{bmatrix},$$

$$\widetilde{\varphi}_{99} = \begin{bmatrix} I & \widetilde{M}_3 \\ * & -\varepsilon_{10}^{-1} \end{bmatrix},$$

$$\widetilde{\Pi}_{12} = -a_2 R_1,$$

$$\widetilde{\Pi}_{13} = \begin{bmatrix} 0 & \widetilde{B}_{fj} D_i \\ 0 & \widetilde{B}_{fj} D_i \end{bmatrix},$$

$$\widetilde{\Pi}_{15} = -a_3 R_1,$$

$$\widetilde{\Pi}_{18} = \begin{bmatrix} -\widetilde{B}_{fj} \\ -\widetilde{B}_{fj} \end{bmatrix},$$

$$\widetilde{\Pi}_{19} = \begin{bmatrix} P_{11} B_i \\ P_{12}^T B_i \end{bmatrix},$$

$$\widetilde{\Pi}_{22} = -Q_2 - a_1 R_1 - a_1 R_2,$$

$$\widetilde{\Pi}_{23} = -a_2 R_2,$$

$$\widetilde{\Pi}_{25} = -a_3 R_1,$$

$$\widetilde{\Pi}_{26} = -a_3 R_2,$$

$$\widetilde{\Pi}_{33} = \begin{bmatrix} \overline{\varphi}_{11} & \widetilde{N}_{21}^T & \overline{\varphi}_{13} & 0 & \overline{\varphi}_{15} & 0 \\ * & \overline{\varphi}_{22} & 0 & 0 & 0 & 0 \\ * & * & \overline{\varphi}_{33} & \overline{\varphi}_{34} & 0 & 0 \\ * & * & * & -\varepsilon_5^{-1} & 0 & 0 \\ * & * & * & * & \overline{\varphi}_{55} & \overline{\varphi}_{56} \\ * & * & * & * & * & -\varepsilon_8^{-1} \end{bmatrix},$$

$$\overline{\varphi}_{11} = -(1-d)\, Q_1 + \lambda D_i^T \psi D_i - 2a_1 R_2,$$

$$\overline{\varphi}_{13} = \begin{bmatrix} 0 & 0 \\ \alpha_1 D_i^T \widetilde{B}_{fj}^T & \alpha_1 D_i^T \widetilde{B}_{fj}^T \end{bmatrix},$$

$$\overline{\varphi}_{15} = \begin{bmatrix} 0 & 0 \\ \alpha_2 D_i^T \widetilde{B}_{fj}^T & \alpha_2 D_i^T \widetilde{B}_{fj}^T \end{bmatrix},$$

$$\overline{\varphi}_{22} = -\left(\varepsilon_2^{-1} + \alpha_1^2 \varepsilon_5^{-1} + \alpha_2^2 \varepsilon_8^{-1}\right)^{-1},$$

$$\overline{\varphi}_{33} = -2\delta P + \delta^2 R_1,$$

$$\overline{\varphi}_{34} = \overline{\varphi}_{56} = \begin{bmatrix} \widetilde{B}_{fj} M_{2j} \\ \widetilde{B}_{fj} M_{2j} \end{bmatrix},$$

$$\overline{\varphi}_{55} = -2\delta P + \delta^2 R_2,$$

$$\widetilde{\Pi}_{88} = \begin{bmatrix} (\lambda-1)\psi & \widetilde{N}_{22}^T & \widetilde{\kappa}_{13} & 0 & \widetilde{\kappa}_{15} & 0 \\ * & \widetilde{\kappa}_{22} & 0 & 0 & 0 & 0 \\ * & * & \widetilde{\kappa}_{33} & \widetilde{\kappa}_{34} & 0 & 0 \\ * & * & * & -\varepsilon_6^{-1} & 0 & 0 \\ * & * & * & * & \widetilde{\kappa}_{55} & \widetilde{\kappa}_{56} \\ * & * & * & * & * & -\varepsilon_9^{-1} \end{bmatrix},$$

$$\widetilde{\kappa}_{13} = \begin{bmatrix} -\alpha_1 B_{fj}^T & -\alpha_1 B_{fj}^T \end{bmatrix},$$

$$\widetilde{\kappa}_{15} = \begin{bmatrix} -\alpha_2 B_{fj}^T & -\alpha_2 B_{fj}^T \end{bmatrix},$$

$$\widetilde{\kappa}_{22} = -\left(\varepsilon_3^{-1} + \alpha_1^2 \varepsilon_6^{-1} + \alpha_2^2 \varepsilon_9^{-1}\right)^{-1},$$

$$\widetilde{\kappa}_{33} = -2\delta P + \delta^2 R_1,$$

$$\widetilde{\kappa}_{34} = \widetilde{\kappa}_{56} = \begin{bmatrix} -\widetilde{B}_{fj} M_{2j} \\ -\widetilde{B}_{fj} M_{2j} \end{bmatrix},$$

$$\widetilde{\kappa}_{55} = -2\delta P + \delta^2 R_2,$$

$$\widetilde{\Pi}_{99} = \begin{bmatrix} -\gamma^2 I & \overline{\kappa}_{12} & \overline{\kappa}_{13} \\ * & \overline{\kappa}_{22} & 0 \\ * & * & \overline{\kappa}_{33} \end{bmatrix},$$

$$\overline{\kappa}_{12} = \alpha_1 \begin{bmatrix} B_i^T P_{11}^T & B_i^T P_{12} \end{bmatrix},$$

$$\overline{\kappa}_{13} = \alpha_2 \begin{bmatrix} B_i^T P_{11}^T & B_i^T P_{12} \end{bmatrix},$$

$$\overline{\kappa}_{22} = -2\delta P + \delta^2 R_1,$$

$$\overline{\kappa}_{33} = -2\delta P + \delta^2 R_2. \tag{60}$$

Moreover, if inequality is feasible, the filter parameters can be obtained by

$$A_{fj} = P_{12}^{-1} \widetilde{A}_{fj},$$
$$B_{fj} = P_{12}^{-1} \widetilde{B}_{fj}, \tag{61}$$
$$C_{fj} = \widetilde{C}_{fj}.$$

Proof. By applying Schur complement lemma, the matrix inequality conditions (32) in Theorem 8 can be rewritten as the following matrix inequality:

$$\widetilde{\Xi} = \begin{bmatrix} \Gamma_{11} & \Gamma_{12} \\ * & \Gamma_{22} \end{bmatrix} < 0 \tag{62}$$

for

$$\Gamma_{11} = \begin{bmatrix} \Theta_{11} & -a_2R_1 & P\widetilde{B}_{ij1} & 0 & -a_3R_1 \\ * & \Theta_{22} & -a_2R_2 & 0 & -a_3R_1 \\ * & * & \Theta_{33} & -a_2R_2 & 0 \\ * & * & * & -Q_3 - a_1R_2 & 0 \\ * & * & * & * & -a_4R_1 \end{bmatrix},$$

$$\Gamma_{12} = \begin{bmatrix} 0 & 0 & P\widetilde{B}_{ij2} & P\widetilde{B}_{wij} \\ -a_3R_2 & 0 & 0 & 0 \\ -a_3R_2 & -a_3R_2 & 0 & 0 \\ 0 & -a_4R_2 & 0 & 0 \\ 0 & 0 & 0 & 0 \end{bmatrix}, \tag{63}$$

$$\Gamma_{22} = \begin{bmatrix} -a_4R_2 & 0 & 0 & 0 \\ * & -a_4R_2 & 0 & 0 \\ * & * & \Theta_{88} & 0 \\ * & * & * & \Theta_{99} \end{bmatrix},$$

where

$$\Theta_{11} = \begin{bmatrix} \phi_{11} & P\widetilde{M}_1 & \widetilde{N}_1^T & P\widetilde{M}_{21} & -P\widetilde{M}_{21} & \widetilde{N}_3^T & \phi_{17} & \phi_{18} & \widetilde{C}_{ij}^T \\ * & -\varepsilon_1^{-1} & 0 & 0 & 0 & 0 & 0 & 0 & 0 \\ * & * & \phi_{33} & 0 & 0 & 0 & 0 & 0 & 0 \\ * & * & * & -\varepsilon_2^{-1} & 0 & 0 & 0 & 0 & 0 \\ * & * & * & * & -\varepsilon_3^{-1} & 0 & 0 & 0 & 0 \\ * & * & * & * & * & -\varepsilon_{10} & 0 & 0 & 0 \\ * & * & * & * & * & * & \phi_{77} & 0 & 0 \\ * & * & * & * & * & * & * & \phi_{88} & 0 \\ * & * & * & * & * & * & * & * & \phi_{99} \end{bmatrix},$$

$$\phi_{11} = P\widetilde{A}_{ij} + \widetilde{A}_{ij}^TP + Q_1 + Q_2 + Q_3 - a_1R_1,$$

$$\phi_{17} = \alpha_1\widetilde{A}_{ij}^TP,$$

$$\phi_{18} = \alpha_2\widetilde{A}_{ij}^TP,$$

$$\phi_{33} = -\left(\varepsilon_1^{-1} + \alpha_1^2\varepsilon_4^{-1} + \alpha_2^2\varepsilon_7^{-1}\right)^{-1},$$

$$\phi_{77} = \begin{bmatrix} -PR_1^{-1}P & P\widetilde{M}_1 \\ * & -\varepsilon_4^{-1} \end{bmatrix},$$

$$\phi_{88} = \begin{bmatrix} -PR_2^{-1}P & P\widetilde{M}_1 \\ * & -\varepsilon_4^{-1} \end{bmatrix},$$

$$\phi_{99} = \begin{bmatrix} I & \widetilde{M}_3 \\ * & -\varepsilon_{10}^{-1} \end{bmatrix},$$

$$\Theta_{22} = -Q_2 - a_1R_1 - a_1R_2,$$

$$\Theta_{33} = \begin{bmatrix} \overline{\phi}_{11} & \widetilde{N}_{21}^T & \alpha_1\widetilde{B}_{ij1}^TP & 0 & \alpha_2\widetilde{B}_{ij1}^TP & 0 \\ * & \overline{\phi}_{22} & 0 & 0 & 0 & 0 \\ * & * & -PR_1^{-1}P & P\widetilde{M}_{21} & 0 & 0 \\ * & * & * & -\varepsilon_5^{-1} & 0 & 0 \\ * & * & * & * & -PR_2^{-1}P & P\widetilde{M}_{21} \\ * & * & * & * & * & -\varepsilon_8^{-1} \end{bmatrix},$$

$$\overline{\phi}_{11} = -(1-d)\,Q_1 + \lambda D_i^T\psi D_i - 2a_1R_2,$$

$$\overline{\phi}_{22} = -\left(\varepsilon_2^{-1} + \alpha_1^2\varepsilon_5^{-1} + \alpha_2^2\varepsilon_8^{-1}\right)^{-1},$$

$$\Theta_{88} = \begin{bmatrix} (\lambda-1)\,\psi & \widetilde{N}_{22}^T & \alpha_1\widetilde{B}_{ij2}^TP & 0 & \alpha_2\widetilde{B}_{ij2}^TP & 0 \\ * & \widetilde{\kappa}_{22} & 0 & 0 & 0 & 0 \\ * & * & -PR_1^{-1}P & P\widetilde{M}_{22} & 0 & 0 \\ * & * & * & -\varepsilon_6^{-1} & 0 & 0 \\ * & * & * & * & -PR_2^{-1}P & P\widetilde{M}_{22} \\ * & * & * & * & * & -\varepsilon_9^{-1} \end{bmatrix},$$

$$\Theta_{99} = \begin{bmatrix} -\gamma^2I & \alpha_1\widetilde{B}_{wij}^TP & \alpha_2\widetilde{B}_{wij}^TP \\ * & -PR_1^{-1}P & 0 \\ * & * & -PR_2^{-1}P \end{bmatrix}. \tag{64}$$

Then, we define the following matrices:

$$P = \begin{bmatrix} P_{11} & P_{12} \\ * & P_{12} \end{bmatrix},$$

$$\widetilde{A}_{fj} = P_{12}A_{fj},$$

$$\widetilde{B}_{fj} = P_{12}B_{fj}, \tag{65}$$

$$\widetilde{C}_{fj} = C_{fj}.$$

We have

$$P\widetilde{A}_{ij} = \begin{bmatrix} P_{11}A_i & \widetilde{A}_{fj} \\ P_{12}^TA_i & \widetilde{A}_{fj} \end{bmatrix},$$

$$P\widetilde{B}_{ij1} = \begin{bmatrix} 0 & \widetilde{B}_{fj}D_i \\ 0 & \widetilde{B}_{fj}D_i \end{bmatrix},$$

$$P\widetilde{B}_{ij2} = \begin{bmatrix} -\widetilde{B}_{fj} \\ -\widetilde{B}_{fj} \end{bmatrix},$$

$$P\widetilde{B}_{wij} = \begin{bmatrix} P_{11}B_i \\ P_{12}^TB_i \end{bmatrix},$$

$$P\widetilde{M}_1 = \begin{bmatrix} \widetilde{A}_{fj}M_1 \\ \widetilde{A}_{fj}M_1 \end{bmatrix},$$

$$P\widetilde{M}_{21} = \begin{bmatrix} \widetilde{B}_{fj}M_2 \\ \widetilde{B}_{fj}M_2 \end{bmatrix},$$

$$P\widetilde{M}_{22} = \begin{bmatrix} -\widetilde{B}_{fj}M_2 \\ -\widetilde{B}_{fj}M_2 \end{bmatrix}. \tag{66}$$

Moreover, we can derive $-PR^{-1}P \leq -2\delta P + \delta^2 R$ by the inequality $(\delta R - P)R^{-1}(\delta R - P) \geq 0$ for any scalar $\delta > 0$. Then, by applying Schur complement lemma and combining (32) and (62)–(66), we can easily derive inequality (58). This completes the proof. □

5. Simulation Results

In this section, we provide two examples to demonstrate the merit of the nonfragile H_∞ filter based event-triggered scheme.

Example 1. Consider the following fuzzy systems:

$$\begin{aligned} \dot{x}(t) &= \sum_{i=1}^{2} u_i [A_i x(t) + B_i \omega(t)] \\ z(t) &= \sum_{i=1}^{2} u_i C_i x(t) \\ y(t) &= \sum_{i=1}^{2} u_i D_i x(t), \end{aligned} \tag{67}$$

where

$$\begin{aligned} A_1 &= \begin{bmatrix} -2.1 & 0.1 \\ 1 & -2 \end{bmatrix}, \\ A_2 &= \begin{bmatrix} -1.9 & 0 \\ -0.2 & -1.1 \end{bmatrix}, \\ B_1 &= \begin{bmatrix} 1 \\ -0.2 \end{bmatrix}, \\ B_1 &= \begin{bmatrix} 0.3 \\ 0.1 \end{bmatrix}, \\ C_1 &= \begin{bmatrix} 1 & -0.5 \end{bmatrix}, \\ C_2 &= \begin{bmatrix} -0.2 & 0.3 \end{bmatrix}, \\ D_1 &= \begin{bmatrix} 1 & 0 \end{bmatrix}, \\ D_2 &= \begin{bmatrix} 0.5 & 0.6 \end{bmatrix}. \end{aligned} \tag{68}$$

And the parameter uncertainties are given as

$$\begin{aligned} M_{11} &= M_{12} = \begin{bmatrix} 0.05 \\ 0.04 \end{bmatrix}, \\ M_{21} &= M_{22} = 0.01, \\ M_{31} &= M_{32} = \begin{bmatrix} 0.02 \\ 0.01 \end{bmatrix}, \\ N_{11} &= N_{12} = \begin{bmatrix} 0.02 & 0.03 \end{bmatrix}, \\ N_{21} &= N_{22} = 0.05, \\ N_{31} &= N_{32} = \begin{bmatrix} 0.03 & 0.02 \end{bmatrix}, \\ K_g(t) &= \sin(t), \quad g = A, B, C. \end{aligned} \tag{69}$$

The membership function $u_i(t)$, $i = 1, 2$, and the external disturbance signal $\omega(t)$ are given as

$$\begin{aligned} u_1(t) &= \sin\left(t^2\right), \\ u_2(t) &= 1 - u_1(t), \\ \omega(t) &= \frac{1}{1 + 3t^2}. \end{aligned} \tag{70}$$

Consider the interval time-varying delay with $\tau_1 = 0.03$ and $\tau_2 = 0.2$. And choose $\lambda = 0.2$, $d = 0.2$, $\delta = 10$, and sample period $h = 20$ ms. By Theorem 9, we derive the desired nonfragile parameter as follows:

$$\begin{aligned} A_{f1} &= \begin{bmatrix} -17.4366 & 8.5268 \\ 7.7146 & -11.2453 \end{bmatrix}, \\ A_{f2} &= \begin{bmatrix} -16.9835 & 6.2715 \\ 7.0539 & -9.1581 \end{bmatrix}, \\ B_{f1} &= \begin{bmatrix} -0.4852 \\ -0.3343 \end{bmatrix}, \\ B_{f2} &= \begin{bmatrix} 0.2106 \\ -0.1013 \end{bmatrix}, \\ C_{f1} &= \begin{bmatrix} -2.4715 & 0.2475 \end{bmatrix}, \\ C_{f2} &= \begin{bmatrix} -1.8903 & 1.1024 \end{bmatrix}. \end{aligned} \tag{71}$$

Meanwhile, we obtain the event-triggered parameter $\psi = 23.6026$ and the minimum H_∞ performance level is $\gamma = 0.99$.

It is assumed that the initial condition is $x(0) = \begin{bmatrix} 1 & -1 \end{bmatrix}^T$. The responses of the original state and the filter state are shown in Figures 1 and 2. The system output signal is depicted in Figure 3. As depicted in Figure 4, we can clearly see that the response of the filtering error is stable after about three seconds, which demonstrates the effective of the proposed approach. Note that, in Figure 5, the transmission rate is mitigated from 100% to 7%, which demonstrates that the event-triggered scheme is effective in reducing data transmission and saving the network resource.

Example 2. To further demonstrate the effectiveness of the proposed method, in this example, we consider a truck-trailer

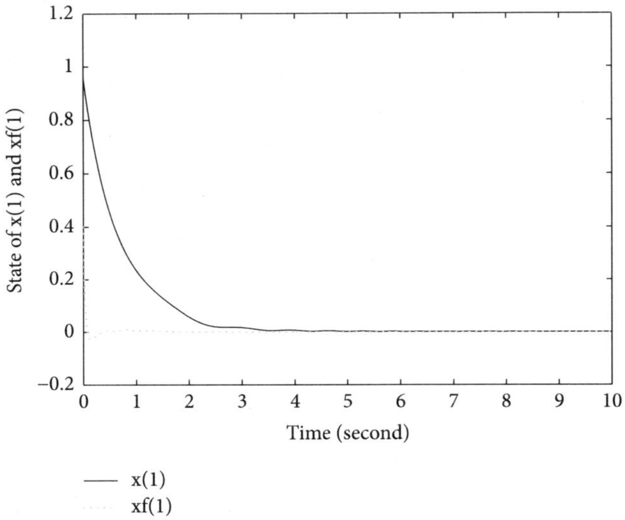

FIGURE 1: State trajectories of x_1.

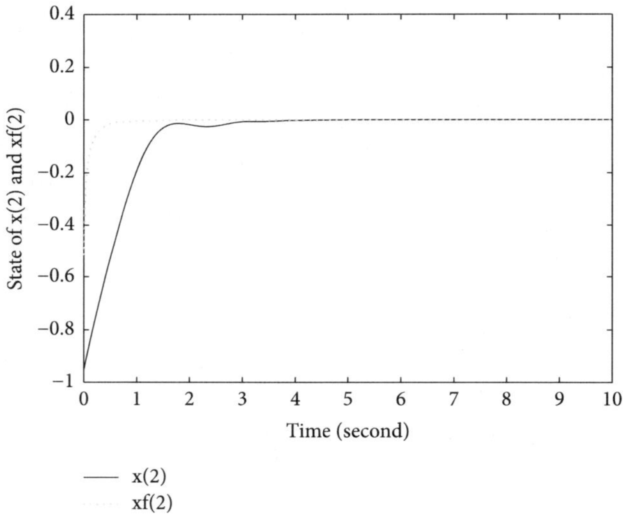

FIGURE 2: State trajectories of x_2.

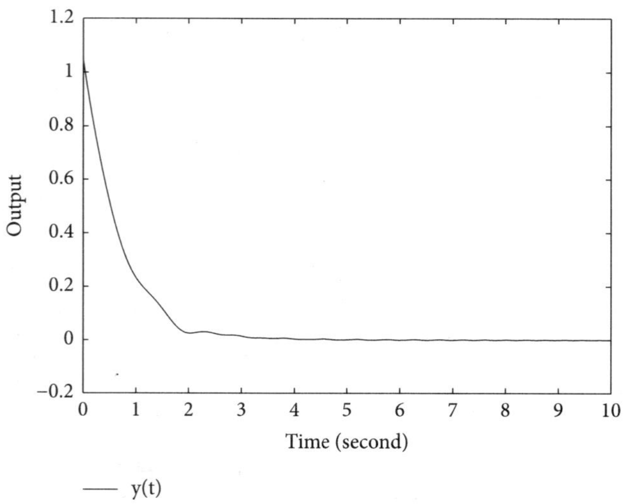

FIGURE 3: The trajectories of $y(t)$.

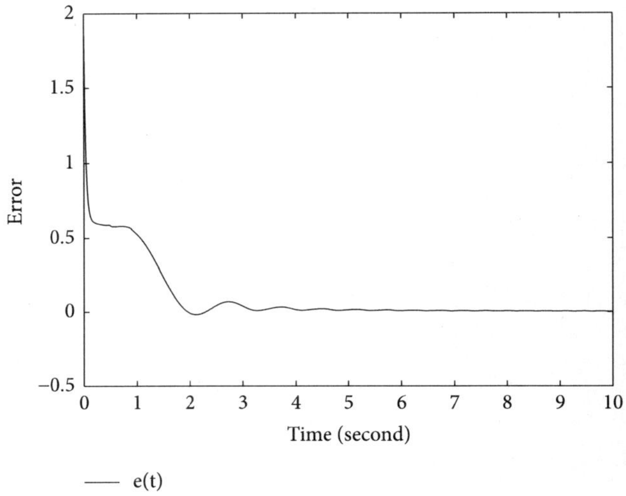

FIGURE 4: The trajectories of estimate error.

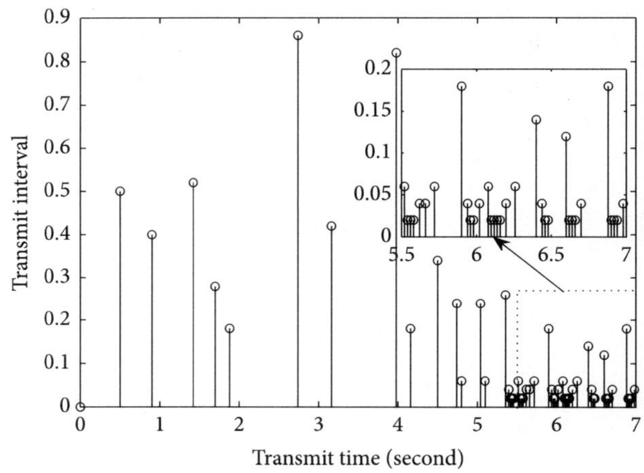

FIGURE 5: Release instant.

system; the practical example is borrowed from [26]. The system parameters are given as follows:

$$A_1 = \begin{bmatrix} -1.6371 & 0.5741 & -0.0157 \\ -0.5091 & 0 & 0 \\ 0.5091 & -4 & 0 \end{bmatrix},$$

$$A_2 = \begin{bmatrix} -1.6086 & 0.5284 & -0.0150 \\ -0.5091 & 0 & 0 \\ 0.0016 & -0.0127 & 0 \end{bmatrix},$$

$$B_1 = \begin{bmatrix} 1 \\ 0.5 \\ 0 \end{bmatrix},$$

$$B_2 = \begin{bmatrix} 1 \\ 0.5 \\ 0 \end{bmatrix},$$

$$C_1 = \begin{bmatrix} 0 & 1 & 0 \end{bmatrix},$$
$$C_2 = \begin{bmatrix} 0 & 1 & 0 \end{bmatrix},$$
$$D_1 = \begin{bmatrix} 4.9 & -2 & 0.04 \end{bmatrix},$$
$$D_2 = \begin{bmatrix} 4.9 & -2 & 0.04 \end{bmatrix}. \quad (72)$$

Then, we suppose that the parameter uncertainties are given by

$$M_{11} = M_{12} = \begin{bmatrix} 0.1 \\ 0.1 \\ 0.21 \end{bmatrix},$$
$$M_{21} = M_{22} = 0.1,$$
$$M_{31} = M_{32} = \begin{bmatrix} -0.2 \\ 0 \\ 0.2 \end{bmatrix}, \quad (73)$$
$$N_{11} = N_{12} = \begin{bmatrix} 0.2 & 0.3 & 0.1 \end{bmatrix},$$
$$N_{21} = N_{22} = 0.5,$$
$$N_{31} = N_{32} = \begin{bmatrix} -0.3 & 0.2 & 0.1 \end{bmatrix},$$
$$K_l(t) = \sin(t), \quad l = A, B, C.$$

We assume that the lower bound and upper bound of the interval time-varying delay are $\tau_1 = 0.01$ and $\tau_2 = 0.25$, respectively. The derivative of the time delay is $d = 0.3$. Furthermore, we suppose that $\lambda = 0.2$ and $\delta = 0.1$. By solving the LMIs in Theorem 9, we obtain minimum performance level $\gamma = 1.1241$, which is less than previously studied $\gamma = 7.0948$ in [26]. Meanwhile, the event-triggered argument $\psi = 2.0069$ is obtained, and the corresponding filter parameters can be derived as follows:

$$A_{f1} = \begin{bmatrix} -8.1449 & 8.2655 & -0.6046 \\ -3.6646 & 3.5043 & -0.2331 \\ -2.3869 & 2.7130 & -1.5149 \end{bmatrix},$$
$$A_{f2} = \begin{bmatrix} -7.6655 & -8.2644 & -0.3039 \\ -3.2074 & 3.1396 & -0.1352 \\ -0.6803 & 0.6193 & -0.9489 \end{bmatrix},$$
$$B_{f1} = \begin{bmatrix} -0.0942 \\ -0.0503 \\ -0.1788 \end{bmatrix}, \quad (74)$$
$$B_{f2} = \begin{bmatrix} -0.0742 \\ -0.0360 \\ -0.2396 \end{bmatrix},$$
$$C_{f1} = \begin{bmatrix} -0.9862 & 0.8778 \end{bmatrix},$$
$$C_{f2} = \begin{bmatrix} -2.6857 & 4.0130 \end{bmatrix}.$$

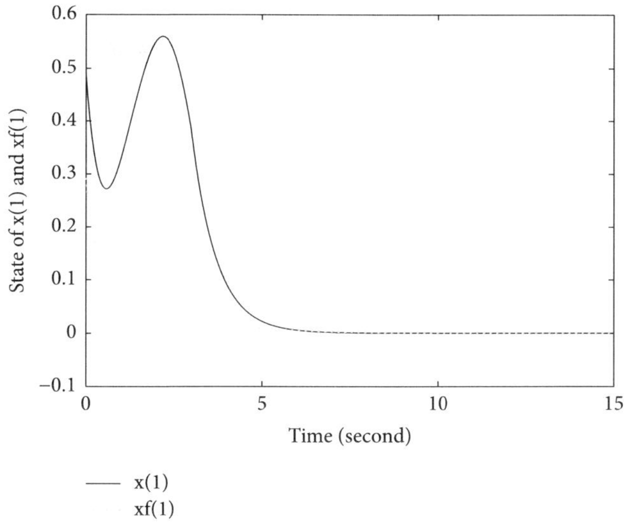

FIGURE 6: State trajectories of x_1.

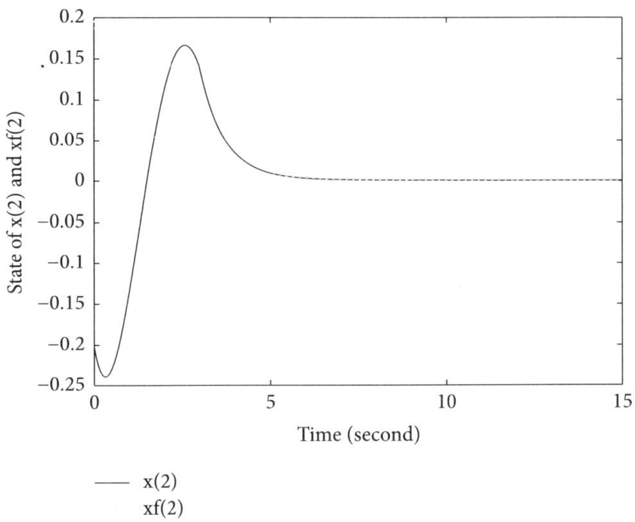

FIGURE 7: State trajectories of x_2.

Figures 6, 7, and 8 show the original state and the filter state; it is clear that the filter state is asymptotically stable. The curve of the system output signal is depicted in Figure 9.

The filtering error response is depicted in Figure 10, which shows that the designed nonfragile H_∞ filter is feasible.

Finally, we choose the sample period $h = 10$ ms; under the event-triggered scheme, it can be seen that Figure 11 shows that only 24 times in all 900 sampling instants are transmitted, which illustrates that the network bandwidth is greatly mitigated.

6. Conclusions

In this paper, the problem of the event-triggered fuzzy nonfragile H_∞ filter design has been investigated for nonlinear NCSs with interval time-varying delay. The resultant filtering error system has been modeled as a system with an

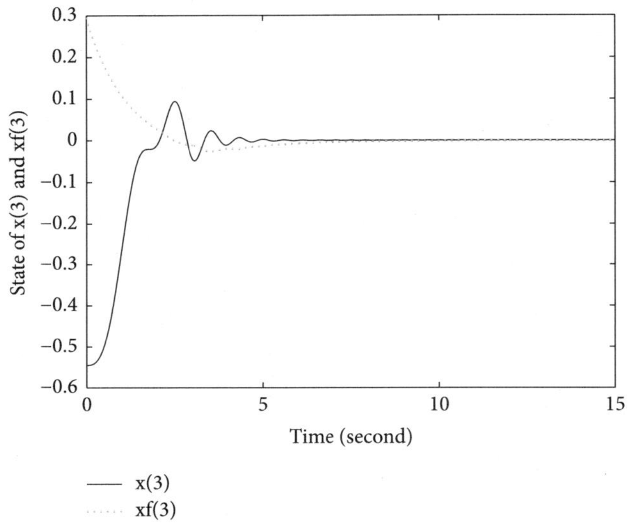

FIGURE 8: State trajectories of x_3.

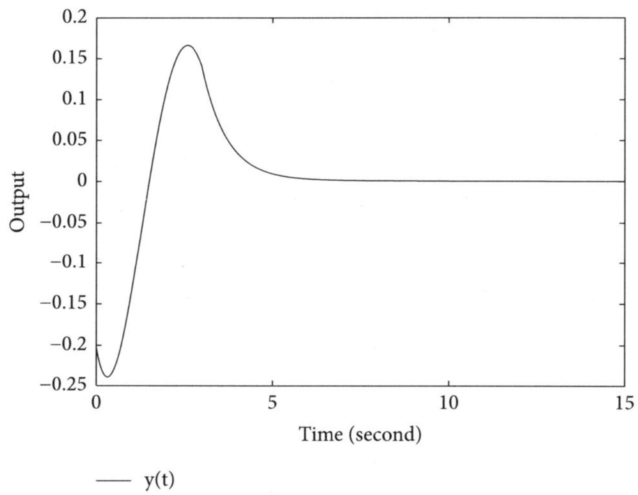

FIGURE 9: The trajectories of $y(t)$.

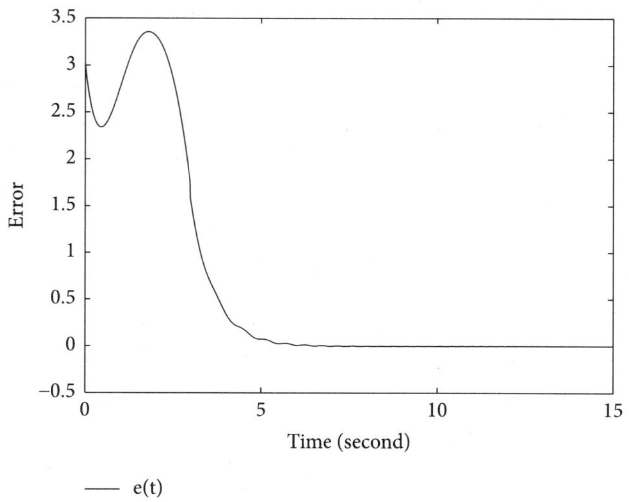

FIGURE 10: The trajectories of estimate error.

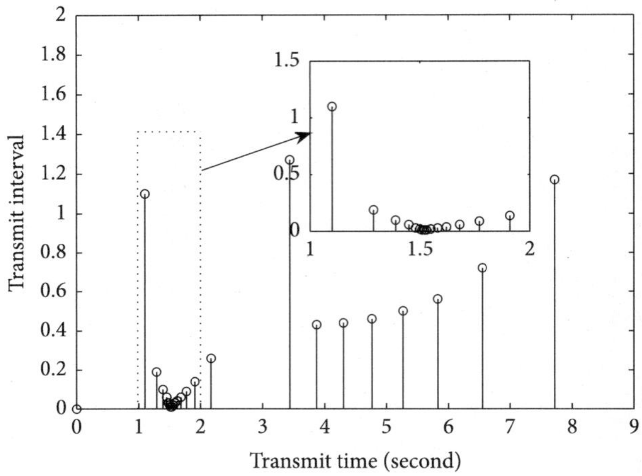

FIGURE 11: Release instant.

interval time-varying delay. By constructing a new Lyapunov-Krasovskii functional and employing Wirtinger inequality, we have obtained a sufficient condition for designing the fuzzy nonfragile H_∞ filter such that the fuzzy filtering error system is asymptotically stable with a prescribed H_∞ performance level. Two numerical examples have been given to demonstrate the effectiveness of the proposed method. Future research includes nonfragile fuzzy fault filter design for nonlinear control systems considering time-varying delays. In addition, type-2 nonfragile fuzzy filter design for NCSs with packet dropout also can be further considered for the future investigation.

Conflicts of Interest

The authors declare that there are no conflicts of interest regarding the publication of this paper.

Acknowledgments

This work was supported in part by Natural Science Foundation of Heilongjiang Province (Grant no. LC2015024) and the Fundamental Research Funds in Heilongjiang Provincial Universities (Grant no. 135109238).

References

[1] Y.-Q. Xia, Y.-L. Gao, L.-P. Yan, and M.-Y. Fu, "Recent progress in networked control systems — A survey," *International Journal of Automation and Computing*, vol. 12, no. 4, pp. 343–367, 2015.

[2] C. Han, L. Wu, H. K. Lam, and Q. Zeng, "Non-fragile control with guaranteed cost of T-S fuzzy singular systems based on parallel distributed compensation," *IEEE Transactions on Fuzzy Systems*, 2013.

[3] C. Han, G. Zhang, L. Wu, and Q. Zeng, "Sliding mode control of T-S fuzzy descriptor systems with time-delay," *Journal of The Franklin Institute*, vol. 349, no. 4, pp. 1430–1444, 2012.

[4] H. K. Lam, "Stability analysis of T-S fuzzy control systems using parameter-dependent Lyapunov function," *IET Control Theory & Applications*, vol. 3, no. 6, pp. 750–762, 2009.

[5] H. K. Lam, "LMI-based stability analysis for fuzzy-model-based control systems using artificial TS fuzzy model," *IEEE Transactions on Fuzzy Systems*, vol. 19, no. 3, pp. 505–513, 2011.

[6] X. Xu, H. Yan, H. Zhang, and F. Yang, "H∞ filtering for T–S fuzzy networked systems with stochastic multiple delays and sensor faults," *Neurocomputing*, vol. 207, pp. 590–598, 2016.

[7] J. An, G. Wen, C. Lin, and R. Li, "New results on a delay-derivative-dependent fuzzy H∞ filter design for T-S fuzzy systems," *IEEE Transactions on Fuzzy Systems*, vol. 19, no. 4, pp. 770–779, 2011.

[8] S. Zhang, Z. Wang, D. Ding et al., "Non-fragile H8 fuzzy filtering with randomly occuring gain variations and channel fadings," *IEEE Transactions on Fuzzy Systems*, vol. 24, no. 33, pp. 505–518, 2016.

[9] D. W. Ding, X. L. Li, Y. Y. Wang, and Z. G. Shi, "Non-fragile H8 fuzzy filtering for discrete-time non-linear systems," *IET Control Theory and Applications*, vol. 7, no. 6, pp. 848–857, 2013.

[10] Z. Li, F. Gu, Y. Q. He, and W. J. Hao, "Nonfragile robust H∞ filter design for a class of fuzzy stochastic systems with stochastic input-to-state stability," *Mathematical Problems in Engineering*, vol. 2015, 10 pages, 2015.

[11] W. P. Heemels, M. C. Donkers, and A. R. Teel, "Periodic event-triggered control for linear systems," *Institute of Electrical and Electronics Engineers Transactions on Automatic Control*, vol. 58, no. 4, pp. 847–861, 2013.

[12] C. Peng, Q.-L. Han, and D. Yue, "To transmit or not to transmit: a discrete event-triggered communication scheme for networked takagi-sugeno fuzzy systems," *IEEE Transactions on Fuzzy Systems*, vol. 21, no. 1, pp. 164–170, 2013.

[13] P. J. Gawthrop and L. Wang, "Event-driven intermittent control," *International Journal of Control*, vol. 82, no. 12, pp. 2235–2248, 2009.

[14] J. Lunze and D. Lehmann, "A state-feedback approach to event-based control," *Automatica*, vol. 46, no. 1, pp. 211–215, 2010.

[15] X. Yin and D. Yue, "Event-triggered tracking control for heterogeneous multi-agent systems with Markov communication delays," *Journal of The Franklin Institute*, vol. 350, no. 5, pp. 1312–1334, 2013.

[16] K. Duan and W. Zhang, "Event-triggered fault-tolerant control for networked systems with dynamic quantiser," *IET Control Theory & Applications*, vol. 10, no. 9, pp. 1088–1096, 2016.

[17] S. Wen, X. Yu, and Z. Zeng, "Event-triggering load frequency control for multi-area power systems," *IEEE Transactions on Industrial Electronics*, vol. 63, no. 2, pp. 1308–1317, 2016.

[18] P. Shi, H. Wang, and C.-C. Lim, "Network-based event-triggered control for singular systems with quantizations," *IEEE Transactions on Industrial Electronics*, vol. 63, no. 2, pp. 1230–1238, 2016.

[19] Z. D. Lu, G. T. Ran, G. L. Zhang, and F. X. Xu, "Event-triggered H∞ fuzzy filtering for networked control systems with quantization and delays," *IEEE Access*, vol. 6, no. 1, pp. 20231–20241, 2018.

[20] C. Peng and J. Zhang, "Even-triggered output-feedback H∞ control for networked control systems with time-varying sampling," *IET Control Theory & Applications*, vol. 9, no. 9, pp. 1384–1391, 2015.

[21] I. R. Petersen, "A stabilization algorithm for a class of uncertain linear systems," *Systems & Control Letters*, vol. 8, no. 4, pp. 351–357, 1987.

[22] L. H. Xie, "Output feedback H∞ control of systems with parameter uncertainty," *International Journal of Control*, vol. 63, no. 4, pp. 741–750, 1996.

[23] Y. Y. Wang, L. H. Xie, and C. E. de Souza, "Robust control of a class of uncertain nonlinear systems," *Systems and Control Letters*, vol. 19, no. 2, pp. 139–149, 1992.

[24] S. Zhou, W. Ren, and J. Lam, "Stabilization for T-S model based uncertain stochastic systems," *Information Sciences*, vol. 181, no. 4, pp. 779–791, 2011.

[25] A. Seuret and F. Gouaisbaut, "On the use of the Wirtinger inequalities for time-delay systems," *IFAC Proceedings Volumes*, vol. 45, no. 14, pp. 260–265, 2012.

[26] Y. J. Liu, B. Z. Guo, and J. H. Park, "Non-fragile H8 filtering for delayed Takagi-Sugeno fuzzy systems with randomly occurring gain variations," *Fuzzy Sets and Systems*, vol. 316, no. 1, pp. 99–116, 2017.

8

An Optimal Hybrid Control Method for Energy-Saving of Chilled Water System in Central Air Conditioning

Yan Zhang[1,2,3] **Xiaoli Chu,**[4] **and Yongqiang Liu**[2]

[1] *Faculty Development and Educational Technology Center, Guangdong University of Finance and Economics, Guangzhou 510320, China*
[2] *School of Electric Power, South China University of Technology, Guangzhou 510640, China*
[3] *Foshan Institute of Modern Services, Guangdong University of Finance and Economics, Foshan 528100, China*
[4] *Center of Network, Guangdong AIB Polytechnic College 510507, China*

Correspondence should be addressed to Yan Zhang; zhangyan@gdufe.edu.cn

Academic Editor: Haiping Du

Chilled water system of central air conditioning is a typical hybrid system; variable frequency behavior with amplitude limited of pumps reflects continuous and discrete dynamic characteristics. The way of energy-saving is variable water volume, via variable frequency behavior of pumps to gain adjustment of power consumption. Facing the situation of the variable frequency pumps with parallel operation, single continuous or discrete modeling cannot reflect the hybrid features. Thus, the control method will have some questions, such as bad energy-saving effect, difficult accurate adjustment of cold capacity, and low running energy efficiency. However, hybrid system modeling can reflect hybrid dynamic behavior of pumps, which is combining continuous and discrete features. The questions of nonlinear and multiparameters can be solved by control method based on hybrid system. Here, an optimum control method is proposed with the principle of the minimum, by setting the minimum power consumption as the performance function in fixed time, which realizes variable control of pumps and accurate adjustment of temperature inside room. At last, it shows the system characteristics and energy-saving affection by hybrid system modeling and the optimum control method.

1. Introduction

Nowadays China is a country with severe energy shortages. Energy conservation and emission reduction, as an important basic national policy of China, have far-reaching guiding significance for the development of industrial technology. At the national level, the "13th Five-Year Plan for Comprehensive Energy Conservation and Emission Reduction" clearly states that vigorously developing energy-saving products and technologies promoting the transformation of high-energy-consuming enterprises or industries and fully implementing energy-saving and emission reduction are the key tasks in the technical field. In building energy consumption, central air-conditioning accounts for 50-70% of total energy consumption [1], followed by lighting and other types of electrical equipment. Energy-saving of air-conditioning systems are not only the key to the success of building energy-saving work, but also the realization of a conservation-oriented society [2].

In addition to the chiller, the energy-saving control of the central air-conditioning is mainly for pumps and fans. The chilled water system is the research object, which is mainly reflected in the variable flow control. At present, the variable flow control objects of the central air conditioning water system mainly include return water temperature [3], supply and return water temperature difference [4, 5], and supply and return water pressure difference [6], as well as temperature difference and differential pressure cascade control [7]. The control methods mainly include switch control, PID control, fuzzy control, adaptive control, optimization control, and decoupling control. J. Mohammed et al. [8] proposed a dual-proportional PID hybrid controller to improve cooling performance and achieve energy saving of chilled water flow. J. Li et al. [9] adopted neural network adaptive control method to monitor the system's work. The state adjusts the parameters of the neural network controller in real time, realizes the intelligent control, and verifies the dynamic performance and steady state performance of the method through simulation. G. Liu et al. [10] combines the technical advantages of fuzzy control and PID algorithm to propose a parameter adaptive fuzzy PID. The algorithm has achieved good test results; M. Koor [11] proposed a predictive control method for finding the best energy efficiency point for the variable frequency pump set with the same performance parameters, considering the pressure and flow factors. Luo et al. [12] designed a system based on PLC fuzzy prediction combined with expert practical experience and control rules, which embodies better control effects, T.Zhao et al. [6]. The number of parallel hydraulic pumps for chilled water systems. Compared with the rate, an online optimization control strategy is proposed, and the performance improvement is demonstrated through experiments. Zhang Yan et al. [13] proposed a kind of central air-conditioning cooling based on hybrid system model. The model of energy saving in the frozen water system, and set the action of the pump group for frequency conversion and switching, and demonstrate the applicability of the model and control strategy through computer simulation; H. Wang et al. [14] put various energy saving such as variable flow, temperature difference and differential pressure. The technology is compared with the corresponding control strategy to analyze the energy saving effect of different methods. J.M. Gordon et al. used a simple thermodynamic model that captures the universal aspects of chiller behavior [15]. Next, they used the simple thermodynamic model to predict the fundamental relation between coefficient of performance and cooling rate for the centrifugal chiller [16].

The traditional model and control method is based on a simple continuous or discrete dynamic model, which cannot accurately reflect the mixed dynamic behavior of continuous features and discrete features. Hybrid systems contain both continuous and discrete characteristics, and have significant advantages in dealing with power system and appliance modeling issues. The central air-conditioning chilled water system is a typical hybrid system. Its energy-saving control is mainly used to change the flow of the water system. It is realized by the frequency conversion or switching action of the pumps group, which reflects the hybridization with the goal of minimum energy consumption. According to the change of the cooling demand, real-time adjustment of the flow of the chilled water system can effectively realize the power saving of the pumps group. Therefore, the modeling and energy-saving control strategies of chilled water systems based on hybrid systems are highly targeted. The nonlinearity, large time lag, and multiparameters of chilled water systems can be modeled and effectively solved by hybrid systems. Thus, it has extremely important theoretical significance.

2. Mathematical Description of Piecewise Affine System

Throughout, the following notation is adopted: $\mathbb{R}$ and $\mathbb{C}$ denote the fields of real and complex number, respectively; $\mathbb{R}^n$, $\mathbb{R}^m$, $\mathbb{R}^l$, and $\mathbb{R}^s$ denote the n, m, l, and s-dimensional real Euclidean spaces; $\mathbb{R}^{n\times n}$ denotes the space of $n \times n$ matrices with real entries; x_i denotes the ith component of the vector x in $\mathbb{R}^n$, respectively; a_{ij} denotes the entry in the (i, j) position of the matrices A or E in $\mathbb{R}^{n\times n}$.

The Piecewise Affine System (PWA) [17] is a class in hybrid systems. The continuous-time PWA model can be described as follows:

$$\begin{aligned} \dot{x}(\tau) &= A_i x(\tau) + B_i u(\tau) + f_i \\ y(\tau) &= C_i x(\tau) + D_i u(\tau) + g_i \end{aligned} \tag{1}$$

which satisfies $\begin{bmatrix} x(\tau) \\ u(\tau) \end{bmatrix} \in \Omega_i$, where $x(\tau) \in \mathbb{R}^n$ and $u(\tau) \in \mathbb{R}^m$, $y(\tau) \in \mathbb{R}^l$, respectively, represent the system's state variables, input variables, and output variables. τ represents a time scalar. f_i and g_i are constant vectors. A_i and C_i represent the state matrices of the system; B_i and D_i are the input matrices of the system. Ω_i denotes a convex polyhedron that represents the state variable of the system and the input variable space, which is a convex partition, namely, $X \times U = \cup\Omega_i$, $\Omega_i \cap \Omega_j = \varnothing$, $i \neq j$. The system switches when the state of the system crosses the boundary. The system consists of N subsystems, $\cup_{i=1}^{N}\Omega_i = \Omega$, and the system switches between subsystems. The switching rules can be state dependent, time dependent, and so on.

3. Mathematical Model of Room Temperature Changes in Air Conditioning Systems

The mathematical model of air-conditioned rooms adopts a common method: based on the dynamic change of the heat capacity in the air-conditioned room, the cold sending air is the control input, and the temperature in the air-conditioned room is the state variable. The heat dissipation device and human constitute a disturbance of heat and the heat transfer medium such as wall surface, ceiling, window, etc. can also affect the indoor temperature to a certain extent, and can be considered as a disturbance or a fixed value. Then this dynamic process conforms to the law of conservation of energy and, according to the adjustment of appropriate parameters, the temperature in the air-conditioned room can be stabilized at the set value, as shown in Figure 1. The dynamic heat balance equation can be expressed as

$$C_z \frac{dt}{d\tau} = q_{sa} c_a (t_{sa} - t) + \sum K_i A_i (t_i - t) + Q \tag{2}$$

where C_z is the heat capacity in the air-conditioned room, J/K; t is the real-time temperature of the air-conditioned room, °C; τ is the time, s; q_{sa} is the air supply volume to the air-conditioned room, kg/s; c_a is the specific heat capacity of the air, J/(kg·K); t_{sa} is the air supply temperature of the air conditioning, °C; K_i is the heat transfer coefficient of the heat transfer medium (wall, window, and roof), W/(m2·K); A_i is the heat transfer medium The area of the wall (wall, window, and roof), m^2; t_i is the temperature of the inner surface of the envelope structure (wall, window and roof), °C; Q is the heat dissipation of the human body and equipment in the air-conditioned room, W.

The cooling capacity produced by the cold units is transferred from the chilled water system to the air system, and the

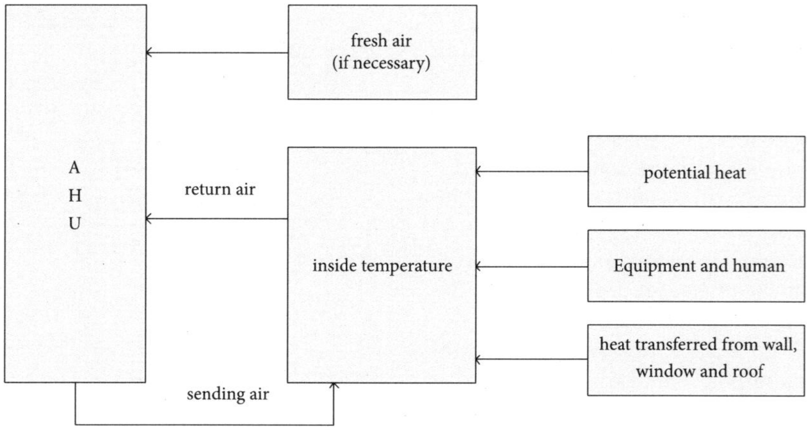

FIGURE 1: Schematic diagram of dynamic thermal equilibrium in air conditioning room.

heat exchange is realized by the cooler in the Air Handling Unit (AHU), and the heat exchange efficiency is set to η. The AHU is also responsible for mixing the indoor return air with the outdoor fresh air, and sending the filtered clean air into the air-conditioned room through the cooler. Among them, the ratio of the fresh air volume q_n to the return air volume q_r is *1-w:w*. According to the requirements of Chinese "Public Building Energy Efficiency Design Standards," there should generally be a fresh air ratio of not less than 15% to ensure indoor air quality. Then the expression of the cooling between the water system and the air system is

$$\gamma\eta Q_{\mathrm{w}} = Q_{\mathrm{a}} = \left(Q_{\mathrm{xr}} + Q_{\mathrm{qr}}\right) \tag{3}$$

By introducing the expressions of the water system carrying the cold capacity and the cooling capacity carried by the air system, the formula (3) can be rewritten as

$$\gamma\eta q_{\mathrm{w}} c_{\mathrm{w}} \Delta t = (1-w)\, q_{\mathrm{sa}} c_{\mathrm{a}} \left(t_{\mathrm{out}} - t_{\mathrm{sa}}\right) + w q_{\mathrm{sa}} c_{\mathrm{a}} \left(t - t_{\mathrm{sa}}\right) + Q_{\mathrm{qr}} \tag{4}$$

where γ is the percentage of cold capacity assigned to an air-conditioned room, $0 \le \gamma \le 100\%$; Q_{w} is the cold capacity carried by the water system, W; Q_{a} is the cold capacity that is sent to the air-conditioned room by the AHU, W; Q_{xr} is the sensible heat in the air-conditioned room, W; Q_{qr} is the potential heat in the room, W; q_w is the flow rate of the water system, kg/s; c_w is the specific heat capacity of water, J/(kg·K); Δt is a constant temperature difference between the supply and return water, °C; t_{out} is the fresh air temperature introduced from the outside, °C.

Bringing (4) into the heat balance equation (2) of the room, we can get

$$C_{\mathrm{z}} \frac{dt}{d\tau} = -\gamma\eta q_{\mathrm{w}} c_{\mathrm{w}} \Delta t - \left[(1-w)\, q_{\mathrm{sa}} c_{\mathrm{a}} + \sum K_i A_i\right] t + (1-w)\, q_{\mathrm{sa}} c_{\mathrm{a}} t_{\mathrm{out}} + \sum K_i A_i T_i + Q + Q_{\mathrm{qr}} \tag{5}$$

Set the state variable $x(\tau) = t(\tau)$ and the input variable $u(\tau) = q_{\mathrm{w}}(\tau)$, use the variable water volume of water system and the constant air volume of air system transfer modes [3, 13], and use the differential equation to describe the system model:

$$\begin{aligned} \dot{x}(\tau) &= Ax(\tau) + Bu(\tau) + H \\ y(\tau) &= x(\tau) \end{aligned} \tag{6}$$

The system parameters are

$$\begin{aligned} A &= -\frac{(1-w)\, q_{\mathrm{sa}} c_a + \sum K_i A_i}{C_z} \\ B &= -\frac{\gamma\eta c_{\mathrm{w}} \Delta t}{C_z} \\ H &= \frac{(1-w)\, q_{\mathrm{sa}} c_{\mathrm{a}} T_{\mathrm{out}} + \sum K_i A_i T_i + Q + Q_{\mathrm{qr}}}{C_z} \end{aligned} \tag{7}$$

where the behaviors of the pumps group show the hybrid characteristics. First, the water system of the air conditioning is dynamic and random, and the factors affecting the system characteristics also show multivariate, dynamic, strong coupling, and nonlinear relationship. Characteristics [18]; second, in variable flow control method, the water flow is generally not less than 50% of the rated value, otherwise the Coefficient of Performance (COP) of the pump will be abruptly attenuated, so the range of variable flow is limited to [50% rated, 100% rated], showing continuous characteristics within this range. If the cold capacity is reduced, the flow control is out of the frequency conversion range, and the amplitude truncation occurs, showing a discrete characteristic. The flow constraints of a single pump and the entire water

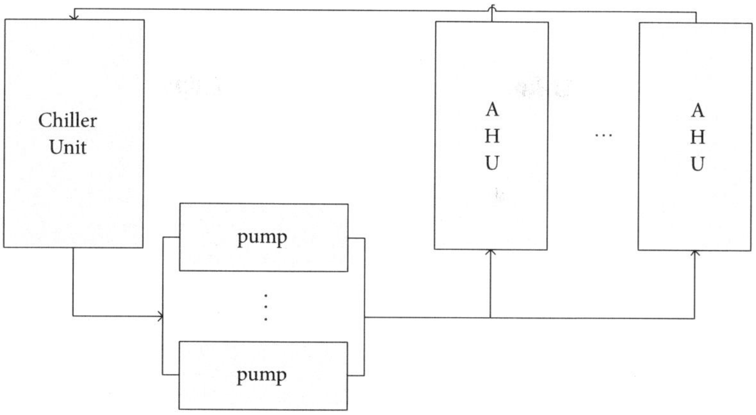

FIGURE 2: Schematic diagram of cold capacity transfer process in chilled water system.

system can be expressed by mathematical descriptions as

$$u_i = \begin{cases} 0.5u_{i0}, & u < 0.5u_{i0} \\ u_{i0}, & u > u_{i0} \\ u_i, & 0.5u_{i0} \le u \le u_{i0} \end{cases}, \quad i = 1, 2, \ldots, N_\mathrm{p} \tag{8a}$$

$$u_\mathrm{s} = \sum_{i=1}^{N_\mathrm{p}} u_i \tag{8b}$$

$$u = \begin{cases} 0.5u_{s0}, & u < 0.5u_{s0} \\ u_{s0}, & u > u_{s0} \\ u, & 0.5u_{s0} \le u \le u_{s0} \end{cases} \tag{9}$$

where u_i is the flow rate of the *i-th* pump as the *i-th* input variable; u_{i0} is the rated flow of the *i-th* pump, which is the constant; N_p is the number of variable-frequency pumps; u_s is the flow of the whole water system; u_{s0} is the rated flow rate of the whole water system. The schematic diagram of the cold capacity of the water system is shown in Figure 2.

4. Hybrid Optimization Control Strategy Based on the Minimum Principle

According to the minimum principle proposed by the former Soviet scholar Pontiac King, the water consumption of the water system in the finite time is set to the performance functional J, and the optimal allowable control u is sought to minimize the performance function. u is a multivariate parameter, which is limited to a closed set and satisfies the inequality constraint $g[x(\tau), u(\tau), \tau] \ge 0$, and g is a r-dimensional continuous vector variable function, $r \le m$.

Because there are many ways to control the frequency of the pumps group, the energy saving effect is also quite different. The method adopts a fixed terminal pressure difference control mode, which sets an electric regulating valve and a differential pressure transmitter on the most unfavorable loop, so that the power of the pump is between one power and three powers [19]. In order to facilitate the calculation, it is approximated that the power (P, W) of the pump and the rotational speed (n, rad/s) are squared, and the flow rate of the pump is proportional to the rotational speed, i.e., $n_i/n_{i0} = q_{\mathrm{w}i}/q_{\mathrm{w}i0}$, where n_i is the rotational speed of a pump, rad/s, n_{i0} is the rated speed of the pump, rad/s; q_{wi} is the flow of a pump, kg/s; q_{wi0} is the rated flow of the pump, kg/s. Let $P_0/q_0^2 = \xi$. Then the performance function is the sum of the energy consumption of the variable frequency pumps, which can be expressed as

$$J = \int_{\tau_o}^{\tau_f} L\left[x(\tau), u(\tau), \tau\right] d\tau = \int_{\tau_o}^{\tau_f} \left(\sum_{i=1}^{N_\mathrm{p}} \xi_i {u_i}^2\right) d\tau \tag{10}$$

where L is a continuous differentiable quantity function; τ_0 and τ_f are initial and terminal moments, respectively, s.

Select the Hamilton function as

$$\begin{aligned} H &= L\left[x(\tau), u(\tau), \tau\right] + \lambda^T f\left[x, u, \tau\right] \\ &= \sum_{i=1}^{N_\mathrm{p}} \xi_i {u_i}^2 + \lambda^T \left(Ax + \sum_{i=1}^{N_\mathrm{p}} Bu_i + H\right) \end{aligned} \tag{11}$$

where λ is the costate vector and $f[\cdot]$ is the state function of the system formula (6), $f[\cdot] = \dot{x}$.

First, suppose the extremum condition and the costate equation are analyzed under the condition of unconstrained control; then the optimal control is discussed according to the actual constraints.

Let

$$\frac{\partial H}{\partial u} = 2\sum_{i=1}^{N_\mathrm{p}} \xi_i u_i + \sum_{i=1}^{N_\mathrm{p}} B\lambda = 0 \tag{12}$$

We get

$$u_i^*(\tau) = -\frac{B\lambda(\tau)}{2\xi_i} \tag{13}$$

which is

$$u^*(\tau) = \sum_{i=1}^{N_\mathrm{p}} u_i(\tau) \tag{14}$$

Then the equation of state can be rewritten as

$$\dot{x} = Ax - \frac{B^2\lambda}{2}\sum_{i=1}^{N_p}\frac{1}{\xi_i} + H \tag{15}$$

The coordination equation is

$$\dot{\lambda} = -\frac{\partial H}{\partial x} = -A\lambda \tag{16}$$

Find the second derivative of the state equation:

$$\begin{aligned}\ddot{x} &= A\dot{x} - \dot{\lambda}\frac{B^2}{2}\sum_{i=1}^{N_p}\frac{1}{\xi_i} + \dot{H} \\ &= A\left(Ax - \frac{B^2\lambda}{2}\sum_{i=1}^{N_p}\frac{1}{\xi_i} + H\right) + \frac{AB^2\lambda}{2}\sum_{i=1}^{N_p}\frac{1}{\xi_i}\end{aligned} \tag{17}$$

We get

$$\ddot{x} = A^2x + AH \tag{18}$$

Integral can be obtained

$$x(\tau) = C_1e^{-A\tau} + C_2e^{A\tau} - \frac{H}{A} \tag{19}$$

where C_1 and C_2 are undetermined coefficients and (19) is substituted into the equation of state (15) and the coordination equation can be obtained as

$$\lambda(\tau) = \frac{4A}{B^2\sum_{i=1}^{N_p}(1/\xi_i)}C_1e^{-A\tau} \tag{20}$$

Furthermore, substituting (20) into (14) for optimal control,

$$u_i^*(\tau) = -\frac{2A}{\xi_iB\sum_{i=1}^{N_p}(1/\xi_i)}C_1e^{-A\tau} \tag{21}$$

Substituting the initial condition $x(\tau_0) = x_0$ and the terminal constraint $N[x(\tau_f), \tau_f] = 0$ (or terminal condition $x(\tau_f) = x_f$), the optimal state trajectory x^* and the optimal performance function J^* are obtained together, where N is a s-dimensional continuous differentiable vector function, $s \leq n$; τ_0 and τ_f are the initial and terminal moments, respectively, s; x_0 and x_f are the initial and terminal values of the state variable, respectively.

The goal proposed in this paper is that the energy consumption is minimum, and the boundary conditions are degenerated into fixed initial and terminal constant values, that is, $x(\tau_0) = x_0$ and $x(\tau_f) = x_f$, respectively. In order to obtain the optimal state trajectory x^* and the optimal performance function J^*, substitute the initial condition $x(\tau_0) = x_0$ and the terminal condition $x(\tau_f) = x_f$ into (19) and determine the values of the undetermined coefficients C_1 and C_2 in the equation of state (19), the coordination equation (20), and the control equation (21).

Considering the limitation of (9) for water system flow, when u_s is used as a single input variable, the optimal control based on the minimum principle is essentially the form of limiting amplifier. However, considering the multi-input variable of multipump and the hybrid dynamic behavior of frequency conversion and truncation, the flow rate of water system is only the sum of the flow rate of multipump, as shown in equation (8a) and (8b), as another constraint condition for input variable. Therefore, in order to achieve the necessary conditions for optimal control, the optimal control u^* must be required to take values from the allowed finite closed set (8a), which is

$$u_i^* = \begin{cases} 0.5u_{i0}, & u_i^* < 0.5u_{i0} \\ u_{i0}, & u_i^* > u_{i0} \\ u_i, & 0.5u_{i0} \leq u_i^* \leq u_{i0} \end{cases} \tag{22}$$

5. Case Analysis

Assume that the building area of air-conditioned room is 120 m^2 and the height is 3 m. The central air-conditioning sends cold air to the indoor temperature and the indoor temperature is required to be within a limited time (3 minutes), and the initial temperature (t_0, 30°C) is reduced to the setting temperature (t_{set}, 26°C), and the goal of power consumption in accordance with the control strategy is minimal. The chilled water system has four variable frequency water pumps with different parameters, and they never leave the work. They transport the cold capacity produced by the cold unit to the AHU for heat exchange and send it into the room in the form of cold air. Assume that the thermodynamic parameters in the air-conditioned room and the parameters and values of the water system pump are shown in Table 1.

Then, (6) can be numerically

$$\begin{aligned} \dot{x}(\tau) &= -0.0002098x(\tau) - 0.003374u(\tau) + 0.005651 \\ y(\tau) &= x(\tau) \end{aligned} \tag{23}$$

The performance functional (10) can be rewritten as

$$\begin{aligned} J = \int_0^{180} &\left(609.6{u_1}^2 + 1567{u_2}^2 + 3858{u_3}^2\right. \\ &\left. + 4542{u_4}^2\right) d\tau \end{aligned} \tag{24}$$

The Hamilton function (11) is updated to

$$\begin{aligned} H &= L[x(\tau), u(\tau), \tau] + \lambda^T f[x, u, \tau] = 609.6{u_1}^2 \\ &+ 1567{u_2}^2 + 3858{u_3}^2 + 4542{u_4}^2 \\ &+ \lambda^T\{-0.0002098x(\tau) \\ &- 0.003374[u_1(\tau) + u_2(\tau) + u_3(\tau) + u_4(\tau)] \\ &+ 0.005651\} \end{aligned} \tag{25}$$

After the solution of the equation of state (formula (19)) is digitized, substitute the initial value (30°C, 0s) and

TABLE 1: Symbols and values of parameters in cold capacity transfer process.

Parameter (Description)	Value
indoor hear capacity, C_z (KJ/K)	469
specific heat of air, c_a (kJ/kg∗k)	1.01
specific heat of water, c_w (kJ/kg∗k)	4.186
temperature of fresh air, t_{out} (°C)	26
setting temperature, t_{set} (°C)	25
indoor initial temperature, t_0 (°C)	30
volume of sending air, q_{sa} (kg/s)	0.623
temperature of walls, windows, and roof, respectively, t_i (°C)	35,36,35
area of walls, windows, and roof, respectively, A_i, (m^2)	80,5,0
heat transfer coefficient of walls, windows, and roof, respectively, K_i, (W/m^2*k)	0.047,0.054,0.051
latent heat load, Q_{qr} (W)	25
random heat load, Q (W)	30
return air rate, w	0.85
transfer efficiency from water system to air system, η	0.9
ratio of cooling capacity allocation, γ	7%
temperature difference, Δt (°C)	6
rated power, respectively, P_0 (kW)	1.69;2.17;4.91;10.6
rated flow, respectively, q_{w0} (L/s)	0.61;0.75;1.77;4.17

TABLE 2: Square ratio of rated power and rated flow.

Number	Rated volume L/s	Rated power kW	ξ
1	4.17	10.6	609.6
2	1.77	4.91	1567
3	0.75	2.17	3858
4	0.61	1.69	4542

the terminal value (26°C, 180s) and find the undetermined coefficients C_1 and C_2 as -51.751 and 54.811. Then the solution to the equation of state can be updated to

$$x(\tau) = -51.751e^{0.0002098\tau} + 54.811e^{-0.0002098\tau} + 26.94 \quad (26)$$

From the costate equation (20), the square ratio of rated power to rated flow (Table 2), the numerical solution of the costate equation is obtained

$$\lambda(\tau) = 1.383300e^{0.0002098\tau} \quad (27)$$

Then the optimal governing equation can be obtained as

$$u_i^*(\tau) = \begin{cases} 3.828e^{0.0002098\tau}, & i = 1 \\ 1.489e^{0.0002098\tau}, & i = 2 \\ 0.605e^{0.0002098\tau}, & i = 3 \\ 0.514e^{0.0002098\tau}, & i = 4 \end{cases} \quad (28a)$$

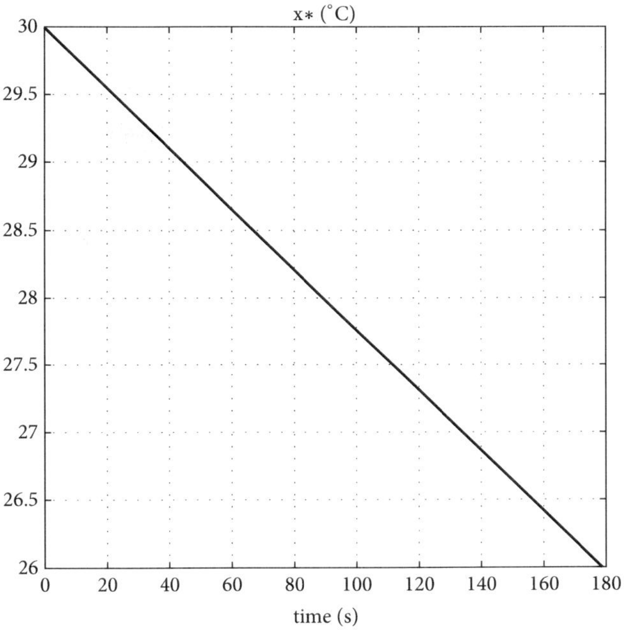

FIGURE 3: Optimal state curve.

The optimal equation for the total flow of the chilled water system is

$$u^*(\tau) = 6.436e^{0.0002098\tau} \quad (28b)$$

The optimal governing equation is verified for a finite time (0-180 s) to satisfy the finite closed set (8a). Then the performance is

$$J^* = 2.808 \times 10^6 (W \cdot s) = 0.78 (kW \cdot h) \quad (29)$$

The optimal curve of this example can be seen in Figures 3–6.

Figure 3 is the optimal state curve showing the requirement that the temperature in the air-conditioned room drops from the initial value (30°C) to the terminal value (26°C) within a limited time (180 seconds) under the action of the optimal control law.

The optimal control equations (28a) and (28b) and Figure 4 show the flow changes of the four pumps under the optimal control law and the change of the total flow of the chilled water system. Under the condition of considering the frequency conversion range, the above flow rate meets the requirements of the limited interval. The total flow of the chilled water system varies from 6.4 to 6.7 L/s in 180 seconds, meeting the amplitude limit requirements of [3.65, 7.3], as shown in Figure 4(b). The flow rate of the pumps (No. 1-4) also meets the variable flow range requirements, as shown in Figure 4(a).

Figure 5 is the curve of the coordination equation, and Figure 6 is the performance function curve, which is the calculation of the total energy consumption of the pumps

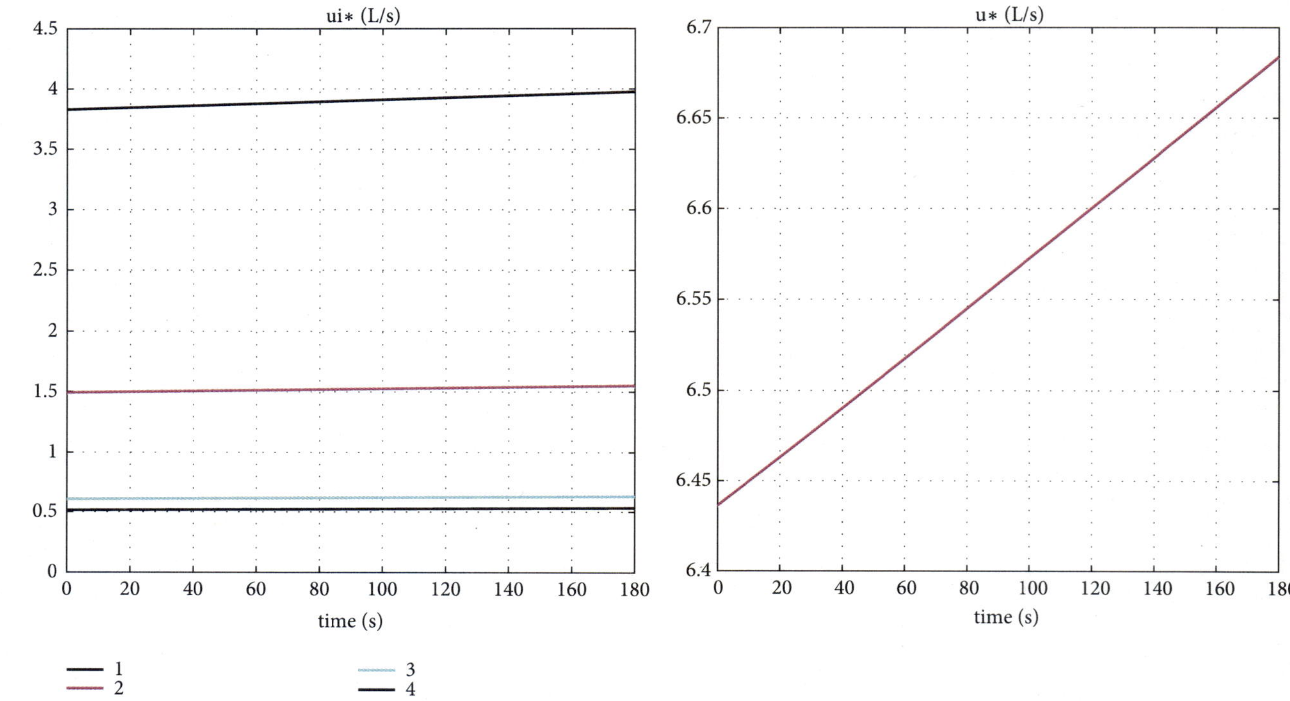

(a) Optimal control curve (u_i)

(b) Optimal control curve (u)

FIGURE 4

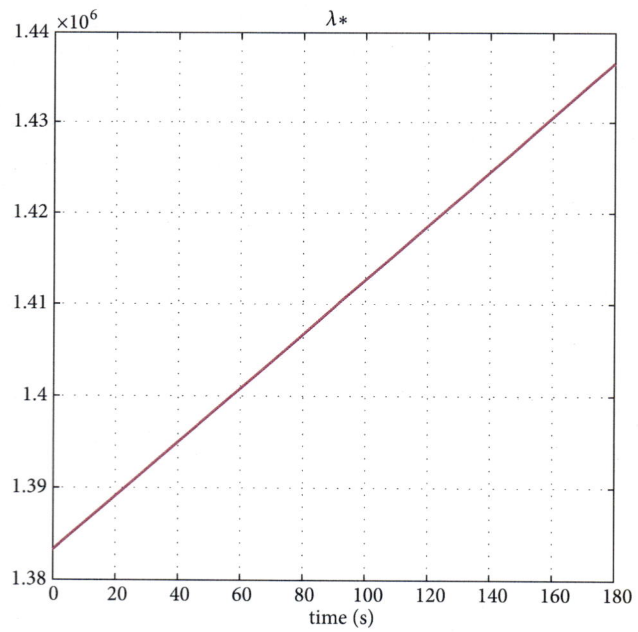

FIGURE 5: Optimal costate curve.

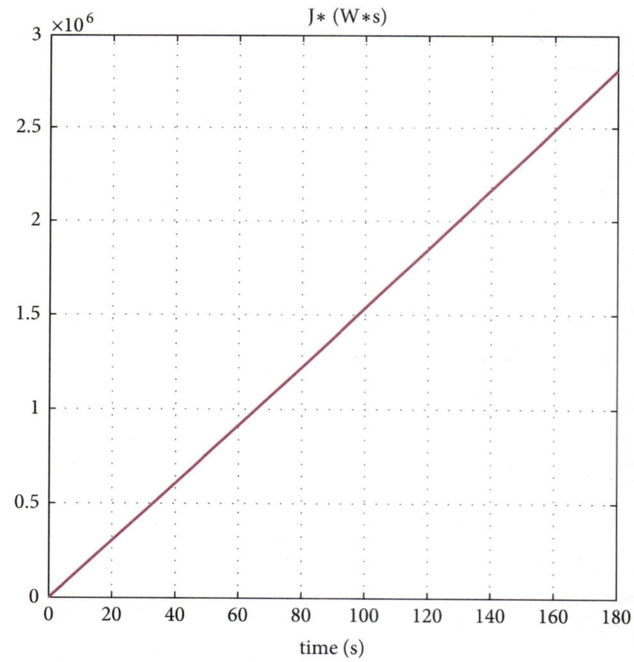

FIGURE 6: Optimal performance curve.

group. Compare the fixed frequency operation of the pumps group, the energy saving situation can be calculated as

$$\varepsilon = 1 - \frac{J^*}{\int_0^{180} \left(\sum_{i=1}^{4} P_{0i}\right) d\tau} = 1 - \frac{2.808 \times 10^6}{3.4866 \times 10^6} = 19.46\% \tag{30}$$

According to the calculation of (30), the optimal control strategy based on the minimum principle of the energy saving rate of 19.46% for the pumps group of the chilled water system within 3 minutes when the indoor temperature drops to the set temperature.

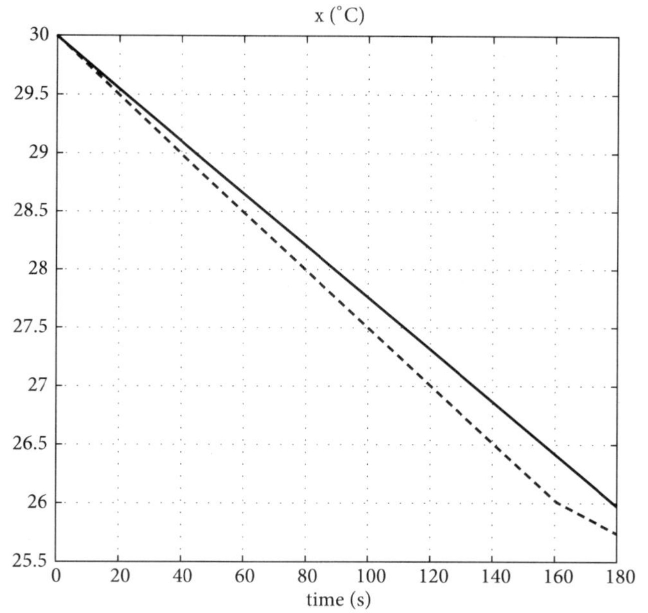

FIGURE 7: Temperature variation (compare with group control strategy).

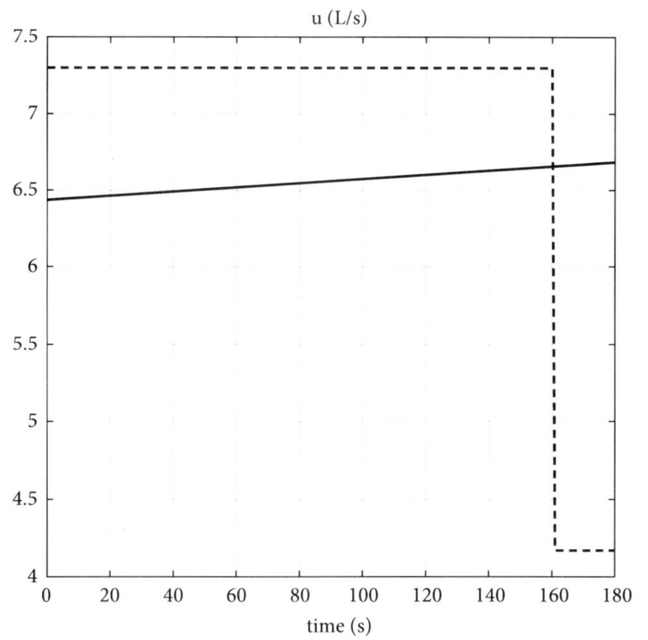

FIGURE 8: Volume variation (compare with group control strategy).

Compare the above method with the pumps group control strategy. The group control strategy is set as follows: 4 pumps are all fixed frequency operation, which are put into use at the initial time. After reaching the set temperature range, the pumps are sequentially discharged according to the switching rules. Let the water pump state stable range be 26 ± 0.5°C. If the temperature is more than 26.5°C for several minutes, put one pump; if the temperature is less than 25.5°C for several minutes, exit one pump. Considering the different pump power, for the sake of energy saving, the high-power

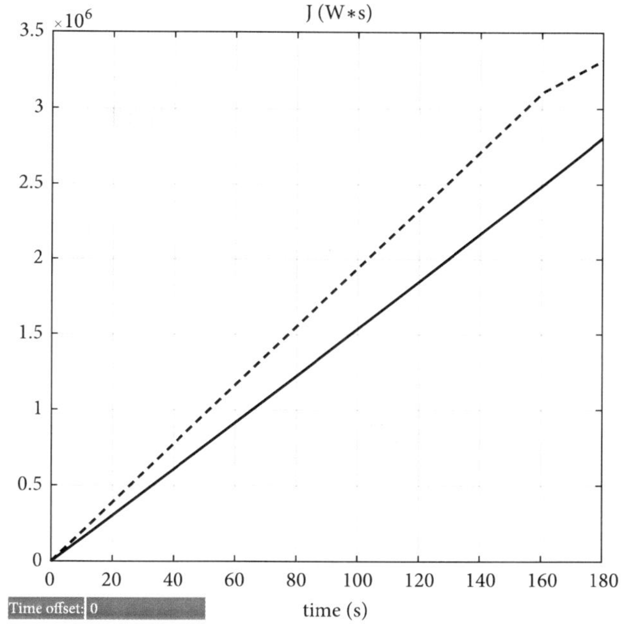

FIGURE 9: Energy consumption (compare with group control strategy).

pump is preferred when withdrawing, and the low-power pump is preferred when cutting in.

The flow of the chilled water system shall not be less than 50% of the rated flow rate, then, the No.1 pump must be fixed and not withdrawn. Other pumps exit in turn, or in turn.

Considering that this numerical case takes 180 seconds as the simulation time, it is assumed that the group control strategy is set to all the pumps Nos. 2, 3, and 4 out of work when the indoor temperature reaches the set value, to meet the constraint minimum flow operation. It can be seen that, in Figure 7, the group control strategy (dashed line) reaches the set temperature of 26°C at about 160 seconds, and then the temperature continues to drop in the stable range. In Figure 8, under the group control strategy, the flow of the water system has changed drastically due to the withdrawal of multiple pumps (dotted line) in the vicinity of 160 seconds. In comparison, the control method (solid line) of this paper is more accurate than the group control strategy, and the curve is smoother.

Figure 9 shows the comparison of energy consumption during the simulation time. It can be seen that the solid line is the energy consumption of the control method in this paper. The value is 2.808 × 106W.s, and the energy consumption value (dashed line) under the group control strategy is 3.316 × 106W.s, and the energy saving rate increased by 15.32% compared with the group control strategy. Compared with the group control strategy, the energy saving effect is obviously superior.

6. Conclusion

The chilled water system of central air conditioning is a multiparameter complex structural system. The continuous and discrete dynamic behaviors of the pumps group determine the mixed characteristics of mathematical modeling. The minimum energy consumption is the goal for the energy-saving control strategy based on hybrid system. The method based on the minimum principle, the optimal control law is established, the frequency conversion regulation of the pumps group is realized, and the optimal state trajectory and optimal association are obtained. The trajectory of the state and the applicability of the method are further verified by numerical examples and computer simulation. The method shows a more accurate indoor temperature regulation effect, a clearer pump group flow control process, and an objective function calculation result. Due to the variety of energy-saving control methods for chilled water systems, the adjustment effects are not the same, and there are also large differences in energy-saving effects. The control strategy of this paper adopts idealized approximation results in water pipe network resistance and pump energy efficiency. Therefore, there are certain limitations. In actual engineering cases, it is still necessary to correct and improve through a large number of measured data.

Conflicts of Interest

The authors declare that they have no conflicts of interest.

Acknowledgments

The research was supported by the National Natural Science Foundation of China (No. 61104181). The simulation work was supported by Dr. Mo Lili (ORCID ID: 0000-0002-8604-5562) from South China University of Technology.

References

[1] L.-J. Zhang, *Fuzzy Neural Network Control Principle and Engineering Application*, Harbin Institute of Technology Press, Harbin, 2004.

[2] Y. Z. Kang and B. Hua, "Technical and Economic Analysis of District Cooling System," *Gas & Heat*, vol. 27, no. 11, pp. 79–82, 2007.

[3] Y. Zhang, Y. Liu, and Y. Liu, "Stability analysis for a type of multiswitching system with parallel structure," *Mathematical Problems in Engineering*, vol. 2018, Article ID 3834601, 16 pages, 2018.

[4] G.-Q. Zhang, Y.-Z. Xu, J. Han, T. Liu, and H.-P. Wu, "Research on optimal economic temperature difference between the chilled water supply and return model of a district-cooling system," *Hunan Daxue Xuebao/Journal of Hunan University Natural Sciences*, vol. 42, no. 9, pp. 128–133, 2015.

[5] Y. Zhang, X. L. Chu, and Y. Q. Liu, "Cold Transmitting Control and Terminal Cold Adjusting of District Cooling System," *Journal of South China University of Technology*, vol. 45, no. 10, pp. 26–33, 2017.

[6] Z. Tianyi, Z. Jili, and M. Liangdong, "On-line optimization control method based on extreme value analysis for parallel variable-frequency hydraulic pumps in central air-conditioning systems," *Building and Environment*, vol. 47, no. 1, pp. 330–338, 2012.

[7] X.-K. Xiao, X.-Q. Jin, P.-J. Liu, and Z.-M. Du, "Research on Optimal Control of Chilled Water Temperature and Head of Secondary Pump in VWV Systems," *Fluid Machinery*, vol. 23, no. 5, pp. 42–45, 2004.

[8] J. A.-K. Mohammed, F. M. Mohammed, and M. A.-S. Jabbar, "Investigation of high performance split air conditioning system by using Hybrid PID controller," *Applied Thermal Engineering*, vol. 129, pp. 1240–1251, 2018.

[9] J.-J. Li, Q.-F. Sheng, and B. Ma, "Temperature and Humidity Adaptive Control in Centre Air-conditioning System Based," *Instrument Technique & Sensor*, vol. 28, no. 1, pp. 55–57, 2007.

[10] L. Gang, Y. Jiao, and S. H. Jia, "Application of Parameter Self-adaptive Fussy PID Control in Central-conditioner Saving-energy System," *Power Electronics*, vol. 41, no. 11, pp. 66-67, 2007.

[11] M. Koor, A. Vassiljev, and T. Koppel, "Optimal pump count prediction algorithm for identical pumps working in parallel mode," in *Proceedings of the 12th International Conference on Computing and Control for the Water Industry, CCWI 2013*, pp. 951–958, Italy, September 2013.

[12] L. Jian and L. Wang, "Application of the Fuzzy Prediction Based on PLC in Variable Water Flow System for Heat Pump Air-conditioning," *Electric Drive*, vol. 40, no. 5, pp. 53–55, 2010.

[13] Y. Zhang, Y. Liu, and Y. Liu, "A Hybrid Dynamical Modelling and Control Approach for Energy Saving of Central Air Conditioning," *Mathematical Problems in Engineering*, vol. 2018, Article ID 6389438, 12 pages, 2018.

[14] H. D. Wang, "Analysis and Comparison of the Control Methods of Variable Frequency Chilled Water Pumps in Air Conditioning Systems," *Construction Machinery for Hydraulic Engineering & Power Station*, 2004.

[15] J. M. Gordon and K. C. Ng, "Predictive and diagnostic aspects of a universal thermodynamic model for chillers," *International Journal of Heat and Mass Transfer*, vol. 38, no. 5, pp. 807–818, 1995.

[16] J. M. Gordon, K. C. Ng, and H. T. Chua, "Centrifugal chillers: Thermodynamic modelling and a diagnostic case study," *International Journal of Refrigeration*, vol. 18, no. 4, pp. 253–257, 1995.

[17] L. F. Rodrigues, *Dissertation Abstracts International*, Stanford University, 2002.

[18] L. Jinping, "Experimental research on energy saving effects of variable frequency chilled water system of air conditioning," *Heating Ventilating & Air Conditioning*, vol. 40, no. 3, pp. 68–72, 2010.

[19] Y. Sun, "Variable water flow control in air conditioning systems for energy efficiency," *Hv & Ac*, vol. 35, no. 10, pp. 90–92, 2001.

9

A Four-Dimensional Hyperchaotic Finance System and its Control Problems

Lin Cao

School of Business and Administration, Qilu University of Technology, Jinan 250353, China

Correspondence should be addressed to Lin Cao; caolinqlu@163.com

Academic Editor: Sundarapandian Vaidyanathan

The construction and several control problems of a new hyperchaotic finance system are investigated in this paper. Firstly, a new four-dimensional hyperchaotic finance system is introduced, based on which a new hyperchaos is then generated by setting parameters. And the qualitative analysis is numerically studied to confirm the dynamical processes, for example, the bifurcation diagram, Poincaré sections, Lyapunov exponents, and phase portraits. Interestingly, the obtained results show that this new system can display complex characteristics: chaotic, hyperchaotic, and quasiperiodic phenomena occur alternately versus parameters. Secondly, three single input adaptive controllers are designed to realize the control problems of such system: stabilization, synchronization, and coexistence of antisynchronization and complete synchronization, respectively. It is noted that the designed controllers are simpler than the existing ones. Finally, numerical simulations are provided to demonstrate the validity and the effectiveness of the proposed theoretical results.

1. Introduction

It is well known that the classical chaotic attractor was firstly found by Lorenz in 1963 [1]. As a most significant system in nonlinear dynamical systems, chaos systems and their relative problems have attracted a lot of consideration from all kinds of researchers in many fields of science, and many significant results have been obtained in the past few decades; for details, see [2–6] and the references therein. From then on, chaos becomes a hot topic for a broad class of applications in physics, electrical engineering, secure communication, and many other fields [4–6]. As we know that the positive Lyapunov exponent of the chaotic systems is a critical prerequisite in these applications. The three-dimensional chaotic system only has one positive Lyapunov exponent; thus it is typically difficult to meet some certain requirements. For instance, the messages which are masked by the simple chaotic systems are easy to decipher in secure communication [4, 5]. Different from the ordinary chaos, the hyperchaotic system has at least two positive Lyapunov exponents and thus has more prominent advantages due to its higher dimensions and more unpredictable behaviors. Accordingly, the hyperchaotic systems have been investigated extensively, and various control techniques and approaches have been developed and utilized [7–9]. It should be pointed out that the above-mentioned works are based on the hyperchaotic models with complex structure. However, such models are difficult to be verified in reality. This motivates the author to construct a simple system that can display rich dynamical characteristics.

It is well known that the OGY method [10] for chaos control was firstly observed in 1990. Meanwhile, the PC method [11] for chaos synchronization firstly achieved chaos synchronization in 1990. Since then, a lot of researchers who are from all kinds of scientific fields had to pay an increasing interest to investigate chaos systems and their relative problems in the past years, and many significant theoretical results and experiments conclusions have been published; for details please see [12–27] and the references therein. In recent years, several typical kinds of chaos synchronization have been identified, for example, complete synchronization (or synchronization) [14, 15], phase synchronization [16, 17], lag synchronization [18, 19], generalized synchronization [20, 21], antisynchronization [22, 23], projective synchronization [24, 25], coexistence of antisynchronization and complete synchronization [26, 27], and simultaneous synchronization and antisynchronization [27]. However, for the chaotic or hyperchaotic systems, there are still some important questions needed to be further investigated, such as the complexity of the controllers and the existence of some control problems, which partly motivates the present work.

For the economic systems, the chaotic behavior in those systems was first found in 1985 [28]. It is noted that chaotic behavior in economics implies that the considered economic system has an inherent indefiniteness. Thus, it is of important value to study the chaotic finance system in order to achieve a stable economic growth. A novel chaotic finance system [3] was presented in 2001. There are four subblocks which construct the system model, that is, money, production, labor force, and stock. This system model is described by three state variables of the time variations: x_1 stands for the interest rate; x_2 and x_3 are the investment demand and the price index, respectively. Since this chaotic finance system is proposed, many works have been done for this finance system [29, 30]. Recently, a novel hyperchaotic finance system [31] is presented in 2010 and some important results have been obtained. However, there are some limitations in the existing results. For example, in the stabilization problem and the complete synchronization problem of such system, the designed controllers are too complicated to be used in applications. On the other hand, the coexistence of antisynchronization and complete synchronization problem of this hyperchaotic finance system has not been investigated so far. Therefore, this new hyperchaotic finance system needs to be further investigated.

Motivated by the above discussions, a four-dimensional hyperchaotic finance system is presented, which can generate double-wing chaotic and hyperchaotic attractors with three equilibrium points. In comparison with the most existing results, this model has simple structure and can display complex dynamics: chaotic, hyperchaotic, and quasiperiodic phenomena occur alternately. Furthermore, some basic dynamic properties of the new hyperchaotic finance system regarding equilibria, dissipation, Lyapunov exponent, Lyapunov dimension, bifurcation diagram, and Poincaré sections are investigated. Then, three control problems: stabilization, synchronization, and coexistence of antisynchronization and complete synchronization are derived with simple yet physically implementable controllers. Finally, numerical simulations are provided to demonstrate effectiveness and the validity of the proposed theoretical results. In conclusion, the main contributions of this paper are given as follows:

(1) A new hyperchaotic finance system is firstly introduced, based on which a new hyperchaos is then generated by setting the parameters.

(2) Dynamic properties of the new generated hyperchaotic finance system are investigated extensively in Section 3, which are important and interesting. It should be pointed out that these dynamic properties results are different from the existing results.

(3) Three control problems of the hyperchaotic finance system are investigated extensively in Section 4. It is noted that the obtained results in this paper have some advantages over the exiting results. In particular, it should be pointed out that the coexistence of antisynchronization and complete synchronization in the two hyperchaotic finance systems is studied, which is a new result.

The rest of this paper is organized as follows. In Section 2, a new hyperchaotic model is introduced, based on which a new hyperchaos is generated by setting the parameters. In Section 3, the dynamic properties of such hyperchaotic finance system are investigated. In Section 4, several control problems of such hyperchaotic finance system are studied extensively, followed by the conclusions in Section 5.

2. Model Formulation

A dynamic model of finance has been reported in [29–31], which is composed of four subblocks, production, money, stock, and labor force, and expressed by four first-order differential equations. The model describes the time variations of four state variables: the interest rate x, the investment demand y, the price exponent z, and the average profit margin w. It is well known that the factors affecting the interest rates are related not only to investment demand and price index, but also to the average profit margin. And the average profit margin and interest rate are proportional. Some important results have been obtained. In the next, the four-dimensional hyperchaotic finance system is expressed as follows:

$$\begin{aligned}\dot{x} &= z + (y - a)x + w \\ \dot{y} &= 1 - by - x^2 \\ \dot{z} &= -x - cz \\ \dot{w} &= -dxy - kw,\end{aligned} \tag{1}$$

where $(x, y, z, w)^T \in R^4$ is the state vector, a is the saving, b is the per investment cost, c is the elasticity of demands of commercials, and a, b, c, d, k are positive constant parameters.

In order to generate a new hyperchaos, we set $a = 0.9$, $b = 0.2$, $c = 1.5$, $d = 0.2$, and select k as a governing parameter according to the two criterions in [7]. The dynamical properties of the hyperchaotic system (1) are firstly studied in Section 3, including symmetry, dissipation, equilibrium point, Lyapunov exponent, Lyapunov dimension, bifurcation diagram, and Poincaré sections. Then, three control problems of such system are investigated in Section 4, that is, stabilization, complete synchronization, and coexistence of antisynchronization and complete synchronization.

3. Dynamic Properties

3.1. Symmetry. Since the hyperchaotic system (1) is invariant under the coordinate transformation: $(x, y, z, w) \rightarrow (-x, y, -z, -w)$, the system (1) is symmetric with respect to z-axis, which implies that all values of such system parameters are under reflection about y-axis.

3.2. Dissipation. The divergence of this four-dimensional hyperchaotic system (1) is described as

$$\nabla V = \frac{\partial \dot{x}}{\partial x} + \frac{\partial \dot{y}}{\partial y} + \frac{\partial \dot{z}}{\partial z} + \frac{\partial \dot{w}}{\partial w} = -(a+b+c+k)$$
$$= -(2.8+k) = -u. \quad (2)$$

When $u > 0$, that is, $k > -2.8$, thus the system (1) is a dissipative system. It results in that volume element $V_0e^{-ut} \to 0$ as $t \to \infty$. Therefore, all the trajectories of this four-dimensional hyperchaotic system (1) are ultimately embedded into an chaotic attractor.

3.3. Equilibria. The equilibria of the system (1) are found by solving the following nonlinear equations:

$$\begin{aligned} z + (y-a)x + w &= 0, \\ 1 - by - x^2 &= 0, \\ -x - cz &= 0, \\ -dxy - kw &= 0. \end{aligned} \quad (3)$$

After computation, it results in

$$\begin{aligned} E_1^* &= \left(0, \frac{1}{b}, 0, 0\right), \\ E_2^* &= \left(\frac{\sqrt{\Delta_1}}{\sqrt{\Delta_2}}, -\frac{\Delta_3}{\Delta_2}, -\frac{\Delta_1}{c\sqrt{\Delta_2}}, -\frac{d(1+ac)\sqrt{\Delta_1}}{\Delta_2}\right), \\ E_3^* &= \left(-\frac{\sqrt{\Delta_1}}{\sqrt{\Delta_2}}, -\frac{\Delta_3}{\Delta_2}, \frac{\Delta_1}{c\sqrt{\Delta_2}}, \frac{d(1+ac)\sqrt{\Delta_1}}{\Delta_2}\right), \end{aligned} \quad (4)$$

where $\Delta_1 = -cd-bk+ck-abck$, $\Delta_2 = -cd+ck$, $m = -k-ack$. As an example, set $k = 0.17$, the equilibria are given as follows:

$$\begin{aligned} E_1^* &= (0, 5, 0, 0), \\ E_2^* &= (1.666, -8.87778, -1.11067, 17.4004), \\ E_3^* &= (-1.666, -8.87778, 1.11067, -17.4004). \end{aligned} \quad (5)$$

Linearizing the system (1), it results in

$$J = \begin{pmatrix} y-a & x & 1 & 1 \\ 2x & -b & 0 & 0 \\ -1 & 0 & -c & 0 \\ -dy & -dx_1 & 0 & -k \end{pmatrix}. \quad (6)$$

For $J_1 = J(E_1^*)$, the eigenvalues can be calculated as $\lambda_1 = 4.1544$, $\lambda_2 = -0.4480$, $\lambda_3 = -1.2764$, $\lambda_4 = -0.2000$, which implies that E_1^* is a saddle-focus, thus it is unstable. About $J_2 = J(E_2^*)$ and $J_3 = J(E_3^*)$, the eigenvalues are $\lambda_1 = -10.0559$, $\lambda_2 = -1.5869$, $\lambda_3 = -0.0025 + 0.2871i$, $\lambda_4 = -0.0025 - 0.2871i$, $\lambda_1 = -10.0559$, $\lambda_2 = -1.5869$, $\lambda_3 = -0.0025+0.2871i$, $\lambda_4 = -0.0025-0.2871i$, respectively, which imply that E_2^* and E_3^* are stable points.

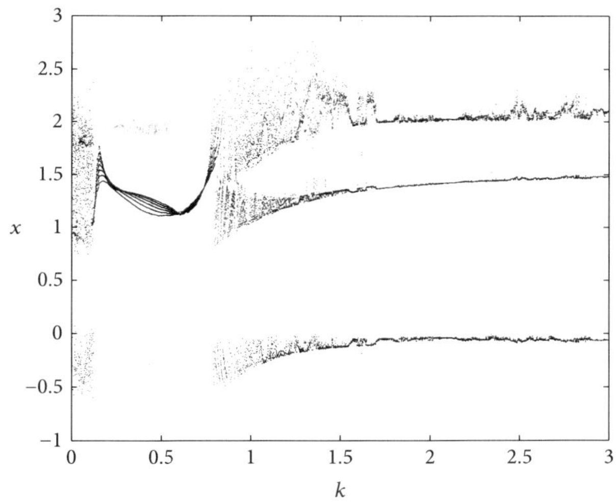

FIGURE 1: Bifurcation diagram of the system (1) with initial conditions (1, 2, 0.5, 0.5) versus parameter k.

3.4. Bifurcation Diagram. Bifurcation diagram is a useful method to show the dynamical processes of a system with respect to a parameter. In this subsection, the parameter k is chosen as a control parameter. Figure 1 displays the bifurcation diagram of the variable x versus the parameter k which is from 0 to 3. It is interesting that the evolution procedure of the system (1) through chaotic, hyperchaotic, and periodic orbit, as well as period doubling route to chaos, which displays complex dynamic properties with the increasing of the parament k. In order to show the dynamics of the system (1) clearly, we test the time course of the system (1) by selecting several parameters k. It can be seen from Figure 2, the state trajectory of the system (1) shows chaos and quasperiodic routes to chaos with the increase of parameter k, which is consistent with above results.

3.5. Lyapunov Exponent and Poincaré Sections. For the chaotic systems, Lyapunov exponent is not only an important index to distinguish chaotic attractor, but also a quantitative description of the sensitivity about the initial values. In addition, Poincaré section is also a critical approach to analyze chaotic system; for example, if there exists a closed curve, the movement is quasicycle; if there only exists one fixed point or a few discrete points, the movement is cycle; and if there exists a collection of distribution points along a segment of a line or a curve, the movement is chaos. Here, the Lyapunov exponent and Poincaré sections of the system (1) with $k = 3$ are investigated, respectively.

Figure 3 shows the Lyapunov exponent diagram of the hyperchaotic finance system (1). It can be seen that there exist always two positive Lyapunov exponents, which means that the proposed finance system (1) is hyperchaotic. Meanwhile,

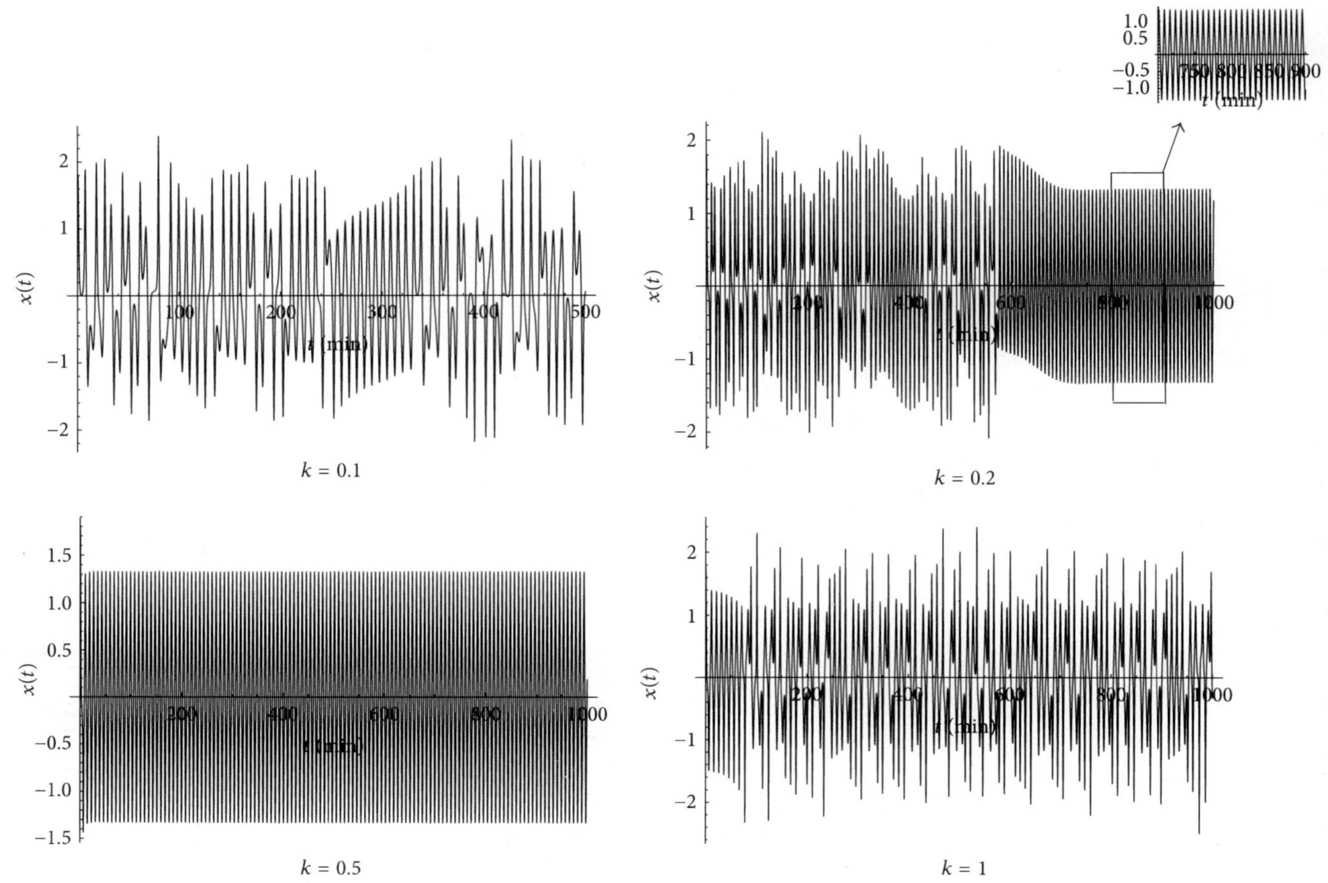

FIGURE 2: Time course of the system (1) with several fixed k values.

the Poincaré sections in various planes verify the hyperchaotic behavior of the system (1) as shown in Figure 4. With the following initial conditions: $x(0) = 5$, $y(0) = 2$, $z(0) = -6$, $w(0) = 4$, Figure 5 describes the 3D projection of the strange attractor of the system (1) on (x, y, z)-space, (x, y, w)-space, (y, z, w)-space, and (x, w, z)-space, respectively. In the next, the time course of x, y, z, w is shown with time t with different initial conditions: $x(0) = 1$, $y(0) = 2$, $z(0) = 0.5$, $w(0) = 0.5$, and $x(0) = 1.01$, $y(0) = 2$, $z(0) = 0.5$, $w(0) = 0.5$, respectively. It can be seen, from Figure 6, the motions of the system (1) do not remain similar for long when they are started with very similar initial conditions, which implies that this system is extraordinarily sensitive to initial values.

3.6. Lyapunov Dimension. By Wolf algorithm, the Lyapunov exponents of this four-dimensional hyperchaotic finance system (1) are presented as follows:

$$\begin{aligned} L_1 &= 0.034432, \\ L_2 &= 0.018041, \\ L_3 &= 0, \\ L_4 &= -1.1499. \end{aligned} \tag{7}$$

In addition, the Lyapunov dimension of the system (1) is derived:

$$\begin{aligned} D_L &= j + \frac{1}{\left|L_{j+1}\right|}\sum_{i=1}^{j} L_i = 3 + \frac{0.034432 + 0.018041}{1.1499} \\ &= 3.045633 \end{aligned} \tag{8}$$

which is fractional.

4. Three Control Problems of the New Hyperchaotic Finance System

In this section, three control problems, stabilization, complete synchronization, and coexistence of antisynchronization and complete synchronization, of the new hyperchaotic finance system (1) are investigated extensively, and some new results are presented.

4.1. Stabilization of the New Hyperchaotic Finance System by a Single Input Controller. For convenience, let $X = (x_1, x_2, x_3, x_4)^T = (x, y, z, w)^T$; then the system (1) is rewritten

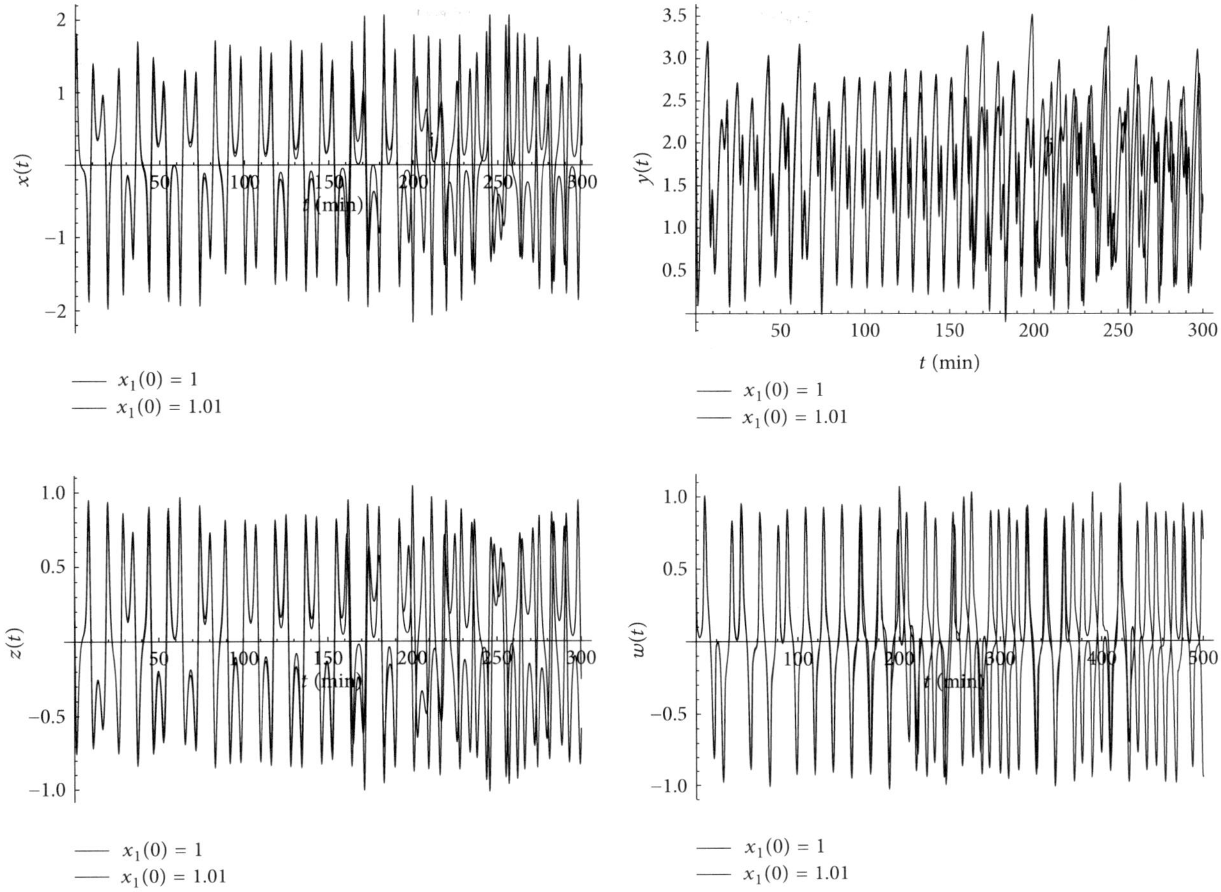

FIGURE 3: Lyapunov exponent of system (1) with varying time t.

as

$$\begin{aligned} \dot{x}_1 &= f_1(X) = x_3 + (x_2 - a)\, x_1 + x_4 \\ \dot{x}_2 &= f_2(X) = 1 - bx_2 - x_1^2 \\ \dot{x}_3 &= f_3(X) = -x_1 - cx_3 \\ \dot{x}_4 &= f_4(X) = -dx_1x_2 - kx_4, \end{aligned} \tag{9}$$

where

$$\begin{aligned} a &= 0.9, \\ b &= 0.2, \\ c &= 1.5, \\ d &= 0.2, \\ k &= 0.17. \end{aligned} \tag{10}$$

Remark 1. Since system (9) is hyperchaotic, there exist four positive constants: λ_i, $i = 1, 2, 3, 4$, satisfying

$$x_i f_i(X) \le \lambda_i x_i^2, \quad i = 1, 2, 3, 4. \tag{11}$$

According to Section 3.3, $E_1^* = (0, 5, 0, 0)$ is an equilibrium point of the hyperchaotic finance system (9). Next, the stabilization problem of system (9) will be studied, that is, designing a physically implementable controller u to force the states of such system to the equilibrium point E_1^*.

Consider system (9). Obviously, if $x_1 = 0$, then the following subsystem

$$\begin{aligned} \dot{x}_2 &= 1 - bx_2 \\ \dot{x}_3 &= -cx_3 \\ \dot{x}_4 &= -kx_4 \end{aligned} \tag{12}$$

is globally asymptotically stable (GAS) with respect to E_1^*.

Thus, the controller is designed as

$$u = \begin{pmatrix} u_1 \\ u_2 \\ u_3 \\ u_4 \end{pmatrix} = \begin{pmatrix} k_1 x_1 \\ 0 \\ 0 \\ 0 \end{pmatrix}, \tag{13}$$

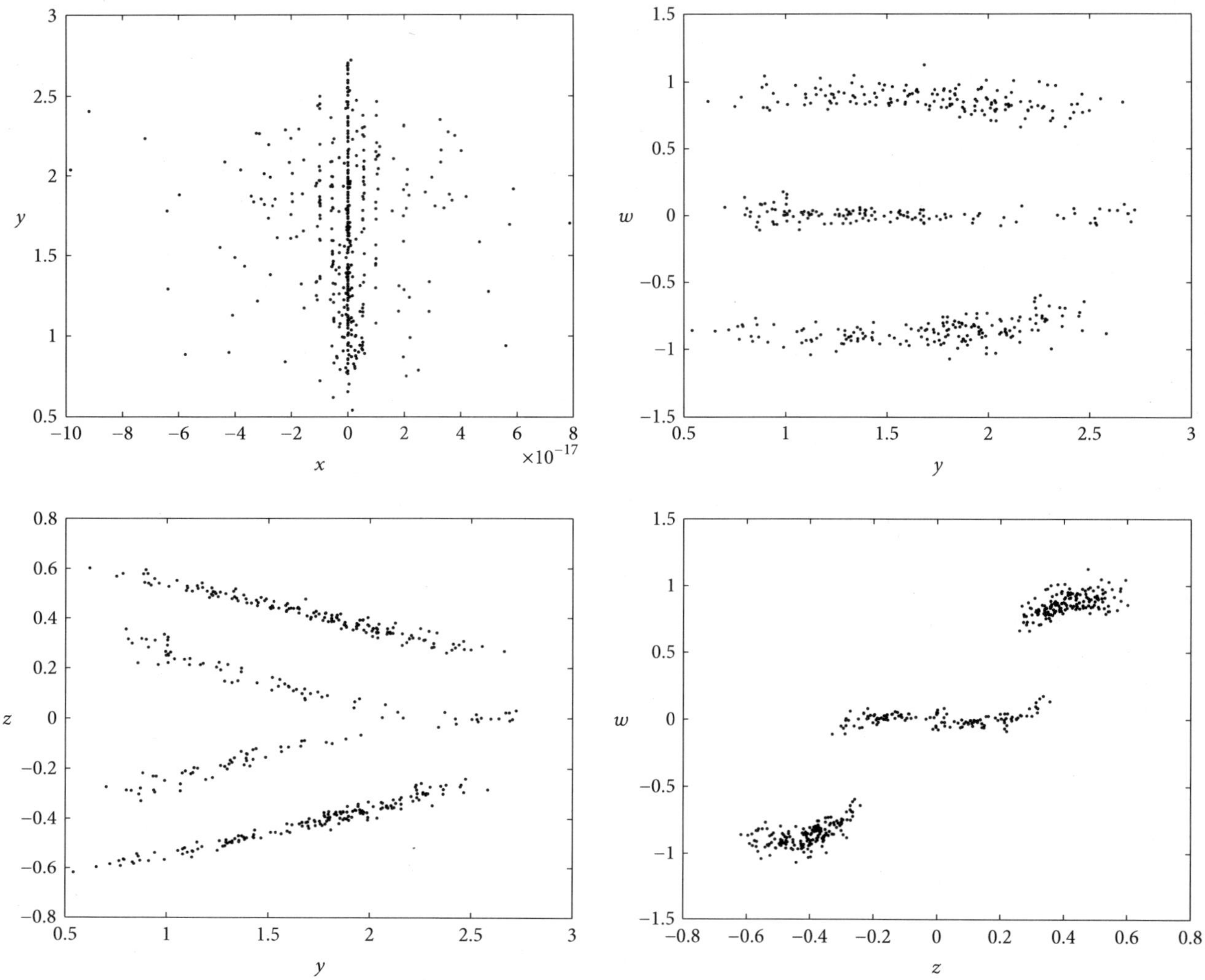

FIGURE 4: Poincaré sections of system (1).

and controlled system (9) is presented as

$$\begin{aligned}\dot{x}_1 &= f_1(X) + u_1 = x_3 + (x_2 - a)x_1 + x_4 + k_1 x_1\\ \dot{x}_2 &= f_2(X) + u_2 = 1 - bx_2 - x_1^2\\ \dot{x}_3 &= f_3(X) + u_3 = -x_1 - cx_3\\ \dot{x}_4 &= f_4(X) + u_4 = -dx_1x_2 - kx_4,\end{aligned} \tag{14}$$

where the dynamic gain k_1 is evolved by the following law:

$$\dot{k}_1 = -\gamma x_1^2, \tag{15}$$

where $\gamma > 0$ which is chosen in advance.

Remark 2. In [30], the obtained controller is given as

$$u = \begin{pmatrix} u_1 \\ u_2 \\ u_3 \\ u_4 \end{pmatrix} = \begin{pmatrix} -k_2x_1 \\ -k_2y_1 \\ 0 \\ -k_2z_1 \end{pmatrix}. \tag{16}$$

Obviously, controller (16) is more complicated than controller (13).

For convenience, systems (14) and (15) are called the auxiliary system, and a candidate Lyapunov function is introduced as follows:

$$V(x, k_1) = \frac{1}{2}X^TX + \frac{1}{2\gamma}(k_1 + L_1)^2, \tag{17}$$

where

$$L_1 > M_1 \sup_{x_1 \neq 0} \frac{X^TX}{x_1^2}, \quad M_1 = \max_{i=1}^{4} \lambda_i. \tag{18}$$

Then, the following conclusion is presented.

Theorem 3. *For any initial values, the orbits $X(t)$ of the system (14) converge to the equilibrium point E_1^* as $t \to \infty$; that is, the stabilization of system (9) is achieved by the above controller (13).*

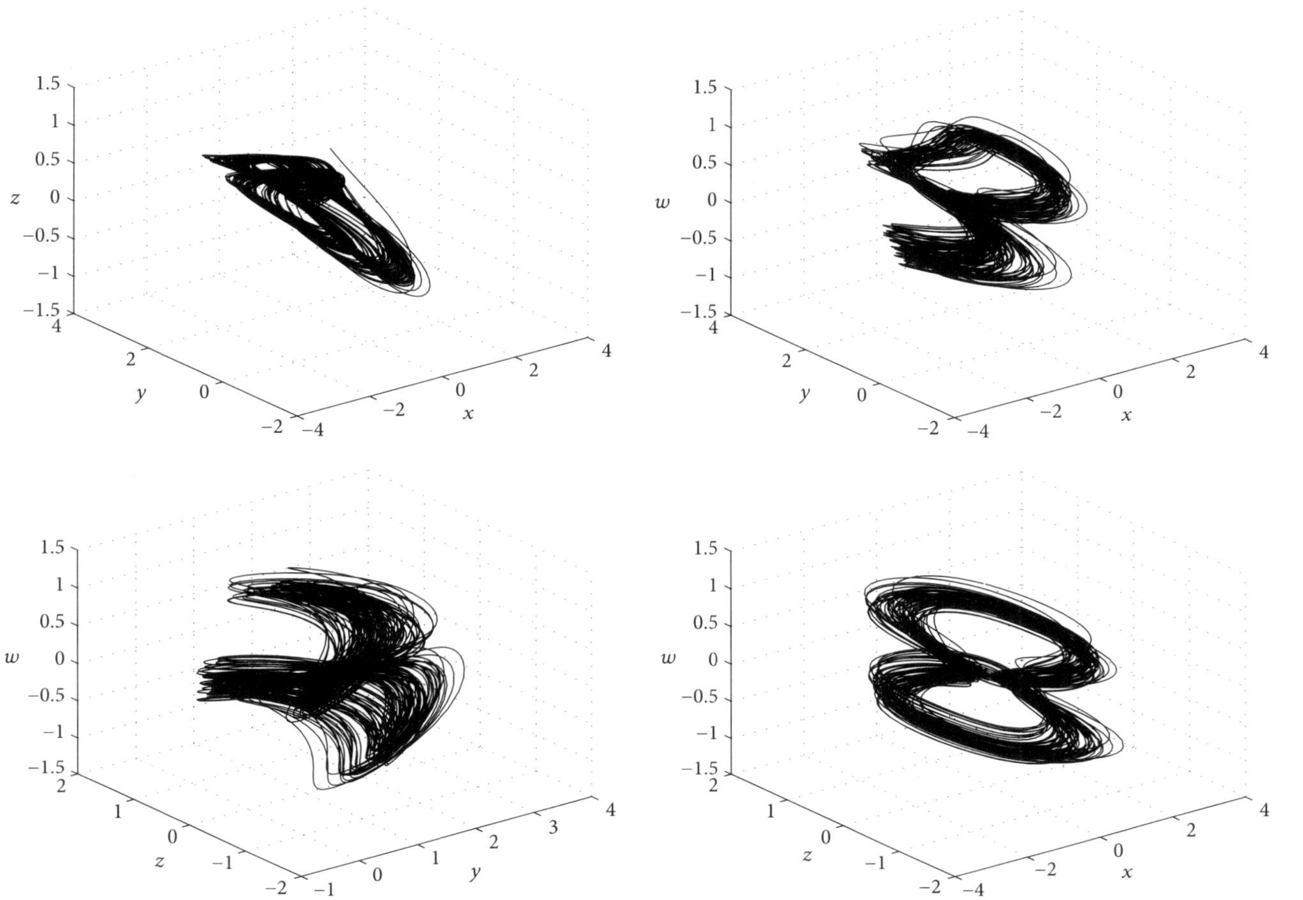

FIGURE 5: 3D projection of the strange attractor of system (1) on (x, y, z)-space, (x, y, w)-space, (y, z, w)-space, and (x, w, z)-space, respectively.

Proof. The time derivative of V with respect to the auxiliary system is given as follows:

$$\begin{aligned}\dot{V} &= X^T\dot{X} + \frac{1}{\gamma}(k_1 + L_1)\dot{k}_1 \\ &= x_1(f_1(X) + k_1x_1) + \sum_{i=2}^{4} x_i f_i(X) - (k_1 + L_1)x_1^2 \\ &= \sum_{i=1}^{4} x_i f_i(X) - L_1x_1^2 \le M_1 X^T X - L_1 x_1^2 < 0.\end{aligned} \quad (19)$$

Thus the conclusion is proved, which completes the proof. □

Numerical simulation is carried out with the following initial conditions: $x_1(0) = 1$, $x_2(0) = 2$, $x_3(0) = -4$, $x_4(0) = -2$ and $k_1(0) = -1$, $\gamma = 1$. Figure 7 displays that system (9) is stabilized to its equilibrium point E_1^*.

4.2. Complete Synchronization of the Two Hyperchaotic Finance Systems by a Single Input Controller. In this subsection, the complete synchronization of two hyperchaotic finance systems with different initial conditions are investigated, and some new results are presented in the following.

Make system (9) the master system; then the corresponding uncontrolled (i.e., $u = 0$) slave system with variable Y being given as

$$\begin{aligned}\dot{y}_1 &= f_1(Y) = y_3 + (y_2 - a)y_1 + y_4 \\ \dot{y}_2 &= f_2(Y) = 1 - by_2 - y_1^2 \\ \dot{y}_3 &= f_3(Y) = -y_1 - cy_3 \\ \dot{y}_4 &= f_4(Y) = -dy_1y_2 - ky_4,\end{aligned} \quad (20)$$

where $Y = (y_1, y_2, y_3, y_4)^T$ is the state vector and a, b, c, d, k are given as (10).

Let $e = (e_1, e_2, e_3, e_4)^T = Y - X$; then the uncontrolled (i.e., $u = 0$) error system is described as

$$\begin{aligned}\dot{e}_1 &= F_1(X, e) = -ae_1 + x_1e_2 + x_2e_1 + e_1e_2 + e_3 + e_4 \\ \dot{e}_2 &= F_2(X, e) = -be_2 - 2x_1e_1 - e_1^2 \\ \dot{e}_3 &= F_3(X, e) = -e_1 - ce_3 \\ \dot{e}_4 &= F_4(X, e) = -d(e_1x_2 + e_2x_1 + e_1e_2) - ke_4.\end{aligned} \quad (21)$$

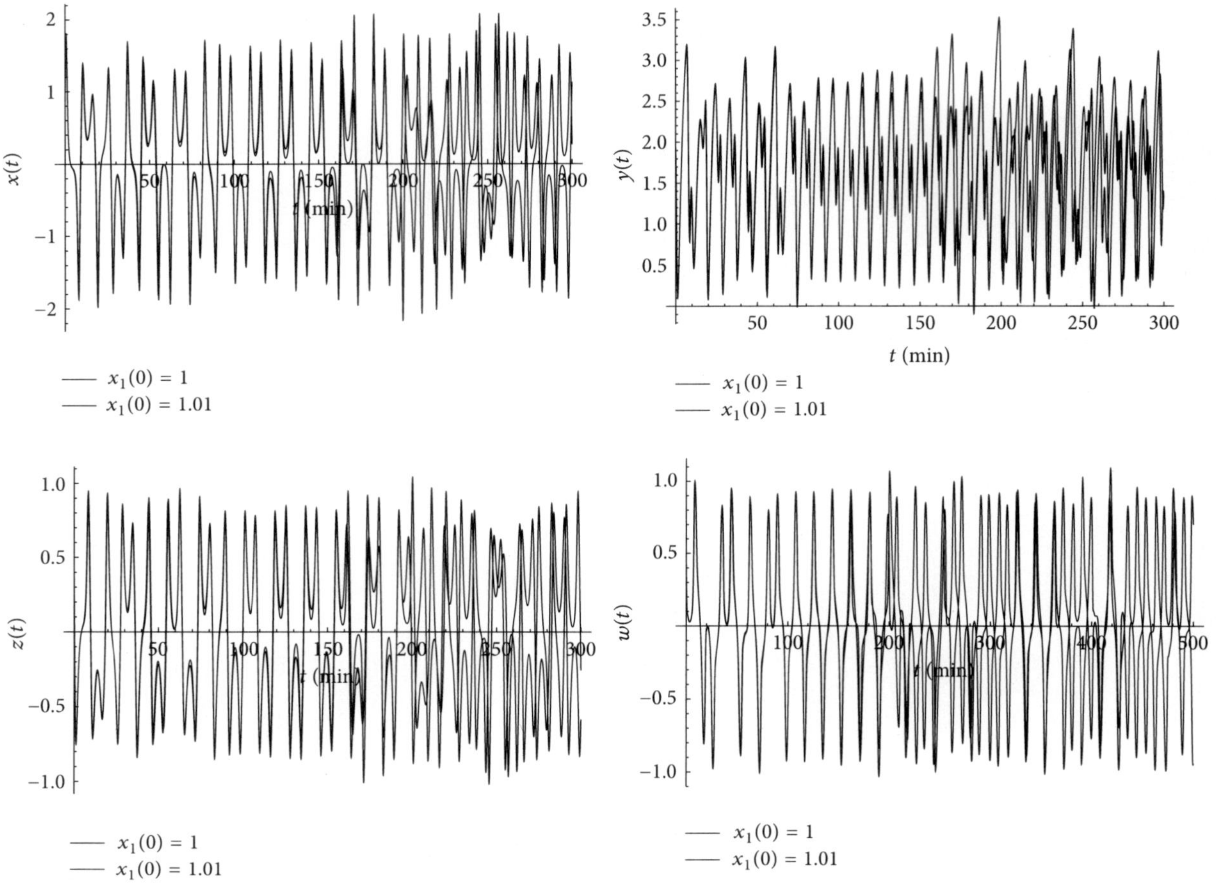

FIGURE 6: Time course of x, y, z, w of system (1) with similar initial values.

Remark 4. Since system (9) is hyperchaotic, there exist four positive constants: l_i, $i = 1, 2, 3, 4$, such that

$$e_i F_i (X, e) \leq l_i e_i^2, \quad i = 1, 2, 3, 4. \tag{22}$$

Consider system (21). Obviously, if $e_1 = 0$, then the following subsystem

$$\begin{aligned}\dot{e}_2 &= -be_2\\ \dot{e}_3 &= -ce_3\\ \dot{e}_4 &= -de_2 x_1 - ke_4\end{aligned} \tag{23}$$

is GAS with respect to origin.

Thus, the designed controller is given as

$$u = \begin{pmatrix} u_1 \\ u_2 \\ u_3 \\ u_4 \end{pmatrix} = \begin{pmatrix} k_2 e_1 \\ 0 \\ 0 \\ 0 \end{pmatrix}, \tag{24}$$

and the controlled error system is described as

$$\begin{aligned}\dot{e}_1 &= F_1 (X, e) + u_1\\ &= -ae_1 + x_1 e_2 + x_2 e_1 + e_1 e_2 + e_3 + e_4 + k_2 e_2\\ \dot{e}_2 &= F_2 (X, e) + u_2 = -be_2 - 2x_1 e_1 - e_1^2\\ \dot{e}_3 &= F_3 (X, e) + u_3 = -e_1 - ce_3\\ \dot{e}_4 &= F_4 (X, e) + u_4 = -d (e_1 x_2 + e_2 x_1 + e_1 e_2) - ke_4,\end{aligned} \tag{25}$$

where the dynamic gain k_2 is evolved by the following law:

$$\dot{k}_2 = -\gamma e_1^2, \tag{26}$$

where $\gamma > 0$ which is chosen in advance.

Similarly, systems (25) and (26) are called the auxiliary system, and a candidate Lyapunov function is presented as follows:

$$V (e, k_2) = \frac{1}{2} e^T e + \frac{1}{2\gamma} (k_2 + L_2)^2, \tag{27}$$

where

$$L_2 > M_2 \sup_{e_1 \neq 0} \frac{e^T e}{e_1^2}, \quad M_2 = \max_{i=1}^{4} l_i. \tag{28}$$

Next, a conclusion is obtained.

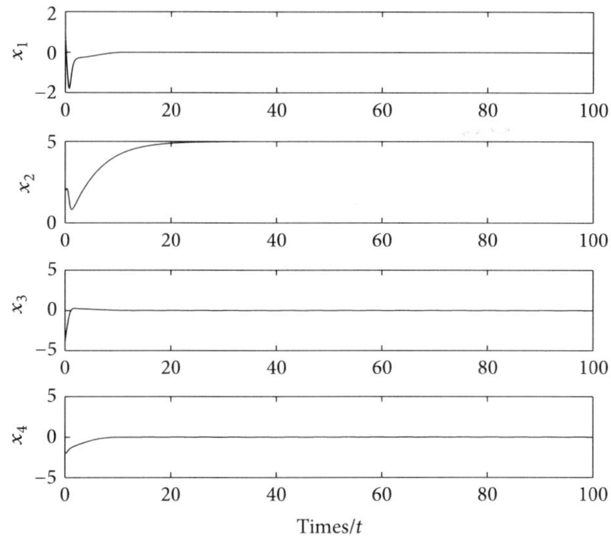

FIGURE 7: The figure shows that the orbit $X(t)$ of the system (14) converges to $(0, 5, 0, 0)^T$ as $t \to \infty$.

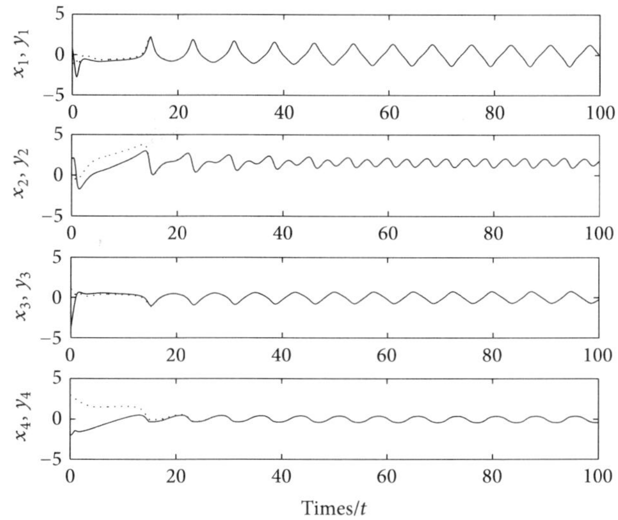

FIGURE 8: The figure shows that the states, x_1, x_2, x_3, x_4, synchronize the states, y_1, y_2, y_3, y_4, respectively.

Theorem 5. *For any initial values, the orbits $e(t)$ of the error system (25) converge to origin as $t \to \infty$, that is, master system (9) synchronizes slave system (20) by the above controller (24).*

Proof. The time derivative of V with respect to the auxiliary system is given as follows:

$$\begin{aligned} \dot{V} &= e^T \dot{e} + \frac{1}{\gamma}(k_2 + L_2)\dot{k}_2 \\ &= e_1 (F_1(X,e) + k_2 e_1) + \sum_{i=2}^{4} e_i F_i(X,e) \\ &\quad - (k_2 + L_2) e_1^2 = \sum_{i=1}^{4} e_i F_i(X,e) - L_2 e_1^2 \\ &\le M_2 e^T e - L_2 e_1^2 < 0. \end{aligned} \tag{29}$$

Thus the result is obtained, which completes the proof. □

Similarly, numerical simulation is carried out with the following initial conditions: $x_1(0) = 1$, $x_2(0) = 2$, $x_3(0) = -4$, $x_4(0) = -1$, $y_1(0) = -2$, $y_2(0) = -1$, $y_3(0) = 1$, $y_4(0) = 1$, and $k_1(0) = -1$, $\gamma = 1$. Figure 8 shows that the complete synchronization of two hyperchaotic finance systems with different initial conditions being achieved by the above controller (24).

4.3. Coexistence of Antisynchronization and Complete Synchronization in the Two Hyperchaotic Finance Systems. In this section, the coexistence of antisynchronization and complete synchronization in the two hyperchaotic finance systems is investigated, and some new results are obtained in the following.

According to the results in [26], we should set

$$\begin{aligned} E_1 &= y_1 + x_1, \\ e_2 &= y_2 - x_2, \\ E_3 &= y_3 + x_3, \\ E_4 &= y_4 + x_4. \end{aligned} \tag{30}$$

Then, the uncontrolled (i.e., $u = 0$) combination (including the sum E and the error e_2) system is described as follows:

$$\begin{aligned} \dot{E}_1 &= G_1(X, E, e_2) \\ &= -aE_1 + E_3 + E_1 e_2 + x_2 E_1 - x_1 e_2 + E_4 \\ \dot{e}_2 &= G_2(X, E, e_2) = -be_2 - E_1^2 + 2x_1 E_1 \\ \dot{E}_3 &= G_3(X, E, e_2) = -E_1 - cE_3 \\ \dot{E}_3 &= G_4(X, E, e_2) = -d(E_1 e_2 + x_2 E_1 - x_1 e_2) - kE_4, \end{aligned} \tag{31}$$

where $E = (E_1, E_3, E_4)^T$.

Remark 6. Since system (9) is hyperchaotic, there exist four positive constants: n_i, $i = 1, 2, 3, 4$ satisfying

$$E_i G_i(X, E, e_2) \le n_i E_i^2, \quad i = 1, 3, 4, \quad e_2 G_2(X, E, e_2) \le n_2 e_2^2. \tag{32}$$

Consider the combination system (31). Obviously, if $E_1 = 0$, then the following subsystem

$$\begin{aligned} \dot{e}_2 &= -be_2 \\ \dot{E}_3 &= -E_1 - cE_3 \\ \dot{E}_4 &= -2x_1 d e_2 - kE_4 \end{aligned} \tag{33}$$

is GAS with respect to origin.

Therefore, the designed controller is expressed as

$$u = \begin{pmatrix} u_1 \\ u_2 \\ u_3 \\ u_4 \end{pmatrix} = \begin{pmatrix} k_3 E_1 \\ 0 \\ 0 \\ 0 \end{pmatrix}, \tag{34}$$

and the controlled combination system is presented as follows:

$$\begin{aligned} \dot{E}_1 &= G_1 (X, E, e_2) \\ &= -aE_1 + E_3 + E_1 e_2 + x_2 E_1 - x_1 e_2 + E_4 + k_3 E_1 \\ \dot{e}_2 &= G_2 (X, E, e_2) = -be_2 - E_1^2 + 2x_1 E_1 \\ \dot{E}_3 &= G_3 (X, E, e_2) = -E_1 - cE_3 \\ \dot{E}_3 &= G_4 (X, E, e_2) = -d (E_1 e_2 + x_2 E_1 - x_1 e_2) - kE_4, \end{aligned} \tag{35}$$

where the dynamic gain k_3 is evolved by the following law:

$$\dot{k}_3 = -\gamma E_1^2, \tag{36}$$

where $\gamma > 0$ which is chosen in advance.

Similarly, systems (35) and (36) are called the auxiliary system, and a candidate Lyapunov function is introduced as follows:

$$V (E, e_2, k_3) = \frac{1}{2} \left(e_2^2 + E^T E \right) + \frac{1}{2\gamma} (k_3 + L_3)^2, \tag{37}$$

where

$$L_3 > M_3 \sup_{E_1 \neq 0} \frac{e_2^2 + E^T E}{E_1^2}, \quad M_3 = \max_{i=1}^{4} n_i. \tag{38}$$

Next, a conclusion is presented as follows.

Theorem 7. *For any initial values, the orbits $(E(t), e_2(t))^T$ of the combination system (35) converge to origin as $t \to \infty$, which implies that the coexistence of antisynchronization and complete synchronization in the two new hyperchaotic finance systems is achieved by the above controller (34), that is, states x_1, x_3, x_4 antisynchronize states y_1, y_3, y_4, while rest state x_2 synchronizes state y_2, respectively.*

Proof. The time derivative of V with respect to the auxiliary system is given as follows:

$$\begin{aligned} \dot{V} &= E^T \dot{E} + e_2 \dot{e}_2 + \frac{1}{\gamma} (k_3 + L_3) \dot{k}_3 \\ &= \sum_{i=1,\neq 2}^{4} E_i G_i (X, E, e_2) + e_2 G_2 (X, E, e_2) - L_3 E_1^2 \\ &\leq M_3 \left(E^T E + e_2^2 \right) - L_3 E_1^2 < 0. \end{aligned} \tag{39}$$

Thus the conclusion holds, which completes the proof. □

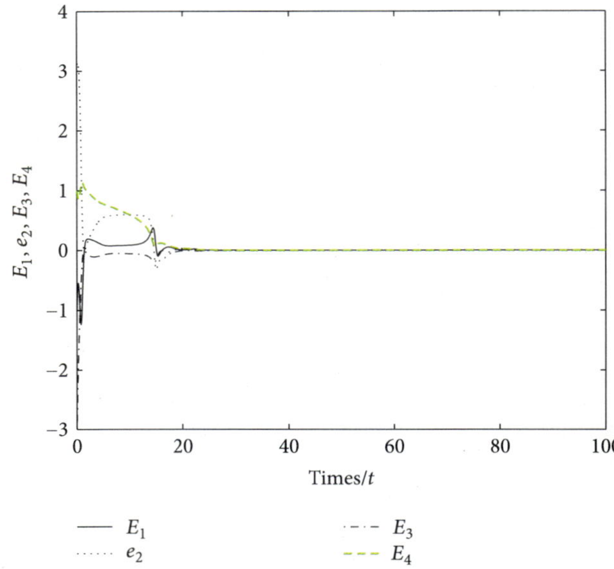

FIGURE 9: The figure shows that the orbits $(E(t), e_2(t))^T$ of the combination system (35) converge to origin as $t \to \infty$.

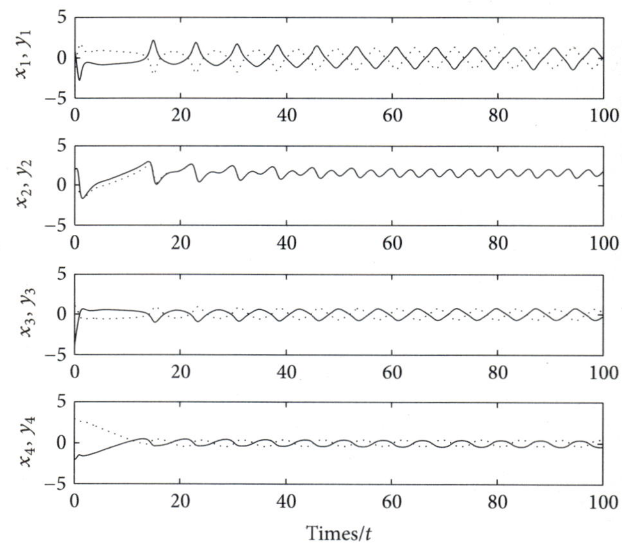

FIGURE 10: The figure shows that states x_1, x_3, x_4 antisynchronize states y_1, y_3, y_4, while state x_2 synchronizes state y_2, respectively.

Similarly, numerical simulation is carried out with the following initial conditions: $x_1(0) = 1$, $x_2(0) = 2$, $x_3(0) = -4$, $x_4(0) = -1$, $y_1(0) = -2$, $y_2(0) = -1$, $y_3(0) = 1$, $y_4(0) = 1$, and $k_1(0) = -1$, $\gamma = 1$. Figure 9 displays that the orbits $(E(t), e_2(t))^T$ of the combination system (35) converge to origin as $t \to \infty$, which implies that the coexistence of antisynchronization and complete synchronization in the two new hyperchaotic finance systems is achieved by the above controller (34). Figure 10 shows that the states x_1, x_3, x_4 antisynchronize the states y_1, y_3, y_4, while state x_2 synchronizes state y_2, respectively.

5. Conclusions

In this paper, a new four-dimensional hyperchaotic finance system has been introduced, based on which a new hyperchaos has been generated. This system has simple algebraic structure and can display complex characteristics: chaotic, hyperchaotic, and quasiperiodic phenomena occur alternately versus parameters. Then, the dynamic properties including symmetry, dissipation, equilibrium point, Lyapunov exponent, bifurcation diagram, and Poincaré sections are studied. Furthermore, three control problems of such system have been achieved by three single input controllers, respectively. Compared to the existing ones, the designed controllers in this paper are simpler. Finally, the validity and the effectiveness of the obtained theoretical results have been verified by the provided numerical simulations.

Conflicts of Interest

The author declares that there are no financial and personal relationships with other people or organizations that can inappropriately influence this work; there is no professional or other personal interest of any nature or kind in any product, service, and/or company that could be construed as influencing the position presented in, or the review of, this article. The author declares that there are no conflicts of interest regarding the publication of this paper.

Acknowledgments

The author would like acknowledge the School of Business Administration at Andrews University for providing wonderful research environment. In particular, the help of Dr. Ralph Trecartin, Dr. Jerry Chi, and the colleagues during the current research is highly appreciated. This paper is supported by the Social Science Foundation of Shandong Province (17CLYJ43).

References

[1] E. N. Lorenz, "Deterministic nonperiodic flow," *Journal of the Atmospheric Sciences*, vol. 20, no. 2, pp. 130–141, 1963.

[2] O. E. Rössler, "An equation for continuous chaos," *Physics Letters A*, vol. 57, no. 5, pp. 397-398, 1976.

[3] J. H. Ma and Y. S. Chen, "study for the bifurcation topological structure and the global complicated character of a kind of nonlinear finance system (I)," *Applied Mathematics and Mechanics*, vol. 22, no. 11, pp. 1240–1251, 2001.

[4] H.-P. Ren, M. S. Baptista, and C. Grebogi, "Wireless communication with chaos," *Physical Review Letters*, vol. 110, no. 18, Article ID 184101, 2013.

[5] G. Bhatnagar and Q. M. Jonathan Wu, "A novel chaos-based secure transmission of biometric data," *Neurocomputing*, vol. 147, no. 1, pp. 444–455, 2015.

[6] L. Liu and R. Guo, "Control problems of Chen–Lee system by adaptive control method," *Nonlinear Dynamics*, vol. 87, no. 1, pp. 503–510, 2017.

[7] O. E. Rossler, "An equation for hyperchaos," *Physics Letters A*, vol. 71, no. 2-3, pp. 155–157, 1979.

[8] S. Mukherjee, S. K. Palit, S. Banerjee, A. W. A. Wahab, M. R. K. Ariffin, and D. K. Bhattacharya, "Computing two dimensional Poincaré maps for hyperchaotic dynamics," *Applied Mathematics and Computation*, vol. 301, pp. 140–154, 2017.

[9] A. Mohammadzadeh and S. Ghaemi, "Synchronization of uncertain fractional-order hyperchaotic systems by using a new self-evolving non-singleton type-2 fuzzy neural network and its application to secure communication," *Nonlinear Dynamics*, vol. 88, no. 1, 2017.

[10] E. Ott, C. Grebogi, and J. A. Yorke, "Controlling chaos," *Physical Review Letters*, vol. 64, no. 11, pp. 1196–1199, 1990.

[11] L. M. Pecora and T. L. Carroll, "Synchronization in chaotic systems," *Physical Review Letters*, vol. 64, no. 8, pp. 821–824, 1990.

[12] D. Auerbach, C. Grebogi, E. Ott, and J. A. Yorke, "Controlling chaos in high-dimensional systems," *Physical Review Letters*, vol. 69, no. 24, pp. 3479–3482, 1992.

[13] J. Sieber, O. E. Omel'Chenko, and M. Wolfrum, "Controlling unstable chaos: stabilizing chimera states by feedback," *Physical Review Letters*, vol. 112, Article ID 054102, 2014.

[14] D. Huang, "Simple adaptive-feedback controller for identical chaos synchronization," *Physical Review E: Statistical, Nonlinear, and Soft Matter Physics*, vol. 71, Article ID 037203, 2005.

[15] R. Guo, "A simple adaptive controller for chaos and hyperchaos synchronization," *Physics Letters A*, vol. 372, no. 34, pp. 5593–5597, 2008.

[16] M. G. Rosenblum, A. S. Pikovsky, and J. Kurths, "Phase synchronization of chaotic oscillators," *Physical Review Letters*, vol. 76, no. 11, pp. 1804–1807, 1996.

[17] A. S. Pikovsky, M. G. Rosenblum, G. V. Osipov, and J. Kurths, "Phase synchronization of chaotic oscillators by external driving," *Physica D: Nonlinear Phenomena*, vol. 104, no. 3-4, pp. 219–238, 1997.

[18] M. G. Rosenblum, A. S. Pikovsky, and J. Kurths, "From phase to lag synchronization in coupled chaotic oscillators," *Physical Review Letters*, vol. 78, no. 22, pp. 4193–4196, 1997.

[19] S. Taherion and Y.-C. Lai, "Observability of lag synchronization of coupled chaotic oscillators," *Physical Review E: Statistical Physics, Plasmas, Fluids, and Related Interdisciplinary Topics*, vol. 59, no. 6, pp. R6247–R6250, 1999.

[20] N. F. Rulkov, M. M. Sushchik, L. S. Tsimring, and H. D. I. Abarbanel, "Generalized synchronization of chaos in directionally coupled chaotic systems," *Physical Review E: Statistical, Nonlinear, and Soft Matter Physics*, vol. 51, no. 2, pp. 980–994, 1995.

[21] H. D. I. Abarbanel, N. F. Rulkov, and M. M. Sushchik, "Generalized synchronization of chaos: the auxiliary system approach," *Physical Review E*, vol. 53, no. 5, article 4528, 1996.

[22] S. Hammami, M. Benrejeb, M. Feki, and P. Borne, "Feedback control design for Rössler and Chen chaotic systems anti-synchronization," *Physics Letters A*, vol. 374, no. 28, pp. 2835–2840, 2010.

[23] L. Ren and R. W. Guo, "A necessary and sufficient condition of anti-synchronization for chaotic systems and its applications," *Mathematical Problems in Engineering*, vol. 2015, Article ID 434651, pp. 1–7, 2015.

[24] R. Mainieri and J. Rehacek, "Projective synchronization in three-dimensional chaotic systems," *Physical Review Letters*, vol. 82, no. 15, pp. 3042–3045, 1999.

[25] D. Xu, "Control of projective synchronization in chaotic systems," *Physical Review E: Statistical, Nonlinear, and Soft Matter Physics*, vol. 63, Article ID 027201, pp. 1–4, 2001.

[26] L. Ren, R. Guo, and U. E. Vincent, "Coexistence of synchronization and anti-synchronization in chaotic systems," *Archives of Control Sciences*, vol. 26, no. 1, pp. 69–79, 2016.

[27] R.-W. Guo, "Simultaneous synchronization and anti-synchronization of two identical new 4D chaotic systems," *Chinese Physics Letters*, vol. 28, no. 4, Article ID 040205, 2011.

[28] J.-M. Grandmont, "On endogenous competitive business cycles," *Econometrica*, vol. 53, no. 5, pp. 995–1045, 1985.

[29] H. Yu, G. Cai, and Y. Li, "Dynamic analysis and control of a new hyperchaotic finance system," *Nonlinear Dynamics*, vol. 67, no. 3, pp. 2171–2182, 2012.

[30] G. Cai, P. Hu, and Y. Li, "Modified function lag projective synchronization of a financial hyperchaotic system," *Nonlinear Dynamics*, vol. 69, no. 3, pp. 1457–1464, 2012.

[31] W. Wu and Z. Chen, "Hopf bifurcation and intermittent transition to hyperchaos in a novel strong four-dimensional hyperchaotic system," *Nonlinear Dynamics*, vol. 60, no. 4, pp. 615–630, 2010.

10

Research on Integrated Guidance and Control of Distributed Cooperation of Multi-Interceptor with State Coupling

Xiang Liu and Xiaogeng Liang

School of Automation, Northwestern Polytechnical University, Xi'an 710072, China

Correspondence should be addressed to Xiang Liu; lx861002@163.com

Academic Editor: Ai-Guo Wu

With the aim of achieving cooperative target interception by using multi-interceptor, a distributed cooperative control algorithm of the multi-interceptor with state coupling is proposed based on the IGC (integrated guidance and control) method. Considering the coupling relationship between the pitch and ya w channels, a state coupling "leader" IGC model is established, an FTDO (finite-time disturbance observer) is designed for estimating the unknown interference of the model, and the "leader" controller is designed according to the adaptive dynamic surface sliding-mode control law. Secondly, the cooperative control strategy of the multi-interceptor is designed with the "leader-follower" distributed network mode to obtain the speed in the three directions of the interceptor in air and transform them to the general flight speed, trajectory inclination angle, and trajectory deflection instruction by using the transformational relation of kinematics. Finally, the "follower" controller is designed with the FTDO and dynamic surface sliding-mode control. The designed multi-interceptor distributed cooperative IGC algorithm with state coupling has good stability according to the simulation results of two different communication topologies.

1. Introduction

With the rapid development of antimissile technology, it is getting difficult for a single interceptor to break through the defense and intercept targets efficiently, thus making it hard for an interceptor to adapt to the demands of future battlefield scenarios. Therefore, an interceptor with cooperative target interception capability would be more suitable for the future. Multi-interceptor can realize functional complementation through information interaction and sharing, which can not only considerably enhance defense penetration and counterforce of the interceptor but also finish tasks that cannot be achieved by a single interceptor [1].

With respect to cooperative guidance and control of multi-interceptors, the authors in [2, 3] proposed a guidance law with controllable attack time and angle-of-attack constraint and applied it to the salvo attack of anti-ship missiles. Based on this idea, researchers subsequently introduced some other guidance and control methods, including sliding-mode control [4, 5], optimal control [6], differential game [7], and dynamic surface control [8]. This group of methods relies on specifying the attack time before launching to achieve coordination. No information exchange occurs between missiles during flight; hence, these methods apparently have temporal limitations. With the progress in consensus of multiagent systems, researchers have begun to use the consensus theory to study the cooperative guidance and control of multi-interceptors. Using the coordination strategy under the cooperative guidance framework, the authors in [9] adjusted missile trajectories, such that the coordination variable of each missile can approach the expected coordination variable and realize cooperative guidance. The authors in [10] applied the "leader-follower" formation control to cooperative guidance of multi-interceptors by putting forward an analogous "leader-follower" cooperative guidance framework. The authors in [11, 12] explored the guidance and control law of the "leader-follower" topology with angle constraint and topology switch present. By constructing an integrated cost function for multiple missiles, the authors in [13] designed a cooperative guidance law for multiple missiles intercepting a maneuvering target. However, the application of this integrated cost function was faced with multiple constrains because each missile required the global information of all the participating partners.

The interceptor guidance and control system is a highly dynamic, strong-coupling, varying, and uncertain multivariant system featuring complicated dynamic characteristics. Therefore, the integrated guidance and control (IGC) design method can allocate the control ability of interceptors more properly. It mainly generates control power according to the relative motive between the interception targets and the interceptors and then drives the interceptors to chase the targets. Moreover, it cannot only stabilize flight attitude but also enhance guidance precision [14]. In recent years, many researchers worldwide conducted studies focusing on the design method of IGC. The authors in [15, 16] designed an IGC control law with the sliding control mode and back-stepping control algorithm. The sliding-mode control method has been widely used in the design of IGC of aircraft [17], missile [18], and unmanned helicopter [19, 20]. In existing papers, most of them were designed in a single channel [21, 22], regardless of the coupling between channels. The authors in [23, 24] designed IGC algorithms in three dimensions, but designing the controller is difficult when establishing the model with a high order.

According to the above literature review, the guidance loop and control loop of the multi-interceptor cooperative guidance and control have been studied separately by experts. However, external disturbance and its coupling relation during the multi-interceptor flight cannot be ignored. In the meantime, multi-interceptor needs to communicate during its flight to finish cooperative control. Thus, unsmooth local communication should be considered while designing the controller. In light of this, a distributed cooperative control strategy was introduced on top of the integrated guidance and control (IGC) method by considering the coupling between the interception pitch and yaw channels, a design method of cooperative IGC of the multi-interceptor with state coupling of the "leader-follower" distributed topology structure is proposed. The design of the "leader" and "follower" control algorithm using the dynamic surface sliding-mode and finite-time disturbance observer can effectively enhance the stability of the interceptor during the flight and furthermore ensure the target to be hit by the "leader" and "follower" simultaneously following the distributed cooperative controlling strategy. The proposed method can enhance the stability of cooperative guidance and control of the multi-interceptor.

2. "Leader" IGC Model of Interceptor with State Coupling

According to the relative motion relation between the interceptor "leader" and target [25, 26], the relative motion model of the interceptor "leader" and target is established as follows:

$$\begin{aligned} r\ddot{q}_\varepsilon &= -2\dot{r}\dot{q}_\varepsilon - r\dot{q}_\beta^2 \sin q_\varepsilon \cos q_\varepsilon - a_{t\varepsilon} - a_{m4\varepsilon} \\ \ddot{r} &= r\dot{q}_\varepsilon^2 + r\dot{q}_\beta^2 \cos^2 q_\varepsilon + a_{tr} - a_{m4r} \\ -r\ddot{q}_\beta \cos q_\varepsilon &= 2\dot{r}\dot{q}_\beta \cos q_\varepsilon - 2r\dot{q}_\varepsilon \dot{q}_\beta \sin q_\varepsilon + a_{t\beta} - a_{m4\beta} \end{aligned} \tag{1}$$

In the equation, q_β and q_ε denote the elevation angle and horizontal sight angle of the "leader" and target, respectively; $a_{m4\varepsilon}$ and $a_{m4\beta}$ denote the longitudinal and lateral motion acceleration of the "leader," respectively; $a_{t\varepsilon}$ and $a_{t\beta}$ denote the longitudinal and lateral motion acceleration of the target, respectively; r represents the relative distance between the "leader" and target.

The kinetic model of the interceptor "leader" can be indicated as follows:

$$\begin{aligned} \dot{\alpha} &= \omega_z + \omega_y \tan\beta \sin\alpha - \frac{qSC_y^\alpha \alpha \cos\alpha}{mV_m \cos\beta} + d_\alpha \\ \dot{\beta} &= \omega_y \cos\alpha + \frac{qSC_y^\alpha \alpha \sin\alpha \sin\beta}{mV_m} + \frac{qSC_z^\beta \beta \cos\beta}{mV_m} \\ &\quad + d_\beta \\ \dot{\omega}_z &= \frac{qS\bar{L}m_z^\alpha \alpha}{J_z} + \frac{qS\bar{L}m_z^{\omega_z}\omega_z}{J_z} + \frac{M_z}{J_z} + d_{\omega_z} \\ \dot{\omega}_y &= \frac{qS\bar{L}m_y^\beta \beta}{J_y} + \frac{qS\bar{L}m_y^{\omega_y}\omega_y}{J_y} + \frac{M_y}{J_y} + d_{\omega_y} \end{aligned} \tag{2}$$

$$\begin{aligned} a_{m3\varepsilon} &= \frac{qSC_y^\alpha \alpha}{m} \\ a_{m3\beta} &= \frac{qSC_z^\beta \beta}{m} \end{aligned} \tag{3}$$

In the equation, S is the reference area of the "leader;" $\bar{L}$ is the reference length of the "leader;" m is the mass of the "leader;" α and β are the attack angle and sideslip angle, respectively; ω_z and ω_y are the pitch angular velocity and yaw rate, respectively; and $d_\alpha, d_{\omega_z}, d_\beta$, and d_{ω_y} are the disturbance and uncertain disturbance of the various links of the system. J_z and J_y are the rotational inertia; $C_y^\alpha, C_z^\beta, m_z^\alpha, m_z^{\omega_z}, m_y^\beta$, and $m_y^{\omega_y}$ are the related aerodynamic force and torque coefficient, respectively; M_z and M_y are the pitch moment and yawing moment of the "leader," respectively; and $a_{m3\varepsilon}$ and $a_{m3\beta}$ are the longitudinal and lateral acceleration of the "leader."

Assuming that the sight angle of the interceptor in the terminal guidance stage changes slightly and the included angle of the sight angle and velocity direction of the interceptor are relatively small, let $a_{m3\varepsilon} = a_{m4\varepsilon}$ and $a_{m3\beta} = a_{m4\beta}$. By defining $x_1 = [\dot{q}_\varepsilon, \dot{q}_\beta]^T$, $x_2 = [\alpha, \beta]^T$, and $x_3 = [\omega_z, \omega_y]^T$, the nonlinear model of the "leader" IGC of the interceptor with state coupling can be obtained, according to (1), (2), and (3).

$$\begin{aligned} \dot{x}_1 &= f_1(x_1) + g_{11}(x_1)x_2 + g_{12}(x_1)a_t \\ \dot{x}_2 &= f_2(x_2) + g_2(x_2)x_3 + d_2 \\ \dot{x}_3 &= f_3(x_2, x_3) + g_3 u + d_3 \\ y &= x_1 \end{aligned} \tag{4}$$

In the equation,

$$f_1(x_1) = \left[-\frac{2\dot{r}}{r}\dot{q}_\varepsilon - \dot{q}_\beta^2 \sin q_\varepsilon \cos q_\varepsilon, -\frac{2\dot{r}}{r}\dot{q}_\beta + 2\dot{q}_\varepsilon\dot{q}_\beta \tan q_\varepsilon\right]^T,$$

$$f_2(x_2) = \left[-\frac{qSC_y^\alpha \alpha \cos\alpha}{mV_m \cos\beta}, \frac{qSC_y^\alpha \alpha \sin\alpha \sin\beta}{mV_m} + \frac{qSC_z^\beta \beta \cos\beta}{mV_m}\right]^T,$$

$$f_3(x_2, x_3) = \left[\frac{qS\bar{L}m_z^\alpha \alpha}{J_z} + \frac{qS\bar{L}m_z^{\omega_z}\omega_z}{J_z}, \frac{qS\bar{L}m_y^\beta \beta}{J_y} + \frac{qS\bar{L}m_y^{\omega_y}\omega_y}{J_y}\right]^T$$

$$g_{11}(x_1) = \begin{bmatrix} -\frac{qSC_y^\alpha}{mr} & 0 \\ 0 & \frac{qSC_z^\beta}{mr\cos q_\varepsilon} \end{bmatrix},$$

$$g_{12}(x_1) = \begin{bmatrix} \frac{1}{r} & 0 \\ 0 & \frac{1}{r\cos q_\varepsilon} \end{bmatrix},$$

$$g_2(x_2) = \begin{bmatrix} 1 & \tan\beta \sin\alpha \\ 0 & \cos\alpha \end{bmatrix},$$

$$g_3 = \mathrm{diag}\left\{J_z^{-1}, J_y^{-1}\right\},$$

$$u = \left[M_z, M_y\right]^T,$$

$$a_t = \left[a_{t\varepsilon}, -a_{t\beta}\right]^T,$$

$$d_2 = \left[d_\alpha, d_\beta\right]^T,$$

$$d_3 = \left[d_{\omega_z}, d_{\omega_y}\right]^T \quad (5)$$

The unknown disturbances d_2 and d_3 in the "leader" IGC model are assumed to be continuously differentiable and the first-order derivative is bounded. $d_i < N$, $i = 2, 3$, N is a positive constant.

3. Design of Interceptor "Leader" Controller

3.1. Design of Finite-Time Disturbance Observer. Aiming at the uncertainty a_t, d_2, and d_3 included in the system model (4), an FTDO (finite-time disturbance observer) is designed for estimating unknown disturbances, in order to eliminate the impact of unknown disturbances on the leader system of the interceptor. Define $v_r = \dot{r}$, $v_\varepsilon = r\dot{q}_\varepsilon$, $v_\beta = r\dot{q}_\beta \cos q_\varepsilon$, and

$$\begin{aligned}\dot{v}_\varepsilon &= \dot{r}\dot{q}_\varepsilon + r\ddot{q}_\varepsilon = -\dot{r}\dot{q}_\varepsilon - r\dot{q}_\beta^2 \sin q_\varepsilon \cos q_\varepsilon + a_{t\varepsilon} - a_{m4\varepsilon} \\ &= -\frac{v_r v_\varepsilon}{r} - \frac{v_\beta^2 \tan q_\varepsilon}{r} + a_{t\varepsilon} - a_{m4\varepsilon}\end{aligned} \quad (6)$$

$$\begin{aligned}\dot{v}_\beta &= \dot{r}\dot{q}_\beta \cos q_\varepsilon + r\ddot{q}_\beta \cos q_\varepsilon - r\dot{q}_\varepsilon \dot{q}_\beta \sin q_\varepsilon \\ &= -\dot{r}\dot{q}_\beta \cos q_\varepsilon + r\dot{q}_\varepsilon \dot{q}_\beta \sin q_\varepsilon - a_{t\beta} + a_{m4\beta} \\ &= -\frac{v_r v_\beta}{r} + \frac{v_\varepsilon v_\beta \tan q_\varepsilon}{r} - a_{t\beta} + a_{m4\beta}\end{aligned} \quad (7)$$

Define $w_1 = [v_\varepsilon, v_\beta]^T$, and according to (6) and (7),

$$\dot{w}_1 = h_1(w_1) + h_2(w_1)x_2 + a_t \quad (8)$$

In the equation,

$$h_1(w_1) = \begin{bmatrix} -\frac{v_r v_\varepsilon}{r} - \frac{v_\beta^2 \tan q_\varepsilon}{r} \\ -\frac{v_r v_\beta}{r} + \frac{v_\varepsilon v_\beta \tan q_\varepsilon}{r} \end{bmatrix},$$

$$h_2(w_1) = \begin{bmatrix} -\frac{qSC_y^\alpha \alpha}{m} & 0 \\ 0 & \frac{qSC_z^\beta \beta}{m} \end{bmatrix} \quad (9)$$

The following FTDO is designed to estimate the acceleration a_t of the target, and

$$\begin{aligned}\dot{z}_{10} &= v_{10} + h_1(w_1) + h_2(w_1)x_2, \\ \dot{z}_{11} &= v_{11}, \\ \dot{z}_{12} &= v_{12} \\ v_{10} &= -\lambda_{10}\left[L_1^{1/3}\left|z_{10} - w_1\right|^{2/3} \mathrm{sgn}(z_{10} - w_1)\right] + z_{11} \\ v_{11} &= -\lambda_{11}\left[L_1^{1/2}\left|z_{11} - v_{10}\right|^{1/2} \mathrm{sgn}(z_{11} - v_{10})\right] + z_{12} \\ v_{12} &= -\lambda_{12}\left[L_1\left|z_{12} - v_{11}\right|^{q_1/p_1} \mathrm{sgn}(z_{12} - v_{11})\right] \\ z_{10} &= \widehat{w}_1, \\ z_{11} &= \widehat{a}_t, \\ z_{12} &= \widehat{\dot{a}}_t\end{aligned} \quad (10)$$

In the equation $w_1 = [w_{11}, w_{12}]^T$, $z_{10} = [z_{101}, z_{102}]^T$, $z_{11} = [z_{111}, z_{112}]^T$, $z_{12} = [z_{121}, z_{122}]^T$, $v_{10} = [v_{101}, v_{102}]^T$, $v_{11} = [v_{111}, v_{112}]^T$, $v_{12} = [v_{121}, v_{122}]^T$, $L_1 = [L_{11}, L_{12}]^T$, $\widehat{a}_t$ is the estimated acceleration a_t of the target, the estimated value w_1 and $\dot{a}_t$ is $\widehat{w}_1$ and $\widehat{\dot{a}}_t$, λ_{10}, λ_{11}, and λ_{12} are the coefficients to be designed for the disturbance observer, q_1 and p_1 are the terminal coefficients, respectively, and $0 < p_1 < q_1$.

It can be learnt according to [27] that appropriate parameters can guarantee that the FTDO error system is steady in finite time. The estimated error of the acceleration a_t of the target is defined as $e_{11} = z_{11} - a_t$.

Similarly, the disturbance d_2 and d_3 of the second subsystem and third subsystem is estimated, and

$$
\begin{aligned}
\dot{z}_{20} &= v_{20} + f_2(x_2) + g_2(x_2)x_3, \\
\dot{z}_{21} &= v_{21}, \\
\dot{z}_{22} &= v_{22} \\
v_{20} &= -\lambda_{20}\left[L_2^{1/3}\left|z_{20} - x_2\right|^{2/3}\operatorname{sgn}(z_{20} - x_2)\right] + z_{21} \\
v_{21} &= -\lambda_{21}\left[L_2^{1/2}\left|z_{21} - v_{20}\right|^{1/2}\operatorname{sgn}(z_{21} - v_{20})\right] + z_{22} \\
v_{22} &= -\lambda_{22}\left[L_2\left|z_{22} - v_{21}\right|^{q_2/p_2}\operatorname{sgn}(z_{22} - v_{21})\right] \\
z_{20} &= \hat{x}_2, \\
z_{21} &= \hat{d}_2, \\
z_{22} &= \dot{\hat{d}}_2
\end{aligned}
\tag{11}
$$

$$
\begin{aligned}
\dot{z}_{30} &= v_{30} + f_3(x_2, x_3) + g_3 u, \\
\dot{z}_{31} &= v_{31}, \\
\dot{z}_{32} &= v_{32} \\
v_{30} &= -\lambda_{30}\left[L_3^{1/3}\left|z_{30} - x_3\right|^{2/3}\operatorname{sgn}(z_{30} - x_3)\right] + z_{31} \\
v_{31} &= -\lambda_{31}\left[L_3^{1/2}\left|z_{31} - v_{30}\right|^{1/2}\operatorname{sgn}(z_{31} - v_{30})\right] + z_{32} \\
v_{32} &= -\lambda_{32}\left[L_3\left|z_{32} - v_{31}\right|^{q_3/p_3}\operatorname{sgn}(z_{32} - v_{31})\right] \\
z_{30} &= \hat{x}_3, \\
z_{31} &= \hat{d}_3, \\
z_{32} &= \dot{\hat{d}}_3
\end{aligned}
\tag{12}
$$

In the equation, the estimated value of disturbance d_2 and d_3 is $\hat{d}_2$ and $\hat{d}_3$, respectively, and the estimated error is $e_{21} = z_{21} - d_2$ and $e_{31} = z_{31} - d_3$, respectively.

3.2. Design of Adaptive Dynamic Sliding-Mode Controller. Because the interceptor IGC model is an unmatched and uncertain system, and aiming at the state coupling IGC model (4) and FTDO estimated value (10)–(12), the "leader" control algorithm is designed by taking advantage of the adaptive dynamic sliding-mode control law.

(1) The command signal of the first subsystem of (4) is defined as x_{1d}. In order to realize the guidance goal, the sight angular velocity should be removed. According to the design method of dynamic surface sliding-mode control, the first dynamic error surface is defined as follows:

$$s_1 = -g_{11}^{-1}(x_1)(x_1 - x_{1d}) \tag{13}$$

Taking the derivative of s_1, the dynamic equation of error is given by

$$
\begin{aligned}
\dot{s}_1 &= g_{11}^{-2}(x_1)\dot{g}_{11}(x_1)(x_1 - x_{1d}) \\
&\quad - g_{11}^{-1}(x_1)\left[f_1(x_1) + g_{12}(x_1)a_t - \dot{x}_{1d}\right] - x_2
\end{aligned}
\tag{14}
$$

According to the dynamic surface design method and FTDO estimated value $\hat{a}_t$ in (10), the virtual control amount of the first dynamic surface can be obtained as follows:

$$
\begin{aligned}
x_2^* &= g_{11}^{-2}(x_1)\dot{g}_{11}(x_1)(x_1 - x_{1d}) \\
&\quad - g_{11}^{-1}(x_1)\left[f_1(x_1) + g_{12}(x_1)\hat{a}_t - \dot{x}_{1d}\right] + k_1 s_1
\end{aligned}
\tag{15}
$$

In the equation, $k_1 = \operatorname{diag}\{k_{11}, k_{12}\}$ is the positive definite matrix. In the design process, differential blast would occur, while the differential of the virtual control amount x_2^* is taken. In order to avoid the complicated computation process owing to item inflation, x_2^* must be obtained through the first-order low-pass filter, and the virtual control amount of the filter can be obtained as follows:

$$
\begin{aligned}
\tau_2\dot{\bar{x}}_2^* + \bar{x}_2^* &= x_2^*, \\
\bar{x}_2^*(0) &= x_2^*(0)
\end{aligned}
\tag{16}
$$

In the equation, $\tau_2 = \operatorname{diag}\{\tau_{21}, \tau_{22}\}$ is the time constant of the filter, and the differential of the virtual control after the error surface filter can be obtained.

$$\dot{\bar{x}}_2^* = -\tau_2^{-1}\left(\bar{x}_2^* - x_2^*\right) \tag{17}$$

(2) The second dynamic error surface is defined as

$$s_2 = x_2 - \bar{x}_2^* \tag{18}$$

Taking the derivative of s_2, the dynamic equation of error can be obtained as follows:

$$\dot{s}_2 = \dot{x}_2 - \dot{\bar{x}}_2^* = f_2(x_2) + g_2(x_2)x_3 + d_2 - \dot{\bar{x}}_2^* \tag{19}$$

Similar to the first dynamic surface design method, the estimated FTDO value $\hat{d}_2$ is substituted in (11). Thus, the virtual control of the second dynamic surface can be obtained as follows:

$$x_3^* = g_2^{-1}(x_2)\left[-f_2(x_2) - \hat{d}_2 + \dot{\bar{x}}_2^* - k_2 s_2\right] \tag{20}$$

In the equation, $k_2 = \operatorname{diag}\{k_{21}, k_{22}\}$ is the positive definite matrix. Similarly, by obtaining x_3^* through the first-order low-pass filter, the virtual control amount of the filter can be obtained as follows:

$$
\begin{aligned}
\tau_3\dot{\bar{x}}_3^* + \bar{x}_3^* &= x_3^*, \\
\bar{x}_3^*(0) &= x_3^*(0)
\end{aligned}
\tag{21}
$$

In the equation, $\tau_3 = \operatorname{diag}\{\tau_{31}, \tau_{32}\}$ is the time constant of the filter. The differential of virtual control after the error surface filter can be obtained as follows:

$$\dot{\bar{x}}_3^* = -\tau_3^{-1}\left(\bar{x}_3^* - x_3^*\right) \tag{22}$$

(3) The third dynamic error surface is defined as follows:

$$s_3 = x_3 - \overline{x}_3^* \quad (23)$$

Taking the derivative of s_3, the dynamic equation of error can be obtained as follows:

$$\dot{s}_3 = \dot{x}_3 - \dot{\overline{x}}_3^* = f_3(x_2, x_3) + g_3 u + d_3 - \dot{\overline{x}}_3^* \quad (24)$$

To guarantee the convergence velocity of the interceptor "leader," an adaptive sliding-mode reaching law is designed:

$$\dot{s} = -k_a \left|\frac{\dot{r}}{r}\right| s - k_b |s|^{\partial} \operatorname{sgn}(s) \quad (25)$$

In the equation, $k_a > 0$, $k_b > 0$, and $\dot{r}$ denotes the change in relative distance between the "leader" and target.

According to (24) and (25) and estimated FTDO $\hat{d}_3$ of (12), the adaptive dynamic surface sliding-mode control law of the interceptor "leader" is given by

$$u = g_3^{-1}\left[-f_3(x_2, x_3) - \hat{d}_3 + \dot{\overline{x}}_3^* - k_3\left|\frac{\dot{r}}{r}\right| s_3 - k_4 |s_3|^{\partial} \operatorname{sgn}(s_3)\right] \quad (26)$$

In the equation, $k_3 = \operatorname{diag}\{k_{31}, k_{32}\}$, $k_4 = \operatorname{diag}\{k_{41}, k_{42}\}$, and $\partial = \operatorname{diag}\{\partial_{11}, \partial_{12}\}$ are positive definite matrices, and $0 < \partial < 1$.

3.3. Stability Analysis

Theorem 1. *Consider the integrated guidance and control system for the "leader" (equation (4)). If the convergence rate is calculated using (25), the disturbance values of the system (see (4)) are estimated using (10)-(12), and filter equations (16) and (21) are implemented; then finally under the dynamic surface sliding-mode control law (see (26)), imposing the constraint for ensuring the system (see (4)) output error converging into the adjacent area of the origin, an arbitrary adjacent area of the origin can be obtained with the appropriate design parameter determined.*

To prove:

Let us assume that the estimated error of FTDO system meets

$$|e_{11}| < N_1, \quad |e_{21}| < N_2, \quad |e_{31}| < N_3 \quad (27)$$

In the equation, N_1, N_2, and N_3 are positive constants. The filter error is defined as follows:

$$y_2 = \overline{x}_2^* - x_2^*, \quad y_3 = \overline{x}_3^* - x_3^* \quad (28)$$

Taking the derivative of y_2 and y_3, the dynamic error of the filter can be obtained as follows:

$$\dot{y}_2 = -\tau_2^{-1} y_2 - \dot{x}_2^*, \quad \dot{y}_3 = -\tau_3^{-1} y_3 - \dot{x}_3^* \quad (29)$$

According to (13)–(23) and (28),

$$\begin{aligned} x_1 &= -g_{11}(x_1) s_1 + x_{1d} \\ x_2 &= s_2 + \overline{x}_2^* = s_2 + y_2 + x_2^* \\ x_3 &= s_4 + \overline{x}_3^* = s_3 + y_3 + x_3^* \end{aligned} \quad (30)$$

According to (4), (11)–(21), and (26)–(28),

$$\begin{aligned} \dot{s}_1 &= g_{11}^{-2}(x_1)\dot{g}_{11}(x_1)(x_1 - x_{1d}) \\ &\quad - g_{11}^{-1}(x_1)\left[f_1(x_1) + g_{12}(x_1) a_t - \dot{x}_{1d}\right] - x_2 \\ &= -s_2 - y_2 - k_1 s_1 + \tilde{e}_{11} \end{aligned} \quad (31)$$

In the equation, $\tilde{e}_{11} = g_{11}^{-1}(x_1) g_{12}(x_1) e_{11}$. Let us assume that $|\tilde{e}_{11}| < \tilde{N}_1$, where $\tilde{N}_1$ is a positive constant.

$$\begin{aligned} \dot{s}_2 &= \dot{x}_2 - \dot{\overline{x}}_2^* = f_2(x_2) + g_2(x_2) x_3 + d_2 - \dot{\overline{x}}_2^* \\ &= g_2(x_2)(s_3 + y_3) - k_2 s_2 - e_{21} \end{aligned} \quad (32)$$

$$\begin{aligned} \dot{s}_3 &= \dot{x}_3 - \dot{\overline{x}}_3^* = f_3(x_2, x_3) + g_3 u + d_3 - \dot{\overline{x}}_3^* \\ &= -k_3\left|\frac{\dot{r}}{r}\right| s_3 - k_4 |s_3|^{\partial} \operatorname{sgn}(s_3) - e_{31} \end{aligned} \quad (33)$$

According to Young's equation and (30)–(33),

$$\begin{aligned} s_1^T \dot{s}_1 &= s_1^T(-s_2 - y_2 - k_1 s_1 + \tilde{e}_{11}) \\ &\le s_1^T\left(\frac{3}{2}I - k_1\right)s_1 + \frac{1}{2}s_2^T s_2 + \frac{1}{2}y_2^T y_2 + \frac{1}{2}\tilde{N}_1^2 \end{aligned} \quad (34)$$

$$\begin{aligned} s_2^T \dot{s}_2 &= s_2^T\left[g_2(x_2)(s_3 + y_3) - k_2 s_2 - e_{21}\right] \\ &\le s_2^T\left(\frac{1}{2}I + g_2^2 - k_2\right)s_2 + \frac{1}{2}s_3^T s_3 + \frac{1}{2}y_3^T y_3 \\ &\quad + \frac{1}{2}N_2^2 \end{aligned} \quad (35)$$

$$\begin{aligned} s_3^T \dot{s}_3 &= s_3^T\left(-k_3\left|\frac{\dot{r}}{r}\right| s_3 - k_4 |s_3|^{\partial} \operatorname{sgn}(s_3) - e_{31}\right) \\ &\le s_3^T\left(\frac{1}{2}I - k_3 + \frac{3}{2}k_4\right)s_3 + \frac{1}{2}N_3^2 + \frac{1}{2}k_4 \end{aligned} \quad (36)$$

It can be learnt that variables and their differential in the system model are bounded, and there are continuous functions $\tilde{z}_2$ and $\tilde{z}_3$, where $\tilde{z}_2 > 0$ and $\tilde{z}_3 > 0$, enabling variables $\dot{x}_2^*$ and $\dot{x}_3^*$ to meet

$$|\dot{x}_2^*| \le \tilde{z}_2, \quad |\dot{x}_3^*| \le \tilde{z}_3 \quad (37)$$

According to Young's equation and (28)–(29) and (37):

$$y_2^T \dot{y}_2 = y_2^T \left(-\tau_2^{-1} y_2 - \dot{x}_2^*\right) \le y_2^T \left(\frac{1}{2} I - \tau_2^{-1}\right) y_2 + \frac{1}{2} \overline{z}_2^2 \quad (38)$$

$$y_3^T \dot{y}_3 = y_3^T \left(-\tau_3^{-1} y_3 - \dot{x}_3^*\right) \le y_3^T \left(\frac{1}{2} I - \tau_3^{-1}\right) y_3 + \frac{1}{2} \overline{z}_3^2 \quad (39)$$

According to the state coupling IGC nonlinear system model (4), a Lyapunov function is selected:

$$V = \frac{1}{2} \left(s_1^T s_1 + s_2^T s_2 + s_3^T s_3 + y_2^T y_2 + y_3^T y_3\right) \quad (40)$$

Taking the derivative of (40),

$$\begin{aligned} \dot{V} &= s_1^T \dot{s}_1 + s_2^T \dot{s}_2 + s_3^T \dot{s}_3 + y_2^T \dot{y}_2 + y_3^T \dot{y}_3 \\ &\le s_1^T \left(\frac{3}{2} I - k_1\right) s_1 + \frac{1}{2} s_2^T s_2 + \frac{1}{2} y_2^T y_2 + \frac{1}{2} \widetilde{N}_1^2 \\ &\quad + s_2^T \left(\frac{1}{2} I + g_2^2 - k_2\right) s_2 + \frac{1}{2} s_3^T s_3 + \frac{1}{2} y_3^T y_3 \\ &\quad + \frac{1}{2} N_2^2 + s_3^T \left(\frac{1}{2} I - k_3 + \frac{3}{2} k_4\right) s_3 + \frac{1}{2} N_3^2 + \frac{1}{2} k_4 \\ &\quad + y_2^T \left(\frac{1}{2} I - \tau_2^{-1}\right) y_2 + \frac{1}{2} \overline{z}_2^2 + y_3^T \left(\frac{1}{2} I - \tau_3^{-1}\right) y_3 \\ &\quad + \frac{1}{2} \overline{z}_3^2 \end{aligned} \quad (41)$$

The design parameters meet the following rules:

$$\begin{aligned} k_1 &\ge \frac{3}{2} I + \frac{1}{2} \kappa I, \\ k_2 &\ge I + g_2^2 + \frac{1}{2} \kappa I, \\ k_3 &\ge I + \frac{3}{2} k_4 + \frac{1}{2} \kappa I \\ \tau_2^{-1} &\ge I + \frac{1}{2} \kappa I, \\ \tau_3^{-1} &\ge I + \frac{1}{2} \kappa I \end{aligned} \quad (42)$$

In the equation, κ is a constant, and $\kappa > 0$. Therefore,

$$\dot{V} \le -\kappa V + \widehat{A} \quad (43)$$

In the equation, $\widehat{A} = (1/2)\widetilde{N}_1^2 + (1/2)N_2^2 + (1/2)N_3^2 + (1/2)\overline{z}_2^2 + (1/2)\overline{z}_3^2 + (1/2)k_4$.

According to (43),

$$V(t) \le \frac{\{[\kappa V(0) - \widehat{A}] e^{-\kappa t} + \widehat{A}\}}{\kappa} \quad (44)$$

s_1, s_2, s_3, y_2, and y_3 are consistent and eventually bounded. Thus, large parameters k_1, k_2, k_3, and k_4, as well as small parameters τ_2 and τ_3 are selected, to make the value of κ sufficiently large and $\widehat{A}/\kappa$ sufficiently small to ensure control precision.

Remark 2. Theoretically, the final boundaries of error surfaces s_1, s_2, and s_3 and filter errors y_2 and y_3 will become smaller with the increasing design parameters k_1, k_2, k_3, and k_4 and the decreasing τ_2 and τ_3. This change leads to a higher controlling precision. However, in reality, using too large parameters (k_1, k_2, k_3, and k_4) and too small parameters (τ_2 and τ_3) will result in an input saturation for the interceptor control system. The nonlinear behavior of the saturated system results in a higher requirement of overload exceeding the available overload. Therefore, the angle of attack and the sideslip angle of the interceptor exceed the allowable range leading to a reduced controlling performance of the system. Furthermore, the physical constraints of the low-pass filter prohibit parameters τ_2 and τ_3 from being too small. Therefore, the parameters of the control algorithm should be properly selected by combining practical situations.

4. Distributed Network Synchronization Strategy

4.1. Design of Cooperative Control Strategy Based on the Distributed Network. Based on the principle of time consistency of a multiagent system, the multi-interceptor cooperative control strategy is designed to ensure that all interceptors hit the targets at the same time. In the cooperative system of the multi-interceptor, the state information of other interceptors can be obtained through information interaction, for realizing time consistency, and such information interaction can be described using the graph theory. Assuming that each interceptor is a communication node, the information exchange among interceptors can be indicated as $\overline{G} = \{\overline{V}, \overline{E}, \overline{A}\}$, where $\overline{V} = \{\overline{v}_i,\ i = 1, 2, \ldots, n\}$ denotes the set of interceptor nodes and $\overline{E}$ denotes the lines between the interceptor nodes. The weighted coefficient matrix is indicated as $\overline{A} = [\overline{a}_{ij}] \in \mathbb{R}^{n\times n}$; $\overline{a}_{ij} > 0$ implies that the interceptor node i and node j can exchange information. However, if $\overline{a}_{ij} = 0$, information cannot be exchanged. $\overline{L}$ denotes the Laplace matrix of the undirected graph $\overline{G}$, among which the elements satisfy

$$\begin{aligned} \overline{l}_{ii} &= \sum_{j=1, j\ne i}^{n} \overline{a}_{ij} \\ \overline{l}_{ij} &= -\overline{a}_{ij}, \quad j \ne i \end{aligned} \quad (45)$$

$\overline{B} = \mathrm{diag}\{\overline{b}_1, \overline{b}_2, \ldots, \overline{b}_j\}$ denotes whether the interceptors can obtain the state information of the leader, $\overline{b}_i > 0$, $i \in \{1, 2, 3, \ldots, j\}$ indicates that the interceptors can obtain the state information of the leader, and $\overline{b}_i = 0$, $i \in \{1, 2, 3, \ldots, j\}$ indicates that the interceptors cannot obtain the state information of the leader.

Based on the "leader-follower" topology structure, the distributed cooperative control strategy of the multi-interceptor is designed as follows:

$$v_i = \overline{k}_{i1} \left[\sum_{j=1}^{n} \overline{a}_{ij} \left(\varsigma_j - \varsigma_i\right) + \overline{b}_i \left(\varsigma_0 - \varsigma_i\right)\right] + \dot{\varsigma}_0 \quad (46)$$

In the equation, $\varsigma_0 = [x_0, y_0, z_0]^T$ denotes the position of the "leader," $\varsigma_i = [x_i, y_i, z_i]^T$, $i \in \{1, 2, 3, \ldots, j\}$ denotes the position of the interceptors, $v_i = \dot{\varsigma}_i$, $i \in \{1, 2, 3, \ldots, j\}$ denotes the velocity of the interceptors, $\overline{k}_{i1} = \mathrm{diag}\{\overline{k}_{i11}, \overline{k}_{i12}, \overline{k}_{i13}\}$ is a constant, and $\overline{k}_{i1} > 0$.

Theorem 3. *If the "follower" state of the interceptor can converge to the "leader" state following the cooperative control strategy (see (46)), such cooperative strategy is then considered to be successful.*

To prove:

Lemma 4. *Laplace matrix $\overline{M}$ has the following properties:*

(1) If $\overline{G}$ is connected, the characteristic value of $\overline{M}$ is $\lambda_{min}(\overline{M}) > 0$, and it is called the algebraic connectivity of the network-connected graph. The larger the value of $\lambda_{min}(\overline{M})$, the more connected the network.

*(2) One of the characteristic values of $\overline{M}$ is 0, and its corresponding characteristic vector is **1**.*

The error variable is defined as $e_i = \varsigma_i - \varsigma_0$, and

$$\dot{e}_i = \dot{\varsigma}_i - \dot{\varsigma}_0 = k_{i1}\left(\sum_{j=1}^{n} \overline{a}_{ij}\left(e_j - e_i\right) + \overline{b}_i e_i\right) \tag{47}$$

The Lyapunov function is defined as

$$V = \frac{1}{2} e^T \left(\overline{M} + \overline{B}\right) e \tag{48}$$

In the equation, $e = [e_1, e_2, \ldots, e_n]^T$.

$\overline{k}_i = \min\{\overline{k}_{i1}\}$, and taking the derivative of the above equation,

$$\begin{aligned}\dot{V} &= \dot{e}^T\left(\overline{M} + \overline{B}\right) e = -\sum_{i=1}^{n}\left[\sum_{j=1}^{n} a_{ij}\left(e_j - e_i\right) - \overline{b}_i e_i\right]\dot{e}_i \\ &= -\sum_{i=1}^{n} \overline{k}_{i1}\left[\sum_{j=1}^{n} a_{ij}\left(e_j - e_i\right) - \overline{b}_i e_i\right] \\ &\quad \cdot\left(\sum_{j=1}^{n} \overline{a}_{ij}\left(e_j - e_i\right) + \overline{b}_i e_i\right) = -\sum_{i=1}^{n}\overline{k}_{i1} \\ &\quad \cdot\left[\sum_{j=1}^{n} a_{ij}\left(e_j - e_i\right) - \overline{b}_i e_i\right]^2 \\ &\le -\overline{k}_i \sum_{i=1}^{n}\left[\sum_{j=1}^{n} a_{ij}\left(e_j - e_i\right) - \overline{b}_i e_i\right]^2\end{aligned} \tag{49}$$

Let $V(e) \neq 0$. Then, according to the above equation,

$$\begin{aligned}&\frac{\sum_{i=1}^{n}\left[\sum_{j=1}^{n} a_{ij}\left(e_j - e_i\right) - \overline{b}_i e_i\right]^2}{V(e)} \\ &= \frac{e^T\left(\overline{M} + \overline{B}\right)^T\left(\overline{M} + \overline{B}\right) e}{(1/2)\left(e^T\left(\overline{M} + \overline{B}\right)^T e\right)} \ge 2\lambda_{\min}\left(\overline{M} + \overline{B}\right)\end{aligned} \tag{50}$$

According to (49)–(50),

$$\dot{V}(t) \le -\overline{k}_i\left[2\lambda_{\min}\left(\overline{M} + \overline{B}\right)\right] \tag{51}$$

Therefore, $V(t)$ is convergent in finite time, namely, the convergence state from the "follower" to the "leader"; it can realize cooperative guidance and control of the multi-interceptor.

4.2. Implementation of Distributed Network Cooperative Control Strategy. In order to implement the distributed network synchronization strategy, instructions provided to the synchronization strategy should be traced for each interceptor "follower." The motion relation of the interceptors involved in cooperative interception is given as follows:

$$\begin{aligned}\dot{x}_i &= V_{mi}\cos\theta_{mi}\cos\varphi_{mvi} \\ \dot{y}_i &= V_{mi}\sin\theta_{mi} \\ \dot{z}_i &= -V_{mi}\cos\theta_{mi}\sin\varphi_{mvi}\end{aligned} \tag{52}$$

In the equation, $\dot{x}_i$, $\dot{y}_i$, and $\dot{z}_i$ are the velocity components of the ith interceptor in the inertial frame, and θ_{mi} and φ_{mvi} are the trajectory inclination angle and trajectory deflection angle of the ith interceptor.

According to the distributed network synchronization strategy (46), the velocity reference instruction of the interceptor is given by

$$\begin{aligned}\overline{V}_{mxi} &= \overline{k}_{i1}\left(\sum_{j=1}^{n}\overline{a}_{ij}\left(x_j - x_i\right) + \overline{b}_i\left(x_m - x_i\right)\right) + \dot{x}_m \\ \overline{V}_{myi} &= \overline{k}_{i1}\left(\sum_{j=1}^{n}\overline{a}_{ij}\left(y_j - y_i\right) + \overline{b}_i\left(y_m - y_i\right)\right) + \dot{y}_m \\ \overline{V}_{mzi} &= \overline{k}_{i1}\left(\sum_{j=1}^{n}\overline{a}_{ij}\left(z_j - z_i\right) + \overline{b}_i\left(z_m - z_i\right)\right) + \dot{z}_m\end{aligned} \tag{53}$$

According to (52), the total velocity, trajectory inclination angle, and trajectory deflection angle of the interceptor can be obtained as follows:

$$\begin{aligned}\overline{V}_{mi}^{*} &= \sqrt{\left(\overline{V}_{mxi}\right)^2 + \left(\overline{V}_{myi}\right)^2 + \left(\overline{V}_{mzi}\right)^2} \\ \overline{\theta}_{mi}^{*} &= \arcsin\left(\frac{\overline{V}_{myi}}{\overline{V}_{mi}}\right) \\ \overline{\varphi}_{vmi}^{*} &= -\arctan\left(\frac{\overline{V}_{mzi}}{\overline{V}_{mxi}}\right)\end{aligned} \tag{54}$$

To obtain the differential of the total velocity and trajectory inclination angle of the tractor $\dot{\overline{x}}$, the signal is obtained through the filter. Let $\overline{x}$ and $\overline{x}_1^*$ be the postfiltering instruction and prefiltering instruction, respectively. Then,

$$\ddot{\overline{x}} = -2\zeta_n\omega_n\dot{\overline{x}} - \omega_n^2\overline{x} + \omega_n^2\overline{x}_1^* \tag{55}$$

In the equation, ζ_n and ω_n are the damping and bandwidth of the filter, respectively.

5. Design of Interceptor "Follower" Controller

The instructions provided by the cooperative control strategy can be transformed into velocity, trajectory inclination angle, and trajectory deflection angle instruction. In order to track the command signal of interceptor "follower" in the cooperative network, the "follower" controller adopts the dynamic surface sliding-mode control algorithm. Assuming that the velocity of the interceptor "follower" is controllable, the flight velocity can be indicated as follows:

$$\dot{V}_m = \frac{\cos\alpha_i \cos\beta_i}{m} P_i - g\sin\theta_{mi} \tag{56}$$

In the equation, P_i is the motor power, m is the quality of "follower", α_i and β_i are the attack angle and sideslip angle, respectively, θ_{mi} is the trajectory inclination angle, and g is the gravitational acceleration.

According to (56), the error surface is defined as follows:

$$s_v = V_m - \overline{V}_{mi} \tag{57}$$

In the equation, $\overline{V}_{mi}$ is the reference velocity command of the "follower" after filtering. Taking the derivative of s_v,

$$\dot{s}_v = \frac{\cos\alpha_i \cos\beta_i}{m} P_i - g\sin\theta_{mi} - \dot{\overline{V}}_{mi} \tag{58}$$

In the equation, $\dot{\overline{V}}_{mi}$ is the differential of total velocity after filtering.

To ensure that the velocity of the "follower" can track the system command rapidly, the following sliding-mode reaching law is adopted:

$$\dot{s} = -k_a s - k_b |s|^{\partial} \operatorname{sgn}(s) \tag{59}$$

According to (56)–(59), the thrust of the "follower" can be obtained as follows:

$$P_i = \frac{m}{\cos\alpha_i \cos\beta_i}\left[\dot{\overline{V}}_{mi} + g\sin\theta_{mi} - k_{v1}s_v - k_{v2}|s_v|^{\partial_v}\operatorname{sgn}(s_v)\right] \tag{60}$$

In the equation, $k_{v1} > 0$, $k_{v2} > 0$, $0 < \partial_v < 1$.

By defining $x_{i1} = [\theta_{mi}, \varphi_{vmi}]^T$, $x_{i2} = [\alpha_i, \beta_i]^T$, and $x_{i3} = [\omega_{zi}, \omega_{yi}]^T$, the state coupling kinetic equations of the ith "follower" can be indicated as follows:

$$\begin{aligned}
\dot{x}_{i1} &= g_{i1}x_{i2} + d_{i1}\\
\dot{x}_{i2} &= f_{i2}(x_{i2}) + g_{i2}(x_{i2})x_{i3} + d_{i2}\\
\dot{x}_{i3} &= f_{i3}(x_{i2}, x_{i3}) + g_{i3}u_i + d_{i3}
\end{aligned} \tag{61}$$

In the equation,

$$\begin{aligned}
f_{i2}(x_{i2}) &= \left[-\frac{qSC_y^{\alpha_i}\alpha_i\cos\alpha_i}{mV_{mi}\cos\beta_i}, \frac{qSC_y^{\alpha_i}\alpha_i\sin\alpha_i\sin\beta_i}{mV_{mi}} + \frac{qSC_z^{\beta_i}\beta_i\cos\beta_i}{mV_{mi}}\right]^T,\\
f_{i3}(x_{i2}, x_{i3}) &= \left[\frac{qS\overline{L}m_z^{\alpha_i}\alpha_i}{J_z} + \frac{qS\overline{L}m_z^{\omega_{zi}}\omega_{zi}}{J_z}, \frac{qS\overline{L}m_y^{\beta_i}\beta_i}{J_y} + \frac{qS\overline{L}m_y^{\omega_{yi}}\omega_{yi}}{J_y}\right]^T\\
g_{i1} &= \begin{bmatrix} -\dfrac{qSC_y^{\alpha_i}}{m} & 0 \\ 0 & \dfrac{qSC_z^{\beta i}}{m}\end{bmatrix},\\
g_{i2}(x_{i2}) &= \begin{bmatrix} 1 & \tan\beta_i\sin\alpha_i \\ 0 & \cos\alpha_i\end{bmatrix},\\
g_{i3} &= \operatorname{diag}\{J_z^{-1}, J_y^{-1}\},\\
u_i &= [M_{zi}, M_{yi}]^T,\\
d_{i1} &= [d_{\theta i}, d_{\varphi_{vi}}]^T,\\
d_{i2} &= [d_{\alpha i}, d_{\beta i}]^T,\\
d_{i3} &= [d_{\omega_{zi}}, d_{\omega_{yi}}]^T
\end{aligned} \tag{62}$$

According to (10)–(12), the following form of FTDO is designed for evaluating the disturbance d_{i1}, d_{i2}, and d_{i3} to (61),

$$\begin{aligned}
\dot{z}_{i10} &= v_{i10} + g_{i1}x_{i2},\\
\dot{z}_{i11} &= v_{i11},\\
\dot{z}_{i12} &= v_{i12}\\
v_{i10} &= -\lambda_{i10}\left[L_{i1}^{1/3}|z_{i10} - x_{i1}|^{2/3}\operatorname{sgn}(z_{i10} - x_{i1})\right] + z_{i11}\\
v_{i11} &= -\lambda_{i11}\left[L_{i1}^{1/2}|z_{i11} - v_{i10}|^{1/2}\operatorname{sgn}(z_{i11} - v_{i10})\right] + z_{i12}\\
v_{i12} &= -\lambda_{i12}\left[L_{i1}|z_{i12} - v_{i11}|^{q_{i1}/p_{i1}}\operatorname{sgn}(z_{i12} - v_{i11})\right]\\
z_{i10} &= \hat{x}_{i1},
\end{aligned}$$

$$z_{i11} = \hat{d}_{i1},$$
$$z_{i12} = \dot{\hat{d}}_{i1} \tag{63}$$

In the equation, the estimated value of d_{i1} is $\hat{d}_{i1}$, and the estimated error is $e_{i11} = z_{i11} - d_{i1}$.

Similarly, the estimated values of d_{i2} and d_{i3} are $\hat{d}_{i2}$ and $\hat{d}_{i3}$, respectively, and the estimated errors are $e_{i12} = z_{i12} - d_{i2}$ and $e_{i13} = z_{i13} - d_{i3}$.

To ensure that the "follower" can track the command signal of the cooperative control strategy rapidly and guarantee steady flight attitude, the "follower" controller is designed with the dynamic surface sliding-mode control law, and according to the FTDO estimated value and state coupling kinetic equation (61).

(1) The first dynamic error surface is defined as follows:

$$s_{i1} = g_{i1}^{-1}(x_{i1} - x_{i1d}) \tag{64}$$

In the equation, $x_{i1d} = [\overline{\theta}_{mi}, \overline{\varphi}_{vmi}]^T$ is the instruction of the trajectory inclination angle and trajectory deflection angle after filtering. Taking the derivative of s_{i1},

$$\dot{s}_{i1} = x_{i2} + g_{i1}^{-1}(d_{i1} - \dot{x}_{i1d}) \tag{65}$$

According to the dynamic surface sliding-mode control method and FTDO estimated value $\hat{d}_{i1}$, the virtual control of the first dynamic surface is selected as

$$x_{i2}^{*} = g_{i1}^{-1}(-\hat{d}_{i1} + \dot{x}_{i1d}) - k_{i1}s_{i1} \tag{66}$$

In the equation, $\dot{x}_{i1d}$ is the differential of the trajectory inclination angle and trajectory deflection angle after filtering, and $k_{i1} = \mathrm{diag}\{k_{i11}, k_{i12}\}$ is the positive definite matrix. The value of x_{i2}^{*} is obtained through the first-order low-pass filter, and the virtual control after filtering and its differential are given by

$$\tau_{i2}\dot{\overline{x}}_{i2}^{*} + \overline{x}_{i2}^{*} = x_{i2}^{*},$$
$$\overline{x}_{i2}^{*}(0) = x_{i2}^{*}(0) \tag{67}$$
$$\dot{\overline{x}}_{i2}^{*} = -\tau_{i2}^{-1}(\overline{x}_{i2}^{*} - x_{i2}^{*})$$

In the equation, $\tau_{i2} = \mathrm{diag}\{\tau_{i21}, \tau_{i22}\}$ is the time constant of the filter.

(2) The second dynamic error surface is defined by

$$s_{i2} = x_{i2} - \overline{x}_{i2}^{*} \tag{68}$$

In the equation, $\overline{x}_{i2}^{*}$ is the command signal after filtering. Taking the derivative of s_{i2},

$$\dot{s}_{i2} = f_{i2}(x_{i2}) + g_{i2}(x_{i2})x_{i3} + d_{i2} - \dot{\overline{x}}_{i2}^{*} \tag{69}$$

According to the dynamic surface sliding-mode control method and FTDO estimated value $\hat{d}_{i2}$, the virtual control of the first dynamic surface is given by

$$x_{i3}^{*} = g_{i2}^{-1}(x_{i2})\left[-f_{i2}(x_{i2}) - \hat{d}_{i2} + \dot{\overline{x}}_{i2}^{*} - k_{i2}s_{i2}\right] \tag{70}$$

In the equation, $\dot{\overline{x}}_{i2}^{*}$ is the instruction differential after filtering, and $k_{i2} = \mathrm{diag}\{k_{i21}, k_{i22}\}$ is the positive definite matrix. The value of x_{i3}^{*} is obtained through the first-order low-pass filter, and the virtual control after filtering and its differential can be obtained as follows:

$$\tau_{i3}\dot{\overline{x}}_{i3}^{*} + \overline{x}_{i3}^{*} = x_{i3}^{*},$$
$$\overline{x}_{i3}^{*}(0) = x_{i3}^{*}(0) \tag{71}$$
$$\dot{\overline{x}}_{i3}^{*} = -\tau_{i3}^{-1}(\overline{x}_{i3}^{*} - x_{i3}^{*})$$

In the equation, $\tau_{i3} = \mathrm{diag}\{\tau_{i31}, \tau_{i32}\}$ is the time constant of the filter.

(3) The third dynamic error surface is defined as follows:

$$s_{i3} = x_{i3} - \overline{x}_{i3}^{*} \tag{72}$$

In the equation, $\overline{x}_{i3}^{*}$ is the command signal after filtering. Taking the derivative of s_{i3},

$$\dot{s}_{i3} = f_{i3}(x_{i2}, x_{i3}) + g_{i3}u_i + d_{i3} - \dot{\overline{x}}_{i3}^{*} \tag{73}$$

According to the sliding-mode reaching law (see (59)), and FTDO estimated value $\hat{d}_{i3}$, the "follower" dynamic surface sliding-mode control law is designed as follows:

$$u_i = g_{i3}^{-1}\left[-f_{i3}(x_{i2}, x_{i3}) - \hat{d}_{i3} + \dot{\overline{x}}_{i3}^{*} - k_{i3}s_{i3} - k_{i4}|s_{i3}|^{\partial_{i3}}\operatorname{sgn}(s_{i3})\right] \tag{74}$$

In the equation $\dot{\overline{x}}_{i3}^{*}$ is the instruction differential after filtering, $k_{i3} = \mathrm{diag}\{k_{i31}, k_{i32}\}$ and $k_{i4} = \mathrm{diag}\{k_{i41}, k_{i42}\}$ are the positive definite matrices, $\partial_{i3} = \mathrm{diag}\{\partial_{i31}, \partial_{i32}\}$, and $0 < \partial_{i3} < 1$.

It can be learnt by referring to (27)–(44) that the stability of the control algorithm of the interceptor "follower" can be guaranteed by selecting appropriate parameters.

6. Simulation Verification

To verify the effectiveness of the distributed cooperative IGC algorithm of the multi-interceptor with state coupling designed in this study, it is assumed that the flight velocity of the interceptor "leader" remains the same. According to the global communication topology structure shown in Figure 1 and the local communication topology structure shown in Figure 2, Figure 1 assumes that the "leader" can communicate with the remaining three "followers," while the "followers" can communicate with each other. Figure 2 assumes that the "leader" can only communicate with "follower 1," while "followers" can communicate with each other. The initial conditions of interceptor "leader," "follower," and target are listed in Table 1.

Focusing on the two communication topology structures shown in Figures 1 and 2, a simulation study is conducted for the cooperative IGC algorithm of the multi-interceptor with state coupling designed in this study. It is assumed

TABLE 1: Initial conditions of the leader, follower, and target.

No.		Parameter	Value	Parameter	Value	Parameter	Value	Parameter	Value
1	Leader	x_{m0}	0m	y_{m0}	0m	z_{m0}	0m	V_{m0}	800 m/s
2	Follower1	x_{m10}	600m	y_{m10}	0m	z_{m10}	500m	V_{m10}	800 m/s
3	Follower2	x_{m20}	800m	y_{m20}	0m	z_{m20}	1000m	V_{m20}	800 m/s
4	Follower3	x_{m30}	400m	y_{m30}	0m	z_{m30}	-500m	V_{m30}	800 m/s
5	Target	x_{t0}	4000m	y_{t0}	5000m	z_{m40}	3000m	V_{t0}	400 m/s

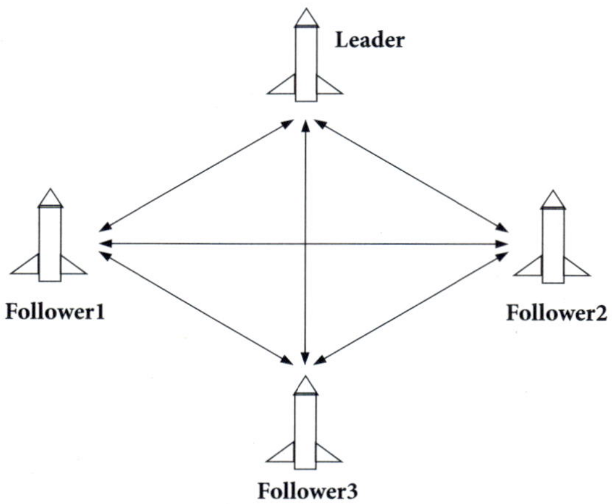

FIGURE 1: Global communication topology structure of interceptor "leader" and "follower".

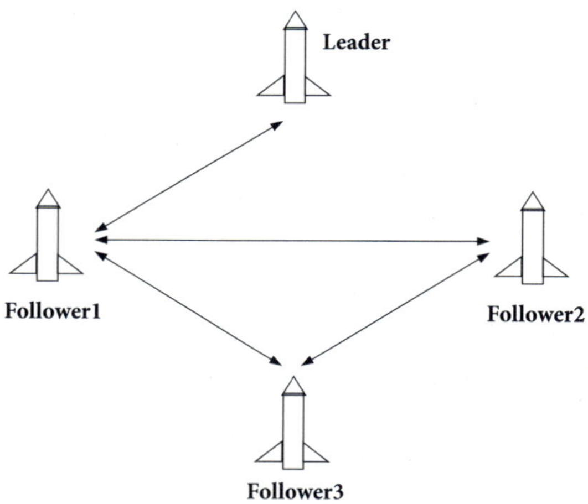

FIGURE 2: Local communication topology structure of the interceptor "leader" and "follower".

that the disturbance of the system is $d_2 = d_3 = d_{i1} = d_{i2} = d_{i3} = 0.02\sin(t)$, and the intercepted target has a linear acceleration of $a_{t\varepsilon} = a_{t\beta} = 5m/s^2$. The comparison of the simulation results of the cooperative IGC algorithm of the multi-interceptor with state coupling in the global and local communication topology is given, as shown in Figures 3–11.

Figures 3 and 4 show the motion trail of the interceptor "leader," "follower," and target in the local and global communication topologies. It can be seen that the motion trail of the interceptor "follower" in two different communication

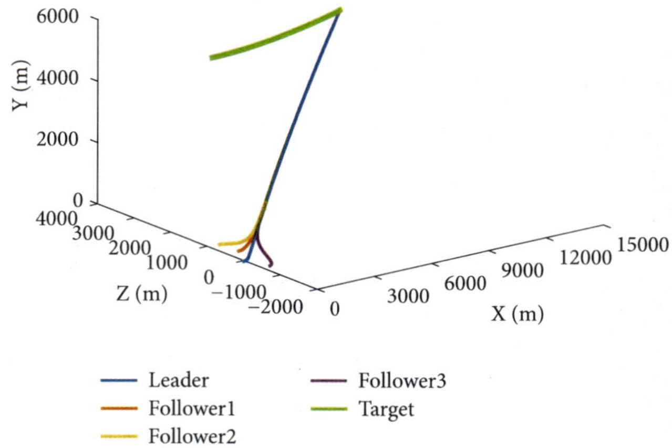

FIGURE 3: Motion trails of the interceptor "leader," "follower," and target in the global communication topology: X denotes the horizontal motion trail of the interceptor and target, Y denotes the longitudinal motion distance of the interceptor and target, and Z denotes the lateral motion trail of the interceptor and target.

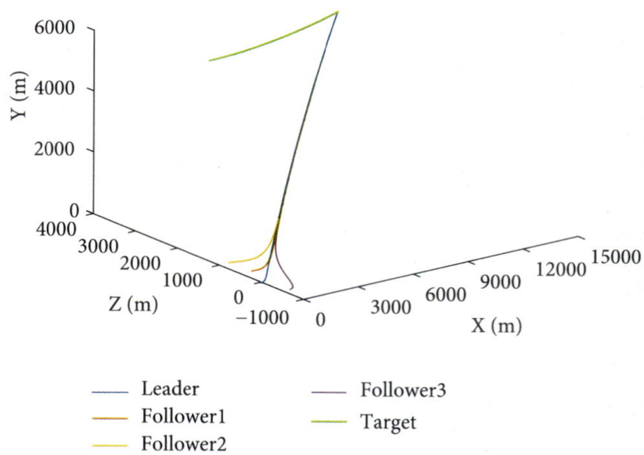

FIGURE 4: Motion trails of the interceptor "leader," "follower," and target in the local communication topology. X denotes the horizontal motion trail of the interceptor and target, Y denotes the longitudinal motion distance of the interceptor and target, and Z denotes the lateral motion trail of the interceptor and target.

topologies is gradually consistent with that of the "leader." Eventually, the "leader" and "follower" hit the target at the same time. The motion trail curve is smooth, showing short interception duration, fast convergence speed, and good stability.

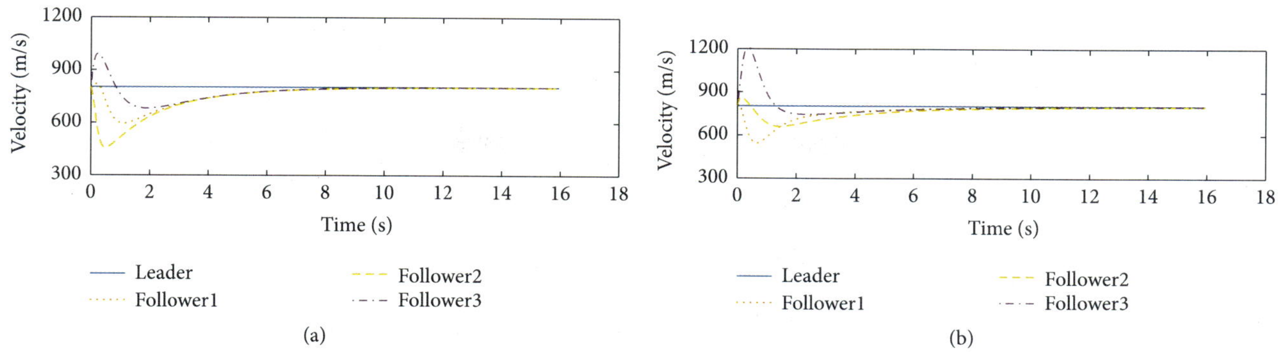

FIGURE 5: Interceptor velocity curve in the global and local communication topologies: (a) velocity curve in the global communication topology; (b) velocity curve in the local communication topology.

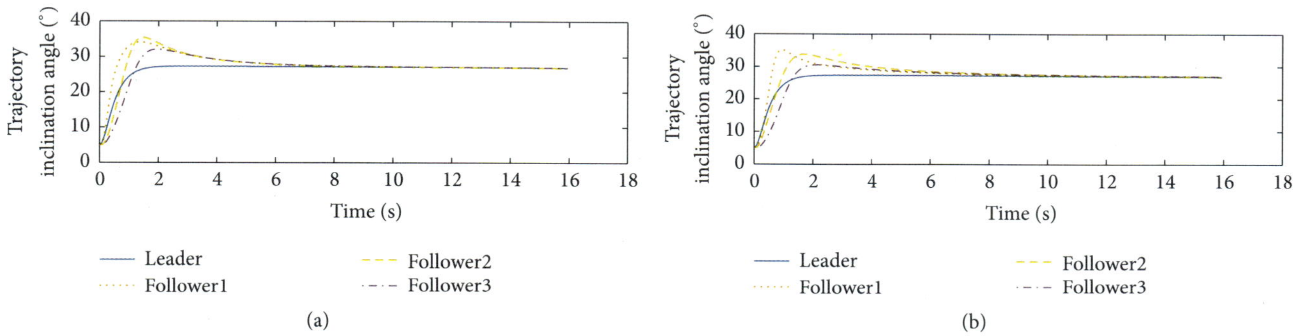

FIGURE 6: Trajectory inclination angle curve in the global and local communication topologies: (a) trajectory inclination angle curve in the global communication topology; (b) trajectory inclination angle curve in the local communication topology.

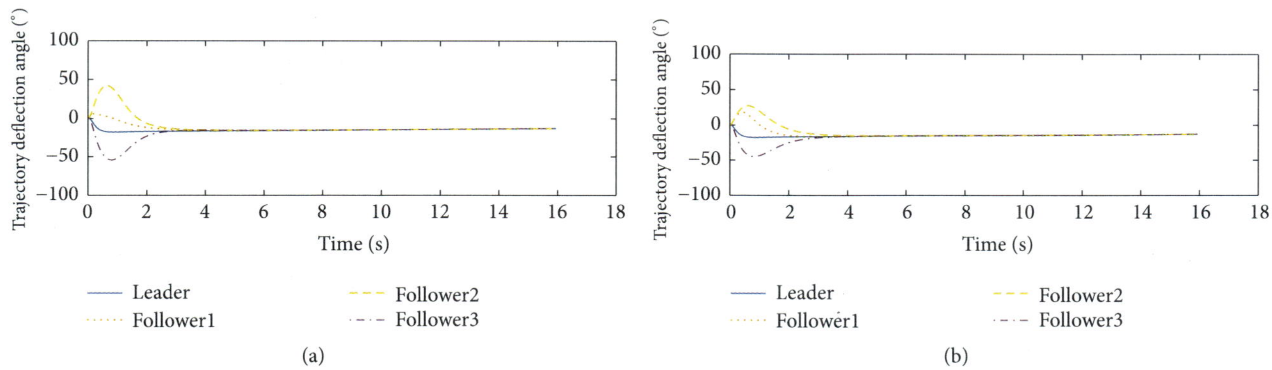

FIGURE 7: Trajectory deflection angle curve in the global and local communication topology: (a) trajectory deflection angle curve in global communication topology; (b) trajectory deflection angle curve in local communication topology.

Figures 5(a) and 5(b) show the velocity curve of the interceptor in the global and local topologies, from which it can be seen that the interceptor "follower" features a considerable overstriking property in the initial stage due to the lack of "leader." The convergence rate is slower than that in the global topology, but it can eventually reach the steady state of the "leader." The convergence process changes smoothly, and it shows good robustness to external disturbance. Similarly, it is clear from Figures 6–11 that the interceptor "follower" in the global and local communication topologies realizes the tracking of control instruction of the "leader." The convergence process is relatively smooth, and it also shows good robustness to external disturbance.

It can be seen from the simulation results of the global and local communication topologies that the distributed cooperative IGC algorithm of the multi-interceptor with state coupling designed in this study completes the tracking the instructions of the cooperative control strategy in the two different topologies and eventually achieves cooperative target interception.

The cooperative IGC algorithm of the multi-interceptor with state coupling proposed in this study and the traditional

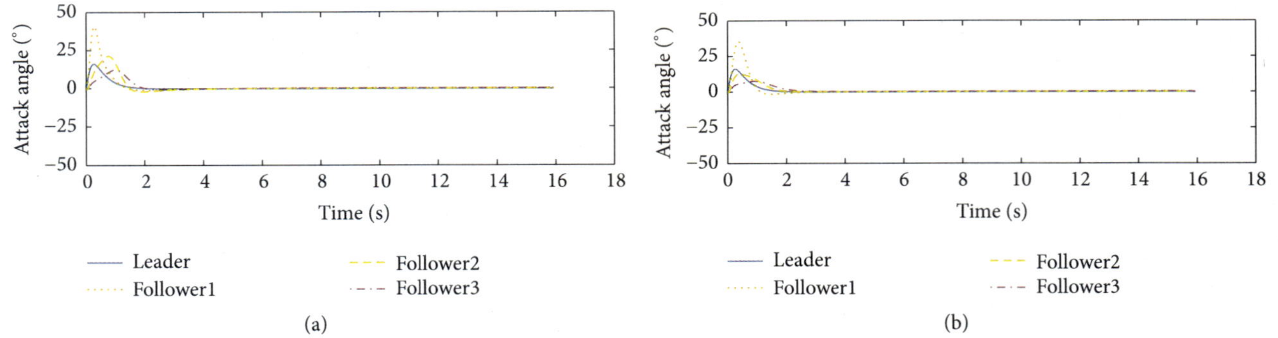

FIGURE 8: Attack angle curve in the global and local communication topologies: (a) attack angle curve in the global communication topology; (b) attack angle curve in the local communication topology.

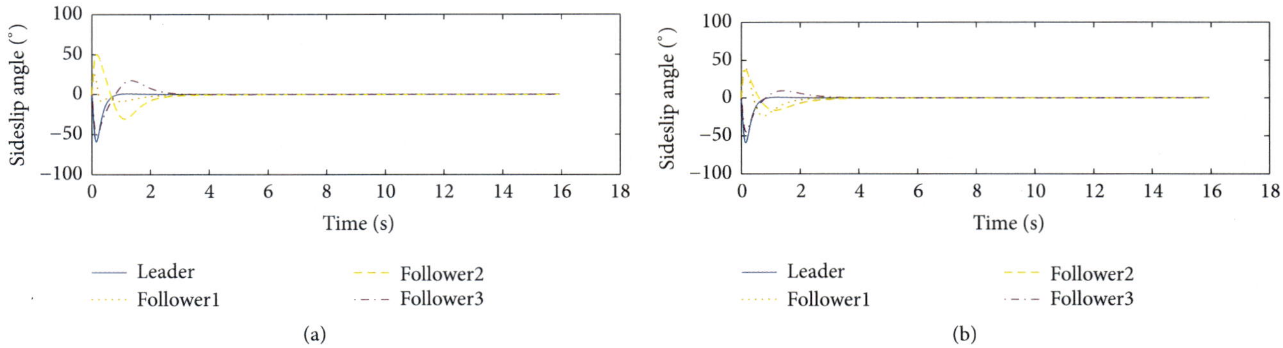

FIGURE 9: Sideslip angle curve in the global and local communication topologies: (a) sideslip angle curve in the global communication topology; (b) sideslip angle curve in the local communication topology.

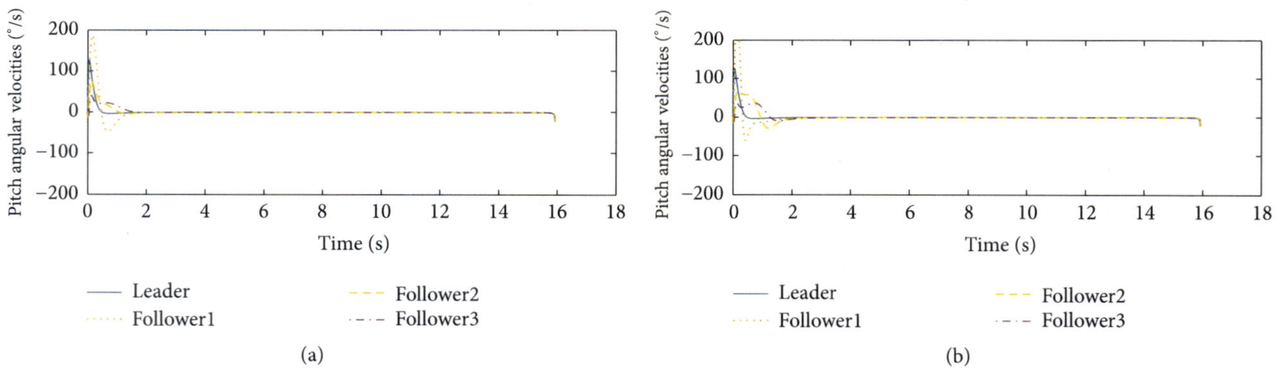

FIGURE 10: Pitch angular velocities curve in the global and local communication topologies: (a) pitch angular velocities curve in the global communication topology; (b) pitch angular velocities curve in the local communication topology.

method for multi-interceptor cooperation without state coupling are compared through numerical simulations. Figures 12 and 13 show the comparison results.

Figures 12 and 13 present the comparison of the attack and sideslip angle curves associated with the cooperative IGC algorithm of the multi-interceptor with state coupling and the traditional method for the multi-interceptor cooperation without state coupling. The convergence process associated with the cooperative IGC algorithm of the multi-interceptor with state coupling appeared to be smoother compared to that associated with the traditional method for the multi-interceptor cooperation without state coupling. The section of the curve after t = 4s particularly revealed more stable angle and sideslip angle curves using the state-coupled design methods. In contrast, fluctuations in the attack and sideslip angle curves were observed for the traditional design methods without state coupling. In other words, the cooperative IGC algorithm of the multi-interceptor with state coupling exhibited better resistance to interference and allowed for a more stable control of the interceptor

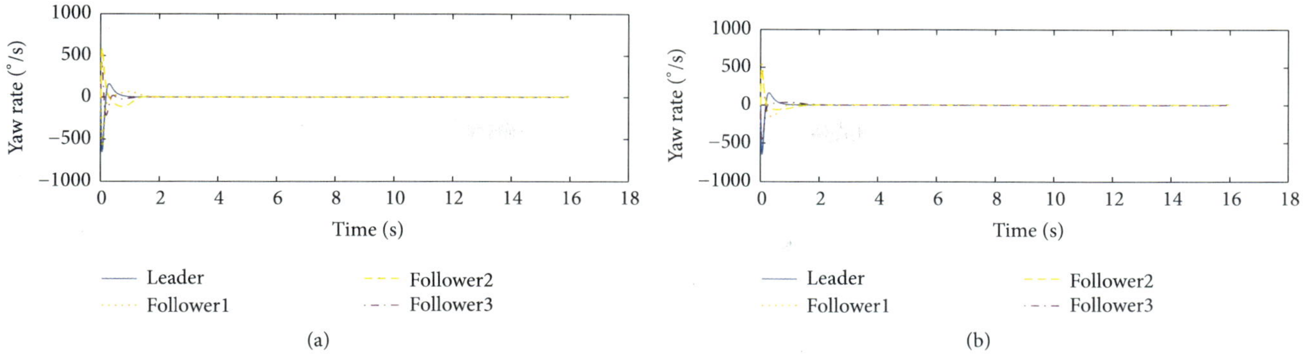

FIGURE 11: Yaw rate curve in the global and local communication topologies: (a) yaw rate curve in the global communication topology; (b) yaw rate curve in the local communication topology.

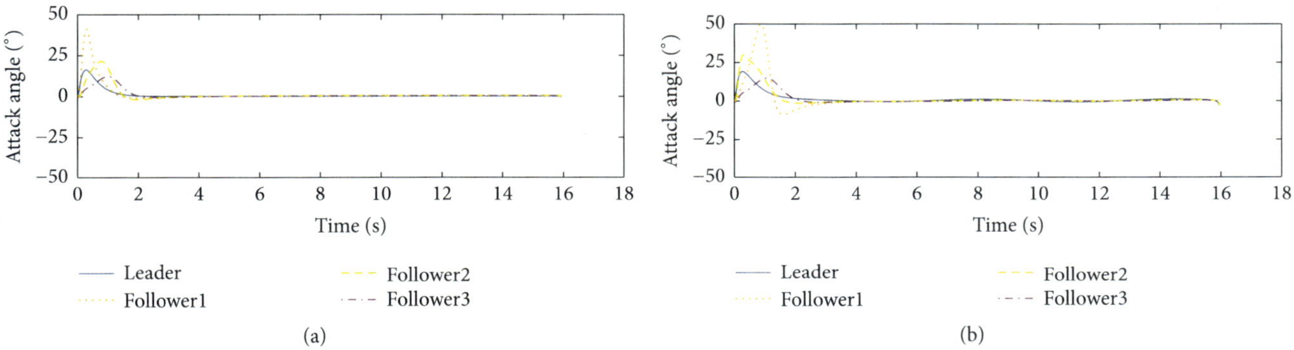

FIGURE 12: Comparison of the interceptor's attack angle curves in the design methods with and without state coupling: (a) attack angle curve in the design method with state coupling; (b) attack angle curve in the design method without state coupling.

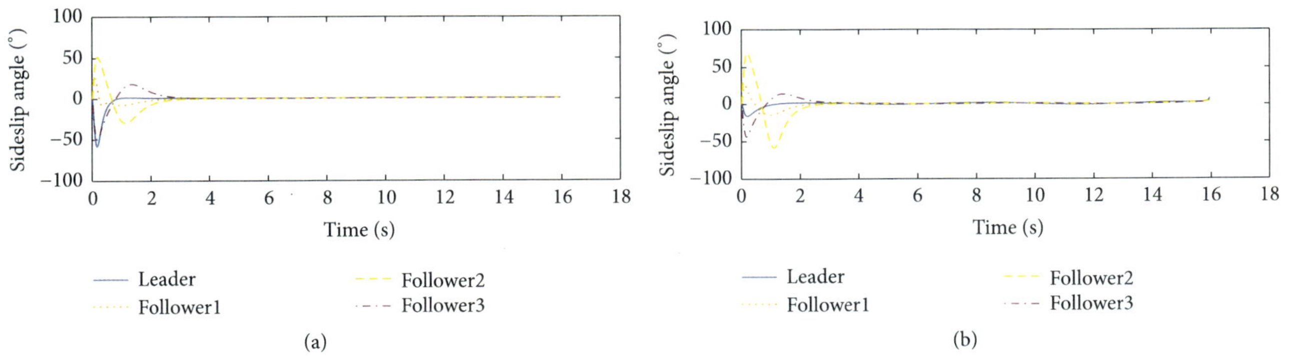

FIGURE 13: Comparison of the interceptor's sideslip angle curves in the design methods with and without state coupling: (a) sideslip angle curve in the design method with state coupling; (b) sideslip angle curve in the design method without state coupling.

compared to the traditional method without state coupling.

7. Conclusions

This study focused on the cooperative target interception by multi-interceptor and designed cooperative IGC algorithm of the multi-interceptor with state coupling "leader-follower" structure. The algorithm is designed by considering the coupling relation between the pitch and yaw channels of the interceptor. Further, this study combines the IGC method and introduces the distributed cooperative control strategy. The interceptor "leader" and "follower" control algorithm is designed separately by employing the dynamic surface sliding-mode control law and FTDO. The distributed cooperative control strategy guarantees that the "leader" and "follower" can hit the targets at the same time. The algorithm displays ideal trajectory characteristics in the simulation verification, and it can realize the cooperative interception of targets in both the global and local communication topologies. Furthermore, the study provides a design method for the cooperative target interception of the multi-interceptor, with certain engineering values.

Notations

V_m, V_{mi}: Interceptor velocity
V_t: Target velocity
α, α_i: Attack angle
β, β_i: Sideslip angle
ω_y, ω_{yi}: Yaw rate
ω_z, ω_{zi}: Pitch angular velocity
$\varphi_{vm}, \varphi_{vmi}$: Trajectory deflection angle of interceptor
ρ: Air density
θ_m, θ_{mi}: Trajectory inclination angle of interceptor
θ_t: Trajectory inclination angle of target
$C_y^{\alpha}, C_y^{\alpha_i}$: Contribution to lift due to angle of attack α
$C_z^{\beta}, C_z^{\beta_i}$: Contribution to yaw force due to sideslip angle β
g: Acceleration due to gravity
J_y: Moment of inertia around the yaw axis
J_z: Moment of inertia around the pitch axis
L: Reference length
m: Interceptor mass
$m_z^{\alpha}, m_z^{\alpha_i}$: Contribution to pitch moment due to angle of attack α
$m_y^{\beta}, m_y^{\beta_i}$: Contribution to yaw moment due to sideslip angle β
$m_y^{\omega_y}, m_y^{\omega_{yi}}$: Contribution to yaw moment due to yaw rate ω_y
$m_z^{\omega_z}, m_z^{\omega_{zi}}$: Contribution to pitch moment due to pitch rate ω_y
M_y, M_{yi}: Yaw moment
M_z, M_{zi}: Pitch moment
q: Dynamic pressure
q_β: Elevation angle
q_ε: Horizontal sight angle
r: Relative distance
S: Reference area
v_r: Relative velocity
v_β: Tangential relative velocity normal to yaw line-of-sight(YLOS)
v_ε: Tangential relative velocity normal to pitch line-of-sight(PLOS)
P_i: Motor power
$a_{m4\varepsilon}, a_{m4\beta}$: Longitudinal and lateral motion acceleration
$a_{t\varepsilon}, a_{t\beta}$: Longitudinal and lateral motion acceleration of the target
$a_{m3\varepsilon}, a_{m3\beta}$: Longitudinal and lateral acceleration.

Conflicts of Interest

The authors declare that there are no conflicts of interest regarding the publication of this paper.

Acknowledgments

The authors gratefully acknowledge the suggestions and help by the Professor Xiaogeng Liang and Northwestern Polytechnical University. This work was supported by the Aeronautical Science Foundation of China [Grant no. 2016ZC12005].

References

[1] J. B. Zhao and S. X. Yang, "Review of muti-missile cooperative guidance," *Acta Aeronautica et Astronautica Sinica*, vol. 38, no. 1, pp. 1–13, 2017.

[2] I.-S. Jeon, J.-I. Lee, and M.-J. Tahk, "Homing guidance law for cooperative attack of multiple missiles," *Journal of Guidance, Control, and Dynamics*, vol. 33, no. 1, pp. 275–280, 2010.

[3] J.-I. Lee, I.-S. Jeon, and M.-J. Tahk, "Guidance law to control impact time and angle," *IEEE Transactions on Aerospace and Electronic Systems*, vol. 43, no. 1, pp. 301–310, 2007.

[4] N. Harl and S. N. Balakrishnan, "Impact time and angle guidance with sliding mode control," *IEEE Transactions on Control Systems Technology*, vol. 20, no. 6, pp. 1436–1449, 2012.

[5] D. Cho, H. J. Kim, and M.-J. Tahk, "Nonsingular sliding mode guidance for impact time control," *Journal of Guidance, Control, and Dynamics*, vol. 39, no. 1, pp. 61–68, 2016.

[6] M. Nikusokhan and H. Nobahari, "Closed-form optimal cooperative guidance law against random step maneuver," *IEEE Transactions on Aerospace and Electronic Systems*, vol. 52, no. 1, pp. 319–336, 2016.

[7] V. Shaferman and T. Shima, "Cooperative differential games guidance laws for imposing a relative intercept angle," *Journal of Guidance, Control, and Dynamics*, vol. 40, no. 10, pp. 2465–2480, 2017.

[8] X. Wang, Y. Zheng, and H. Lin, "Integrated guidance and control law for cooperative attack of multiple missiles," *Aerospace Science and Technology*, vol. 42, pp. 1–11, 2015.

[9] S. R. Kumar and D. Ghose, "Cooperative Rendezvous Guidance using Sliding Mode Control for Interception of Stationary Targets," *IFAC Proceedings Volumes*, vol. 47, no. 1, pp. 477–483, 2014.

[10] J. Zhao and S. Yang, "Integrated cooperative guidance framework and cooperative guidance law for multi-missile," *Chinese Journal of Aeronautics*, vol. 31, no. 3, pp. 546–555, 2018.

[11] T. Lyu, Y. Lyu, and C. Li, "Cooperative guidance with impact angle constraint based on leader-follower strategy," *Advances in the Astronautical Science*, vol. 160, pp. 4009–4025, 2017.

[12] Q. Zhao, J. Chen, X. Dong, Q. Li, and Z. Ren, "Cooperative guidance law for heterogeneous missiles intercepting hypersonic weapon," *Hangkong Xuebao/Acta Aeronautica et Astronautica Sinica*, vol. 37, no. 3, pp. 936–948, 2016.

[13] V. Shaferman and T. Shima, "Cooperative optimal guidance laws for imposing a relative intercept angle," *Journal of Guidance, Control, and Dynamics*, vol. 38, no. 8, pp. 1395–1408, 2015.

[14] P. K. Menon and E. J. Ohlmeyer, "Integrated design of agile missile guidance and autopilot systems," *Control Engineering Practice*, vol. 9, no. 10, pp. 1095–1106, 2001.

[15] Y. B. Shtessel and C. H. Tournes, "Integrated higher-order sliding mode guidance and autopilot for dual-control missiles," *Journal of Guidance, Control, and Dynamics*, vol. 32, no. 1, pp. 79–94, 2009.

[16] M. Cross and Y. B. Shtessel, "Integrated Guidance Navigation and Control Using High-Order Sliding Mode Control for a Missile Interceptor," in *Proceedings of the 2018 AIAA Guidance, Navigation, and Control Conference*, Kissimmee, Florida.

[17] S. Shamaghdari, S. K. Nikravesh, and M. Haeri, "Integrated guidance and control of elastic flight vehicle based on robust MPC," *International Journal of Robust and Nonlinear Control*, vol. 25, no. 15, pp. 2608–2630, 2015.

[18] Z. Zhu, D. Xu, J. Liu, and Y. Xia, "Missile guidance law based on extended state observer," *IEEE Transactions on Industrial Electronics*, vol. 60, no. 12, pp. 5882–5891, 2013.

[19] T. Yamasaki, S. N. Balakrishnan, and H. Takano, "Integrated guidance and autopilot design for a chasing UAV via high-order sliding modes," *Journal of The Franklin Institute*, vol. 349, no. 2, pp. 531–558, 2012.

[20] H. Zhou, H. Zhao, H. Huang, and X. Zhao, "Integrated guidance and control design of the suicide UCAV for terminal attack," *Journal of Systems Engineering and Electronics*, vol. 28, no. 3, pp. 546–555, 2017.

[21] B. Panchal, K. Subramanian, and S. E. Talole, "Robust Integrated Guidance and Control Design for Tactical Missiles," in *Proceedings of the 2018 AIAA Guidance, Navigation, and Control Conference*, Kissimmee, Florida.

[22] S. H. Seyedipour, M. F. Jegarkandi, and S. Shamaghdari, "Nonlinear integrated guidance and control based on adaptive backstepping scheme," *Aircraft Engineering and Aerospace Technology*, vol. 89, no. 3, pp. 415–424, 2017.

[23] F.-K. Yeh, "Design of nonlinear terminal guidance/autopilot controller for missiles with pulse type input devices," *Asian Journal of Control*, vol. 12, no. 3, pp. 399–412, 2010.

[24] J. Wang, L. Liu, T. Zhao, and G. Tang, "Integrated guidance and control for hypersonic vehicles in dive phase with multiple constraints," *Aerospace Science and Technology*, vol. 53, pp. 103–115, 2016.

[25] Huibo Zhou, Shenmin Song, Junhong Song, and Jing Niu, "Design of Second-Order Sliding Mode Guidance Law Based on the Nonhomogeneous Disturbance Observer," *Journal of Control Science and Engineering*, vol. 2014, pp. 1–10, 2014.

[26] Bhavnesh Panchal and S. E. Talole, "Generalized ESO and Predictive Control Based Robust Autopilot Design," *Journal of Control Science and Engineering*, vol. 2016, pp. 1–12, 2016.

[27] Y. B. Shtessel, I. A. Shkolnikov, and A. Levant, "Smooth second-order sliding modes: missile guidance application," *Automatica*, vol. 43, no. 8, pp. 1470–1476, 2007.

11

Nonlinear Predictive Control System for Stiction Compensation in Electropneumatic Control Valves

M. E. U. J. Araújo,[1,2] J. R. T. Gadelha,[2] W. M. Santos,[2] A. L. Maitelli,[2] and F. M. U. Araújo[2]

[1]*Department of Informatics, Federal Institute of Paraíba, N-720, 58015-435 João Pessoa, PB, Brazil*
[2]*Department of Automation and Computing, CT, Federal University of Rio Grande do Norte, 59078-970 Natal, RN, Brazil*

Correspondence should be addressed to M. E. U. J. Araújo; marcio.ugulino@ifpb.edu.br

Academic Editor: Petko Petkov

This paper presents the implementation of a system that deals with static friction (stiction) in electropneumatic control valves, one of the most common nonlinearities that causes problems such as limit cycles and consequently wear of the valve and its moving parts, as well as losses in production and maintenance costs. This system is composed of a nonlinear predictive controller with adjustable constraints and an online database for estimation of the stiction parameters. The predictive controller uses constraints on the valve speed during its excursion, as well as constraints on the control signal to bring the valve to the desired position and slip it when necessary. The strategy adopted also showed robustness, being able to cope with changes in the spring and stiction parameters, which caused mismatch between the model and the controller and consequently loss of performance or even instability.

1. Introduction

Electropneumatic valves are common in the petrochemical area and are essential in many other industrial processes on fluid control. The positional control of these valves usually presents a challenge, since they present nonlinearities as dead zones, hysteresis, and static and dynamic friction. Among these nonlinearities, we can mention the occurrence of the grip and slip (stick-slip) phenomenal on the valve stem due to static friction (stiction) between its moving parts. This phenomenon may compromise position control that causes a regime error and even generates periodic oscillations, also known as limit cycles, which force the process variables to exceed their limits of operation, safety, and efficiency, causing a premature degradation not only in the valve, but in all components of the control mesh [1] which results in losses in productivity. 30% of industrial processes present oscillations are caused by problems in control valves [2]. The average cost of maintaining a valve is between US$ 400.00 and US$ 2,000.00 [3] and a production halt for maintenance routines may cause significant financial losses. Thus, the importance of the study and elaboration of viable solutions for the problems of grip in control valves are evident.

Stick-slip phenomenon needs to be represented by a mathematical model that approaches its physical characteristics. Moreover, accurate models that allow the identification and quantification of stiction are critical for their compensation by means of a model-based control strategy. However, a problem associated with its modeling and subsequent control resides in the fact that stiction presents stochastic characteristics [4]. Due to this stochastic nature, at each time, different control signal values are needed to overcome static friction in the same specific position, and different values of minimum speed are required so that the valve does not grip during opening or closing procedure, being not compatible with traditional model-based controllers. In addition to the problems inherent to the modeling and control of electropneumatic valves, the change of its parameters over time (caused by the wear of its mechanical components) makes the control and maintenance of production rates even more challenging, increasing the importance of periodic reidentification processes. These parameter changes can generate mismatch problems between the nominal model and the actual behavior of the plant, which may cause losses on the control loop efficiency, as well as nonfeasibility problems in the optimization process when a predictive controller is used.

PID controllers are the most used to control valves and also in a wide variety of industrial processes that include them, due to their robustness and satisfactory performance [5]. However, the inherent complexity of the stiction and other nonlinearities of control valves significantly reduce the performance of PID controllers, rendering them unable to drive the system to a desired position, which causes oscillations in the process control plant. These facts justify the rise of several approaches that treat stiction compensation and that preferably deal with stress constraint. Furthermore, other relevant characteristics in the performance of controllers, such as the control effort, may not be contemplated in cases where the control valves are gripped. An alternative is the adoption of Model-Based Predictive Controllers (MPC), since they contemplate these characteristics and their efficiency in the control of systems with Multiple Inputs and Multiple Outputs (MIMO), unlike the PID that is used in Single Input and Single Output (SISO).

MPC emerged in the late 70s [6, 7] and refer to a family of controllers that use the explicit process model to predict the future behavior of the dynamic system [8]. Initially designed and implemented in power plants and oil refineries, its use extends to several different areas in the present day. This can be partially justified by the difficulty of the classic controllers in dealing with the complexity, with operational constraints, MIMO systems, and nonlinearities of the current systems. These controllers use the system linear model to estimate the output trajectory N steps ahead, into the future, as well as the control signal trajectory M steps ahead, into the future. These signals are calculated by the minimization of an objective function defined in an optimization problem that may or may not consider the system constraints. The solution of the optimization problem can be obtained by linear programming, quadratic programming, min–max method, and multicriteria optimization [9]. MPC are heavily dependent on the type of model used to represent the dynamics of the process, which, in some cases, may be a disadvantage. Linear models are commonly described in the form of impulse response, step response, transfer function, and the state space model [10]. In MPC, state space models are most commonly used compared to others.

Although many industrial processes can be controlled by controllers based on linear models, there are processes that have many nonlinearities, which makes these controllers unsatisfactory. Therefore, the search for solutions based on nonlinear models has become attractive, such as neural controllers, fuzzy controllers, and predictive controllers based on nonlinear models (NMPC). This work proposes the use of NMPC, since it has a direct relation with the mathematical model that represents the system. On the other hand, the use of NMPC is a challenge due to the wide variety of nonlinearities present in physical systems. Some alternatives to deal with nonlinearities can be found in the literature, such as controllers based on the Hammerstein Model, Volterra Model, Wiener Model, and bilinear models. But for this, it is necessary to represent the system or process by one of these specific models, which becomes a limitation, because the nonlinearity in these models is static and stiction is stochastic parameter. Another alternative is the NMPC using successive linearization [13]. However, this approach requires considerable computational effort due to some computationally complex mathematical operations. In this paper, a nonlinear predictive controller based on a state space model structured in a similar to [13] was modified for the proposed compensation methodology in position controller system of the valve.

Kayihan and Doyle [14] proposed the use of an input-output linearization (IOL) scheme for friction compensation in pneumatic valves. However, it can only be applied to valves with an internal controller. Hägglund [15] describes a compensation technique for static friction in pneumatic valves, called Knocker. A similar proposal to Knocker was presented by Ivan and Lakshminarayanan [16], where the compensator received the denomination Constant Reinforcement (CR). These two methods use PID controllers and may not reach the above-mentioned operational constraint requirements, in addition to generating a control signal with reasonable variability and in some cases taking a long stabilization time.

This paper proposes the implementation of a nonlinear, iterative, predictive control system for stiction compensation. This system is able to lead the valve to different setpoints, which reduces the oscillations presented in the previous papers. For this, the basic principle of the proposed system is not to try to compensate stiction, but to use it as an advantage by designing a nonlinear predictive controller that incorporates minimum and maximum speed restrictions on the valve excursion; it decelerates and forces the valve to hold as close as the desired setpoint. In order to do this, a table that is updated with the excursion speed and control signal necessary to initiate slip phase was used. This update occurs every time the valve is grasped.

For better control dynamics, the concept of variable constraints was incorporated into the predictive controller, as well as the need to store the last ten (10) values in the stiction table to each setpoint aiming at calculating the medium value to compensate the stochastic character of the stiction [4]. Based on these parameters, the adopted predictive control strategy proved itself robust, capable of dealing with variations in valve parameters value, such as changes of 20% in the spring constant and the stiction value which causes mismatch between controller model and valve model.

This paper is organized as follows: Section 2 presents the linear displacement valve model and its electropneumatic control. In Section 3, a mathematical model of the friction force on the diaphragm-spring electropneumatic valves stem is presented. Section 4 presents brief explanation of the nonlinear predictive control based on the state space representation model. In Section 5, the proposed method is presented. In Section 6, simulation responses of the control methodology proposed are showed in five different situations. Finally, Section 7 draws conclusions and proposals for future work are suggested.

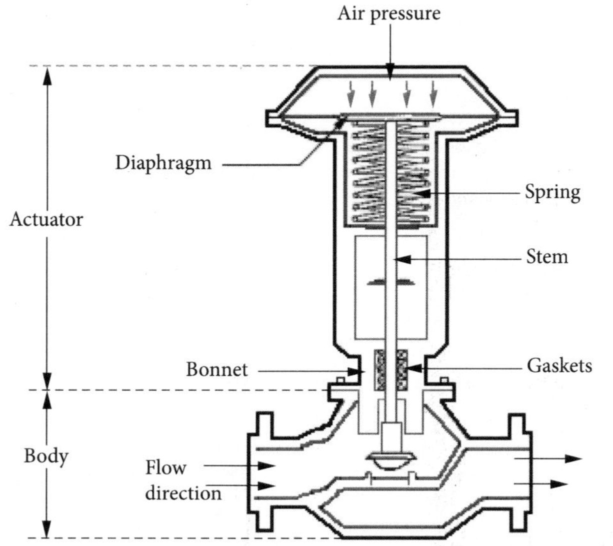

FIGURE 1: Body and actuator of a control valve with pneumatic actuator (adapted from [11]).

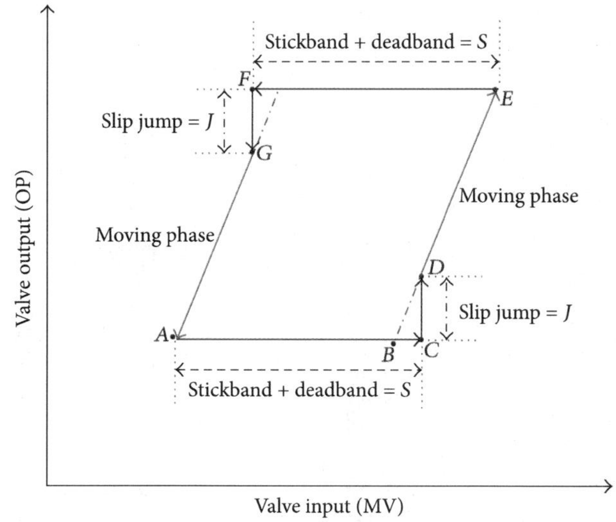

FIGURE 2: Standard signature curve of a valve with stiction (adapted from [12]).

2. Valves

Valves are devices that are able to restrict or allow the passage of a fluid in response to a control signal. Valves can be classified as manual, self-regulatory, and controlled, which use an auxiliary force to operate, and its operation is driven by a control signal. This auxiliary force comes through the combined action of the actuator and the positioner. Similarly, there are several types of actuators and positioners. The types of actuators include the handwheel, manual lever, electric motor, pneumatic, solenoid, hydraulic piston, and self-actuation, where more than 90% of these are pneumatic actuators [15]. They all have a particular composition due to their type. This paper focuses on electropneumatic linear displacement control valves which consist basically of two main sets: the actuator and the body. The actuator, in this case, converts the electrical signal into a corresponding pressure to move valve stem. The conversion is done through a linear relationship between current and pressure, which, in turn, uses air pressure on one or both sides of a diaphragm to provide the force to position the valve.

The body assembly is basically divided into the following subsets: body, internal components, bonnet, and lower flange. In some valve types, the body and bonnet form one set simply called the body. In other valve types, there is no lower flange; hence, it makes the whole valve body the part that comes into direct contact with the fluid. The valve types are classified according to their body types. These valves are divided into the following:

(I) *Linear displacement*: conventional globe, three-way globe, cage globe, angle globe, diaphragm, and guillotine.

(II) *Rotary displacement*: butterfly, ball, shutter eccentric, and sphere segment.

One of the key elements of a valve is the box set of gaskets that serves to seal against process fluid leakage. Figure 1 represents the body and actuator of a linear displacement control valve (globe type).

Two problems related to the difference between the atmospheric and internal valve pressures may arise in cases when the gaskets box is not correctly sealed. If internal pressure is higher, it can cause leakage from the valve's internal fluid to the environment. In those cases where these valves are used in flammable or toxical fluid pipes, the gaskets must be firmly pressed in order to prevent any leakage. This may cause an increase in stiction and consequently the problems associated with this phenomenon.

The desired characteristics of the material used in the composition of gaskets are elasticity, low coefficient of friction, pressure, temperature, and resistance to corrosion caused by the process fluid. The most common materials used in the construction of gaskets are Teflon and impregnated asbestos.

Stiction is the most common type of nonlinearity in electropneumatic control valves [1]. The sealing rings (also known as gaskets) are the principal responsible component in the valve for the phenomenon.

The stiction phenomenon can be better explained by the valve input-output behavior, as illustrated in Figure 2. In the absence of stiction, the valve moves along the dotted line trace that passes through the origin. Any controller output signal (OP) would result in the same change in the value of the manipulated variable (MV). However, for a stuck valve, static and dynamic friction components must be taken into account. The input-output can therefore be described by four components: deadband, stiction band, slip jump (J), and mobile phase.

Let us assume that in point A, Figure 2, the valve stem is stopped. Shortly thereafter there was a reversal in the movement direction where MV is constant, as long as the valve is stuck by the action of static friction force (F_s). The deadband range (AB) exists due to the presence of Coulomb friction (F_c) (a constant force that operates in the opposite direction of the movement). Point BC corresponds to the band stiction (J), where as in point C the MV value is abruptly changed (stiction jump-J), due to the applied force and consequently valve stem sliding. After that, MV varies progressively (movement phase); it is only opposed by the dynamic frictional force (F_v). When the speed of the valve stem is too low, it grasps again and the movement inversion does not occur. The band formed by the deadband and the stiction band will be referred to as S. In an attempt to overcome the gripping, the actuator may jump around the desired position that causes the temporary oscillatory behavior of the manipulated variable. This behavior of the manipulated variable is also propagated to the process variable which, in turn, is how the stiction disturbance contributes to the instability of the process and, consequently, to the loss of performance of the control loop.

As discussed above, friction is a force that directly influences the valve dynamics. Its modeling is fundamental, in order to detect and quantify the stiction force to controller design that focuses on its compensation.

3. Modeling

Several models have been proposed to represent stiction dynamics in pneumatic valves using phenomenological approaches. Reference [17] presents several of these models. Since then, new models have been created using the input-output relationship. References [12, 18–20] presented new models, where [18] presents one parameter based model, and the other works are based on two-parameter model.

3.1. Valve Model and Friction Force Model. Eight friction models were selected by [2] (physical and based on process operating data models) in order to conduct a performance comparison among them: [21], the seven-parameter models of [22], Siskin model [23], the one parameter model of [18], [12] model, [19] model, and [20] model. In conclusion, the best representations of friction in a control valve are Karnopp, Siskin, and Kano models, since they had been approved in all tests.

In this paper, the physical friction model of Karnopp was selected, since it adequately represents the phenomenon of stiction, as concluded in the work of [2], and it models friction as a static function of speed. In this way, the position and speed of the stem can be manipulated directly by the predictive controller, which is the basis of the methodology proposed by this paper.

In order to obtain the valve model, a force balance is applied to the valve stem according to Newton's second law. It is assumed that the input signal is the signal defined by the controller, converted to pressure measurement unit, and the system output is the valve stem position (usually normalized between 0 and 100%), as shown in

$$m\ddot{x}(t) = F_{\text{pressure}} - F_{\text{spring}} - F_{\text{friction}} - F_{\text{fluid}} - F_{\text{settlement}}, \tag{1}$$

where

(i) m is the mobile mass parts (obturator + stem);

(ii) the position of valve is $x(t)$;

(iii) $F_{\text{pressure}} = S_a P$ is the force applied by the pressure on the diaphragm actuator; S_a is the diaphragm area and P is the pressure on the actuator;

(iv) $F_{\text{spring}} = K_m x(t)$ is the force applied by the actuator spring and K_m is the spring elastic constant;

(v) F_{fluid} is the force applied by the process fluid in the valve obturator;

(vi) F_{friction} is the total friction force on stem;

(vii) $F_{\text{settlement}}$ is a settlement external force (necessary to lock the valve on the seat) of the shutter in the seat of the valve; it only acts on the end of the stem excursion.

Reference [14] assumed that the force of the fluid F_{fluid} can be neglected, since it is very small compared to the other forces involved. But [14] does not present results to prove that F_{fluid} can, in fact, be neglected. Reference [24] conducted laboratory experiments to confirm such hypothesis. The settlement force $F_{\text{settlement}}$ also will be neglected since it acts only in the end of the valve actuation range, when there is no further stem movement made; only the shutter is being forced against the seat.

Therefore, (1) can be written as

$$m\ddot{x}(t) = F_{\text{pressure}} - F_{\text{spring}} - F_{\text{friction}}. \tag{2}$$

The main difference between the most commonly used physical models for control valve lies in the way that the friction force is calculated. Therefore, in [2], a study for the application of different friction models to a control valve is conducted. There are more complex models in the literature as described in [22] using seven parameters. These complex models, in general, have parameters in their model of difficult access. Thus, it is important to select a model that presents simplicity and, at the same time, adequately describes the friction.

In this paper, the Karnopp model in [21] is adopted, since it easily deals with the null speed detection problem, an important point in valve modeling, through the definition of a minimum speed DV. The model used is represented by equations below:

$$F_{\text{friction}} = \begin{cases} F_{\text{friction}}(\dot{x}(t)), & \text{if } |\dot{x}(t)| \geq DV, \\ F_e, & \text{if } |\dot{x}(t)| < DV,\ |F_e| \leq F_s, \\ F_s \cdot \operatorname{sign}(F_e), & \text{if } |\dot{x}(t)| < DV,\ |F_e| > F_s. \end{cases} \tag{3}$$

F_{friction} is the total friction force defined as

$$F_{\text{friction}}(\dot{x}(t)) = \gamma \operatorname{sign}(\dot{x}(t)) + F_v \dot{x}(t),$$
$$\gamma = \left[F_c + (F_s - F_c)\, e^{(\dot{x}(t)/V_s)^2}\right], \tag{4}$$

where

$$\operatorname{sign}(\dot{x}(t)) \begin{cases} 1, & \text{if } \dot{x}(t) > 0, \\ 0, & \text{if } \dot{x}(t) = 0, \\ -1, & \text{if } \dot{x}(t) < 0. \end{cases} \tag{5}$$

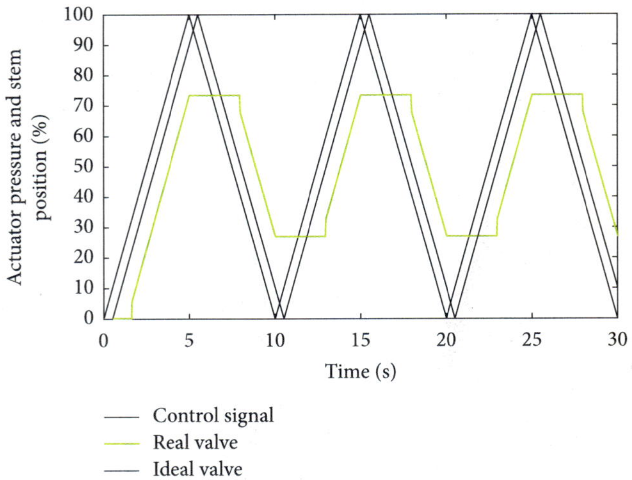

FIGURE 3: Valve responses to an input of a triangular signal.

TABLE 1: Nomenclatures in mathematical Karnopp model.

Parameter	SI base units	Description
F_e	N	The sum of the external forces: $F_{\text{pressure}} - F_{\text{spring}}$
F_c	N	Coulomb friction coefficient
F_s	N	Static friction coefficient
F_v	N	Viscous friction coefficient
V_s	m/s	Stribeck speed
$x(t)$	m	Stem position
$\dot{x}(t)$	m/s	Stem speed

A presence of static friction (stiction) causes a reduced control valve working range. In many cases, the valve stem signals are normalized to a corresponding value between the MV limit ranges. This way, when the friction is absent, the stem position curve behaves as a replica of the controller signal curve, which can be observed in the red and blue color curves, respectively. However, in the green color curve, in Figure 3, the actual behavior of the stem is observed, where it presents a smaller amplitude than the blue curve, due to the phenomenon of the stick-slip caused by the presence of the friction. In Figure 3, the triangular input signal and the outputs are displayed as time works for a valve which is considered in perfect condition and, in the case of the green curve, the friction action is present. The application of a triangular signal to the valve is a commonly used method when defining the valve signature curve.

Table 1 describes the nomenclature to be adopted in the Karnopp model equation and its respective units of measure.

As previously mentioned, valves are elements that present many nonlinearities and a stochastic parameter (stiction). Beyond that, the parameters values of the valve can suffer variations causing mismatching between the dynamic system and the phenomenological model.

One of the parameters that suffers a great degradation due to the oscillations caused by the stiction phenomenon is the spring constant, which tends to decrease when compared to new valves. This forces a periodic reidentification process of the model parameters and consequently controller parameters recalculation.

Valves that operate in external environments or with fluids that suffer changes with temperature may represent a challenge, since these temperature changes affect the spring elastic constant and the friction in the gaskets, due to thermal expansion.

The use of a PID controller forces the system into an oscillatory cycle. This is caused by the integrative action component of the controller when the valve grasps in a wrong position. In this situation, the integrative action keeps increasing the control input value until the valve starts moving. However, when this happens, the valve usually slips, and its position jumps over the reference, completing the cycle.

3.2. Valve Model in the State Space Representation. From the total friction force model described by (4) and from equation of the valve model, we have

$$m\ddot{x}(t) = F_{\text{pressure}} - F_{\text{spring}} - F_v\dot{x}(t) - \left[F_c + (F_s - F_c)e^{(\dot{x}(t)/V_s)^2}\right]\text{sign}(\dot{x}(t)), \tag{6}$$

where forces F_{pressure} and F_{spring} are, respectively, defined by

$$F_{\text{pressure}} = S_a P, \qquad F_{\text{spring}} = K_m x(t), \tag{7}$$

where S_a is the diaphragm area, P is the pressure in diaphragm, and K_m is the elastic constant of the spring. From (7) in (6)

$$m\ddot{x}(t) = S_a P - K_m x(t) - F_v\dot{x}(t) - \left[F_c + (F_s - F_c)e^{(\dot{x}(t)/V_s)^2}\right]\text{sign}(\dot{x}(t)). \tag{8}$$

Parameters inherent to the actuator such as maximum pressure ($P_{\max}$), minimum pressure ($P_{\min}$), maximum current ($I_{\max}$), and minimum current ($I_{\min}$) are used so that a linear relationship can be established in the pressure conversion in the diaphragm of valve P, as a function of the current signal $u(t)$ in mA. The dynamics in a linear converter can be neglected because it has a faster dynamic in relation to the system.

$$K_p = \frac{\Delta P}{\Delta I} = \frac{P_{\max} - P_{\min}}{I_{\max} - I_{\min}}, \tag{9}$$

$$P = K_p u(t). \tag{10}$$

From the substitution of (10) in (8), we reach

$$m\ddot{x}(t) = S_a K_p u(t) - K_m x(t) - F_v\dot{x}(t) - \left[F_c + (F_s - F_c)e^{(\dot{x}(t)/V_s)^2}\right]\text{sign}(\dot{x}(t)). \tag{11}$$

Finally, in order to define the position and the speed of the stem as state variables, as well as by algebraically manipulating equation (11), it gives a state space model representation, based only on the total frictional force, that is, the classic model found in the literature to represent the frictional force.

$$\begin{gathered} x_1(t) = x(t), \\ x_2(t) = \dot{x}(t) \\ \Downarrow \\ \dot{x}_1(t) = \dot{x}(t), \\ \dot{x}_2(t) = \ddot{x}(t), \end{gathered} \tag{12}$$

$$\begin{aligned} \dot{X}(t) &= AX(t) + Bu(t) + K_{nL}\operatorname{sign}(x_2(t)), \\ y(t) &= CX(t), \end{aligned} \tag{13}$$

where A, B, C, K_{nL}, and $X(t)$ are matrices defined by

$$\begin{aligned} X(t) &= \begin{bmatrix} x_1(t) \\ x_2(t) \end{bmatrix}, \\ \dot{X}(t) &= \begin{bmatrix} \dot{x}_1(t) \\ \dot{x}_2(t) \end{bmatrix}, \\ A &= \begin{bmatrix} 0 & 1 \\ -\dfrac{K_m}{m} & -\dfrac{F_v}{m} \end{bmatrix}, \\ B &= \begin{bmatrix} 0 \\ \dfrac{S_a K_p}{m} \end{bmatrix}, \\ C &= \begin{bmatrix} 0 & 1 \end{bmatrix}, \\ K_{nL} &= \begin{bmatrix} 0 \\ -\dfrac{1}{m}\left[F_c + (F_s - F_c)\, e^{(x_2/V_s)^2}\right] \end{bmatrix}. \end{aligned} \tag{14}$$

The adoption of the classical friction force model in place of the Karnopp model in the state space representation is due to the fact that some model-based controllers, such as predictive controllers, require the model to be represented in state space, which is difficult to obtain from the Karnopp model. In addition, this representation also needs to be in discrete time that implies the use of discretization techniques for its practical application in the control of electropneumatic valves.

4. Nonlinear MPC-NMPC

Systems that can be represented in the following manner are considered in this work:

$$\begin{aligned} x(k+1) &= f(x(k), u(k)) \cdot x(k) + g(x(k), u(k)) \\ &\quad \cdot u(k) \\ y(k) &= Cx(k), \end{aligned} \tag{15}$$

where $x \in R^n$ is the state vector, $u \in R^m$ is the input vector, $y \in R^P$ is the output vector, and $k \in \mathbb{N}$ is the sampling time. It is assumed that the controlled system and state space variables are subject to the following linear constraints:

$$\begin{aligned} Gx(k) &\le \rho, \\ Hu(k) &\le \mu. \end{aligned} \tag{16}$$

The predictive control scheme defines, at each time step k, an input trajectory $\hat{u}(k)$, $k = 0, 1, \ldots, H_u$, that satisfies the following nonlinear optimization problem:

$$\begin{aligned} \min_{\hat{x}(k+i), \hat{u}(k+j)} \quad & J(k) \\ & = \sum_{i=1}^{H_p} \left\| r(k+i) - C\hat{x}(k+i) \right\|_Q \\ & \quad + \sum_{H_u-1}^{j=0} \left\| \hat{u}(k+j) \right\|_R, \\ \text{Subj. to} \quad & \hat{x}(k+1+i) \\ & = f(\hat{x}(k+i), \hat{u}(k+i))\, \hat{x}(k+i) \\ & \quad + g(\hat{x}(k+i), \hat{u}(k+i))\, \hat{u}(k+i), \\ & \hat{x}(k) = x(k), \\ & \hat{u}(k+j+1) = \hat{u}(k+j), \\ & G\hat{x}(k+i) \le \rho, \\ & H\hat{u}(k+i) \le \mu, \end{aligned} \tag{17}$$

where the objective function $J(k)$ is the performance index of the finite prediction horizon; $r(k+i)$ is the future reference trajectory to be followed by the output $y(k) = Cx(k)$; H_p is the prediction horizon, it is constant, and $H_p \in \mathbb{N}$; H_u is the control horizon, it is constant, and $H_u \in \mathbb{N}$; Q is the tracking error weighting matrix; R is the control effort weighting matrix (being Q and R positive definite); $\hat{x}$ is the state vector predicted by the model in (15), while in (17) $\hat{x}(k) = x(k)$; and $\hat{u}$ is the input vector to be determined in a manner that optimizes the performance index.

The $J(k)$ minimization results in the control sequence $\hat{u}(k+j)$ that optimizes the predicted behavior of the system in the prediction horizon H_p. The mathematical model does not represent with perfect accuracy the behavior of real systems, the occurrence of disturbances, and the presence of noise, among other difficulties that arise in practical applications, and the predictive control strategy employs the movable horizon principle that consists in the application of the first element $\hat{u}(k \mid k)$ of the control sequence computed and the subsequent recalculation of the optimization problem in the next time step.

Prior knowledge of the full state vector $x(t)$ is required for the solution of the optimization problem which characterizes a full state feedback structure.

The optimization (17) is a nonlinear, nonconvex problem and may depend on many variables that depend on the system dimension and the length of the horizons H_p and H_u. The solution for this problem presents a considerably high computational cost, hindering its application on systems with fast dynamics, where the sampling time is not high enough so that the problem solution can be computed in a feasible time period. However, if the system dynamic is linear, explicit linear equations can be defined in a way that represents the predicted states $\hat{x}(k+i)$ as functions of the inputs variables $\hat{u}(k+j)$. This way, the number of variables is greatly reduced and the problem can be formulated as a quadratic optimization problem, enabling the usage of efficient and fast algorithms. Thus, the methods that have been proposed for the resolution of nonlinear predictive control problems can be roughly divided into two categories: the first one proposes fast algorithms to solve the problem (17) and the second one proposes different ways to linearize the model (15), so that efficient methods for quadratic optimization can be employed. The method employed in this work fits the second category and will be described in the following section.

4.1. Iterative Compensation Algorithm. The one-step state prediction is considered in the following equation:

$$\hat{x}(k+1+i) = f(\hat{x}(k+i), \hat{u}(k+i))\,\hat{x}(k+i) + g(\hat{x}(k+i), \hat{u}(k+i))\,\hat{u}(k+i). \tag{18}$$

It is important to notice that if the values of $\hat{x}(k+i)$ and $\hat{u}(k+i)$ were numerically defined through the prediction horizon $i = 0, \ldots, H_p - 1$, the following time-variant linear state equation would be defined as

$$\hat{x}(k+1+i) = \widetilde{A}_i(\hat{x}, \hat{u})\,\hat{x}(k+i) + \widetilde{B}_i(\hat{x}, \hat{u})\,\hat{u}(k+i). \tag{19}$$

Application of the movable horizon principle in (19) is

$$\begin{aligned}\hat{x}(k+1+i \mid k) &= \widetilde{A}_i(\hat{x}(k+i \mid k), \hat{u}(k+i \mid k))\,\hat{x}(k+i \mid k) \\ &\quad + \widetilde{B}_i(\hat{x}(k+i \mid k), \hat{u}(k+i \mid k))\,\hat{u}(k+i \mid k),\end{aligned} \tag{20}$$

where $\widetilde{A}_i(\hat{x}, \hat{u}) = f(\hat{x}(k+i), \hat{u}(k+i))$ and $\widetilde{B}_i(\hat{x}(k+i \mid k), \hat{u}(k+i \mid k)) = g(\hat{x}(k+i), \hat{u}(k+i))$. However, $\hat{x}(k+i)$ and $\hat{u}(k+i)$ values are not known throughout the horizon, since the values $\hat{x}(k+i)$ depend on the future values $\hat{u}(k+i)$. Nevertheless, in [13], an algorithm that promotes the convergence of states to the correct predicted values is described and defined as follows.

For each sample time k, one has the following.

Step 1. Set up the input vector $U(k) = U^*(k)$.

Step 2. It is considered that $x(k)$ is measured in the plant; it applies $U(k)$ to the model (19) to obtain the prediction vector states $\widehat{X}(k) = [\hat{x}(k+1 \mid k), \hat{x}(k+2 \mid k), \ldots, \hat{x}(k+H_p-1 \mid k), \hat{x}(k+H_p \mid k)]$ over the prediction horizon of H_p.

Step 3. Apply $U(k)$ and $\widehat{X}(k)$ to model state prediction one step ahead as in (20). In order to obtain linear state equations over H_p prediction horizon from current time k, $\hat{x}(k+1+i \mid k) = A(k+i \mid k)\hat{x}(k_i \mid k) + B(k+i \mid k)\hat{u}(k+i \mid k)$. Note that $\widetilde{A}(\hat{x}(k+i \mid k), \hat{u}(k+i \mid k))$ and $\widetilde{B}(\hat{x}(k+i \mid k), \hat{u}(k+i \mid k))$ are functions of the states and inputs along the prediction horizon, while $A_i(k+i \mid k)$ and $B_i(k+i \mid k)$ are matrices numerically defined at the end of Step 3.

Step 4. From the linear model generated in Step 3, define the quadratic optimization problem based on (17) (problem equation optimization with constraints), to meet a new vector prediction of entries, that is, an optimal control signal vector $U^*(k)$.

Step 5. If it is confirmed that the values of $U(k)$ and $U^*(k)$ follow a certain criteria convergence, it is assumed that the algorithm arrived at its end that allows the implementation of the first value $U(k)$ and moving the plant to the next instant k sampling. Otherwise, it returns to Step 1.

If this algorithm converges, then, the values $\hat{x}(k+i)$ and $\hat{u}(k+i)$ predicted by the time-variant linear model would coincide with the ones predicted by the original now linear model. Thus, if the optimization problem has a feasible solution, so there is a guarantee that the constraints (16) will be respected throughout the horizon prediction. This fact represents an advantage over methods based on linearized models, since, as close as these models may represent the original model, they cannot guarantee constraints satisfaction.

5. Proposed Method

The proposed method is based on two steps that continuously occur during the operation of the valve. The first step is responsible for recording, on a table, the valve speed whenever a stop is detected, as well as the control action determined by the controller an instant before a slip occurs. The second process updates the quadratic programming constraints considering the information registered in the first step.

5.1. Learning Process. It consists of the assembly and update of stiction velocities table. Every moment a valve stop is detected, the algorithm records the position and speed of the valve stem (at the moment before your speed becomes null) in the table.

5.2. Stiction Treatment Process. At each instant, the algorithm checks if there is a register that relates the stem speed to the desired valve position. If one or more of these registers exist in the table, we use the mean of these speeds in a mathematical equation to move the speed constraint of the NMPC. As a consequence, the optimization problem calculates a control trajectory that causes the stem to move with the speed required to adhere as close to the desired valve position as possible. On the other hand, if there is no register in the table that relates speed to a valve position, it will be immediately registered in the table. We will refer to this movable constraint of speed as restVel. The absolute error between a current position of the stem and the setpoint of the position control system is another parameter that can

TABLE 2: Stiction table in instant k.

Pos (cm)	$x_2(k_1)$	$x_2(k_2)$	$x_2(k_3)$	Average speed (cm/s)
15	0.01	0.03	0.02	0.02
10	Null	Null	Null	Null
5	0.05	0.04	0.03	0.04

be considered in the equation of the movable constraint of speed. The valve speed constraint must not be too strict when the absolute difference between the controller and the valve is considerable, as this may lead to an infeasible optimization problem. A possible approach for mathematical equation of the movable constraint of speed as a function of the absolute error may be as explained below:

$$\text{restVel} = K \cdot |\text{error}| + v_{\text{Ag}}, \tag{21}$$

where v_{Ag} is the value registered in the table for the speed the valve had before sticking, error is the difference between the current valve position and the controller reference, and K is the weight constant. Note that the lower the value of the variable K is, the faster the value of the speed constraint will reach the value registered in the table for the speed constraint.

In the tests phase, the formulation to update restVel caused some feasibility problems to the solution of the optimization problem for large values of K. The following formulation is assumed, which updates the values in a smoother manner:

$$\text{restVel} = \alpha \cdot \left(-e^{-\beta\cdot|\text{error}|} + 1\right) + v_{\text{Ag}}, \tag{22}$$

where α and β are two tuning parameters and e is the exponential function, α must have a greater value than the estimated maximum speed value for the valve, and the value of β is an arbitrary value that should be large enough so that the valve sticks near the reference.

From the calculated value of restVel, the value of the speed constraints θ is set as

$$-V_{\text{min}} \le \theta \le \text{restVel}, \tag{23}$$

if the valve position value is smaller than the reference value. Otherwise, the speed constraint is defined as

$$-\text{restVel} \le \theta \le V_{\text{max}}, \tag{24}$$

where V_{min} and V_{max} are estimates of minimum and maximum valve speed, respectively.

Whenever there is a need to change the position of the valve, the controller will increase the absolute value of the control action until it reaches the minimum necessary force to detach the valve. So, whenever a slip occurs, the control action Γ applied at the moment to the valve is stored in the table. This information is used to accelerate the detachment of the valve whenever it is stuck again in the same position registered in the table. It is considered that the control action varies slowly until it reaches the value of Γ, and the control action u may be limited by the following restriction,

$$\Gamma \le u, \tag{25}$$

in case the controller reference is greater than the actual position. Otherwise,

$$u \le \Gamma. \tag{26}$$

5.3. Example. As an example, a controller with a reference set to 10 (ten) cm in initial instant k has the configuration shown in Table 2.

For the determined reference, there is no particular registered value for the stiction speed in the table. Thus, the controller does not consider restrictions on the speed variable, forcing the valve into an oscillatory cycle around the reference.

From the values recorded in the table for the case where the reference is set to 5 (five) cm and $v_{\text{Ag}} = 0.04$ cm/s, the algorithm will calculate a new value for the speed constraint in the optimization problem (17). With the v_{Ag} value calculated, setting values for $\alpha = 5$, $\beta = 150$, using (22),

$$\text{restVel} = 5 \cdot \left(-e^{-150\cdot|1-0.03|} + 1\right) + 0.04 = 5.04. \tag{27}$$

Since the valve position value is lower than the reference value, the new speed constraint θ is set as

$$-V_{\text{min}} \le \theta \le 5.04. \tag{28}$$

Thus, the algorithm works with variable constraints on the valve speed variable that increases their strictness as the valve approaches the control reference. In this manner, the stiction compensation scheme aims to force the valve to stop as close as possible to the controller reference, cancelling the oscillatory movement around the setpoint that characterizes the stiction disturbance. The control action values necessary to start a valve movement whenever it is stuck can also be stored in the table. The control action values that initiate valve movement must also be stored in the table whenever it gets stuck. These values are used to update the constraints on the input variable which speeds the valve detachment process.

6. Tests and Results

For simulation purposes, the valve and controller models, as well as the control algorithms and the proposed system, were implemented in the software Matlab®/Simulink. The Karnopp model was used to represent the friction force in the valve model simulation while the classic model was employed in the NMPC. In the discretization of the state space model (13), the differential equations were approximated by finite difference equations, $dx/dt \approx (x(t+T_s) - x(t))/T_s$, and were

TABLE 3: Parameters estimated valve control (adapted from [2]).

Parameter	Value	Measurement unit
m	1.361	Kg
S_a	0.06452	m^2
F_s	1707.7	N
F_c	1423	N
F_y	612.9	N·s/m
K_m	52538	N/m
V_s	0.000254	m/s
DV	0.0001524	m/s
K_p	(0.75 × 6894.757)	Pa/mA

used with sampling time T_s. The result of the discretization can be checked in the equation below:

$$X(k+1) = \overline{A}X(k) + \overline{B}u(k) + \overline{K}_{nL}\operatorname{sign}(x_2(k)), \quad y(k) = CX(k), \tag{29}$$

where $\overline{A}$, $\overline{B}$, C, $\overline{K}_{nL}$, and $X(k)$ are matrices defined by

$$X(k) = \begin{bmatrix} x_1(k) \\ x_2(k) \end{bmatrix}, \quad C = \begin{bmatrix} 0 & 1 \end{bmatrix}, \quad \overline{A} = \begin{bmatrix} 1 & T_s \\ -\frac{K_m}{m}T_s & 1 - \frac{F_v}{m}T_s \end{bmatrix}, \quad \overline{B} = \begin{bmatrix} 0 \\ \frac{S_a K_p}{m}T_s \end{bmatrix}, \tag{30}$$

$$\overline{K}_{nL} = \begin{bmatrix} 0 \\ -\frac{1}{m}T_s\left[F_c + (F_s - F_c)e^{(x_2(k)/V_s)^2}\right] \end{bmatrix}.$$

In the linearization phase, it is necessary that (29) be in the format described by (15). For this, the matrix $\overline{K}_{nL}$ had to be incorporated on transition matrix of states $\overline{A}$ which implies that the initial course of the stem is obligatorily different from zero, $x_1(k) \neq 0$, and the discrete model (29) used in NMPC will be described by

$$X(k+1) = \overline{\Phi}X(k) + \overline{B}u(k), \quad y(k) = CX(k),$$

$$\overline{\Phi} = \begin{bmatrix} 1 & T_s \\ -\frac{\overline{\gamma}\cdot\operatorname{sign}(x_2(k)) + K_m x_1(k)}{x_1(k)\,m}T_s & 1 - \frac{F_v}{m}T_s \end{bmatrix}, \tag{31}$$

$$\overline{\gamma} = \left[F_c + (F_s - F_c)e^{(\dot{x}(k)/V_s)^2}\right].$$

The numerical values of the parameters belonging to the valve model adopted in the simulations are shown in Table 3. Table 4 describes the nomenclature of control strategy adopted, as well as its respective dimensions. Similarly, in Table 5, there is information regarding the tuned predictive controller, where its restrictions are the state variable $x_2(k)$, the stem speed, and the control signal $u(k)$. The values of the α, β, H_p, and H_u tuning parameters of Table 5 were obtained empirically. In order to represent the existing measurement noise in industrial environments, a random signal with a mean power of 2% in relation to the measured signal was incorporated in all simulations in the measurement of the position variable.

TABLE 4: Nomenclatures in mathematics of the control strategy.

Parameter	Description
p	Number of outputs
m	Number of inputs
Q	Error weighting matrix
R	Control action weighting matrix
H_p	Prediction horizon
H_u	Control horizon
α	Tuning parameter
β	Tuning parameter
eye(·)	Identity matrix function

TABLE 5: Predictive controller parameters.

Parameter	Value
H_p	4
H_u	4
Q	$\text{eye}(p(H_p+1) \times p(H_p+1))$
R	$10^{-5} I_m \cdot H_u \times m \cdot H_u$
T_s (sample time)	10^{-4} seconds
α	5
β	150

For the correct operation of the system, the controller must be aggressive enough so that the valve does not stick too far from the desired reference. A more aggressive controller, along with minimum constraint on the input u, will also help to detach the valve when a change of reference is required.

In the following section, 5 simulations results will be presented. These simulations aim to explain the effectiveness of the proposed stiction compensation system when applied under different conditions. The trajectories of the stem position, the control signal, and the speed variable are illustrated.

6.1. NMPC Nonlinear in the Absence of Stiction (F_s = 0 N, F_c = 0 N, F_v = 612.9 N). In Figures 4(a), 5(a), 6(a), 7(a), and 8(a), the blue line represents the position of the stem and the red line represents desired position, that is, the valve *setpoint*.

In the figures, it is possible to observe the absence of friction, and the predictive controller drives the stem position to the correct reference and, due to the aggressive tuning, causes an overshoot and oscillations during the transient system response. The decision to maintain the viscous friction parameter in the model during the simulation is explainable

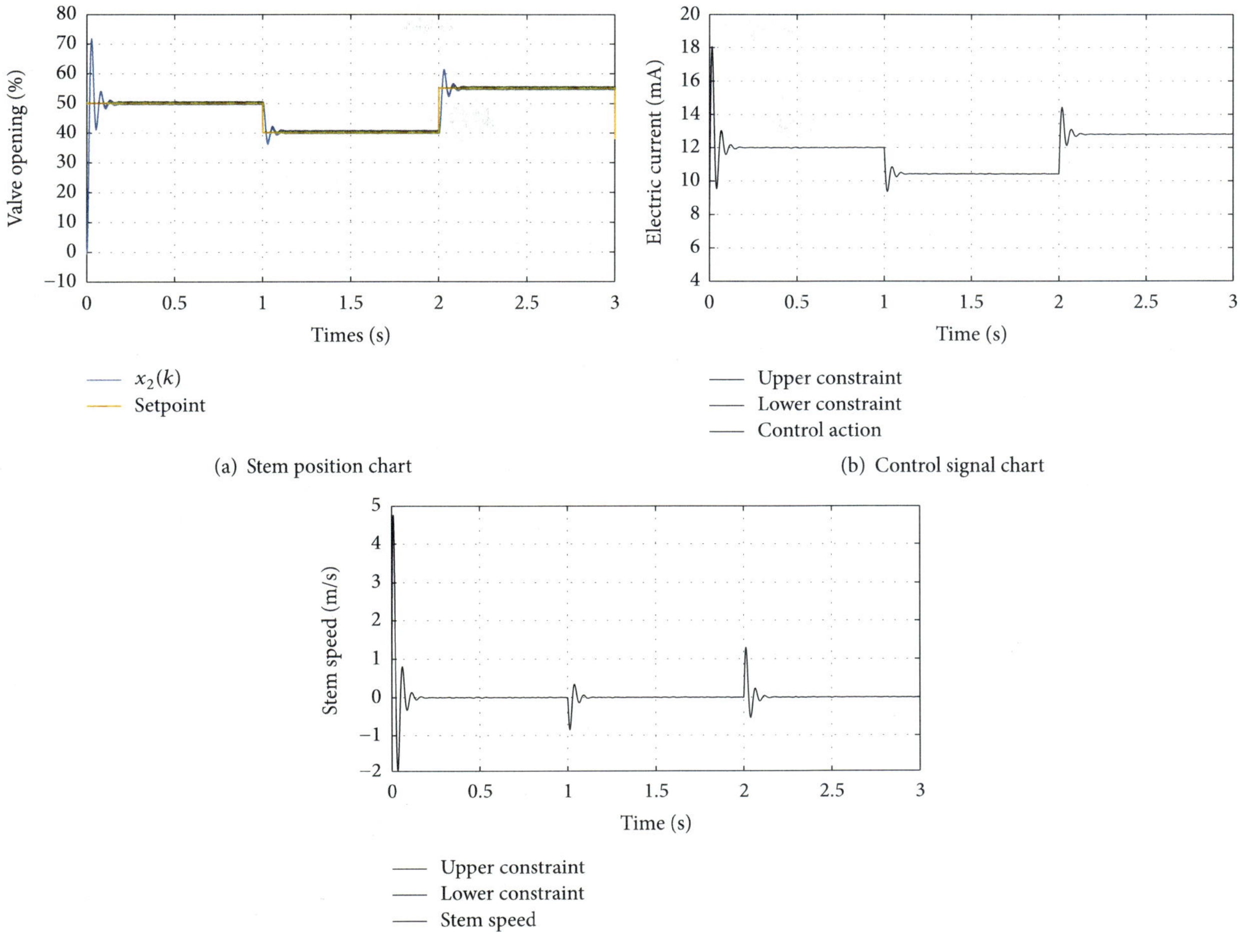

(a) Stem position chart

(b) Control signal chart

(c) Speed chart

FIGURE 4: Simulation 1.

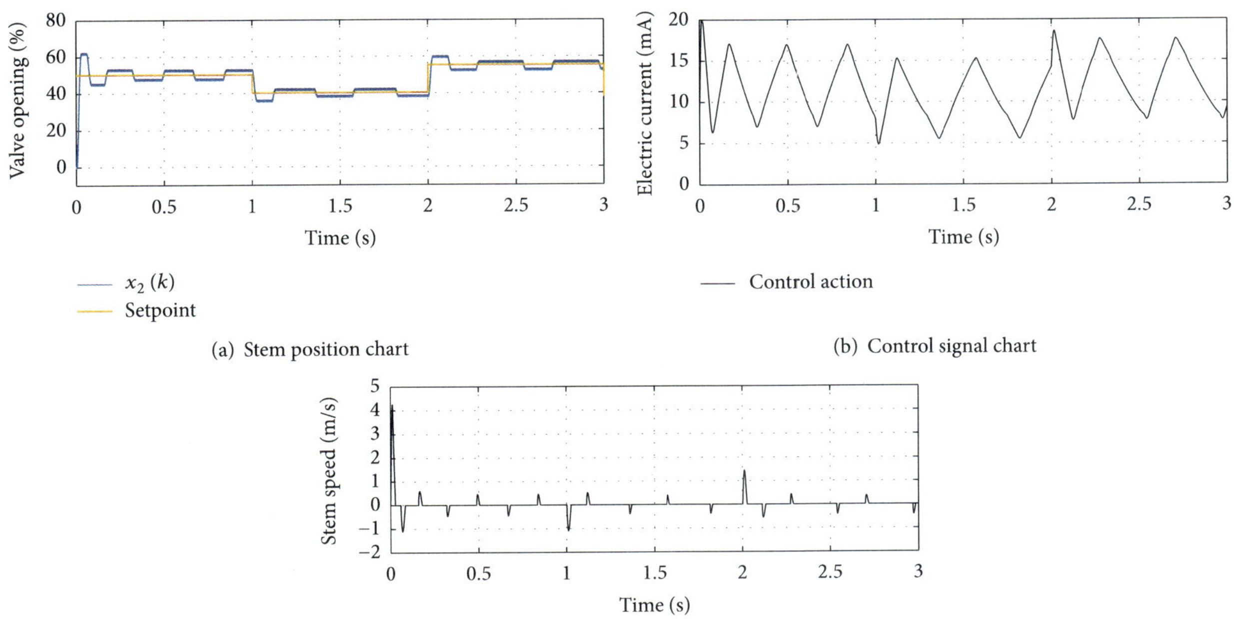

(a) Stem position chart

(b) Control signal chart

(c) Speed chart

FIGURE 5: Simulation 2.

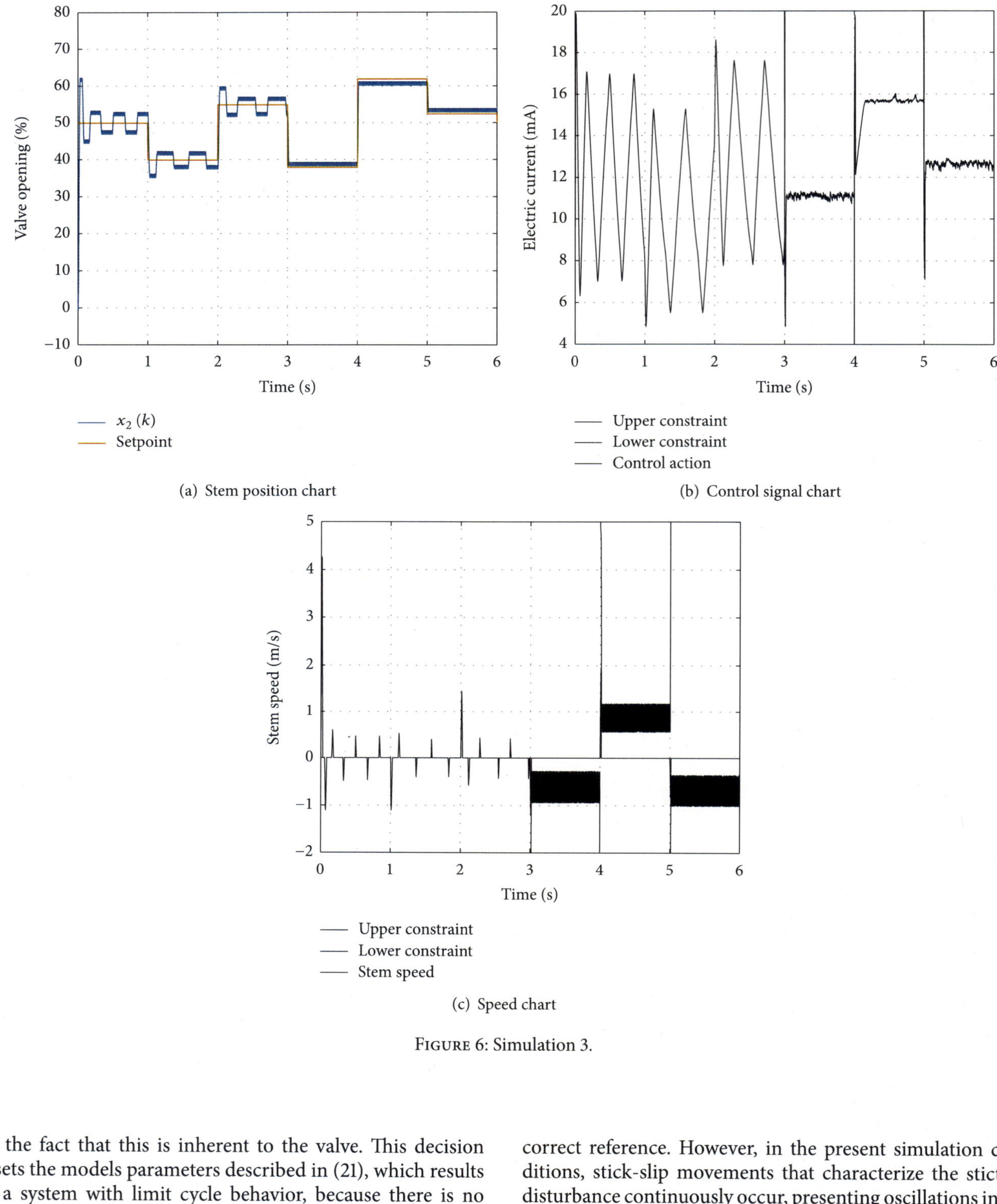

FIGURE 6: Simulation 3.

by the fact that this is inherent to the valve. This decision resets the models parameters described in (21), which results in a system with limit cycle behavior, because there is no damping effect on the valve dynamics.

6.2. NMPC Applied in the Presence of Stiction Disturbance (F_s = 1707.7 N, F_c = 1423 N, F_v = 612.9 N). In simulation 2, it becomes evident that the inclusion of the static and Coulomb friction components in the valve dynamics reduces the standard predictive controller effectiveness. In the previous simulation, the same controller was able to stabilize the frictionless system and drive the valve output to the correct reference. However, in the present simulation conditions, stick-slip movements that characterize the stiction disturbance continuously occur, presenting oscillations in the valve output trajectory (Figure 5(a)), preventing the error from being satisfactorily treated. In Figure 5(b), it is possible to see the oscillatory behavior of the control signal due to the aggressive tuning of the controller plus the presence of the frictional forces in the model. This behavior can also be seen in the speed peaks presented in Figure 5(c), as a result of the inversions of the stem movement. These inversions in the movement are a reflection of the attempts by the controller to bring the valve to the reference position.

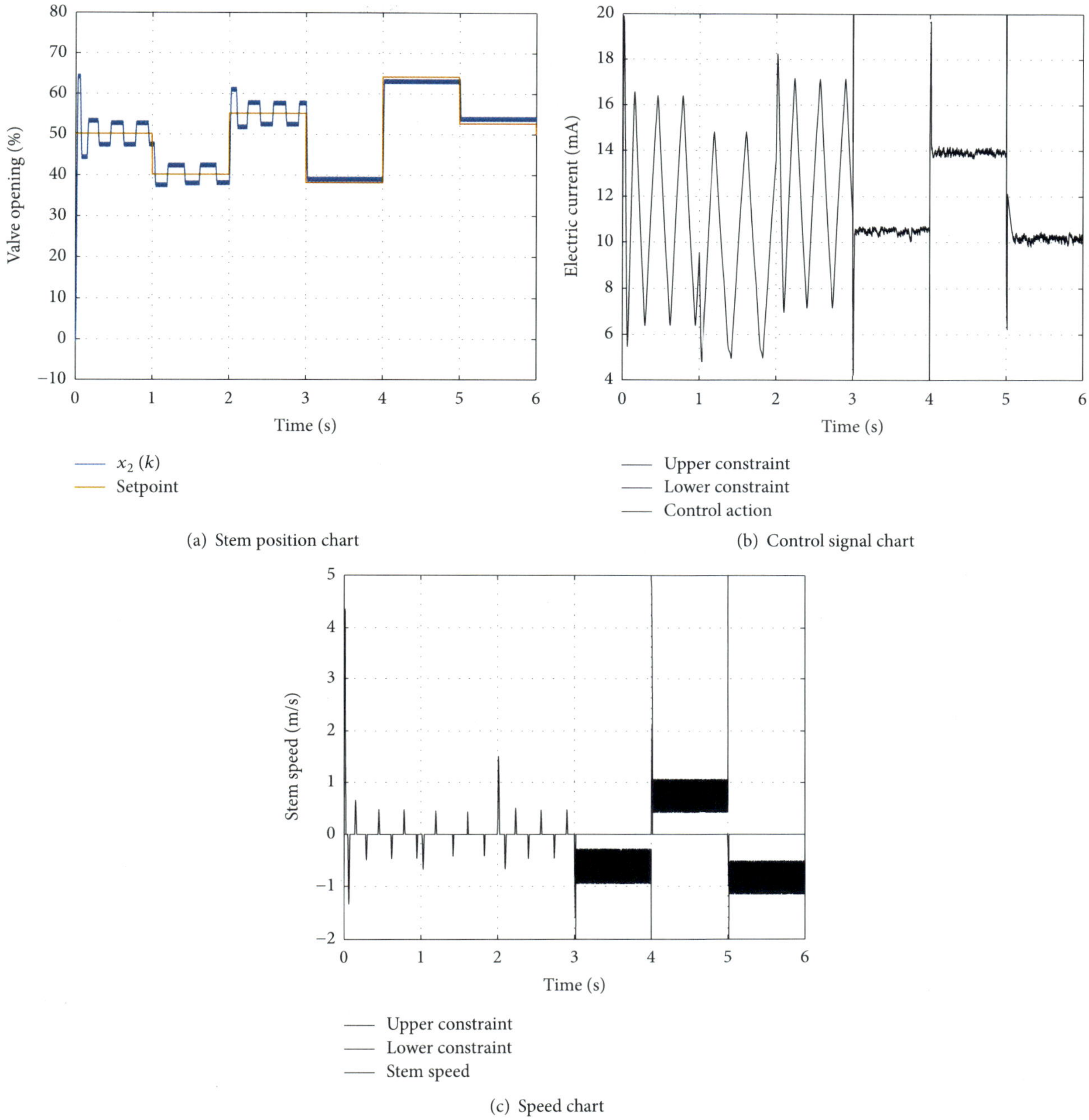

FIGURE 7: Simulation 4.

6.3. NMPC with Proposed Stiction Treatment Strategy (F_s = 1707.7 N, F_c = 1423 N, F_v = 612.9 N). Through the observation of the results presented by simulation 3 in Figures 6(a), 6(b), and 6(c), one can notice that, up until the instant of 3 seconds, the valve has a similar behavior to that seen in simulation 2. This is because the stiction table is empty and the controller does not know the speed that drives the valve to the desired reference, which causes periodical oscillations of the position variable around the reference. Between the 3rd and 4th seconds, the table is not empty and the method of control has learned from 3 reference changes that occurred between the 1st and 2nd seconds and the average speed required to drive the valve to the desired reference. The input signal value needed to release the valve grasped is also registered in the table. This information is used to update the input constraints in the predictive control formulation that forces its detachment potentially sooner. In Figures 6(b) and 6(c), we observe that, from the third second, the control signal and the stem speed stabilize, which become almost constant with peaks in the reference transitions.

The green and blue lines in Figures 6(b) and 6(c) represent minimum and maximum constraints on the controller, respectively, and these constraints vary as a function of the error.

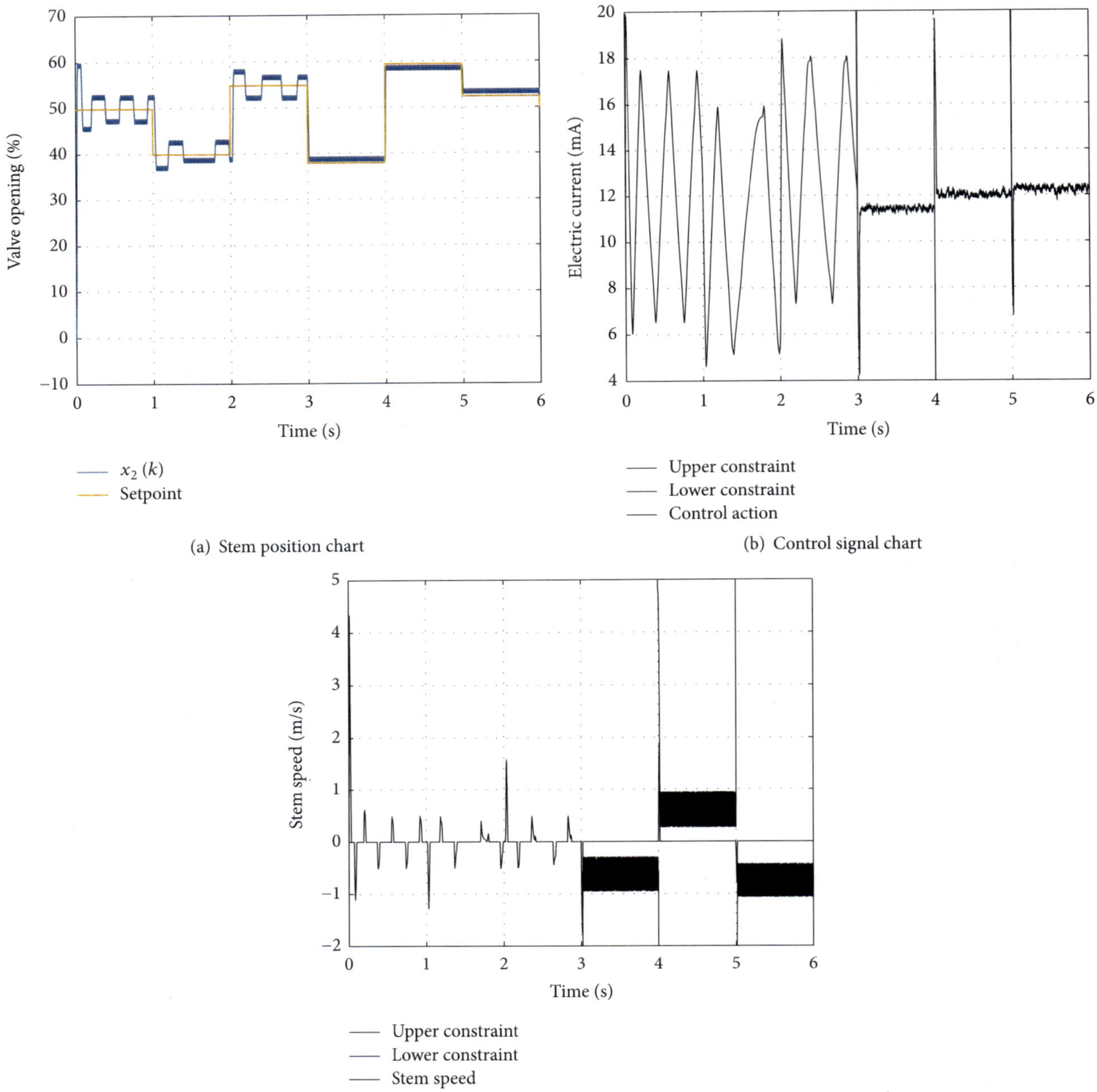

(a) Stem position chart

(b) Control signal chart

(c) Speed chart

FIGURE 8: Simulation 5.

6.4. NMPC Proposed System with Stiction and Degradation in the Spring Constant 10% (F_s = 1707.7 N, F_c = 1423 N, F_v = 612.9 N). Simulation 4 shows that even with a change in the parameter K_m (spring constant), which consequently causes a mismatch between the valve model and the controller model, the system continues to compensate the stiction, since the online learning and the minimum speed (present in the stiction table) define where the valve will stop.

Through the analysis of the position charts in simulation 4, it is possible to notice that the softer spring causes oscillations from 1 to 3 seconds with higher energy when compared to the same chart in simulation 3, since the decrement in the K value has an opposite action to the pressure force relative to the control action u. Even so, the stiction compensation works satisfactorily.

6.5. NMPC Proposed System with Stiction and Increased Friction Parameters by 10% (F_s = 18784.7 N, F_c = 1565.3 N, F_v = 674.19 N). Simulation 5 proves that even with changes in all the parameters related to the stiction disturbance, which consequently cause mismatching between the valve model and the controller model, the system continues to compensate for the stiction since the learning is online and it will be the minimum speed valve in the stiction table which defines where it will stop.

Through the position charts for simulation 5, in Figures 8(a), 8(b), and 8(c), it is possible to conclude that the increase in the stiction parameters causes the oscillations from 1 to 3 seconds on the system to be lower than the simulation 3 since the increase in the friction effect has a contrary action to the pressure force in comparison to the control action. Even under these conditions, the stiction compensation scheme works satisfactorily.

There are two main reasons that steady state error is not null: (1) the adjustment of the α and β constants and (2) the average velocity recorded in the table (v_{Ag}) and parameters of (22). In the case of v_{Ag}, the small value associated with it due to statistical bias, because there are few samples of speed in the table. An alternative solution to this problem is the empirical definition of the slightly larger parameters of α and β and the increase of the statistical information of the speeds as a function of the position of the stem. Furthermore, in Figures 6(c), 7(c), and 8(c), it is possible to observe, from the instant of 3 s, that the upper (green line) and lower (blue line) constraints of x_2 show an irregular behavior. This is due to the fact that the state variable x_2 is a derivative of the position variable x_1, amplifying the measurement noise.

7. Conclusion

This paper presented a system that deals with static friction in electropneumatic valves, one of the most serious problems encountered in these devices, which causes oscillations and consequent wear of the mobile components and production losses. The main idea for the construction of the system is based on a simple principle, which consists of the use of stiction in its favor, instead of trying to compensate it, as presented in previous papers. The strategy adopted uses a nonlinear predictive controller with flexible constraints. The choice of the control strategy is justified by the presence of severe nonlinearities and the presence of the stick-slip phenomenon in the dynamics of the valve, even more so in the latter case which, because of its stochastic nature, requires different values of the control signal to overcome static friction in the same specific position (setpoint), which is not possible with traditional model-based controllers. As an alternative solution to the studied problem, it is proposed to implement a nonlinear and iterative predictive control system for static friction compensation that, associated with a database, reduces the degeneration of the valve positioning system caused by the stick-slip. We describe briefly some of the main linear displacement valve parts used in the industry, as well as their operation and the mathematical models used in the tests (classic model and Karnopp model) of the methodology of control proposed by this paper. In the simulations graphs (Figures 4 and 5), it was observed that the existence of stiction significantly affects the control and that the oscillatory effect was eliminated by the controller presenting a small stationary error (Figures 6, 7, and 8). The controller presented enough robustness to deal with variations in stiction and spring parameters (Figures 7(b) and 8(b)), which may occur in valves exposed to temperature variations and by the natural wear of the spring and sealing gaskets because of us.

Conflicts of Interest

The authors declare that they have no conflicts of interest.

References

[1] M. S. Choudhury, M. Jain, and S. L. Shah, "Stiction definition, modelling, detection and quantification," *Journal of Process Control*, vol. 18, no. 3-4, pp. 232–243, 2008.

[2] C. Garcia, "Comparison of friction models applied to a control valve," *Control Engineering Practice*, vol. 16, no. 10, pp. 1231–1243, 2008.

[3] R. Srinivasan and R. Rengaswamy, "Stiction compensation in process control loops: a framework for integrating stiction measure and compensation," *Industrial & Engineering Chemistry Research*, vol. 44, no. 24, pp. 9164–9174, 2005.

[4] Q. P. He and J. Wang, "Valve stiction quantification method based on a semiphysical valve stiction model," *Industrial & Engineering Chemistry Research*, vol. 53, no. 30, pp. 12 010–12 022, 2014.

[5] A. Ingimundarson and T. Hägglund, "Performance comparison between PID and dead-time compensating controllers," *Journal of Process Control*, vol. 12, no. 8, pp. 887–895, 2002.

[6] J. Richalet, A. Rault, J. L. Testud, and J. Papon, "Model predictive heuristic control," *Automatica*, vol. 14, no. 5, pp. 413–428, 1978.

[7] C. E. Garcia, D. M. Prett, and M. Morari, "Model predictive control: theory and practice a survey," *Automatica*, vol. 25, no. 3, pp. 335–348, 1989.

[8] S. J. Qin and T. A. Badgwell, "A survey of industrial model predictive control technology," *Control Engineering Practice*, vol. 11, no. 7, pp. 733–764, 2003.

[9] J. M. Maciejowski, "Predictive control with constraints," *International Journal of Adaptive Control and Signal Processing*, vol. 17, pp. 261-262.

[10] E. F. Camacho and C. Bordons, *Model Predictive Control*, Springer, London, UK, 2007.

[11] T. Shen, K. Tamura, N. Henmi, and T. Nakazawa, "Robust model following controller applied to positioning of pneumatic control valve with friction," in *Proceedings of the 1998 International Conference on Control Applications*, pp. 512–516, Trieste, Italy.

[12] M. Shoukat Choudhury, N. Thornhill, and S. Shah, "Modelling valve stiction," *Control Engineering Practice*, vol. 13, no. 5, pp. 641–658, 2005.

[13] A. B. Fontes, C. E. Dorea, and M. R. da S. Garcia, "An iterative algorithm for constrained MPC with stability of bilinear systems," in *Proceedings of the 16th Mediterranean Conference on Control and Automation*, pp. 1526–1531, Ajaccio, France, June 2008.

[14] A. Kayihan and F. J. Doyle, "Friction compensation for a process control valve," *Control Engineering Practice*, vol. 8, no. 7, pp. 799–812, 2000.

[15] T. Hägglund, "A friction compensator for pneumatic control valves," *Journal of Process Control*, vol. 12, no. 8, pp. 897–904, 2002.

[16] L. Z. X. Ivan and S. Lakshminarayanan, "A new unified approach to valve stiction quantification and compensation," *Industrial & Engineering Chemistry Research*, vol. 48, no. 7, pp. 3474–3483, 2009.

[17] H. Olsson, *Control Systems with Friction*, Department of Automatic Control Lund Institute of Technology, 1996.

[18] A. Stenman, F. Gustafsson, and K. Forsman, "A segmentation-based method for detection of stiction in control valves," *International Journal of Adaptive Control and Signal Processing*, vol. 17, no. 7-9, pp. 625–634, 2003.

[19] M. Kano, H. Maruta, H. Kugemoto, and K. Shimizu, "Practical model and detection algorithm for valve stiction," *IFAC Proceedings Volumes*, vol. 37, no. 9, pp. 859–864, 2004.

[20] Q. P. He, J. Wang, M. Pottmann, and S. J. Qin, "A curve fitting method for detecting valve stiction in oscillating control loops," *Industrial & Engineering Chemistry Research*, vol. 46, no. 13, pp. 4549–4560, jun 2007.

[21] D. Karnopp, "Computer simulation of stick-slip friction in mechanical dynamic systems," *Journal of Dynamic Systems, Measurement, and Control*, vol. 107, no. 1, pp. 100–103, 1985.

[22] B. Armstrong-Hélouvry, P. Dupont, and C. C. de Wit, "A survey of models, analysis tools and compensation methods for the control of machines with friction," *Automatica*, vol. 30, no. 7, pp. 1083–1138, 1994.

[23] C. Canudas de Wit, H. Olsson, K. J. Astrom, and P. Lischinsky, "A new model for control of systems with friction," *IEEE Transactions on Automatic Control*, vol. 40, no. 3, pp. 419–425, 1995.

[24] B. C. Silva and C. Garcia, "Comparison of stiction compensation methods applied to control valves," *Industrial & Engineering Chemistry Research*, vol. 53, no. 10, pp. 3974–3984, 2014.

12

Fault-Tolerant Control Strategy for Neutral-Point-Clamped Three-Level Inverter

Yanxia Shen, Beibei Miao, Dinghui Wu, and Kader Ali Ibrahim

Engineering Research Center of Internet of Things Technology Applications of Ministry of Education, Jiangnan University, Wuxi 214122, China

Correspondence should be addressed to Yanxia Shen; shenyx@jiangnan.edu.cn and Beibei Miao; 6151920004@vip.jiangnan.edu.cn

Academic Editor: Ahmed Rhif

A fault-tolerant control technique is discussed for the Neutral-Point-Clamped (NPC) three-level inverter, which ensures that the NPC inverter operates normally even under device failures. A two-level leg is added to the NPC inverter; when the device open circuit fault occurs, the load of this faulty phase is connected to the neutral point of this two-level leg through the bidirectional thyristors. An improved Space Vector Pulse Width Modulation (SVPWM) strategy called "addition and subtraction substitution SVPWM" is proposed to effectively suppress fluctuation in capacitor neutral-point voltages by readjusting the sequence and action time of voltage vectors. The fault-tolerant topology in this paper has the advantages of fewer switching devices and lower circuit costs. Experimental results show that the proposed fault-tolerant system can operate in balance of capacitor neutral-point voltages at full output power and the reliability of the inverter is greatly enhanced.

1. Introduction

Compared with the conventional two-level inverters, multilevel inverters have the advantages of lower voltage stress, better waveform spectrum, less voltage change rate, and waveform distortion [1–3]. Therefore, it is widely used in various electric power conversion and power transmission applications, such as inverter system, low-voltage motor drives, and wind power generation [4–6]. However, there are many switching devices in multilevel inverters, and the control of these switches becomes more complex, which leads to a significant increase in failure rate. Then the reliability of the circuit is greatly deteriorated. If the open fault is not dealt with in time, the voltage stress of certain switch will increase owing to the fluctuation of capacitor neutral-point voltages. What is worse, the entire circuit operates abnormally, causing serious accidents or incalculable economic losses [7–9]. In some applications, higher reliability of the circuit is required that it is even expected that the circuit should keep working for a period under the device failures. So, in recent years, the research of fault-tolerant control for multilevel inverters has attracted wide attention from scholars [10–12].

For three-level inverters, there are a variety of fault-tolerant control methods. The fault-tolerant control of the Active Neutral-Point-Clamped (ANPC) inverter using the redundant switching states has been proposed in [13]. The inverter can operate in balance of capacitor neutral-point voltages with symmetrical three-phase output after the proposed fault-tolerant control method is applied. However, the amplitude of output voltage and current are greatly decreased. A load-type fault-tolerant solution of the ANPC inverter has been proposed in [14, 15]. The basic idea of this method is to keep the faulty leg at O level, which is achieved by connecting the faulty leg to the neutral point of the DC-link capacitor. Similarly, the inverter can only operate at reduced rated power. The fault-tolerant of half-bridge switches and neutral-point switches in T-type inverter has been proposed in [16, 17]. The redundant switching states are used to synthesize the reference vector of the rotational voltage in SVPWM algorithm, which enables the inverter to operate normally in the fault condition. However, for the open fault of half-bridge switches, the DC-link voltage utilization ratio is half of this inverter. The fault-tolerant control of the T-type inverter using the additional leg has

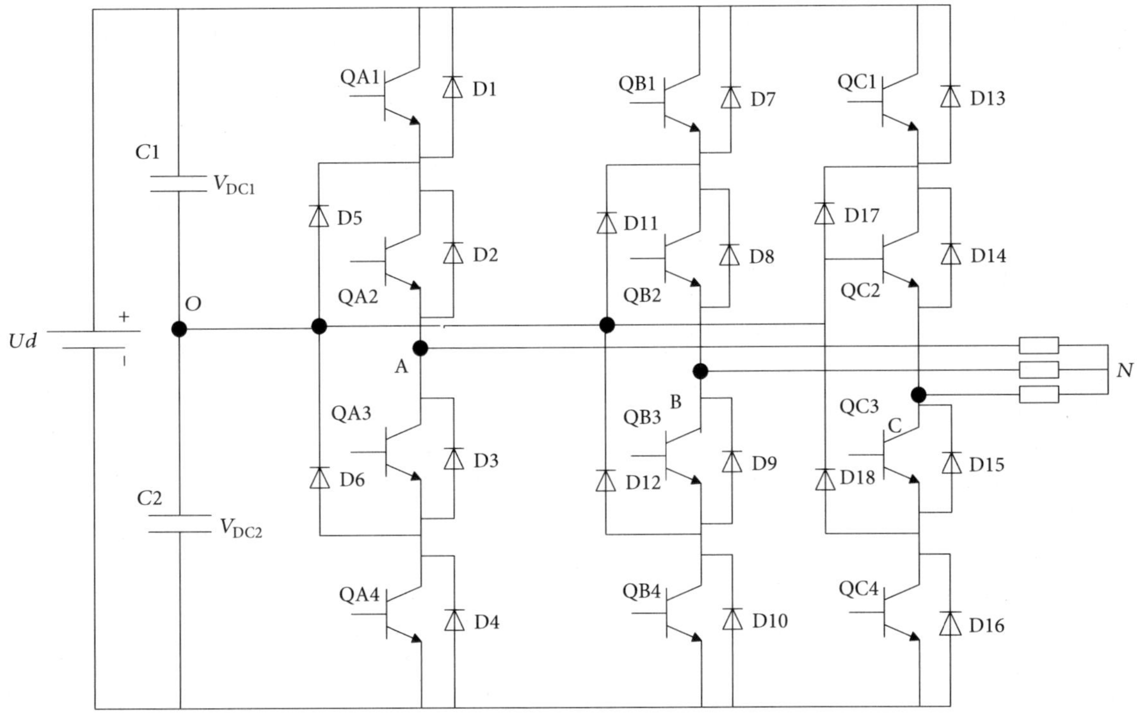

FIGURE 1: The main circuit of NPC three-level inverter.

been proposed in [18]. The one-phase bridge of the T-type inverter is used as the additional leg. Once a fault is detected, the faulty phase is connected to the additional leg by the corresponding bidirectional thyristors. The inverter can output the full-rated power in any fault condition, with higher manufacturing costs and larger size and weight. The eight-switch three-phase inverters (ESTPIs) reconfigured topologies for the NPC inverter have been proposed in [19]. When the device open circuit fault occurs, the corresponding bidirectional thyristors of the faulty leg is triggered, and the faulty leg is connected to the neutral point of the DC-link capacitor. However, the inverter can only operate at half output power; also the amplitude of the output voltage is decreased by half. The fault-tolerant control of the NPC inverter using the asymmetric leg has been proposed in [20]. The faulty leg is isolated by the bidirectional thyristors under device failures, and an asymmetric two-level leg is added to the inverter. To eliminate the low-order harmonic component of the output phase voltages, the bipolar Sinusoidal Pulse Width Modulation (SPWM) method is used to control the asymmetric leg. The inverter can operate in balance of three phases' output without reducing the output power of the original circuit. A software redundancy fault-tolerant control algorithm of the NPC inverter has been proposed in [21]. This method is implemented by using the redundant voltage vectors, with some limitations that apply only to open circuit fault and short circuit fault of outer switches.

When the open circuit fault of inner switches occurs in the NPC three-level inverter, some voltage vectors are lost, and the synthesis of the possible voltage vectors for reference voltage cannot be implemented. Consequently, the reconfiguration of the circuit is required for fault-tolerant control. In order to ensure the NPC inverter can output the full-rated power under device failures, a two-level leg serves as the additional leg for the faulty phase in this paper. When the open circuit fault occurs in the NPC inverter, the faulty leg is connected to the additional leg through the bidirectional thyristors. Both the switching states P and N, generated by modifying the SVPWM algorithm, are employed to control the additional leg. In addition, in order to accurately suppress the fluctuation of the capacitor neutral-point voltages, an improved SVPWM strategy called "addition and subtraction substitution SVPWM" is proposed to readjust the sequence and action time of voltage vectors in SVPWM algorithm.

2. Description of NPC Inverter

Figure 1 shows the main circuit of NPC three-level inverter. There are three switching states for each leg [22], namely, P, O, and N. For instance, the relation between switching states and switching sequence of phase A is listed in Table 1.

In Table 1, "1" indicates that the power device is turned on and "0" indicates turn-off. The symbol "+" represents the current flows from the power supply to the load; on the contrary, the symbol "−" represents the current flows from the load to the power supply.

It can be known from Figure 2 that the NPC inverter has 27 combinations of switching states that correspond to 27 space voltage vectors, divided into four classes according to their magnitudes, namely, zero-, small-, medium-, and large-voltage vectors [23, 24].

TABLE 1: Switching states in NPC three-level inverter.

Switching states	Switching sequence									
	QA1	QA2	QA3	QA4	D1	D2	D3	D4	D5	D6
P+	1	1	0	0	0	0	0	0	0	0
P−	0	0	0	0	1	1	0	0	0	0
O+	0	1	0	0	0	0	0	0	1	0
O−	0	0	1	0	0	0	0	0	0	1
N+	0	0	0	0	0	0	1	1	0	0
N−	0	0	1	1	0	0	0	0	0	0

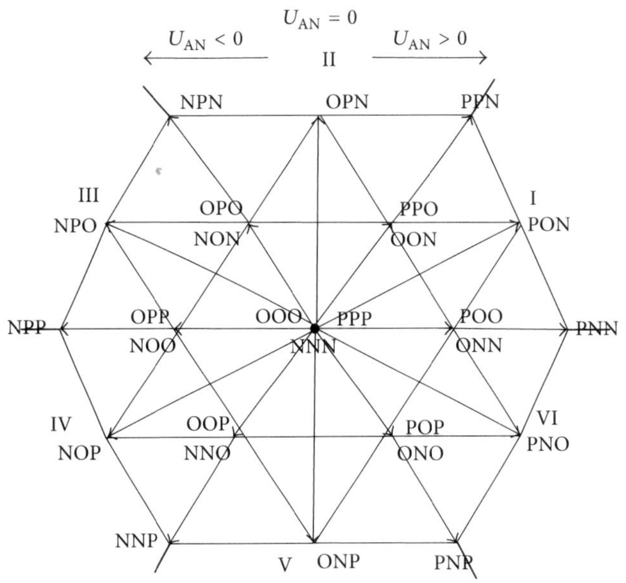

FIGURE 2: Voltage vector distribution of NPC three-level inverter.

3. Effect of the Switching States on Neutral-Point Voltages

When the open circuit fault occurs in NPC three-level inverter, the absence of the switching states leads to the change in the current flow paths. Consequently, an undesirable bridge arm phase voltage is generated, which can bring out the distortion of the load phase current, resulting in the fluctuation of the capacitor neutral-point voltages. The analysis is illustrated with QA1 and QA2 single-switch fault of phase A. The direction of the current flows from the power supply to the load is defined as the positive direction $I_a > 0$.

(1) QA1 Fault. The current can only flow through the switch QA2 and the diode D5 to the load under the switch QA1 fault, while the inverter operates in the P state and the current $I_a > 0$. The load is connected to the neutral point of DC side and the inverter operates in the O state, as shown in Figure 3(a). The solid line represents that the inverter operates normally, whereas the dotted line denotes the inverter under the QA1 fault condition. The bridge arm phase voltage varies from $+1/2U_d$ to zero, which causes the distortion of load phase current. Under these circumstances, all voltage vectors that start with "P" are undesirable, including small-voltage vectors [PPO], [POO], and [POP]; medium-voltage vectors [PON] and [PNO]; and large-voltage vectors [PNN], [PPN], and [PNP], as shown in Figure 4(a). These impossible voltage vectors are replaced by the voltage vectors enclosed with brackets that start with "O"; conversely, the voltage vectors that start with "O" or "N" state in phase A are desirable, because the QA1 fault has no effect on them. That is, the number of the current branches flowing to load from the neutral point of DC side is more than that in the normal state. Therefore, the upper capacitor voltage V_{DC1} is greater than the lower capacitor voltage V_{DC2}.

(2) QA2 Fault. As illustrated in Figure 3(a), one condition is that the QA2 fault occurs in the case that the inverter operates in the P state and the current $I_a > 0$, the current flows through the diodes D3 and D4 to the load, and the inverter operates in the N state. This condition brings out the change in the magnitude of bridge arm phase voltage and current. The other one is that the inverter operates in the O state and the current $I_a > 0$. In the same way, the current flows through the diodes D3 and D4 to load, and the switching state is N, as shown in Figure 3(b). The solid line means the inverter in the normal state and the dotted line indicates the inverter under the QA2 fault condition. Thus, it can be observed from Figure 4(b) that all voltage vectors beginning with "P" and "O" are crossed out and these voltage vectors are replaced by the voltage vectors beginning with "N"; the number of the current branches flowing to load from the neutral point of DC side is less than that in the normal state. Similarly, it can be derived that V_{DC2} is greater than V_{DC1}.

In summary, the device open circuit fault can cause unbalance of the capacitor neutral-point voltages in the NPC inverter.

4. Fault-Tolerant Topology and Working Principle

As shown in Figure 4, when the QA1 open circuit fault occurs, the equivalent stator flux circle can still be synthesized by adjusting the sequence of the available voltage vectors based on vector superposition theorem, but the voltage utilization ratio is half. In contrast, the inverter has no fault tolerance ability in case of the QA2 fault because there is no redundant voltage vectors to synthesize the stator flux circle in the area $U_{AN} > 0$. Therefore, in order to enable the NPC inverter to operate at full output power under the device failures, a

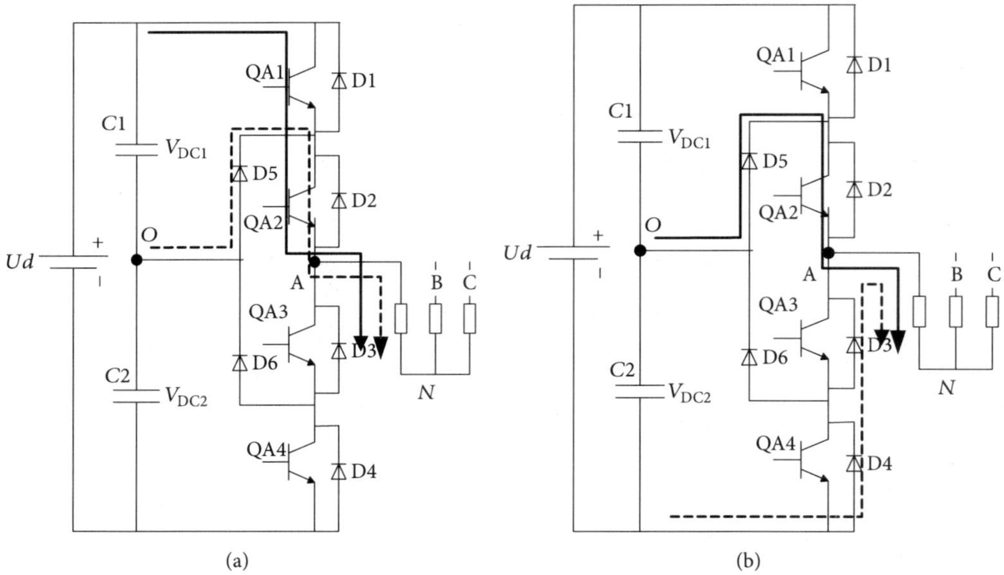

FIGURE 3: Change of current paths under open circuit fault. (a) QA1 fault. (b) QA2 fault.

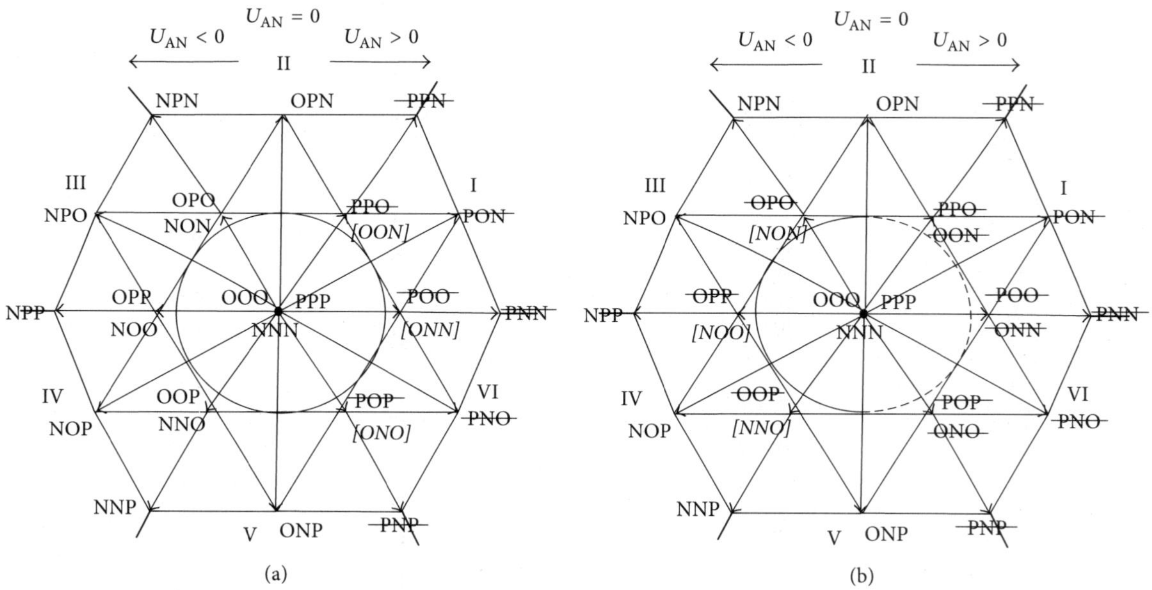

FIGURE 4: Voltage vector distribution under device open fault of phase A. (a) QA1 fault. (b) QA2 fault.

two-level leg is added to the main circuit of the NPC three-level inverter, which consists of two switches Q1 and Q2 and corresponding antiparallel diodes D19 and D20. In addition, six bidirectional thyristors *Trx*1 and *Trx*2 (x = a, b, and c) are used to constitute the fault-tolerant topology of the NPC inverter, as shown in Figure 5.

When the NPC inverter operates normally, *Trx*1 (x = a, b, and c) are connected and *Trx*2 (x = a, b, and c) are disconnected. If the open fault of phase A occurs, the faulty phase is isolated by *Tra*1, and the midpoint O′ of this two-level leg is connected to load of the faulty phase through *Tra*2. Both the switching states P and N, generated by modifying the SVPWM algorithm, are aimed at controlling the two switches Q1 and Q2 of this two-level leg. Obviously, the O state is not available. It is essential to readjust the sequence and action time of voltage vectors. In general, the proposed reconfiguration topology can maintain the neutral-point voltages balance and operate at full output power under any fault condition.

5. SVPWM Strategy of Addition and Subtraction Substitution

After the fault reconstruction topology is adopted, different voltage vector distribution can be obtained, as illustrated in Figure 6.

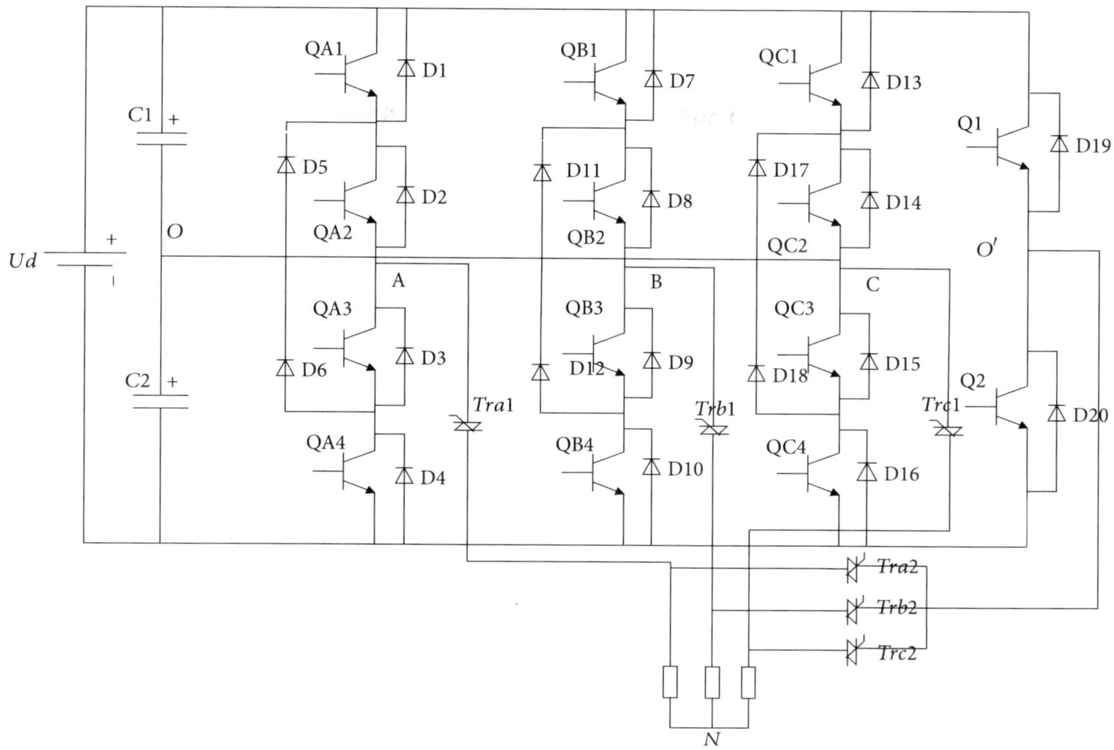

FIGURE 5: Fault-tolerant topology of NPC inverter.

TABLE 2: Substitution law of voltage vectors.

Region	Voltage vectors
I	[OOO] → [PPP], N-type → P-type
II-1, 3, 5	[OOO] → [PPP], N-type → P-type, [OPN] → [PPO]+ [NON]
II-2, 4, 6	[OOO] → [NNN], P-type → N-type, [OPN] → [PPO]+ [NON]
III	[OOO] → [NNN], P-type → N-type,
IV	[OOO] → [NNN], P-type → N-type
V-1, 3, 5	[OOO] → [NNN], P-type → N-type, [ONP] → [POP]+ [NNO]
V-2, 4, 6	[OOO] → [PPP], N-type → P-type, [OPN] → [POP]+ [NNO]
VI	[OOO] → [PPP], N-type → P-type

The equivalent stator flux circle can still be synthesized by adjusting the sequence of the available voltage vectors, and the inverter can output the full-rated power. An improved SVPWM strategy called "addition and subtraction substitution SVPWM" is proposed to effectively suppress the fluctuation of the capacitor neutral-point voltages. The fault-tolerant control of NPC inverters is carried out by adjusting the sequence and operation time of voltage vectors. The research is conducted only taking phase-A fault as an example.

If the phase-A fault occurs, the faulty phase A is replaced with a two-level leg, which is controlled by the switching states P and N. Therefore, the switching states that start with "O" are prohibited, as shown in Figure 6, and these undesirable voltage vectors should be replaced by other voltage vectors of the same size. Namely, N-type voltage vectors ([OON], [ONN], and [ONO]) in the regions I, II-1, II-3, II-5, and VI are replaced, respectively, with P-type voltage vectors ([PPO], [POO], and [POP]). For the purpose of restraining the fluctuation of the capacitor neutral-point voltages effectively, P-type voltage vectors ([OPO], [OPP], and [OOP]) in the opposite regions IV, V-1, V-3, V-5, and III should be replaced, respectively, with N-type voltage vectors ([NON], [NOO], and [NNO]). However, there are no redundant voltage vectors in the medium-voltage vectors ([OPN] and [ONP]) in regions II and V. In this paper, the medium-voltage vectors are replaced by the small-voltage vectors adjacent to them. Table 2 presents the substitution law of voltage vectors in each region.

"Addition and subtraction substitution SVPWM" is simply defined by adding or subtracting a minimum action time (T_{low}) to the three-phase action times (T_a, T_b, and T_c) to realize the substitution of voltage vectors in the SVPWM algorithm. The sequence and action time of voltage vectors are adjusted by replacing the undesirable voltage vectors with the remaining voltage vectors of the same size, where T_{low}

TABLE 3: The sequence and action time of voltage vectors in each region.

Region	Switching sequence	Action time
I-1	POO-PPO-PPP-PPP-PPO-POO	M
I-2	POO-PPO-PPP-PPP-PPO-POO	$M1$
I-3	PON-POO-PPO-PPO-POO-PON	$M1$
I-4	PON-POO-PPO-PPO-POO-PON	$M2$
I-5	PNN-PON-POO-POO-PON-PNN	$M2$
I-6	PON-PPN-PPO-PPO-PPN-PON	$M2$
IV-1	NNN-NNO-NOO-NOO-NN0-NNN	$M2$
IV-2	NNN-NNO-NOO-NOO-NN0-NNN	$M1$
IV-3	NNO-NOO-NOP-NOP-NOO-NNO	$M1$
IV-4	NNO-NOO-NOP-NOP-NOO-NNO	M
IV-5	NOO-NOP-NPP-NPP-NOP-NOO	M
IV-6	NNO-NNP-NOP-NOP-NNP-NNO	M

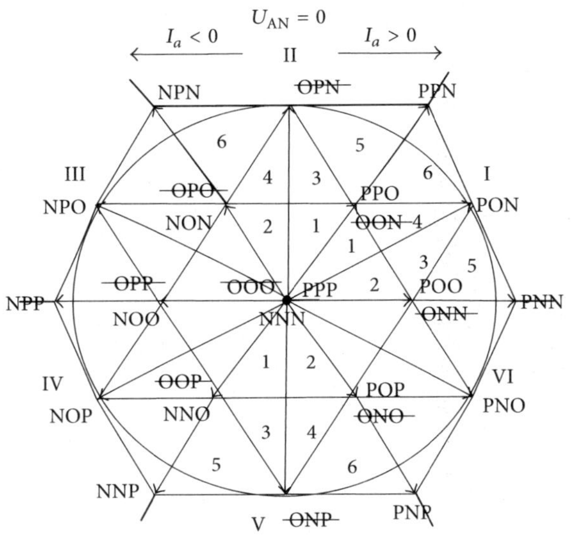

FIGURE 6: Voltage vector distribution after the fault tolerance.

TABLE 4: Experimental parameters.

Dc-link voltage	400 V
Fundamental frequency	$f_n = 50$ Hz
Switching frequency	$f_k = 3$ kHz
Sampling time	$T = 1e-4$
Neutral capacitance	$C1 = C2 = 0.008$ F
RL load	$4\Omega/3$ mH

represents the minimum value among T_a, T_b, and T_c. In this method, the N-type voltage vectors can be substitute for the P-type voltage vectors after T_{low} is added to T_a, T_b, and T_c. In contrast, the P-type voltage vectors take the place of the N-type voltage vectors by subtracting T_{low} to T_a, T_b, and T_c. The process will be explained in two regions I-1 and III-2.

The sequence and action time of voltage vectors in region I-1 in the normal state are shown in Figure 7(a). The undesirable N-type voltage vectors [ONN] should be replaced by the P-type voltage vectors [POO], and this process is achieved by subtracting T_{low} in three-phase action times (T_a, T_b, and T_c), as shown in Figure 7(b), where T_{low} is T_b. Similarly, the impossible voltage vectors [OON] and [OOO] should be changed to [PPO] and [PPP]. Finally, according to the principle that the switching states can only be changed once or unchanged, the redefined sequence and action time of voltage vectors should be rearranged, as shown in Figure 7(c).

The sequence and action time of voltage vectors in region III-2 in the normal state is shown in Figure 8(a). In order to replace the impossible P-type voltage vectors [OPO] with N-type voltage vectors [NON], T_{low} is added in three-phase action times (T_a, T_b, and T_c), where T_{low} is T_c. The redefined sequence and action time of voltage vectors can be obtained as given in Figure 8(b). Therefore, the redefined sequence and action time of voltage vectors in each region listed in Table 3 can be summed up.

$$M = \frac{T1}{2}, \frac{T2}{2}, \frac{T3}{2}, \frac{T3}{2}, \frac{T2}{2}, \frac{T1}{2}$$
$$M1 = \frac{T3}{2}, \frac{T1}{2}, \frac{T2}{2}, \frac{T2}{2}, \frac{T1}{2}, \frac{T3}{2} \quad (1)$$
$$M2 = \frac{T2}{2}, \frac{T3}{2}, \frac{T1}{2}, \frac{T1}{2}, \frac{T3}{2}, \frac{T2}{2}.$$

M, $M1$, and $M2$ represent the action time of voltage vectors corresponding to the different switching sequence.

6. Experiment Verification

In order to validate the improved SVPWM strategy proposed previously, a laboratory prototype of the NPC three-level inverter with fault-tolerant capability is built based on the dSPACE DS1007 platform, as shown in Figure 9. The main experimental parameters are listed in Table 4. Figure 9 shows the overall structure of the experimental platform, including NPC inverter, three-phase load, DS1007, and PC. The main boards used in the experiment are DS5202 and ACMC (AC Motor Control). DS5202 provides 8 channels with maximum sampling frequency of 10 MHz, and EV1048 not only can be

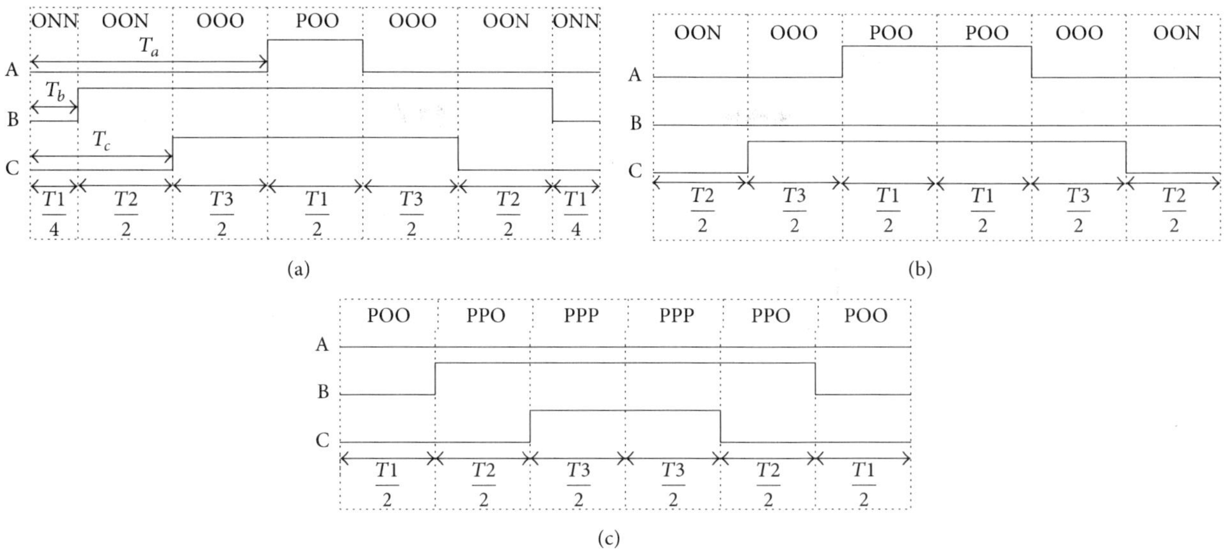

FIGURE 7: The sequence of voltage vectors and its action times in region I-1. (a) Conventional switching sequence. (b) After T_{low} is subtracted. (c) After the switching sequence is rearranged.

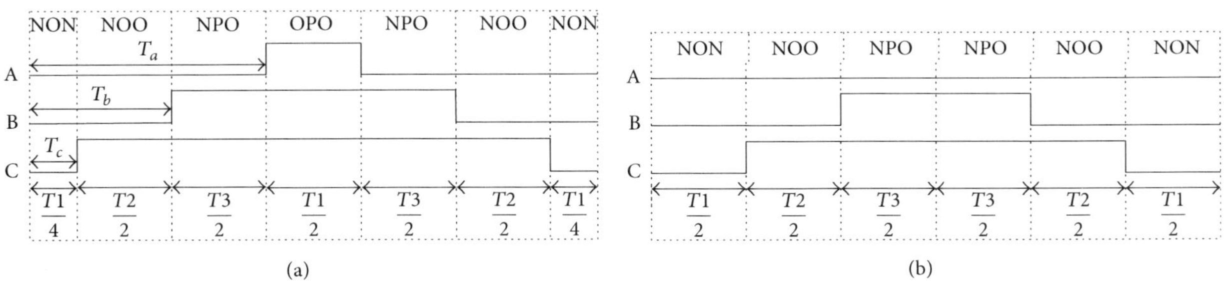

FIGURE 8: The sequence of voltage vectors and its action times in region III-2. (a) Conventional switching sequence. (b) After the switching sequence is rearranged.

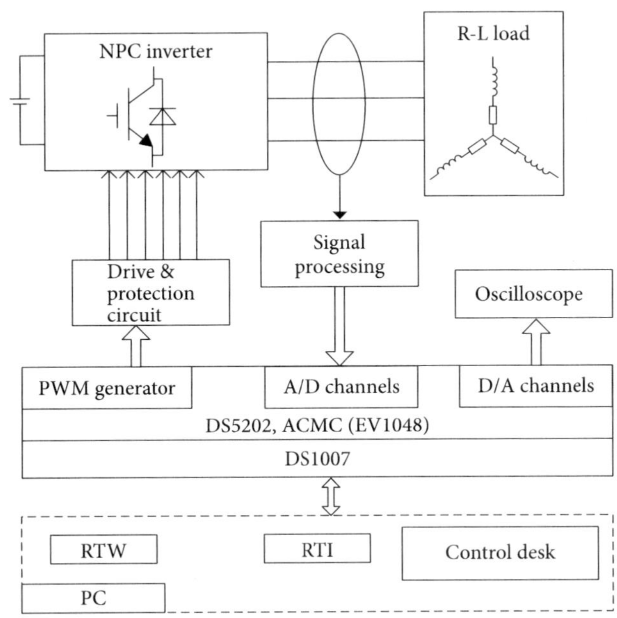

FIGURE 9: Experimental structure diagram.

used for AD sampling but also generate PWM signal to drive the inverter circuit. In the experiment, the QA1 open fault is set up and fault-tolerant control is carried out.

Figure 10 shows the different experimental waveforms of the output voltages, current, and capacitor neutral-point voltages without and with fault-tolerant control in sequence. The waveforms of phase-A bridge arm phase voltage U_{ao} and line-to-line voltage U_{ab} are given in Figure 10(a), respectively. It can be observed from Figure 10(a) that the phase voltage U_{ao} exhibits three levels and the line-to-line voltage U_{ab} shows five levels under the normal condition. When the fault occurs, the amplitude of the phase voltage U_{ao} in positive half period is decreased from $+1/2U_d$ to zero, and a zero level appeared for a long time. Similarly, the amplitude of line-to-line voltage U_{ab} in positive half period is lower than that of the normal condition. The voltage waveforms are the same as that under the normal condition when the proposed method is adopted for fault-tolerant control; namely, U_{ao} is three-level waveform with constant output amplitude and U_{ab} is five-level waveform with constant output amplitude.

Figure 10(b) shows the waveforms of current I_a in phase-A without and with fault-tolerant control, respectively. It can

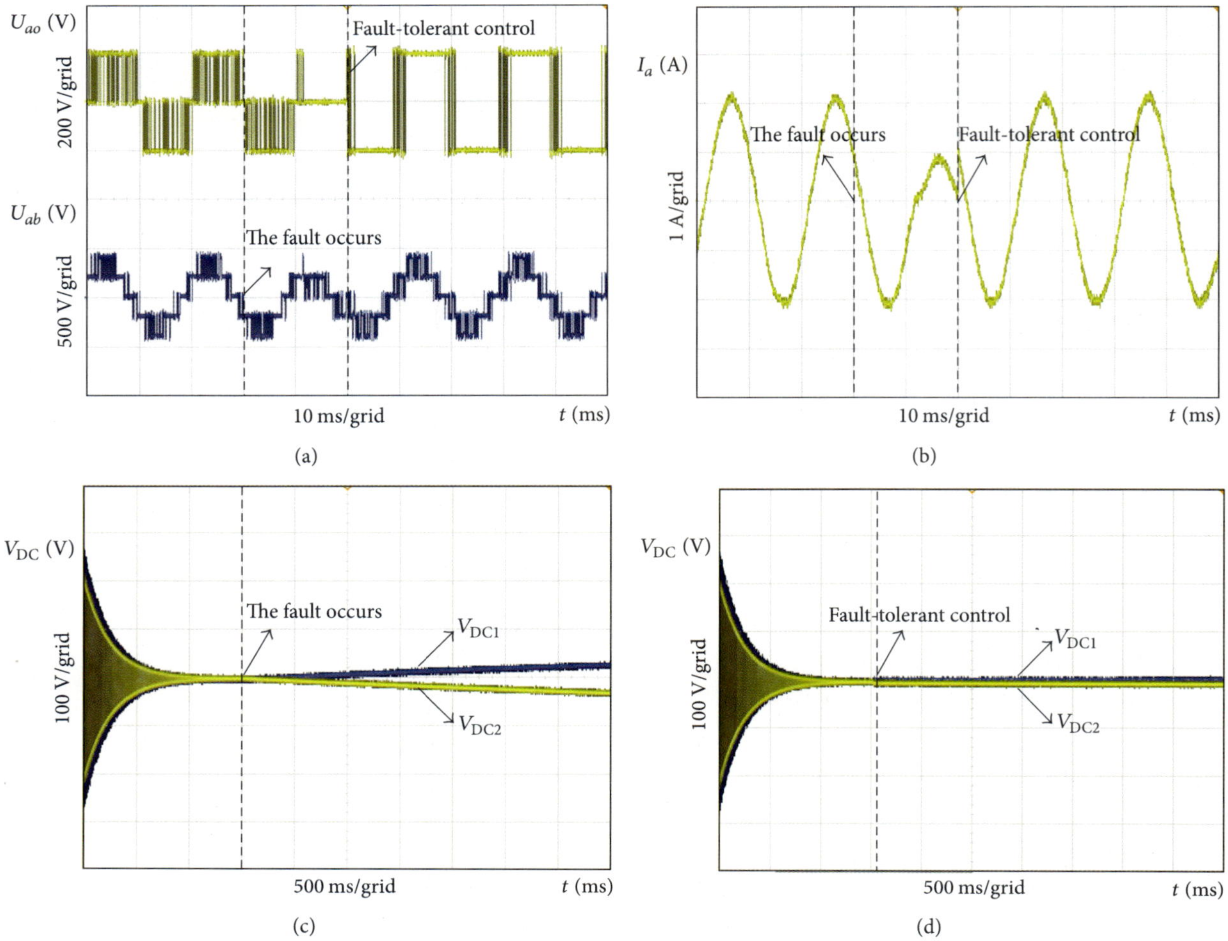

FIGURE 10: Experimental waveforms. (a) U_{ao} and U_{ab}. (b) I_a. (c) Capacitor neutral-point voltages before fault-tolerant control. (d) Capacitor neutral-point voltages after fault-tolerant control.

be seen from Figure 10(b) that the current I_a is sinusoidal before the fault occurs, but when the fault occurs, the amplitude of the load phase current I_a in positive half period is approximately reduced to half of the normal condition, and the waveform is obviously distorted. The load phase current I_a is still sinusoidal with constant amplitude after the proposed fault-tolerant method is applied, and the inverter can operate normally.

Figures 10(c) and 10(d) show the capacitor neutral-point voltages without and with fault-tolerant control, respectively. As shown in Figure 10(c), when the fault occurs, the neutral-point voltage of the capacitor fluctuates obviously, and $V_{DC1} > V_{DC2}$. The fluctuation of the neutral-point voltage of the capacitor is well suppressed by the fault-tolerant control, as shown in Figure 10(d). Obviously, the correctness of the fault-tolerant strategy is proved by experiments.

7. Conclusion

In order to enable the NPC inverter to operate at full output power under the device failures, a two-level leg is taken as the additional leg of the faulty phase in this paper, which is controlled with the two switching states P and N. In addition, an improved SVPWM strategy called "addition and subtraction substitution SVPWM" is proposed to readjust the sequence and action time of voltage vectors in SVPWM algorithm. The main advantages of the proposed fault-tolerant SVPWM strategy include the following:

(1) The proposed fault-tolerant topology has fewer switching devices and lower costs.

(2) The inverter is operated continuously without reduced output power by the proposed method. Therefore, the proposed fault-tolerant SVPWM strategy can be applied for all kinds of applications.

(3) The distortion of capacitor neutral-point voltages is eliminated well after the proposed fault-tolerant method is applied.

(4) The two switches Q1 and Q2 of this two-level leg should withstand the entire input DC voltage after fault-tolerant control. Therefore, the proposed method is more suitable for the NPC inverter with medium- or low-voltage power supply.

The effectiveness and fault-tolerant operation of the proposed SVPWM strategy are demonstrated by both simulation and experimental results under various types of faults.

Conflicts of Interest

The authors declare that they have no conflicts of interest regarding the publication of this paper.

Acknowledgments

This work is supported in part by the National Natural Science Foundation of China (61573167 and 61572237). The authors appreciate the support of the Priority Academic Program Development of Jiangsu Higher Education Institutions as well as The Research Innovation Program for College Graduates of Jiangsu Province, SJLX16_0107.

References

[1] T. Wang, H. Xu, J. Han, E. Elbouchikhi, and M. E. H. Benbouzid, "Cascaded H-bridge multilevel inverter system fault diagnosis using a PCA and multiclass relevance vector machine approach," *IEEE Transactions on Power Electronics*, vol. 30, no. 12, pp. 7006–7018, 2015.

[2] J. He, N. A. O. Demerdash, N. Weise, and R. Katebi, "A fast on-line diagnostic method for open-circuit switch faults in SiC-MOSFET-based T-type multilevel inverters," *IEEE Transactions on Industry Applications*, vol. 53, no. 3, pp. 2948–2958, 2017.

[3] U.-M. Choi, H.-G. Jeong, K.-B. Lee, and F. Blaabjerg, "Method for detecting an open-switch fault in a grid-connected NPC inverter system," *IEEE Transactions on Power Electronics*, vol. 27, no. 6, pp. 2726–2739, 2012.

[4] K. Ma and F. Blaabjerg, "Modulation methods for neutral-point-clamped wind power converter achieving loss and thermal redistribution under low-voltage ride-through," *IEEE Transactions on Industrial Electronics*, vol. 61, no. 2, pp. 835–845, 2014.

[5] M. Schweizer and J. W. Kolar, "Design and implementation of a highly efficient three-level T-type converter for low-voltage applications," *IEEE Transactions on Power Electronics*, vol. 28, no. 2, pp. 899–907, 2013.

[6] O. S. Senturk, S. Munk-Nielsen, R. Teodorescu, L. Helle, and P. Rodriguez, "Power density investigations for the large wind turbines' grid-side press-pack IGBT 3L-NPC-VSCs," in *Proceedings of the 4th Annual IEEE Energy Conversion Congress and Exposition (ECCE '12)*, pp. 731–738, IEEE, Raleigh, NC, USA, September 2012.

[7] J.-S. Lee and K.-B. Lee, "An open-switch fault detection method and tolerance controls based on SVM in a grid-connected t-type rectifier with unity power factor," *IEEE Transactions on Industrial Electronics*, vol. 61, no. 12, pp. 7092–7104, 2014.

[8] D. J. Chen, Y. Z. Ye, and R. Hua, "Fault diagnosis of three level inverter in bullet train based on waveform real-time analysis," *Transactions of China Electrotechnical Society*, vol. 29, no. 6, pp. 106–113, 2014.

[9] S. Yang, D. Xiang, A. Bryant, P. Mawby, L. Ran, and P. Tavner, "Condition monitoring for device reliability in power electronic converters: A review," *IEEE Transactions on Power Electronics*, vol. 25, no. 11, pp. 2734–2752, 2010.

[10] L. L. Jiang, W. Z. Ma, and L. H. Li, "Fault tolerant control strategy for modular multilevel converters," *Power System source*, vol. 38, no. 9, pp. 2497–2503, 2014.

[11] J. Amini and M. Moallem, "A fault-diagnosis and fault-tolerant control scheme for flying capacitor multilevel inverters," *IEEE Transactions on Industrial Electronics*, vol. 64, no. 3, pp. 1818–1826, 2017.

[12] S. Ceballos, J. Pou, J. Zaragoza, E. Robles, J. L. Villate, and J. L. Martín, "Fault-tolerant neutral-point-clamped converter solutions based on including a fourth resonant leg," *IEEE Transactions on Industrial Electronics*, vol. 58, no. 6, pp. 2293–2303, 2011.

[13] J. Li, A. Q. Huang, Z. Liang, and S. Bhattacharya, "Analysis and design of active NPC (ANPC) inverters for fault-tolerant operation of high-power electrical drives," *IEEE Transactions on Power Electronics*, vol. 27, no. 2, pp. 519–533, 2012.

[14] Z. Wang, B. Zhang, Y. Wang, Y. Zhang, and M. Cheng, "Analysis and control of active neutral-point-clamping three-level inverters under fault tolerant operation modes," in *Proceedings of the 18th International Conference on Electrical Machines and Systems (ICEMS '15)*, pp. 2140–2146, Pattaya, Thailand, October 2015.

[15] A. I. L. de Lacerda and E. R. C. da Silva, "Fault tolerant Active Neutral Point Clamped inverter: short-circuit and open-circuit failures," in *Proceedings of the 24th IEEE International Symposium on Industrial Electronics (ISIE '15)*, pp. 1068–1073, Rio de Janeiro, Brazil, June 2015.

[16] U.-M. Choi, F. Blaabjerg, and K.-B. Lee, "Reliability improvement of a T-type three-level inverter with fault-tolerant control strategy," *IEEE Transactions on Power Electronics*, vol. 30, no. 5, pp. 2660–2673, 2015.

[17] U.-M. Choi, K.-B. Lee, and F. Blaabjerg, "Diagnosis and tolerant strategy of an open-switch fault for t-type three-level inverter systems," *IEEE Transactions on Industry Applications*, vol. 50, no. 1, pp. 495–508, 2014.

[18] S. Xu, J. Zhang, and J. Hang, "Investigation of a fault-tolerant three-level T-type inverter system," in *Proceedings of the 7th Annual IEEE Energy Conversion Congress and Exposition (ECCE '15)*, pp. 1632–1638, Montreal, Canada, September 2015.

[19] Y. C. Liu, X. L. Ge, and X. Y. Feng, "Research on the comparison of SVPWM algorithm for two level and three level NPC inverter single bridge fault reconfiguration topology," *Proceedings of the CSEE*, vol. 36, no. 3, pp. 775–783, 2016.

[20] D. J. Chen, Y. Z. Ye, and R. Hua, "Research on Fault Tolerant Topology and control of NPC three level inverter," *Control Engineering of China*, vol. 21, no. 2, pp. 258–267, 2014.

[21] D. B. Matos, J. O. Estima, and A. J. M. Cardoso, "Performance of a synchronous reluctance motor drive with a fault-tolerant three-level neutral point clamped inverter," in *Proceedings of the 22nd International Conference on Electrical Machines (ICEM '16)*, pp. 1152–1159, Lausanne, Switzerland, September 2016.

[22] J.-S. Lee, K.-B. Lee, and F. Blaabjerg, "Open-switch fault detection method of a back-to-back converter using NPC topology for wind turbine systems," *IEEE Transactions on Industry Applications*, vol. 51, no. 1, pp. 325–335, 2015.

[23] Z. Wang, Y. Z. Wang, J. Chen, and M. Cheng, "Fault-Tolerant Control of NPC three-level inverters-fed double-stator-winding PMSM drives based on vector space decomposition," *IEEE Transactions on Industrial Electronics*, vol. 64, no. 11, pp. 8446–8458, 2017.

[24] U.-M. Choi and K. B. Lee, "Space vector modulation strategy for neutral-point voltage balancing in three-level inverter systems," *IET Power Electronics*, vol. 6, no. 7, pp. 1390–1398, 2013.

13

Approximate Prediction-Based Control Method for Nonlinear Oscillatory Systems with Applications to Chaotic Systems

Thiago P. Chagas,[1] Pierre-Alexandre Bliman,[2,3] and Karl H. Kienitz[4]

[1]*Universidade Estadual de Santa Cruz (UESC), 45662-900 Ilhéus, BA, Brazil*
[2]*Lab. J.-L. Lions UMR CNRS, Inria, UPMC University Paris 06, Sorbonne Universités, 7598 Paris, France*
[3]*Fundação Getúlio Vargas, Rio de Janeiro, RJ, Brazil*
[4]*Instituto Tecnológico de Aeronáutica (ITA), 12228-900 São José dos Campos, SP, Brazil*

Correspondence should be addressed to Thiago P. Chagas; thchagas@gmail.com

Academic Editor: Sundarapandian Vaidyanathan

The approximate Prediction-Based Control method (aPBC) is the continuous-time version of the well-known Prediction-Based Chaos Control method applied to stabilize periodic orbits of nonlinear dynamical systems. The method is based on estimating future states of the free system response of continuous-time systems using the solution from the Runge-Kutta implicit method in real time. Some aspects of aPBC are evaluated in the present work, particularly its robustness to low future states estimation precision is exemplified.

1. Introduction

Oscillatory systems are typical in many problems of engineering, biological sciences, physics, economy, and other areas [1–7]. In general, oscillations shall be damped in order to reduce amplitude or stopped avoiding damage, reducing costs, increasing precision, and others [1–4]. These oscillations are mainly divided into periodic and aperiodic. Chaotic aperiodic oscillations are related to unpredictability, disorder, and instability while periodic oscillations are associated with order [8, 9].

Chaotic sets are composed of an infinite number of unstable periodic orbits (UPOs) [10, 11] and the stabilization of one of these orbits leads to periodic oscillation, possibly, reducing amplitude. Different chaos control methods have been developed aiming at stabilizing these orbits with low control effort using the main characteristics of chaos [9].

One classical chaos control method is the Delayed Feedback Control (DFC) proposed by Pyragas (1992) [12]. This is a state-feedback method whose control signal is computed through the difference between delayed and current time measured system states. DFC was initially proposed for continuous-time and has many applications on discrete-time systems [13]. Nevertheless, DFC has a well-known limitation proofed for discrete-time systems, the odd-number limitation [14–17], which means that DFC cannot stabilize orbits with an odd number of Floquet multipliers larger than +1. This limitation is questioned in literature for continuous-time systems with counter example [18–20].

Different modifications on DFC have been proposed to overcome the odd-number limitation [13] and one of the most interesting modifications is the Prediction-Based Control (PBC) [21] developed for discrete-time systems. Instead of the delayed state, it uses the future state one period of the target UPO ahead computed along the trajectories of the free system response as reference for the control signal. Developments of PBC are proposed always for discrete-time systems; for example, in [22] a method is proposed for tuning control gain without previous knowledge about the target period-p UPO that leads to fast convergence rate of trajectories to the stabilized periodic orbit; in [23] it is proved that period-2 UPOs can be stabilized by a pulsed control signal reducing the required control effort for stabilization; and in [24] equilibrium points (or period-1 UPOs) are stabilized in the presence of multiplicative or additive noise.

The literature provides some results on a PBC-like strategy applied to continuous-time systems. In [25] it is proposed to use the system Jacobian matrix at each point of the trajectory, instead of system future states, for stabilization of equilibrium points. The results showed that this strategy does not lead, necessarily, to stabilization of free system equilibrium points or UPOs and control effort does not vanish in steady state, two main advantages of chaos control proposed by Ott et al. [26]. In [27], the strategy proposed in [25] is incremented using Neural Networks also for stabilization of equilibrium points.

In fact, PBC has a practical limitation for continuous-time systems if it is applied as proposed for discrete-time systems [21], the need of future state values in real time. In [28], we proposed the approximate Prediction-Based Control (aPBC) with a methodology based on the implicit Runge-Kutta method and state estimation applied to predict future states for the free system in real time based on system model. The authors claim that aPBC is the continuous-time applicable version of PBC because it uses prediction and stabilizes free system UPOs, ideally, vanishing steady state control effort. On the contrary, predicting future states has some drawbacks because the proposed future state prediction scheme requires an increase in the closed-loop system order and consequently computational power for numerical integration. Moreover, it is also subjected to prediction model and real system mismatch and integration method precision for application.

Continuing the aPBC development and study, one of the main questions that remained in [28] is about the robustness of the method for low future state estimation precision. The forced van der Pol (vdP) oscillator (nonautonomous) and Rössler system (autonomous) are used in the present work for numerical examples on the aPBC robustness and trade-off between estimation accuracy and computational cost. We present a methodology and performance indexes that aid finding this trade-off and show results evidencing that it is possible to find a lower bound for precision. Besides that, this work presents an UPO of the forced van der Pol oscillator with one of the Floquet multipliers larger than +1 that is stabilized by the aPBC and could not be stabilized by the DFC with constant control gain following the procedures proposed in [18–20]. The optimization procedure proposed in [29] was also applied to the DFC without success. This is evidence that the aPBC maintain the advantage of discrete-time PBC of not being sensitive to the odd-number limitation.

The paper is divided as follows. The aPBC is reviewed in Section 3, generalizing its proposition. In Section 4, aPBC application is presented using orthogonal collocation method [30, 31] as implicit Runge-Kutta method. In Section 5, the forced van der Pol (vdP) oscillator and Rössler system and their chaotic behavior are presented. Both systems are used for numerical examples on the aPBC robustness and trade-off between estimation accuracy and computational cost in Section 6. The example of an UPO of the forced vdP oscillator that could not be stabilized by DFC and is stabilized by aPBC is shown in Section 6.1.4.

2. Problem Statement

Consider the following continuous-time dynamical system:

$$\dot{x}(t) = f(t, x(t), u(t)), \quad x(0) \text{ given}, \tag{1}$$

where $t \in \mathbb{R}^+$, $x : \mathbb{R}^+ \to \mathbb{R}^n$, $u : \mathbb{R}^+ \to \mathbb{R}^m$, $n, m \in \mathbb{N}$, and $f : \mathbb{R}^+ \times \mathbb{R}^n \times \mathbb{R}^m \to \mathbb{R}^n$ is a T-periodic function with respect to time t; that is, by definition

$$f(t+T, x, u) = f(t, x, u), \quad \forall t \in \mathbb{R}^+,\ \forall x \in \mathbb{R}^n,\ \forall u \in \mathbb{R}^m. \tag{2}$$

Moreover, we assume the existence of a T-periodic solution $x^*(t)$ to the free system (1), which is the system obtained by setting $u(t) = 0$, $t \geq 0$. In other words,

$$\begin{aligned} x^*(t+T) &= x^*(t), \quad \forall t \in \mathbb{R}^+, \\ \dot{x}^*(t) &= f(t, x^*(t), 0_m), \quad \forall t \in \mathbb{R}. \end{aligned} \tag{3}$$

We assume that this periodic solution is unstable and can be stabilized by the continuous-time version of the PBC with feedback law defined as

$$u(t, x(t)) = K(t, x(t))\left(\varphi(t+T, t, x(t), 0) - x(t)\right), \tag{4}$$

where $u : \mathbb{R}^+ \times \mathbb{R}^n \to \mathbb{R}^m$ is the control signal, $K : \mathbb{R}^+ \times \mathbb{R}^n \to \mathbb{R}^{m \times n}$ is the control gain, $\varphi(t_1, t_0, x, 0)$ is the value at time t_1 of the state of (1) with $x(t_0) = x$ and $u(t) = 0$, $t_0 \leq t \leq t_1$. In other words, $\varphi(t_1, t_0, x, 0)$ is the value at time t_1 of the state along the trajectory departing from x at time t_0 of the free system ($u(t) \equiv 0$).

For an appropriate control gain $K(t, x(t))$, x^* is a stable solution of the closed-loop system

$$\dot{x}(t) = f(t, x(t), u(t, x(t))) \tag{5}$$

and control signal u ideally verifies

$$u(t, x^*(t)) = 0 \quad t \geq 0. \tag{6}$$

The general formulation of the continuous-time PBC method can be written as partial differential equation where the solution of system (5) with control signal (4) is the solution of the PDE (7a) and (7b). Consider

$$\frac{\partial X(t,0)}{\partial t} = f(t, X(t,0), K(t, X(t,0))(X(t,T) - X(t,0))), \quad t \geq 0 \tag{7a}$$

$$\frac{\partial X(t,Ts)}{\partial s} = Tf(t+Ts, X(t,Ts), 0), \quad t \geq 0,\ s \in [0,1] \tag{7b}$$

$$X(0,0) = x(0). \tag{8}$$

The function $X : \mathbb{R}^+ \times [0, T] \to \mathbb{R}^n$ is such that $X(t, 0) = x(t)$ and $X(t, T) = \varphi(t + T, t, x(t), 0)$.

Condition (6) ensures zero control effort when the trajectory is on the unstable periodic solution x^* of the free system. However, applying $u(t, x(t))$ in real systems is based on the exact prediction and real-time computation of future states $\varphi(t + T, t, x(t), 0)$. Due to these practical constrains, alternatively, we are also interested in stabilizing an orbit *close to* the UPO x^* of (1) resulting in a low control effort. This task is solved using the aPBC reviewed in the sequel.

3. Principles of the Approximate Prediction-Based Control Method (aPBC)

Computing the future state of the prediction term $\varphi(t + T, t, x(t), 0)$ requires solving the free system ODE at each time t, from time t to $t + T$. This cannot be done exactly in real time and its solution is the basis of the aPBC. This task is divided into two steps, the first one consists in approximating the solution of $\varphi(t + T, t, x(t), 0)$ by an implicit Runge-Kutta (R-K) ODE integration method, and the second is expressing the solution of the R-K method as a state observer that can be integrated by any explicit method in real time.

3.1. Approximation of the Prediction Term: 1st Step. The first step consists in approximating the solution of (7b) by an implicit R-K ODE integration method [32], in order to estimate the prediction term, that is, the terminal value,

$$X(t,T) = x + T\int_0^1 f(t + Ts, X(t,Ts), 0)\,ds, \quad x \text{ given.} \tag{9}$$

In order to estimate $X(t, T)$ given by (9), the state transition map of the free system $\varphi(t_2, t_1, x, 0)$ is first approximated by the operator $\mathbf{z}$ defined by

$$\mathbf{z}(t + T, t, x) = x + T\sum_{i=1}^{N} c_i l_i(t) \tag{10a}$$

$$l_i(t) = f\left(t + Ts_i, x + T\sum_{j=1}^{N} a_{ij} l_j(t), 0\right), \tag{10b}$$

where $l_j : \mathbb{R} \to \mathbb{R}^n$ and $a_{ij}, c_i \in \mathbb{R}$ are weights chosen according to the implicit method used [32]. The approximation $\mathbf{z}(t + T, t, x(t))$ of $X(t, Ts)$, $s \in [0, 1]$, is calculated at the discretization points $s = s_i$, $i = 1, \ldots, N$. N shall be chosen accordingly such that the sums in (10a) and (10b) lead to an adequate approximation of the integral in (9) for the given $f(t, x, u)$.

For simplicity, (10b) is written in the vector form

$$L(t) = F_T(t, x, L(t)), \tag{11}$$

where

$$L(t) = \begin{bmatrix} l_1(t) \\ \vdots \\ l_N(t) \end{bmatrix} \in \mathbb{R}^{nN}, \quad \forall t \geq 0, \tag{12}$$

and $F_T : \mathbb{R} \times \mathbb{R}^n \times \mathbb{R}^{nN} \to \mathbb{R}^{nN}$ is defined by

$$F_T(t, x, L) = \begin{bmatrix} f\left(t + Ts_1, x + \sum_{j=1}^{N} a_{1j} l_j, 0\right) \\ \vdots \\ f\left(t + Ts_N, x + \sum_{j=1}^{N} a_{Nj} l_j, 0\right) \end{bmatrix}, \quad \forall x \in \mathbb{R}^n, \ \forall L \in \mathbb{R}^{nN}. \tag{13}$$

In order to compute $\mathbf{z}(t + T, t, x)$ through (10a), it is necessary to solve the algebraic system of (11) with unknown $L(t) \in \mathbb{R}^N$.

Writing

$$C = \begin{bmatrix} c_1 & \cdots & c_N \end{bmatrix} \tag{14}$$

and closing (5) by

$$\begin{aligned} u(t, x(t)) &= K(t, x(t))(\mathbf{z}(t + T, t, x(t)) - x(t)) \\ &= TK(t, x(t))CL(t) \end{aligned} \tag{15}$$

yield the differential algebraic equation (DAE)

$$\dot{x}(t) = f(t, x(t), TK(t, x(t))CL(t)), \quad x(0) \text{ given} \tag{16a}$$

$$L(t) = F_T(t, x(t), L(t)). \tag{16b}$$

The real-time solution of the DAE (16a) and (16b) requires the computation of $L(t)$, its algebraic term, at each time t. We therefore introduce an observer equation in the sequel, to transform the controlled system into a system of ODEs.

3.2. Approximation of the Prediction Term: 2nd Step. We now approximate (16b) by solving the nN-dimensional ODE (17) whose solution $\widehat{L}(t)$ is an estimation of $L(t)$. The initial value $\widehat{L}(0)$ is intended to be (precisely) computed off-line to provide good tracking quality for $L(t)$.

$$\begin{aligned} &\frac{d}{dt}\left(\widehat{L}(t) - F_T\left(t, x(t), \widehat{L}(t)\right)\right) \\ &\quad + k_o\left(\widehat{L}(t) - F_T\left(t, x(t), \widehat{L}(t)\right)\right) = 0, \\ &\widehat{L}(0) = L(0). \end{aligned} \tag{17}$$

The scalar gain k_o is chosen positive in order that $\widehat{L}(t)$ of (17) tends asymptotically towards the solution $L(t)$ of (16b) when $t \to +\infty$ and typically in such a way that the estimator dynamics are faster than the controlled system dynamics. If the evolution of $\widehat{L}(t)$ may indeed be chosen in order to fulfill (17), convergence does occur.

From (17), we obtain

$$\begin{aligned} &\dot{\widehat{L}}(t) - \left[\partial_1 F_T\left(t, x(t), \widehat{L}(t)\right)\right. \\ &\quad + \partial_2 F_T\left(t, x(t), \widehat{L}(t)\right)\left(\mathbf{1}_N \otimes f(t, x(t), 0)\right) \\ &\quad \left.+ \partial_3 F_T\left(t, x(t), \widehat{L}(t)\right)\left(T(A \otimes I_n)\dot{\widehat{L}}(t)\right)\right] \\ &\quad + k_o\left(\widehat{L}(t) - F_T\left(t, x(t), \widehat{L}(t)\right)\right) = 0, \end{aligned} \tag{18}$$

where $A = (a_{ij})$ and ∂_i is the partial derivative with respect to the ith variable. I_n is the $n \times n$ identity matrix and $\mathbf{1}_N$ is the column vector of dimension N with all elements equal to 1. $\otimes$ is the Kronecker product.

By determining $\dot{\widehat{L}}(t)$ from (18) we get

$$\begin{aligned}\dot{\widehat{L}}(t) &= \left[I_{nN} - T\partial_3 F_T\left(t, x(t), \widehat{L}(t)\right)(A \otimes I_n)\right]^{-1} \\ &\cdot \left[\partial_1 F_T\left(t, x(t), \widehat{L}(t)\right)\right. \\ &+ \partial_2 F_T\left(t, x(t), \widehat{L}(t)\right)\left(\mathbf{1}_N \otimes f(t, x(t), 0)\right) \\ &\left. - k_o\left(\widehat{L}(t) - F_T\left(t, x(t), \widehat{L}(t)\right)\right)\right].\end{aligned} \tag{19}$$

Clearly, solving (19) in order to obtain (17) requires invertibility of the first factor.

We then define $G_T : \mathbb{R}^n \times \mathbb{R}^n \times \mathbb{R}^{nN} \rightarrow \mathbb{R}^{nN}$,

$$\begin{aligned}G_T\left(t, x, \widehat{L}\right) &= \left[I_{nN} - T\partial_3 F_T\left(t, x, \widehat{L}\right)(A \otimes I_n)\right]^{-1} \\ &\cdot \left[\partial_1 F_T\left(t, x, \widehat{L}\right) + \partial_2 F_T\left(t, x, \widehat{L}\right)\left(\mathbf{1}_N \otimes f(t, x, 0)\right)\right. \\ &\left. - k_o\left(\widehat{L} - F_T\left(t, x, \widehat{L}\right)\right)\right].\end{aligned} \tag{20}$$

From (16a), (16b), (17), and (20) and denoting $\widehat{l}_i(t)$ as the components of $\widehat{L}(t)$, the control law yields the following closed-loop system of ODEs:

$$\begin{bmatrix}\dot{x}(t) \\ \dot{\widehat{L}}(t)\end{bmatrix} = \begin{bmatrix} f\left(t, x(t), K(t, x(t))\, TC\widehat{L}(t)\right) \\ G_T\left(t, x(t), \widehat{L}(t)\right)\end{bmatrix} \tag{21}$$

$$x(0) \text{ given}, \quad \widehat{L}(0) = L(0).$$

The solution of (21) is an approximation of the solution of the PDE given in (7a) and (7b) and therefore only an approximate stabilization of the initial orbit is expected or rather the stabilization to an orbit close to the initial one.

The ODE (21) has two types of state components, corresponding to the controlled system dynamics and to the dynamical state controller. Once $\widehat{L}(t)$ stands for a set of unmeasured state variable components, (17) can be interpreted as a state observer. Notice that this estimator introduces dynamical feedback whose state has a dimension equal to the number of points of the R-K method adopted multiplied by the dimension of the initial system to be controlled (nN).

4. aPBC Application Issues

Two aspects of the aPBC shall be considered for application. One is the choice of an implicit R-K method to solve (10a) and (10b) and the other is the control gain tuning for stabilization. Here, we use the orthogonal collocation method with Lagrange polynomials and a constant control gain K computed through the closed-loop monodromy matrix described in the sequel.

4.1. Orthogonal Collocation Method with Lagrange Polynomials. The implicit R-K method given in (10a) and (10b) is a general formulation used for the integration of differential equations whose application depends on the choice of a specific implementation. Herein, using the results in [28], we apply the orthogonal collocation [30–32] as implicit R-K method.

Collocation methods amount to approximate the prediction term by $\mathbf{z}(t + T, t, x(t))$, where $\mathbf{z}(t + Ts, t, x(t))$ is defined on the whole interval $s \in [0, 1]$ by

$$\mathbf{z}(t + Ts, t, x(t)) = \sum_{j=1}^{N} w_j(s)\, m_j(t), \tag{22}$$

$$m_1(t) = x(t), \quad s \in [0, 1], \quad t \geq 0,$$

where $m_j : \mathbb{R}^+ \rightarrow \mathbb{R}^n$, $j = 1, \ldots, N$, is a column vector. The functions $w_j(s)$ are the Lagrange polynomials

$$w_j(s) = \prod_{i=1, i \neq j}^{N} \frac{s - s_i}{s_j - s_i}, \quad j = 1, \ldots, N, \tag{23}$$

attached to the choice of the points $0 = s_1 < s_2 < \cdots < s_{N-1} < s_N = 1$.

The link with the implicit R-K method is as follows.

Theorem 1 (adapted from Theorem 7.7 in [32]). *The collocation method (22) with Lagrange polynomials $w_j(s)$ is equivalent to the N-stage implicit Runge-Kutta method (10a) and (10b) with coefficients*

$$a_{ij} = \int_0^{s_i} w_j(s)\, ds, \qquad c_j = \int_0^{1} w_j(s)\, ds, \qquad i, j = 1, \ldots, N. \tag{24}$$

It is possible to choose the collocation points in order to fulfill the orthogonality relations:

$$\int_0^1 (1 - s)\, s w_i(s)\, w_j(s)\, ds = 0, \qquad i, j = 2, \ldots, N - 1, \; i \neq j. \tag{25}$$

Note that, for each $j = 1, \ldots, N$, $\varphi(t + Ts_j, t, x(t), 0) \approx \mathbf{z}(t + Ts_j, t, x(t)) = m_j(t)$ as $w_j(s_i) = \delta_j^i$ (with δ_j^i, the Kronecker delta symbol). A characteristic of the orthogonal collocation method is that each $m_j(t)$ is an approximation of the state that differs from $l_j(t)$ in (10a) and (10b), which are the derivatives.

Equation (21) is obtained applying the substitutions provided by (22) and Theorem 1 with interpolating times s_j obtained by solving (25).

4.2. Closed-Loop Monodromy Matrix. The aPBC was developed for the general control gain $K(t, x(t))$ of system (21). However, methods to choose the time and state dependent control gain have not yet been studied. Moreover, the control gain is tuned for the PBC method and applied to the aPBC, justified by the precision of the estimation of

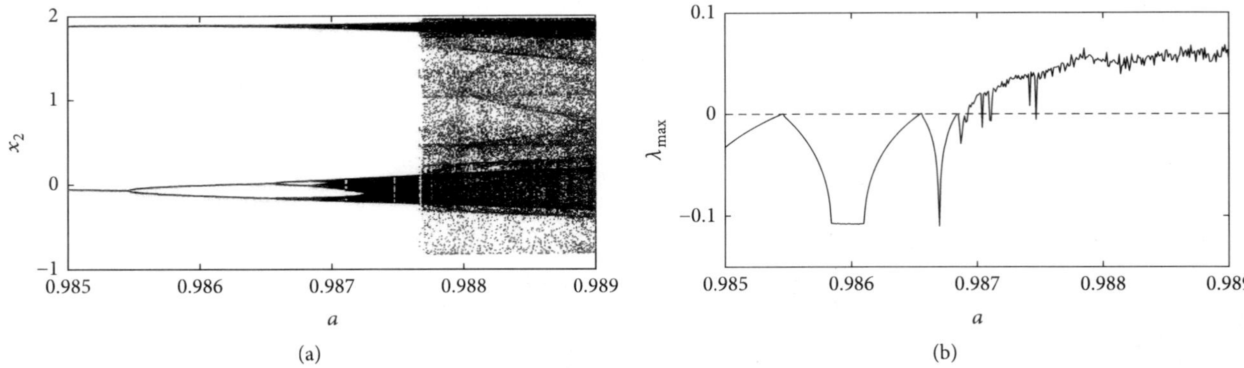

FIGURE 1: Bifurcation diagram of forced van der Pol oscillator for $\omega = 0.45$ and $\eta = 1$ using (a) Poincaré map and (b) largest nonzero Lyapunov exponent.

future states [28]. For simplicity, a constant control gain K is applied that depends upon the ability of computing the closed-loop monodromy matrix of $x^*(t)$. The computation of this matrix requires the integration of the closed-loop system and its variational equation (26) along a trajectory in the vicinity of $x^*(t)$ (linearised dynamics around the periodic orbit) [33, Appendix B]. To integrate this trajectory, the initial condition is chosen close to $x^*(0)$. Integrating (26) over a period yields the corresponding closed-loop monodromy matrix $\Psi(t) = \Phi(t+T,t)$.

$$\begin{aligned}\frac{d\Phi(t,0)}{dt} &= \nabla_x f\left(t,x,K\left(\varphi(t+T,t,x,0)-x\right)\right)\Big|_{x=x^*(t)}\Phi(t,0), \\ & t\in[0,T],\ \Phi(0,0)=I_n,\end{aligned} \tag{26}$$

where

$$\begin{aligned}&\nabla_x f\left(t,x,K\left(\varphi(t+T,t,x,0)-x\right)\right) \\ &\quad = \nabla_x f(t,x,u) \\ &\qquad + K\nabla_u f(t,x,u)\left(\nabla_x\varphi(t+T,t,x,0)-I_n\right).\end{aligned} \tag{27}$$

$\nabla_x\varphi(t+T,t,x,0)$ is the free system monodromy matrix.

Using (26) and (27), we compute the closed-loop monodromy matrix of $x^*(t)$ given gain K. The Floquet multipliers are computed to measure the local stability of the controlled orbit for the chosen K.

In practice, we fix K, compute the monodromy matrix by integrating (26) with an explicit R-K method, and $\varphi(t+T,t,x^*(t),0)$ is computed by integrating the free system over a period T at each step of the integration of (26). After that, we obtain the corresponding Floquet multipliers of the closed-loop system.

5. Case Studies

Two continuous-time systems are used as case study for the aPBC—the forced van der Pol oscillator and Rössler system. In this section we provide a brief analysis about them evidencing their chaotic behavior through bifurcation diagrams and chaotic attractors.

5.1. Forced van der Pol Oscillator. The forced van der Pol (vdP) oscillator is described by the nonlinear state space model (28):

$$\dot{x}(t) = \begin{bmatrix} x_2(t) \\ a\sin(\omega t) - \eta\left(x_1^2(t)-1\right)x_2(t) - x_1(t)\end{bmatrix}, \tag{28}$$

where $x:\mathbb{R}^+\to\mathbb{R}^2$ and $f:\mathbb{R}^+\times\mathbb{R}^2\to\mathbb{R}^2$. In the present work we set $\omega = 0.45$ and $\eta = 1$ and provide bifurcation diagrams by attractor Poincaré map (Figure 1(a)) and largest nonzero Lyapunov exponent (λ_{max}, Figure 1(b)) for $a = [0.985\ \ 0.989]$ evidencing the route to chaos. Poincaré section is defined as usual for nonautonomous systems using time, and in this case

$$\begin{aligned}&S_{vdP} \\ &= \left\{s_{vdP} = (x_1,x_2)\in\mathbb{R}^2 : (t \bmod T) = 0, T = \frac{2\pi}{\omega}\right\},\end{aligned} \tag{29}$$

where (A mod B) is the modulo operation or the remainder of the integer division of A by B.

Form Figure 1 it is possible to note that increasing a we have a series of period-doubling bifurcations leading to two chaotic attractors that are merged by a merging crises bifurcation. Lyapunov exponents are computed for each a evidencing the bifurcations and chaotic attractors. This work follows [29] and adopts $a = 0.988$ with chaotic attractor in continuous-time and Poincaré map shown in Figure 2.

5.2. Rössler System. The Rössler system is described by the nonlinear state space model

$$\dot{x}(t) = \begin{bmatrix} -x_2(t) - x_3(t) \\ x_1(t) + \beta_1 x_2(t) \\ \beta_2 + x_1(t)x_3(t) - \beta_3 x_3(t)\end{bmatrix}, \tag{30}$$

where $x:\mathbb{R}^+\to\mathbb{R}^3$ and $f:\mathbb{R}^+\times\mathbb{R}^3\to\mathbb{R}^3$. Here, $\beta_1 = \beta_2 = 0.2$ and the route to chaos through period-doubling bifurcations for $\beta_3 = [2\ \ 5]$ is presented by attractors Poincaré map (Figure 3(a)) and largest Lyapunov exponent (λ_{max}, Figure 3(b)) bifurcation diagrams. This work follows [34]

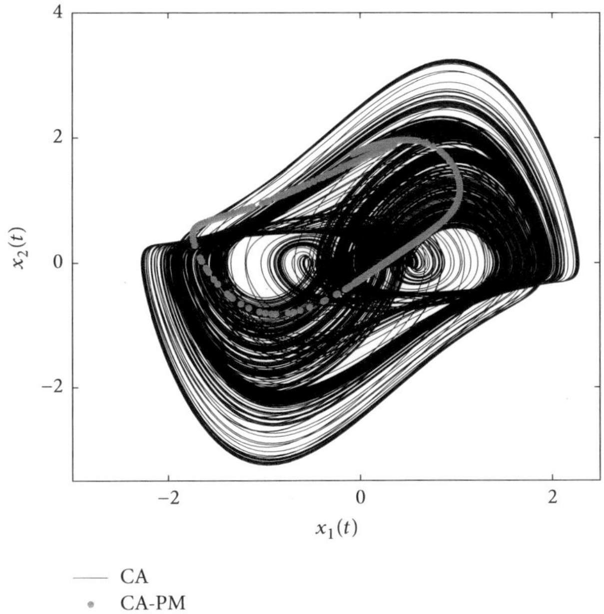

FIGURE 2: Forced van der Pol chaotic attractor in continuous-time (CA) and Poincaré map (CA-PM) for $\omega = 0.45, \eta = 1$, and $a = 0.988$.

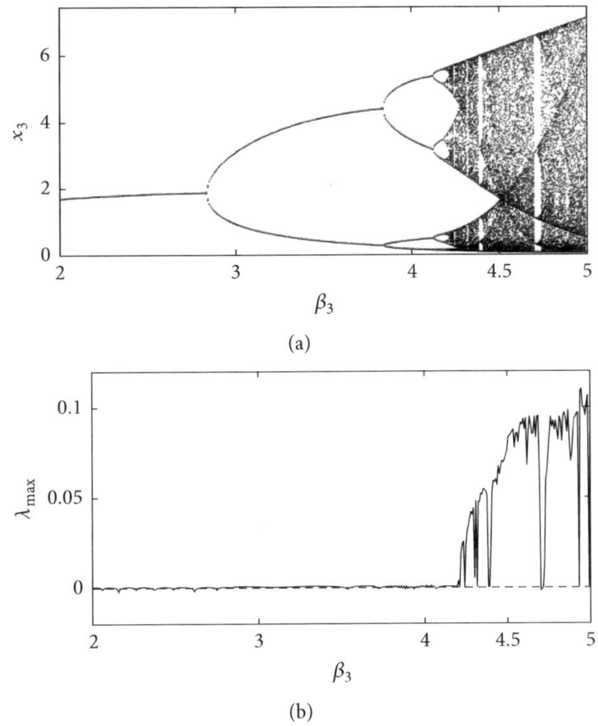

FIGURE 3: Bifurcation diagram of Rössler system for $\beta_1 = \beta_2 = 0.2$ using (a) Poincaré map and (b) largest Lyapunov exponent.

and adopts $\beta_3 = 4.5$ with chaotic attractor in continuous-time and Poincaré map shown in Figure 4. Poincaré section is defined for the Rössler system in

$$S_{Rs} = \left\{ s_{Rs} = (x_1, x_2, x_3) \in \mathbb{R}^3 : x_1 = 0, x_2 \geq 0 \right\}. \quad (31)$$

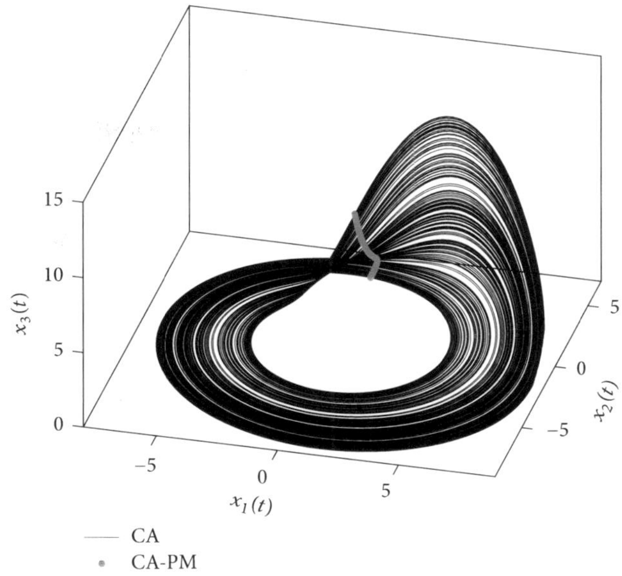

FIGURE 4: Rössler system chaotic attractor in continuous-time (CA) and Poincaré map (CA-PM) for $\beta_1 = \beta_2 = 0.2$ and $\beta_3 = 4.5$.

6. Numerical Results

Numerical results are presented providing illustrative examples about the aPBC characteristics. We use the forced van der Pol oscillator as case study to show an UPO that cannot be stabilized by the DFC method due to the odd-number limitation and is stabilized using the aPBC. Using the same dynamical system, the robustness of the aPBC is evaluated for low-precision future states estimation. The aPBC is applied to the autonomous Rössler system and its robustness is also evaluated. For all the cases, the values N, k_o, and K that characterize the aPBC are tuned prior to application.

6.1. Applying aPBC to the Forced van der Pol Oscillator. In this section, we provide numerical results on the approximated Prediction-Based Control (aPBC) applied to the forced van der Pol (vdP) oscillator according to (32), where $f(t, x(t))$ is defined in (28).

$$\dot{x}(t) = f(t, x(t)) + u(t, x(t)). \quad (32)$$

The state transition map $\varphi(t + Ts_j, t, x(t), 0)$ in (4) is approximated by an implicit R-K method using the operator $\mathbf{z}(t + Ts_j, t, x(t))$ in the s_j, $j = 1, 2, \ldots, N$, discretization points. The implicit R-K method is solved using estimator (17) resulting in the aPBC (21).

Herein, we apply the aPBC using the orthogonal collocation method (see Section 4.1), obtaining $\mathbf{z}(t + Ts_j, t, x(t)) = m_j(t)$, and estimator (17) that leads to $\widehat{m}_j(t) \approx m_j(t)$. The control signal of the closed-loop control method (21) applied to system (32) is

$$u(t, x(t)) = K\left(\widehat{m}_N(t) - x(t)\right) \quad (33)$$

with the constant control gain

$$K = \begin{bmatrix} k & 0 \\ 0 & 0 \end{bmatrix}, \quad k \in \mathbb{R}. \quad (34)$$

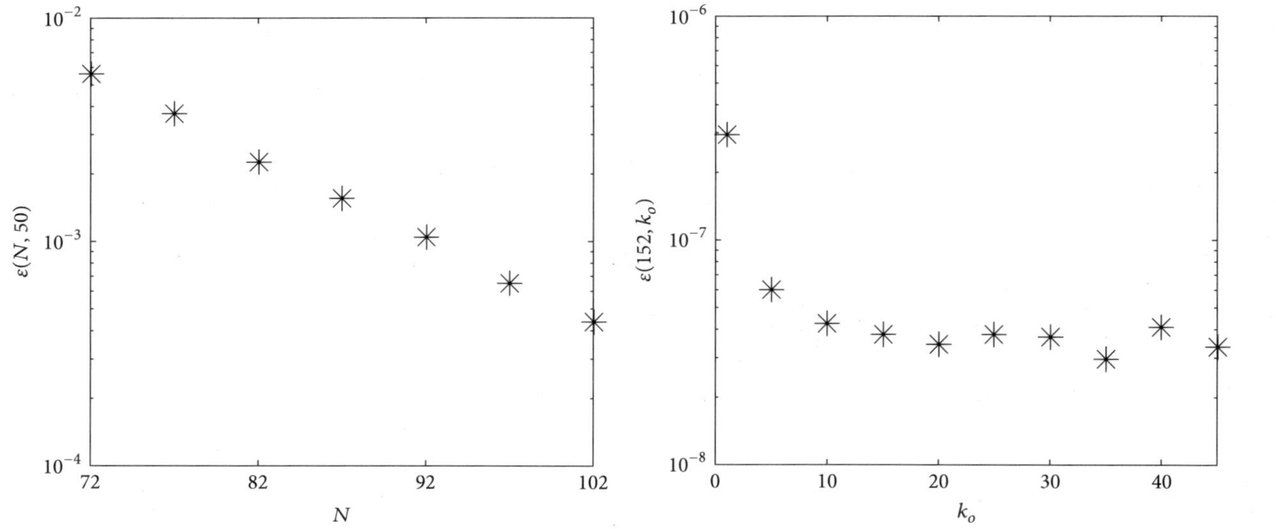

(a) (b)

FIGURE 5: Relative error of the free system response estimated future value using different (a) N and (b) k_o for the forced vdP oscillator.

$u : \mathbb{R}^+ \times \mathbb{R}^2 \to \mathbb{R}^2$ is defined with the same dimension of the free system, but only the first component of the control signal vector is different from zero. This choice of matrix K makes only the first state variable influence on feedback and simplifies its tuning; however, better results can be obtained for a full matrix K.

The control signal is designed by approximating the feedback term, leading to the closed-loop system (21). The solution of the latter depends on the choice of three parameters: the control gain k, the observer gain k_o, and the number of discretization points for the orthogonal collocation method N. The values k_o and N are directly related to the estimation quality and these two parameters will be tuned first.

6.1.1. Tuning the aPBC Parameters k_o and N. In order to evaluate the future state estimation, whose characteristics are related to the parameters k_o and N, we first set $k = 0$. The system is integrated for $t \in [0, T]$ and the points $\widehat{m}_j(t+jT/N)$, $j = 0, \ldots, N$, are collected for different values of k_o and N. One then assesses the convergence by comparison with the solution obtained with $N = 152$ and $k_o = 50$, by computing

$$\varepsilon(N, k_o) = \frac{\left\| \widehat{m}_{j,i}^{(152,50)}(\cdot) - \widehat{m}_{j,i}^{(N,k_o)}(\cdot) \right\|_{L^1(0,T)}}{\left\| \widehat{m}_{j,i}^{(152,50)}(\cdot) \right\|_{L^1(0,T)}}. \quad (35)$$

The superscript on $\widehat{m}_{j,i}^{(N,k_o)}(t)$ indicates the value of N and k_o used in the estimation and the subscript i refers to the state variable of the n-dimensional vector $\widehat{m}_j(t)$.

Figure 5(a) shows $\varepsilon(N, 50)$ for different values of N using initial condition $x(0) = [-0.75\ \ 0.75]^T$, $T = 2\pi/0.45$, and state variable $i = 2$. Note that the collocation parameters s_i and ω_j are previously computed for each N tested. The computation of ω_j for large N requires intense off-line computational burden and reduces the number of values N tested. These values are computed once for all and are independent of $f(t, x(t))$. The values of $N < 72$ lead to integration instability and are not computed.

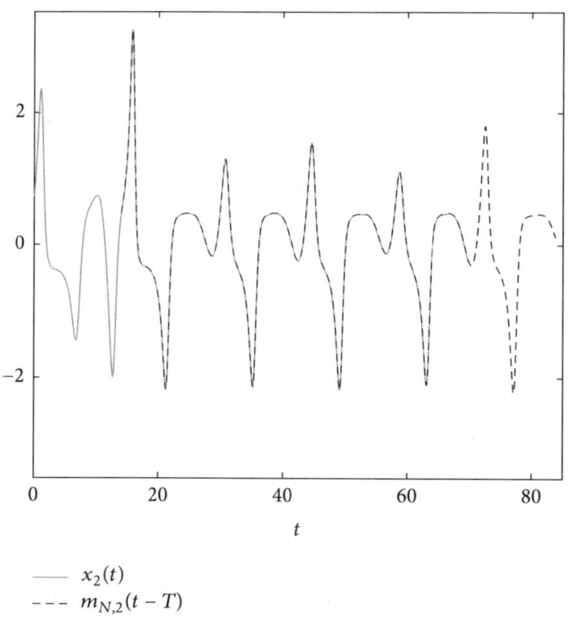

FIGURE 6: Current time and shifted future time trajectories of the forced vdP oscillator for $k = 0$, $k_o = 10$, and $N = 102$.

Repeating the same process, we evaluate $\varepsilon(152, k_o)$ for different values of k_o; see Figure 5(b). Increasing N and k_o results in larger computation effort and trade-off between estimation quality and computation effort should be considered.

Due to the error level shown in Figure 5 for $N = 102$ and $k_o = 10$, we adopt these values in the sequel. Figure 6 shows $x_2(t)$ and $\widehat{m}_{j,2}(t-T)$ (beware of the time shift here). In this case, it is expected that $x(t) = \varphi(t, t-T, x(t-T), 0) \approx \mathbf{z}(t, t-T, x(t-T)) \approx \widehat{m}_N^{(102,10)}(t-T)$ and the proximity of both time-series in Figure 6 allows the use of

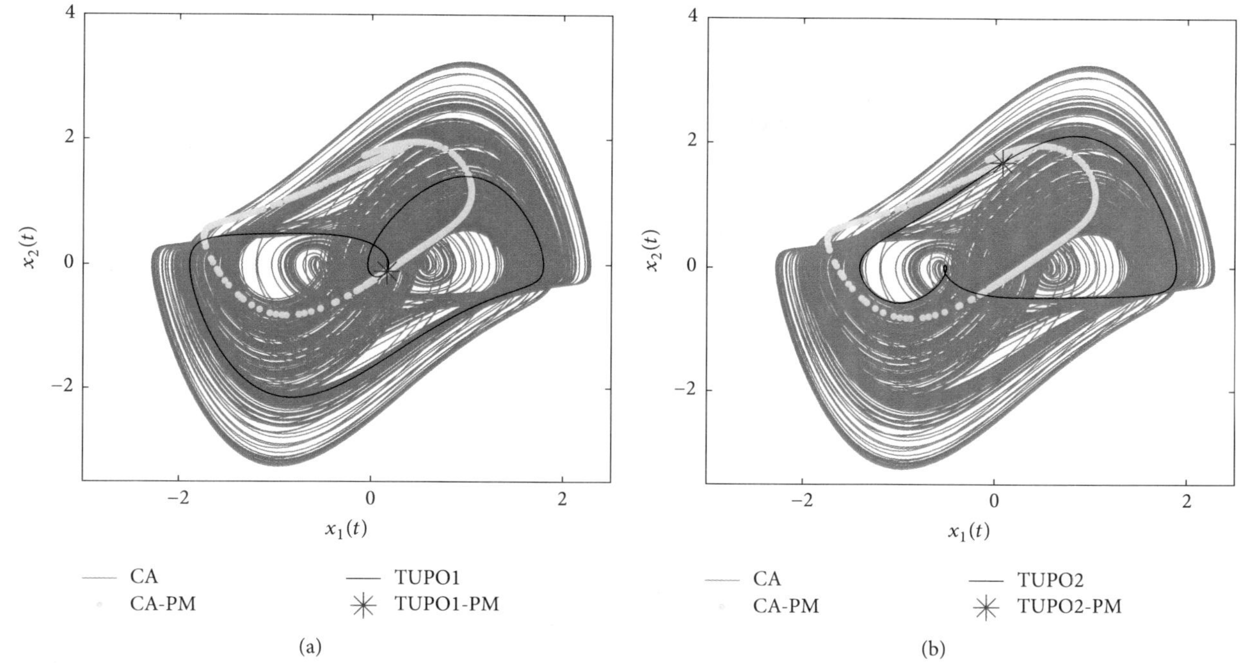

FIGURE 7: Target UPOs (TUPO1 and TUPO2) embedded in the chaotic attractor (CA) of the forced vdP oscillator. The points for the target UPOs and the chaotic attractor in the Poincaré map (PM) are represented by asterisk and gray bullet, respectively.

TABLE 1: Floquet multipliers and one Poincaré map point for $a = 0.988$ of the two target periodic orbits of the vdP oscillator.

	TUPO1	TUPO2
$x^*(t_0)$	$[0.1588445 \quad -0.1106056]^T$	$[0.0771541 \quad 1.6893651]^T$
μ	$[-1.872 \quad -0.048]$	$[4.627 \quad 0.147]$

$m_j(t)$ instead of $\widehat{m}_j^{(102,10)}(t)$. Once the computational cost of the aPBC comes from processing the set of nN ODEs used for prediction with reasonable error tolerance, $\widehat{m}_j^{(102,10)}(t)$ is also a reference for a good trade-off between future state estimation and computation cost.

6.1.2. Target Unstable Periodic Orbits. Two UPOs are chosen as target UPO (TUPO) to be stabilized by the aPBC. TUPO1 was used in [29] for the application of an optimal DFC and in [28] for comparison between aPBC and optimal DFC. Herein, it is used to exemplify the effect of low-precision future state estimation through a small number of collocation points N. TUPO2 is used to exemplify aPBC stabilizing an orbit subjected to the odd-number limitation when applying DFC.

Figure 7 shows the state space of the target periodic orbits, in continuous-time (line) and Poincaré map ($*$), embedded in the chaotic attractor. One point of the orbits with adequate precision for open-loop simulations and their Floquet multipliers are presented in Table 1. Period-1 (where period-1 means one cycle of the forced term or a one-point Poincaré map for the nonautonomous system) orbits of the forced van der Pol oscillator with $\omega = 0.45$ leads to $T = 2\pi/0.45$.

6.1.3. Stabilizing TUPO1 through aPBC for Low-Precision Future State Estimation. In Section 6.1.1, the values of

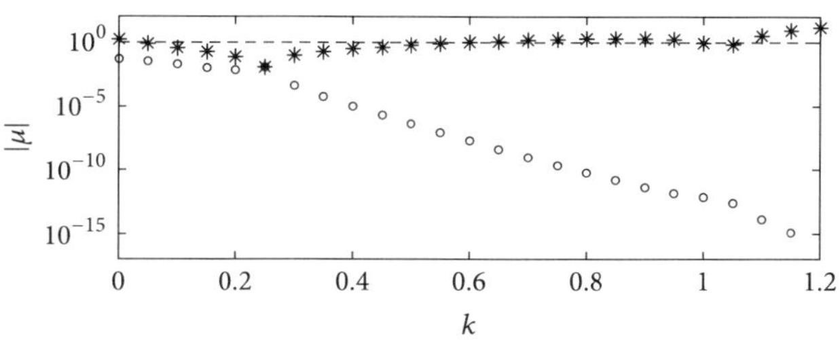

FIGURE 8: Magnitude of the two Floquet multipliers for different values of k for the TUPO1 controlled by the aPBC.

$N = 102$ and $k_o = 10$ that result in a good estimation of future states according to value $\varepsilon(N, k_o)$ have been obtained. Results for these N and k_o and a comparison with DFC are shown in [28]. In the present section, the aPBC is applied to stabilize TUPO1 using $N < 102$ exemplifying the robustness of the method to the use of low-precision future state estimates, aiming at possible experimental applications.

Before applying aPBC to TUPO1, a control gain k that leads to closed-loop stabilization shall be found.

Tuning of k. We use (26), with $x^*(0) = [0.1588445 \quad -0.1106056]^T$, to obtain a value k that stabilizes the orbit of (7a) and (7b). Note that, once the estimation with $N = 102$ and $k_o = 10$ is precise, a stabilizing k for (7a) and (7b) yields stability for a slightly perturbed cycle of (32) with the control signal (33) [28].

Figure 8 shows the magnitude of the two Floquet multipliers $|\mu|$ for different values of k. Stability is achieved when $|\mu_i| \leq 1$, $i = 1, 2$. We choose $k = 0.25$, which results in

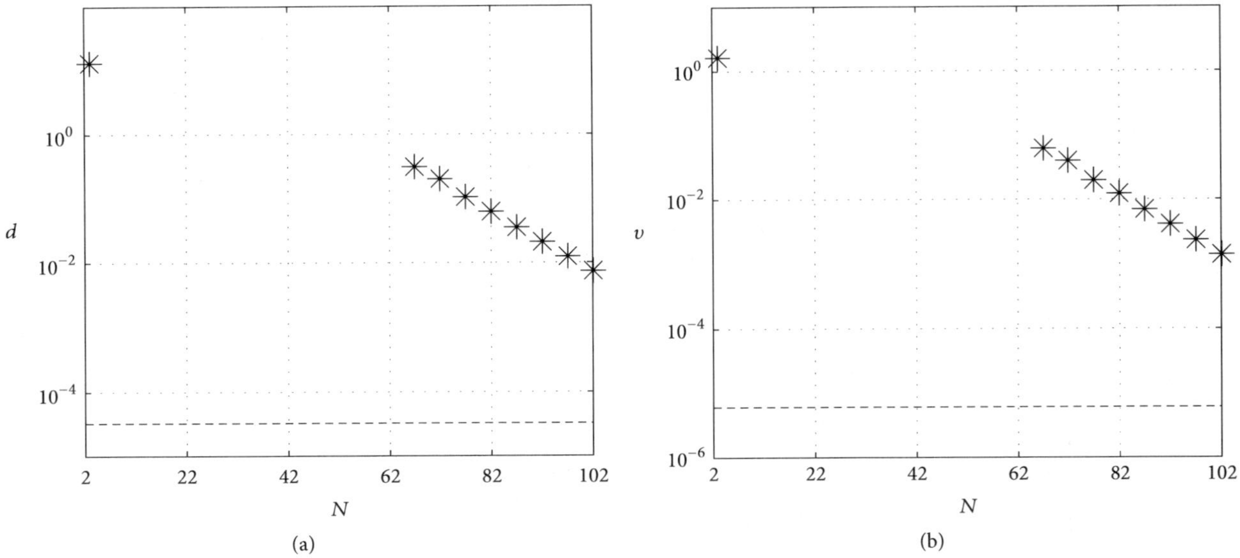

FIGURE 9: Indexes (a) d and (b) υ when applying aPBC to TUPO1 for different N with $k = 0.25$ and $k_o = 10$. Bottom dashed line are the values of (a) d and (b) υ computed with $N = 152$ and $k_o = 50$ used as reference.

$|\mu|_{\max} \approx 0.01213$. All the Floquet multipliers in this case are real numbers.

Applying the aPBC for a Low N. Figure 5(a) shows an increase of $\varepsilon(N, 50)$ when decreasing N for $k = 0$. This value refers to the free system and now we are then interested in characteristics of the closed-loop system, particularly the value of steady state error and control effort as a function of N.

Steady state error index d is the distance (Euclidean norm) between predicted state $\widehat{m}_N(t)$ and current state $x(t)$ for an entire cycle in the close vicinity of target UPO. Steady state control effort index υ is the measure of total external signal (norm-1) necessary to keep the trajectory in close vicinity around one cycle of the target periodic orbit.

$$d = \lim_{t\to+\infty} \left\|\widehat{m}_N(\tau) - x(\tau)\right\|_{L^2(t,t+T)}, \qquad \upsilon = \lim_{t\to+\infty} \left\|u(\tau)\right\|_{L^1(t,t+T)}. \tag{36}$$

Figure 9(a) shows d when stabilizing the TUPO1 by applying the aPBC for different values of N with $k = 0.25$ and $k_o = 10$. Figure 9(b) shows the same for υ.

It can be observed in Figure 9 that even for a reduction in the number of collocation points, stability of (an orbit close to) TUPO1 is achieved for $N \geq 67$. TUPO1 stabilized with the aPBC for $N = 82$ and $N = 67$ and $x(0) = \begin{bmatrix}-1.5 & 1.0\end{bmatrix}^T$ are shown in Figures 10 and 11, respectively. Figures 10(e) and 11(e) show the initial value for the predicted states $m_j(0)$, $j = 1, 2, \ldots, N$, and $\varphi(T, 0, x(0), 0)$. It is observed that $m_j(0)$ is relatively close to $\varphi(T, 0, x(0), 0)$.

The high values observed in Figure 9 for $N = 3$ are explained in Figure 12. It exemplifies a situation wherein a stable orbit different from TUPO1 is achieved due to the reduced N. High amplitude periodic control signal in Figure 12(b) indicates that this stabilized orbit is not an orbit of the original free system. Figure 12(e) shows that $N = 3$ leads to $m_j(0)$ almost unrelated to $\varphi(T, 0, x(0), 0)$ and consequently $u_2(t) = k(\widehat{m}_{N,2}(t) - x_2(t))$ is a high amplitude control signal. According to Figure 12(a), a trajectory with initial condition $x(0)$ on TUPO1 (Poincaré point) diverges from it showing that it is not stable in this case.

It is noticed from the time-series of trajectories in Figures 10(c), 10(d), 11(c), and 11(d) that in certain torsions of the trajectory of the system controlled by the aPBC there is an oscillation on $\widehat{m}_N(t)$. This oscillation is generated because small N leads to instability on the predicted state and it reflects in time-series of control signals in Figures 10(b) and 11(b) and on the values of υ and d in Figure 9.

Another consequence on reducing N is the loss of stability of $\widehat{m}_N(t)$, exemplified in Figure 13 and occurring even after changing the error tolerances of the explicit R-K integrator (the simulations present in this work were performed using the MATLAB® integration routine ode45 and, in this specific case, the routines ode113 and ode15s were also tried, all of them resulting in interruption of the numerical integration). One of the reasons for this loss of stability is that N is the parameter that rules the implicit R-K integration error and Figure 13(d) shows that $m_N(0)$ slightly diverges from $\varphi(T, 0, x(0), 0)$ at time T and, contrasting to example of Figure 12, the trajectories of the new system do not converge to a stable periodic orbit.

6.1.4. Stabilizing TUPO2 through aPBC: An Odd-Number Limitation Example. One of the main purposes when developing PBC for discrete-time systems is the odd-number limitation subjected by the DFC [14, 21]. Herein, we show a case where the continuous-time DFC with a constant control gain is subjected to the odd-number limitation and aPBC is not. TUPO2 has an odd number of real Floquet multipliers larger than +1 (see Table 1) and is used for a numerical example.

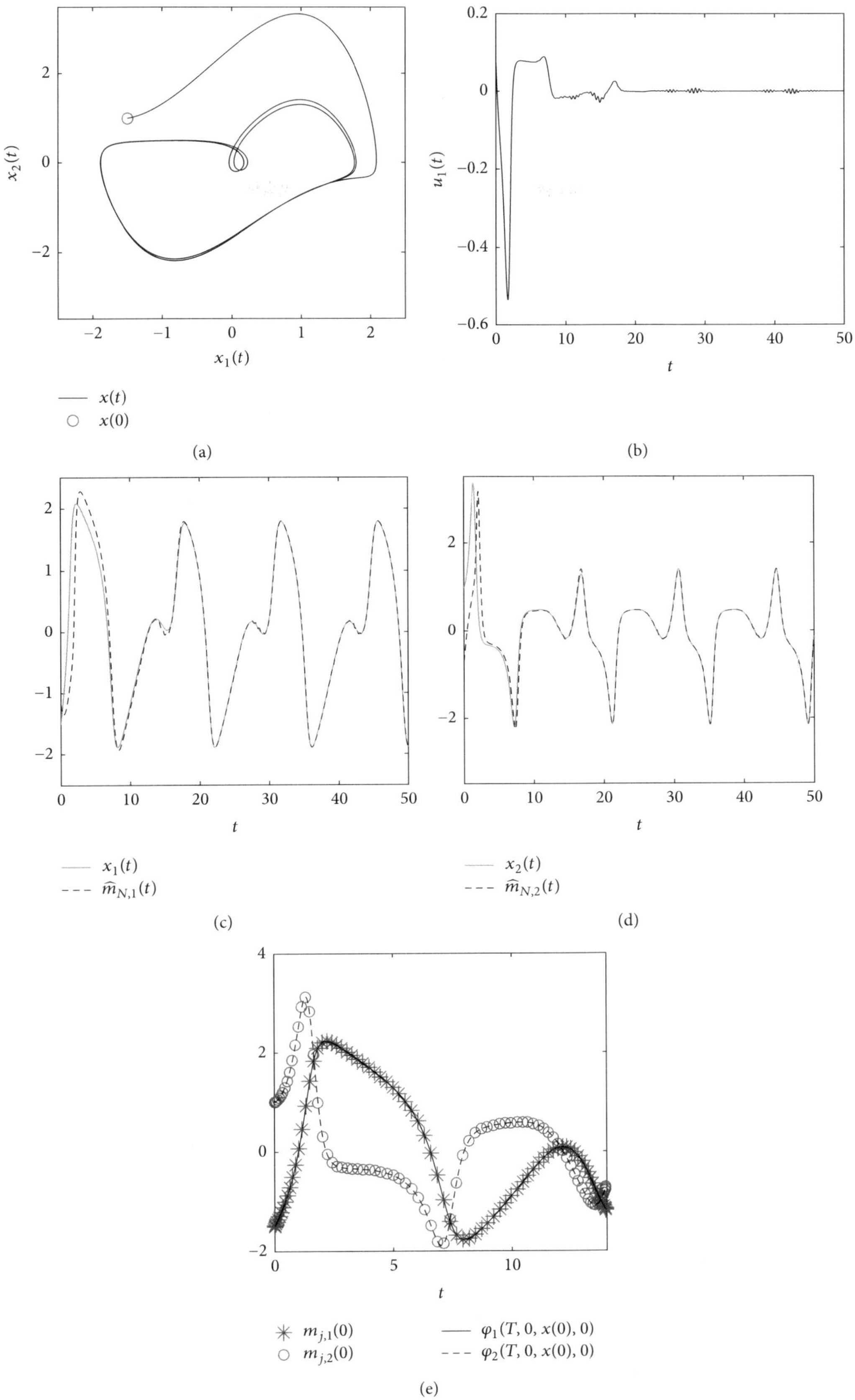

FIGURE 10: aPBC applied with $N = 82$, $k = 0.25$, and $k_o = 10$. (a) Trajectory in state space; (b) time-series of the control signal; ((c) and (d)) time-series of the actual and predicted state variables; (e) time-series of $\varphi(T, 0, x(0), 0)$ and the initial value of the predicted states $m_j(0)$.

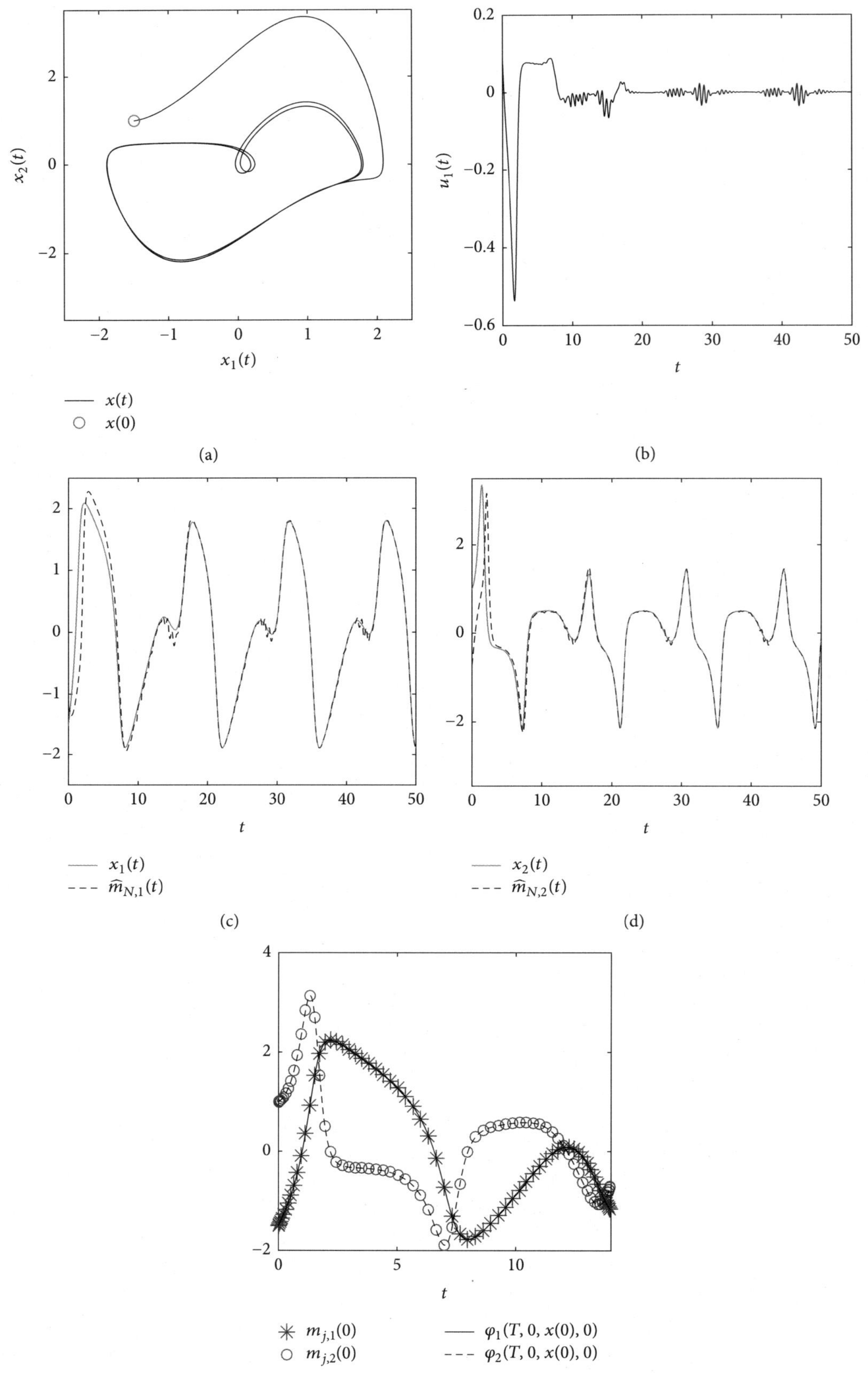

FIGURE 11: aPBC applied with $N = 67$, $k = 0.25$, and $k_o = 10$. (a) Trajectory in state space; (b) time-series of the control signal; ((c) and (d)) time-series of the actual and predicted state variables; (e) time-series of $\varphi(T, 0, x(0), 0)$ and the initial value of the predicted states $m_j(0)$.

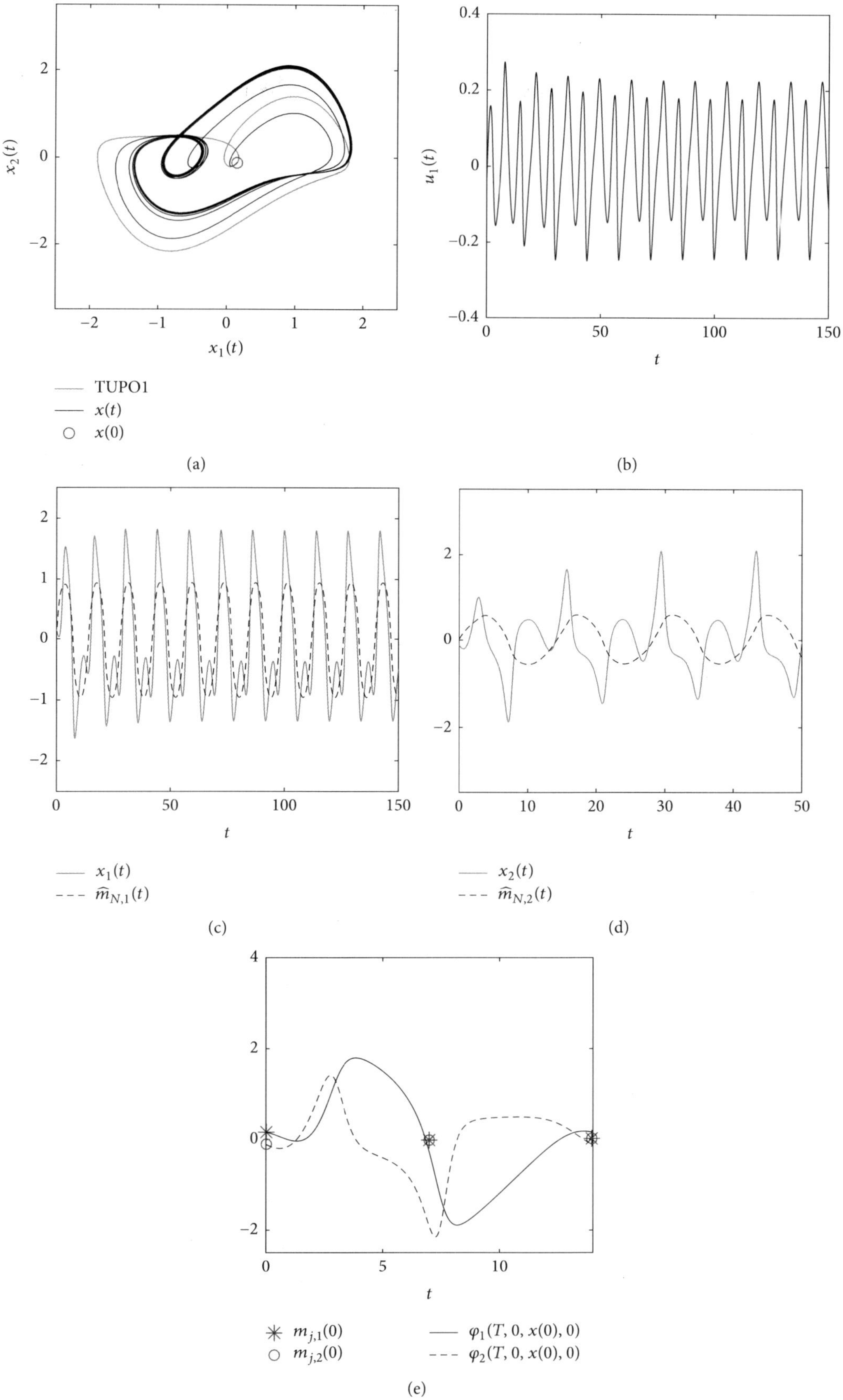

FIGURE 12: aPBC applied with $N = 3$, $k = 0.25$, and $k_o = 10$. (a) Trajectory in state space; (b) time-series of the control signal; ((c) and (d)) time-series of the actual and predicted state variables; (e) time-series of $\varphi(T, 0, x(0), 0)$ and the initial value of the predicted states $m_j(0)$.

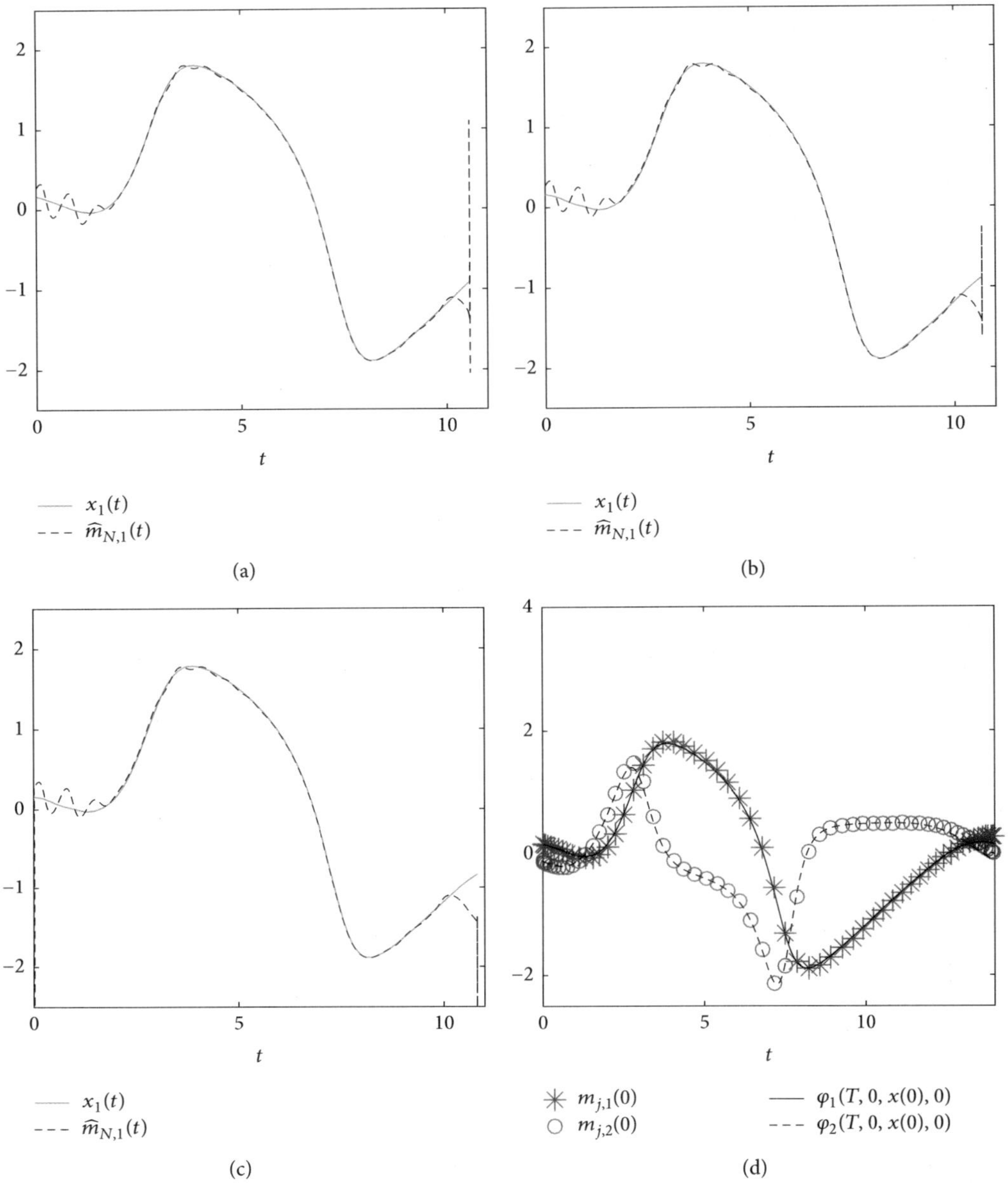

FIGURE 13: Time-series of a trajectory when applying the aPBC with $N = 62$ and (a) $k = 0$ and $k_o = 10$; (b) $k = 0.25$ and $k_o = 10$; (c) $k = 0.25$ and $k_o = 1000$; (d) time-series of $\varphi(T, 0, x(0), 0)$ and the initial value of the predicted states $m_j(0)$.

Applying the DFC. The DFC with control signal

$$u(t, x(t)) = K(x(t-T) - x(t)), \tag{37}$$

$K \in \mathbb{R}^{2\times 2}$, is applied to system (32) with the goal of stabilizing TUPO2. The procedure of finding K for DFC follows [29].

A scalar control gain k is used with K defined in (34) resulting in the root locus chart of Figure 14. This chart contains the free system orbit Floquet multipliers μ marked with ×: in black solid line, the fourth largest magnitude (the closed-loop system is infinite-dimensional when applying DFC; see [29] for details) closed-loop Floquet multipliers of the orbit for positive k and, in dark gray, the largest closed-loop Floquet Multiplier of the orbit for negative k. Arrows indicate increasing $|k|$. It is possible to conclude that for any k there is a Floquet multiplier with magnitude larger than 1 and the orbit is not stabilizable. In addition, all the Floquet multipliers of the orbit tend to +1 when increasing DFC positive scalar gain k. The same chart was done for the other three possible scalars k with equivalent (omitted) results.

Following [29], a matrix gain was tried with values obtained by an optimization process using the MATLAB routines *fminsearch* and *fmincon* [35–38]. This optimization aims at finding a minimum for the largest magnitude Floquet multiplier through changing matrix K. A scan process varying all the values of the matrix gain was also tried. The final trial used a matrix gain composed of a rotational matrix (formed by sines and cosines) multiplied by a real scalar gain equivalent to the gain used in [18–20], varying both the rotation angle and the gain. After performing the three

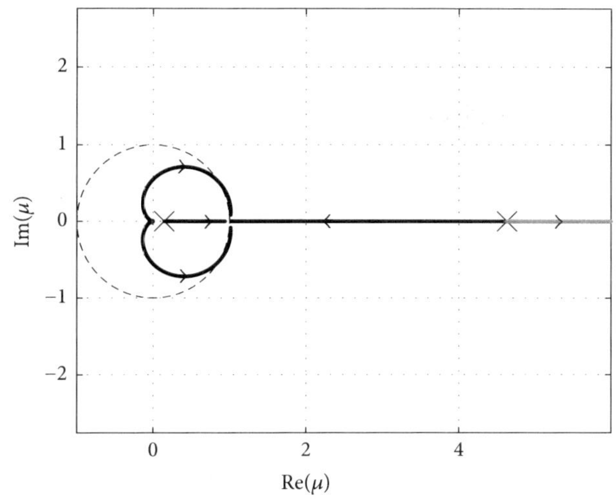

FIGURE 14: Root locus chart for the Floquet multipliers of the TUPO2 using the DFC. In black, the four largest magnitude Floquet multipliers for positive control gain k. In dark gray, the largest magnitude Floquet multiplier for negative control gain k. The Floquet multipliers of the free system orbit are represented by (×) and the unit circle by dashed line in black.

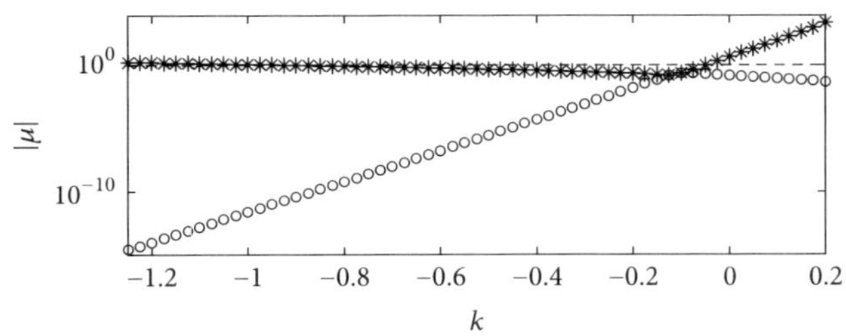

FIGURE 15: Magnitude of the Floquet multipliers for different values of k for the TUPO2 controlled by the aPBC.

methodologies, a matrix control gain that stabilizes TUPO2 was not found.

Applying aPBC and Stabilizing TUPO2. Figure 15 shows the magnitude of the Floquet multipliers as a function of the control gain k for the TUPO2 by applying the aPBC. It was noticed that the aPBC stabilizes orbits with an odd number of real Floquet multipliers larger than +1. One detail is relevant here; the stabilization is achieved only for negative values of the gain k and positive values of k make the largest magnitude Floquet multiplier increase.

TUPO2 stabilized by the aPBC with $k = -0.125$ and initial condition $x(0) = \begin{bmatrix}0.1 & 1.8\end{bmatrix}^T$ is shown in state space in Figure 16(a). Time-series of the control signal is presented in Figure 16(b) with $u_2(t) \rightarrow 0$ as $t \rightarrow +\infty$ indicating stabilization.

6.2. Applying aPBC to the Rössler System. In this section, we provide numerical results on approximated Prediction-Based Control (aPBC) for the Rössler system using control scheme (21) applied to system (30) according to (32). The goals are (i) to show that the method is applicable to autonomous systems without any modification and (ii) to compare the effect of different numbers of collocation points N on stabilization.

aPBC is applied following the same procedure shown in Section 6.1 with control signal $u(t, x(t))$ defined in (33). Herein, we use the constant control gain

$$K = \begin{bmatrix} 0 & 0 & 0 \\ 0 & 0 & 0 \\ 0 & 0 & k \end{bmatrix}, \quad k \in \mathbb{R}. \tag{38}$$

This choice of matrix K makes only the third state variable influence on feedback and simplifies its tuning.

6.2.1. Tuning the aPBC Parameters k_o and N. The procedure shown in Section 6.1.1 is repeated here to characterize the performance of the estimator for different values of the parameters k_o and N for $k = 0$. We used initial condition $x(0) = \begin{bmatrix}0 & -6 & 0.0375\end{bmatrix}^T$ to compute $\varepsilon(N, k_o)$ in (35). The results are shown in Figure 17 for different values of (a) N with $k_o = 50$ ($\varepsilon(N, 50)$) and (b) k_o with $N = 152$ ($\varepsilon(152, k_o)$).

In this case $N < 32$ values lead to numerical integration instability, except when $N = 3$, and are not computed. Herein we notice that less collocation points are necessary to obtain estimation quality equivalent to the one obtained for the vdP oscillator (see Figure 5). It can be seen that for the vdP there is $\varepsilon(102, 50) \approx 4.4 \times 10^{-4}$, while for the Rössler system there is $\varepsilon(52, 50) \approx 4.4 \times 10^{-4}$. Values $\varepsilon(152, k_o)$ for large k_o of the chosen trajectory of the Rössler system also stabilize to a value much smaller than for the vdP oscillator.

Figure 18 shows $x_2(t)$ and $\widehat{m}_{N,2}(t-T)$ (beware of the time shift here) for the Rössler system, which can be compared with Figure 6. Notice that $T = 5.8439698764$ (the period of the TUPO4) for the Rössler system while $T = 2\pi/0.45$ for vdP oscillator, but the reduction on $\varepsilon(152, k_o)$ is much more related to the behavior of the trajectories than the period itself, because collocation points s_i are computed for a normalized time between 0 and 1.

6.2.2. Target Unstable Periodic Orbit. Asenjo et al. [34] present one UPO of the chaotic attractor for $\beta_1 = \beta_2 = 0.2$ and $\beta_3 = 4.5$. We apply the Newton-Raphson method with Poincaré section (31) to refine the position and period of the orbit and call it TUPO4 (Figure 19). One point of TUPO4 and its period T with adequate precision for open-loop simulations and Floquet multipliers μ are presented in Table 2.

TABLE 2: Poincaré map point $x(t_0)$, period T, and Floquet multipliers μ of a periodic orbit (TUPO4) of the Rössler system.

	TUPO4
$x(t_0)$	$\begin{bmatrix}0 & 5.0206004746 & 1.6806432235\end{bmatrix}^T$
T	5.8439698764
μ	$\begin{bmatrix}-1.918 & 1 & -1.866 \times 10^{-11}\end{bmatrix}$

6.2.3. Stabilizing TUPO4 through aPBC. This section is dedicated to the stabilization of TUPO4 using aPBC for different

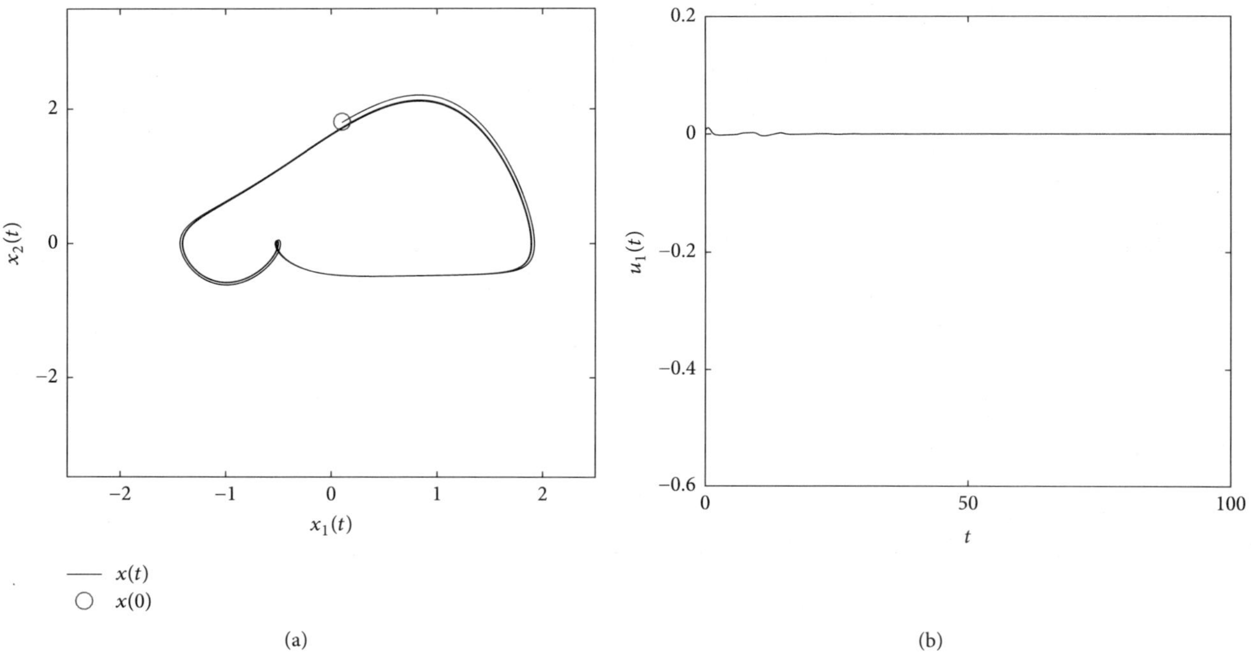

FIGURE 16: TUPO2 of forced vdP oscillator stabilized using aPBC with $k = -0.125$ and $x(0) = [0.1 \quad 1.8]^T$. (a) State space trajectory and (b) time-series of the control effort.

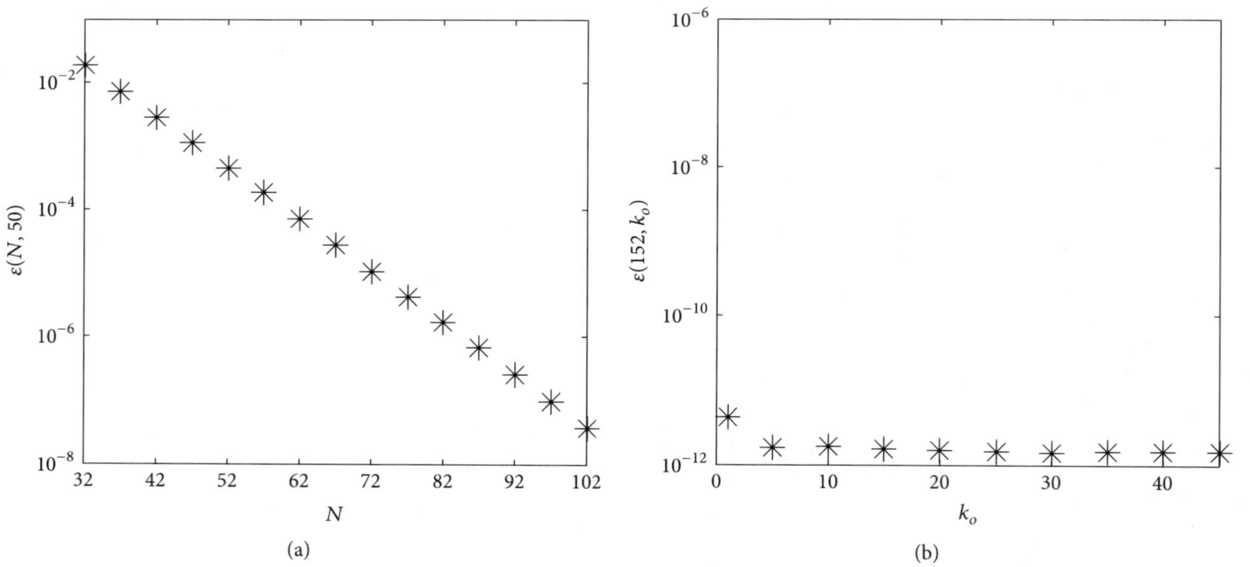

FIGURE 17: Relative error of the free system response estimated future value using different (a) N and (b) k_o for the Rössler system.

values N. From Figure 17, values $N = 102$ and $k_o = 10$ are indicated for an accurate estimation of future states and aPBC will be initially applied for these values. Results for low N are also presented, evidencing the robustness of the controller for low-precision future states estimation in autonomous systems.

The control gain k of (38) shall be tuned for TUPO4 prior to the application.

Tuning of k. Figure 20 shows the magnitude of the three Floquet multipliers $|\mu|$ for different values of k according to the procedure presented in Section 6.1.3. The value $\mu_1 = 1$ is characteristic of autonomous system trajectories and stability is achieved when, for the other Floquet multipliers, $|\mu_i| \leq 1$, $i = 2, 3$. Choosing $k = 1.85$ results in $|\mu_2| \approx 2.994 \times 10^{-3}$ (and $|\mu_3| \approx 2.167 \times 10^{-14}$). All Floquet multipliers in this case are real numbers.

Applying the aPBC for Different Values of N. In this section, we evaluate the characteristics of aPBC for different values N applied to the stabilization of TUPO4. Herein, we use $N \leq 102$, $k_o = 10$, and $k = 1.85$.

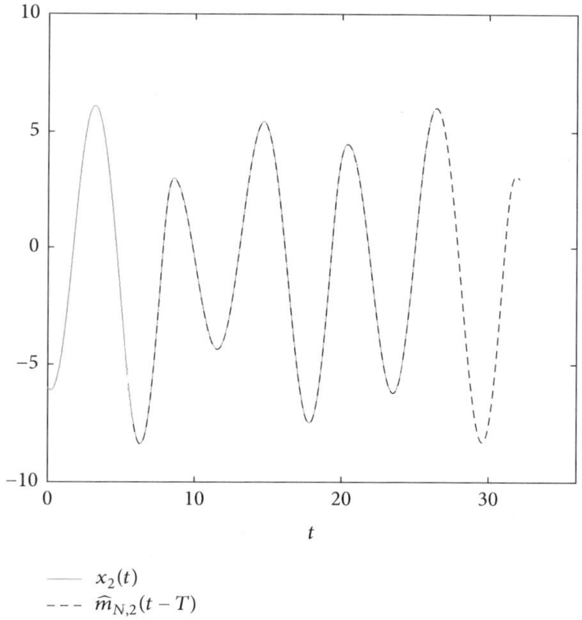

FIGURE 18: Current time and shifted future time trajectories of the Rössler system for $k = 0$, $k_o = 10$, and $N = 102$.

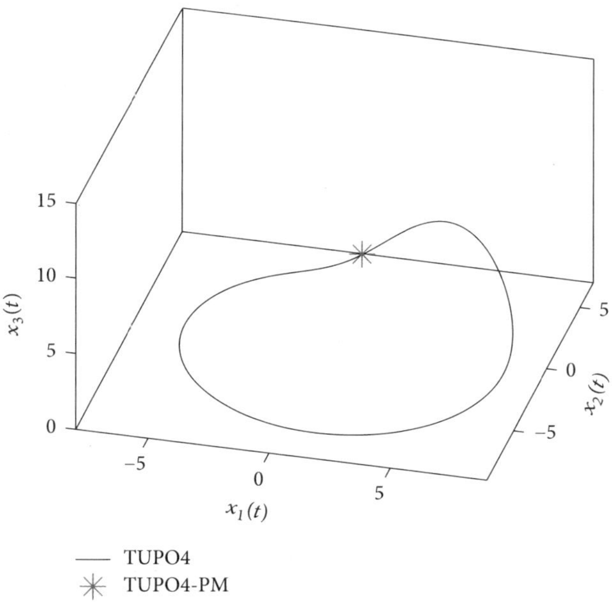

FIGURE 19: TUPO4 of Rössler system in continuous-time (line) and Poincaré map point (∗) for $\beta_1 = \beta_2 = 0.2$ and $\beta_3 = 4.5$.

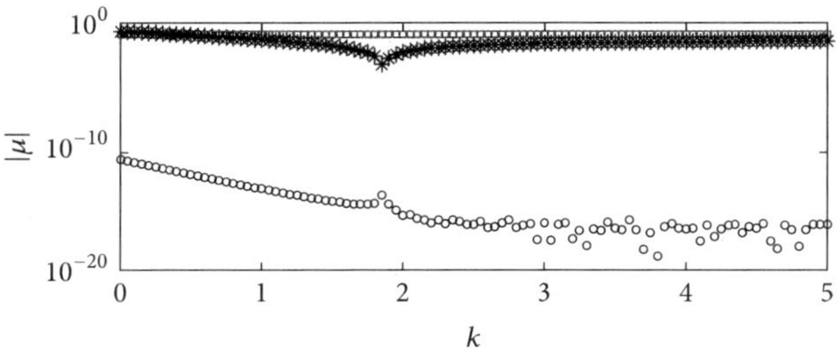

FIGURE 20: Magnitude of the Floquet multipliers for different values of k for the TUPO4 controlled by the aPBC.

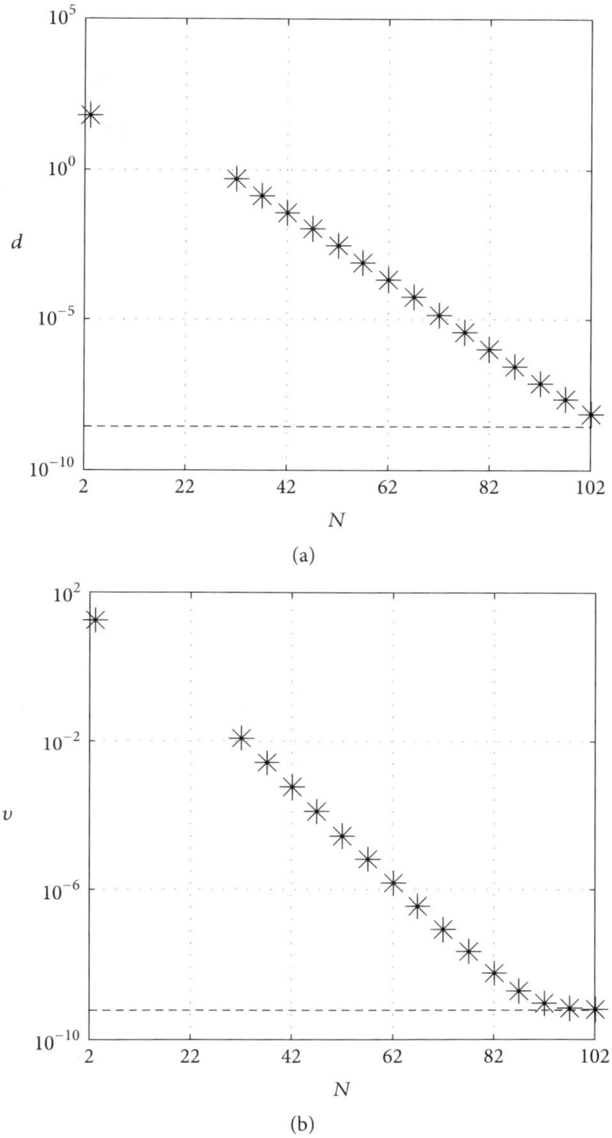

FIGURE 21: Indexes (a) d and (b) υ when applying aPBC to TUPO4 for different N with $k = 1.85$ and $k_o = 10$. The bottom dashed lines are the values of (a) d and (b) υ computed with $N = 152$ and $k_o = 50$ used as reference.

Figure 17(a) shows an increase of $\varepsilon(N, 50)$ observed when decreasing N for $k = 0$. Figure 21 shows d and υ for different values of N with $k = 1.85$ and $k_o = 10$. Values of $3 < N < 32$ were not computed because of instability, the same as observed in Section 6.1.3.

Herein, we observe that $N \geq 102$ leads to a precise estimation once there is a convergence of υ to the value computed using $N = 152$ and $k_o = 50$ (dashed line). Comparing Figure 21 with Figure 9, a great reduction on υ and d is observed for the control method applied to the Rössler system, which is in accordance with what was observed for $\varepsilon(N, 50)$. In fact, N between 47 and 52 leads to a value of d equivalent to the value obtained for $N = 102$ for the vdP oscillator. Considering υ, this reduction is more significant, with $N = 37$ leading to a value of υ close to the value

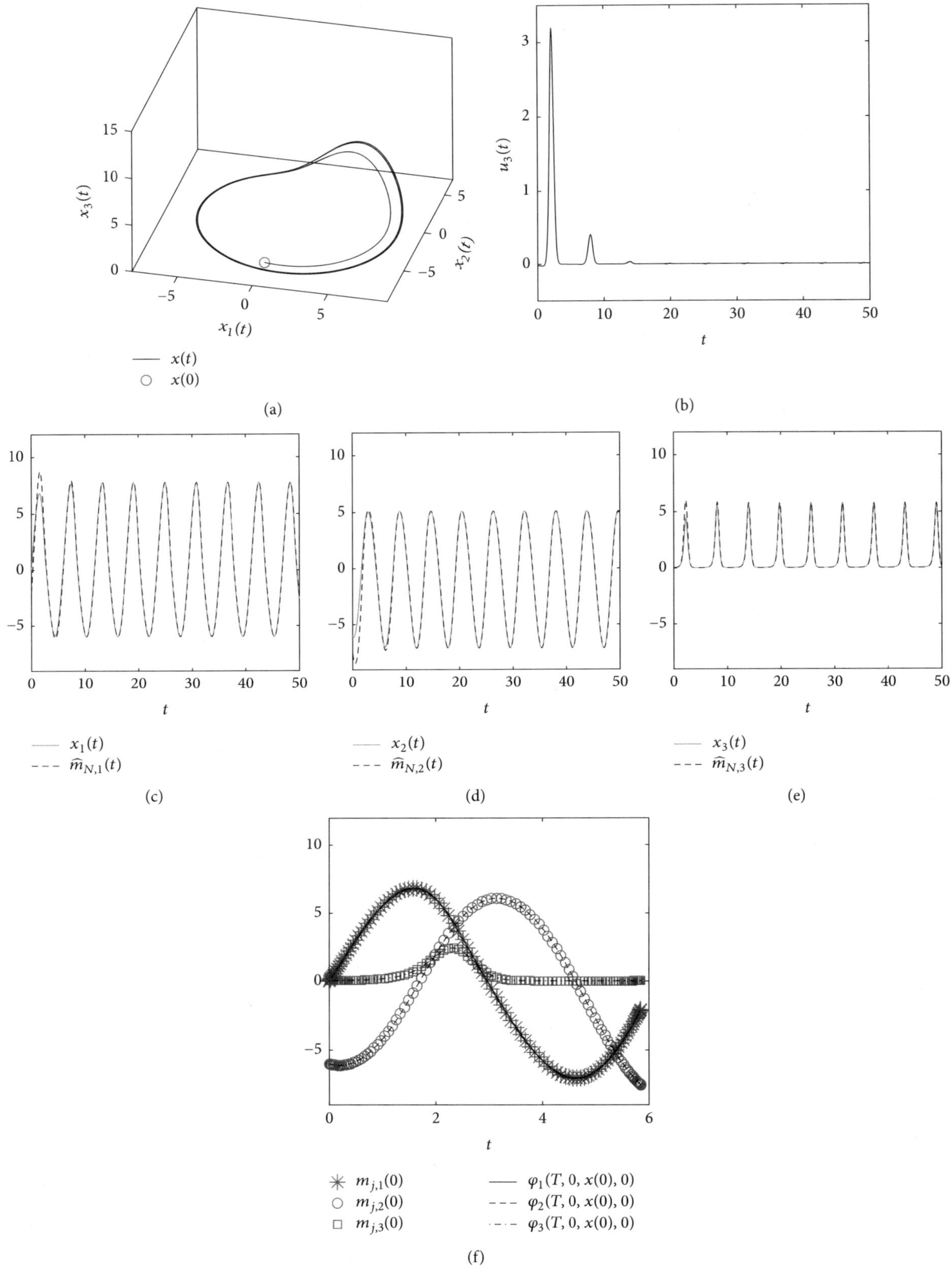

FIGURE 22: aPBC applied to the Rössler system with $N = 102$, $k = 1.85$, and $k_o = 10$. (a) Trajectory in state space; (b) time-series of the control signal; ((c), (d), and (e)) time-series of the actual and predicted state variables; (f) time-series of $\varphi(T, 0, x(0), 0)$ and the initial value of the predicted states $m_j(0)$.

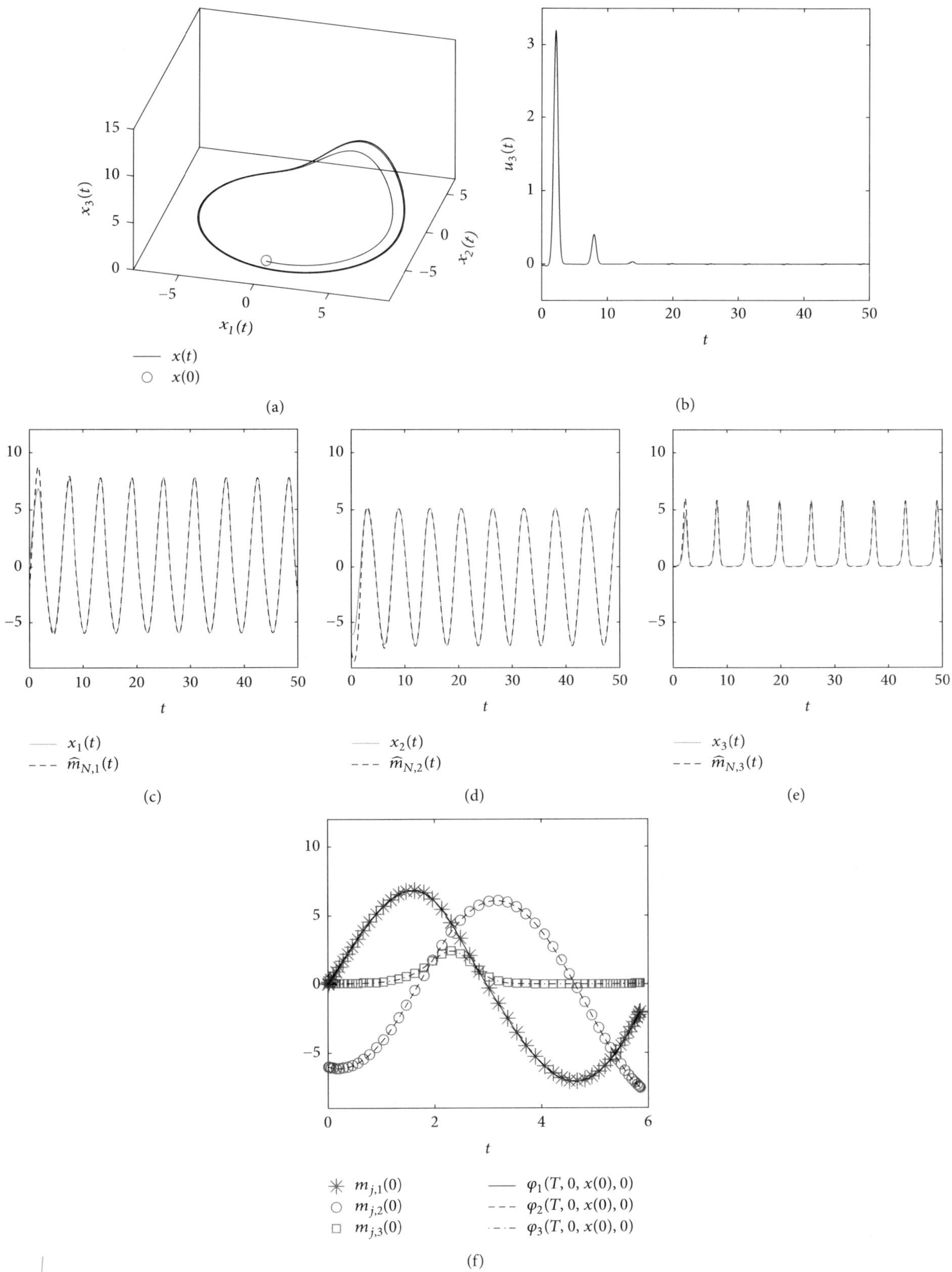

FIGURE 23: aPBC applied to the Rössler system with $N = 52$, $k = 1.85$, and $k_o = 10$. (a) Trajectory in state space; (b) time-series of the control signal; ((c), (d), and (e)) time-series of the actual and predicted state variables; (f) time-series of $\varphi(T, 0, x(0), 0)$ and the initial value of the predicted states $m_j(0)$.

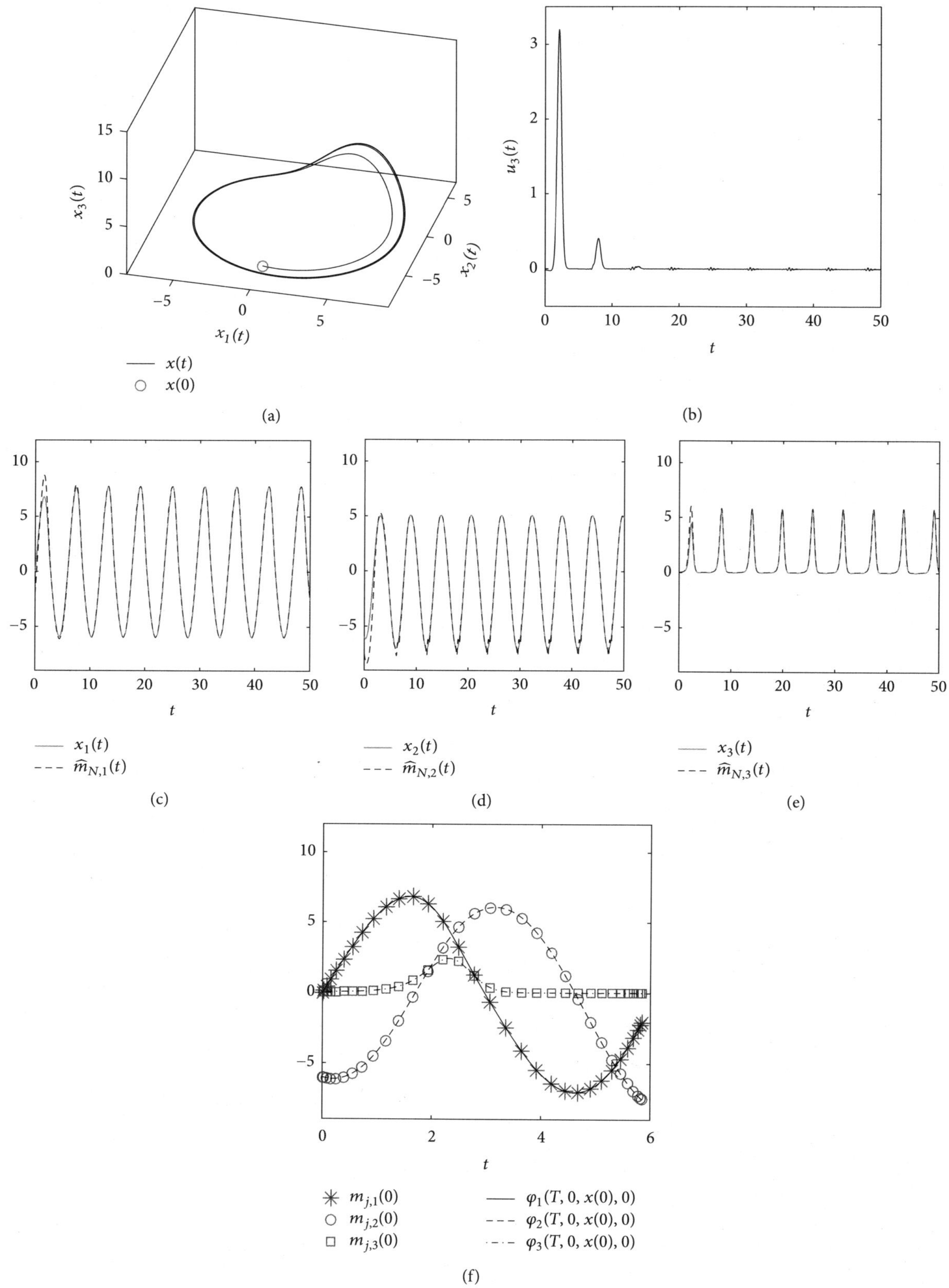

FIGURE 24: aPBC applied to the Rössler system with $N = 32$, $k = 1.85$, and $k_o = 10$. (a) Trajectory in state space; (b) time-series of the control signal; ((c), (d), and (e)) time-series of the actual and predicted state variables; (f) time-series of $\varphi(T, 0, x(0), 0)$ and the initial value of the predicted states $m_j(0)$.

obtained for $N = 102$ for the vdP. This result is important for practical aspects because it shows that, for certain systems, the augmented state of the closed-loop system controlled with the aPBC can be reduced significantly (in comparison with the initial results shown for the vdP oscillator).

Figure 21 shows that stability of TUPO4 is achieved for $N \geq 32$. The TUPO4 stabilized by the aPBC for $N = 102$, $N = 52$, and $N = 32$ is shown in Figures 22, 23, and 24, respectively, using $x(0) = [0\ \ -6\ \ 0.0375]^T$. Deviation of $\widehat{m}_{N,2}(t)$ from $x_2(t)$ is observed in Figure 24(d), which reflects the numerical integration instabilities for $N < 32$. We verify that stabilization is not achieved for $N = 3$, which leads to the high values of υ and d observed in Figure 21.

7. Conclusions

Approximate Prediction-Based Control (aPBC), which is the continuous-time version of Prediction-Based Control (PBC), was reviewed in this work and some of its characteristics were analyzed through numerical simulations. It was shown that continuous-time application of PBC requires future state values and aPBC is based on a methodology for estimating them in real-time.

The main drawback in aPBC is that estimation of future states requires extending the order of the closed-loop system by the number of implicit Runge-Kutta solution points, herein the orthogonal collocation points, times the number of state variables. Applications require real-time explicit integration of these extended equations and reducing them implies reducing the computational cost while reducing estimation precision. Both exemplified cases have shown that there is a lower bound on the number of collocation points to achieve stability, to both target UPO and numerical integration. While comparing examples, it was noticed that the number of collocation points depends upon the trajectory torsions in state space. Future studies should be carried out in order to reduce the number of implicit Runge-Kutta solution points through different orthogonal polynomials or different implementation methods.

Another interesting example is an UPO of a continuous-time system that could not be stabilized by DFC with constant control gain because of the odd-number limitation. The same orbit is stabilized by aPBC with a negative scalar constant gain without any modification on the method. Then, the main justification for the application of PBC in comparison to DFC in discrete-time is also valid for continuous-time.

The method proposed for estimation of open-loop future states of nonlinear systems applied to aPBC may be generalized for other applications. As example, model-based predictive control [39] applied to nonlinear systems requires open-loop future state values to optimize control gain at each time step and this routine may be fastened by using the proposed scheme.

Conflicts of Interest

The authors declare that they have no conflicts of interest regarding the publication of this paper.

Acknowledgments

This project has been funded by CAPES-COFECUB (Project Ma 624/09), FAPESP (grant 2011/17610-0), and Université Paris-Sud, Orsay.

References

[1] A. Mesquita, E. L. Rempel, and K. H. Kienitz, "Bifurcation analysis of attitude control systems with switching-constrained actuators," *Nonlinear Dynamics*, vol. 51, no. 1-2, pp. 207–216, 2008.

[2] Y. C. Fung, *An Introduction to the Theory of Aeroelasticity*, Dover Publications, 2002.

[3] V. P. Zhuravlev and D. M. Klimov, "Theory of the shimmy phenomenon," *Mechanics of Solids*, vol. 45, no. 3, pp. 324–330, 2010.

[4] L. F. R. Turci, E. E. N. Macau, and T. Yoneyama, "Efficient chaotic based satellite power supply subsystem," *Chaos, Solitons & Fractals*, vol. 42, no. 1, pp. 396–407, 2009.

[5] D. T. Schmitt and P. C. Ivanov, "Fractal scale-invariant and nonlinear properties of cardiac dynamics remain stable with advanced age: a new mechanistic picture of cardiac control in healthy elderly," *American Journal of Physiology-Regulatory, Integrative and Comparative Physiology*, vol. 293, no. 5, pp. R1923–R1937, 2007.

[6] G. F. Fussmann, S. P. Ellner, K. W. Shertzer, and J. Hairston N.G., "Crossing the hopf bifurcation in a live predator-prey system," *Science*, vol. 290, no. 5495, pp. 1358–1360, 2000.

[7] A. C.-L. Chian, E. L. Rempel, and C. Rogers, "Complex economic dynamics: chaotic saddle, crisis and intermittency," *Chaos, Solitons & Fractals*, vol. 29, no. 5, pp. 1194–1218, 2006.

[8] R. L. Devaney, *A First Course In Chaotic Dynamical Systems*, Avalon Publishing, 1992.

[9] M. A. F. Sanjuán and C. Grebogi, *Recent Progress in Controlling Chaos*, World Scientific Publishing Company, 2010.

[10] P. Cvitanović, "Invariant measurement of strange sets in terms of cycles," *Physical Review Letters*, vol. 61, no. 24, pp. 2729–2732, 1988.

[11] V. Franceschini, C. Giberti, and Z. M. Zheng, "Characterization of the Lorenz attractor by unstable periodic orbits," *Nonlinearity*, vol. 6, no. 2, pp. 251–258, 1993.

[12] K. Pyragas, "Continuous control of chaos by self-controlling feedback," *Physics Letters A*, vol. 170, no. 6, pp. 421–428, 1992.

[13] K. Pyragas, "Delayed feedback control of chaos," *Philosophical Transactions of the Royal Society A: Mathematical, Physical & Engineering Sciences*, vol. 364, no. 1846, pp. 2309–2334, 2006.

[14] T. Ushio, "Limitation of delayed feedback control in nonlinear discrete-time systems," *IEEE Transactions on Circuits and Systems I: Fundamental Theory and Applications*, vol. 43, no. 9, pp. 815-816, 1996.

[15] S. Yamamoto, T. Hino, and T. Ushio, "Dynamic delayed feedback controllers for chaotic discrete-time systems," *IEEE Transactions on Circuits and Systems I: Fundamental Theory and Applications*, vol. 48, no. 6, pp. 785–789, 2001.

[16] S. Yamamoto, T. Hino, and T. Ushio, "Delayed feedback control with a minimal-order observer for stabilization of chaotic discrete-time systems," *International Journal of Bifurcation and Chaos*, vol. 12, no. 5, pp. 1047–1055, 2002.

[17] J. Zhu and Y.-P. Tian, "Necessary and sufficient conditions for stabilizability of discrete-time systems via delayed feedback control," *Physics Letters A*, vol. 343, no. 1-3, pp. 95–107, 2005.

[18] B. Fiedler, V. Flunkert, M. Georgi, P. Hövel, and E. Schöll, "Refuting the odd-number limitation of time-delayed feedback control," *Physical Review Letters*, vol. 98, no. 11, Article ID 114101, 2007.

[19] W. Just, B. Fiedler, M. Georgi, V. Flunkert, P. Hövel, and E. Schöll, "Beyond the odd number limitation: A bifurcation analysis of time-delayed feedback control," *Physical Review E: Statistical, Nonlinear, and Soft Matter Physics*, vol. 76, no. 2, Article ID 026210, 2007.

[20] B. Fiedler, V. Flunkert, P. Hövel, and E. Schöll, "Beyond the odd number limitation of time-delayed feedback control of periodic orbits," *The European Physical Journal Special Topics*, vol. 191, no. 1, pp. 53–70, 2011.

[21] T. Ushio and S. Yamamoto, "Prediction-based control of chaos," *Physics Letters A*, vol. 264, no. 1, pp. 30–35, 1999.

[22] T. P. Chagas, P.-A. Bliman, and K. H. Kienitz, "New feedback laws for stabilization of unstable periodic orbits," *IFAC Proceedings Volumes*, vol. 43, no. 14, pp. 1005–1010, 2010.

[23] M. P. F. de Córdoba and E. Liz, "Prediction-based control of chaos and a dynamic Parrondo's paradox," *Physics Letters A*, vol. 377, no. 10-11, pp. 778–782, 2013.

[24] E. Braverman, C. Kelly, and A. Rodkina, "Stabilisation of difference equations with noisy prediction-based control," *Physica D: Nonlinear Phenomena*, vol. 326, pp. 21–31, 2016.

[25] A. Boukabou, A. Chebbah, and N. Mansouri, "Predictive control of continuous chaotic systems," *International Journal of Bifurcation and Chaos*, vol. 18, no. 2, pp. 587–592, 2008.

[26] E. Ott, C. Grebogi, and J. A. Yorke, "Controlling chaos," *Physical Review Letters*, vol. 64, no. 11, pp. 1196–1199, 1990.

[27] M. A. Khelifa and A. Boukabou, "Design of an intelligent prediction-based neural network controller for multi-scroll chaotic systems," *Applied Intelligence*, vol. 45, no. 3, pp. 793–807, 2016.

[28] T. P. Chagas, P.-A. Bliman, and K. H. Kienitz, "A new method for stabilizing unstable periodic orbits of continuous-time systems. Application to control of chaos," in *Proceedings of the 51st IEEE Conference on Decision and Control (CDC '12)*, pp. 2146–2151, Maui, Hawaii, USA, December 2012.

[29] T. P. Chagas, B. A. Toledo, E. L. Rempel, A. C. Chian, and J. A. Valdivia, "Optimal feedback control of the forced van der Pol system," *Chaos, Solitons & Fractals*, vol. 45, no. 9-10, pp. 1147–1156, 2012.

[30] B. A. Finlayson, *The Method of Weighted Residuals and Variational Principles, with Application in Fluid Mechanics, Heat And Mass Transfer*, vol. 87 of *Mathematics in Science and Engineering*, Academic Press, 1972.

[31] J. Viladsen and M. L. Michelsen, *Solution of Differential Equation Models by Polynomial Approximation*, Prentice Hall, 1978.

[32] E. Hairer, S. P. Norsett, and G. Wanner, *Solving Ordinary Differential Equation I: Nonstiff Problems*, Springer-Verlag, 2008.

[33] T. S. Parker and L. O. Chua, *Practical Numerical Algorithms for Chaotic Systems*, Springer, 1989.

[34] F. Asenjo, B. A. Toledo, V. Muoz, J. Rogan, and J. A. Valdivia, "Optimal control in a noisy system," *Chaos: An Interdisciplinary Journal of Nonlinear Science*, vol. 18, no. 3, Article ID 033106, 2008.

[35] S. P. Han, "A globally convergent method for nonlinear programming," *Journal of Optimization Theory and Applications*, vol. 22, no. 3, pp. 297–309, 1977.

[36] O. L. Mangasarian, R. R. Meyer, and S. M. Robinson, *Nonlinear Programming 3: Proceedings of The Special Interest Group on Mathematical Programming Symposium*, vol. 3, Academic Press, 1978.

[37] P. E. Gill, W. Murray, and M. H. Wright, *Practical Optimization*, Academic Press, 1981.

[38] M. J. Powell, *A Fast Algorithm for Nonlinearly Constrained Optimization Calculations*, vol. 630 of *Lecture Notes in Mathematics*, Springer, 1978.

[39] E. F. Camacho and C. B. Alba, *Model Predictive Control*, Advanced Textbooks in Control and Signal Processing, Springer-Verlag, 2nd edition, 2007.

14

Detection of Two-Level Inverter Open-Circuit Fault using a Combined DWT-NN Approach

Bilal Djamal Eddine Cherif and Azeddine Bendiabdellah

Diagnostic Group, Laboratory LDEE, Electrical Engineering Faculty, University of Sciences and Technology of Oran, Bir El Djir, Algeria

Correspondence should be addressed to Bilal Djamal Eddine Cherif; cherif.doc84@gmail.com

Academic Editor: Qiang Liu

Three-phase static converters with voltage structure are widely used in many industrial systems. In order to prevent the propagation of the fault to other components of the system and ensure continuity of service in the event of a failure of the converter, efficient and rapid methods of detection and localization must be implemented. This paper work addresses a diagnostic technique based on the discrete wavelet transform (DWT) algorithm and the approach of neural network (NN), for the detection of an inverter IGBT open-circuit switch fault. To illustrate the merits of the technique and validate the results, experimental tests are conducted using a built voltage inverter fed induction motor. The inverter is controlled by the SVM control strategy.

1. Introduction

Several researchers have carried out their investigation in relation to the field of detection and location of faults in static converters and more particularly those related to three-phase power inverters [1]. The treated fault is mainly concerned with the open-circuit fault of an inverter IGBT switch [2]. Most published papers are based on Park's current vectors approach [3]. This approach is based on the trajectory tracking of the phase current vector. In fact, for the case of a healthy state condition of the inverter, the trajectory of these current vectors in the (d-q) frame is a circle. It was found that the circle becomes a semicircle under an open-circuit IGBT switch fault in one of the legs of the inverter. The position of this semicircle in the (d-q) frame makes it possible to identify the faulty IGBT switch [4]. Another paper used the mean value of the phase currents in Park's frame for the extraction of the open-circuit fault angle of each IGBT switch [1–5]; unfortunately this method presents an inconvenience as it depends on the load. To overcome the problem, some authors suggested the normalized DC current method which is fundamentally based on the DC component of the current and the first-order harmonic coefficients of the alternating current (AC) [6]. Some detection techniques mentioned above are briefly discussed in [7, 8].

Others researchers used stator current as key parameter for fault diagnosis purpose because it does not require costly sensors. This technique is widely termed as Motor Current Signature Analysis (MCSA) technique [9–11]. For steady state constant load conditions, the Fast Fourier Transform (FFT) algorithm has been used for different induction motor fault diagnosis purpose. The (FFT) algorithm is able to diagnose bearing fault for full load conditions but not for the no-load or light load condition [7]. Therefore, some researchers proposed short-term Fourier transform (STFT) and wavelet transform (WT) methods for different fault diagnosis purpose of the induction motor by taking into consideration the quantity of time (that is for nonstationary applications). However, the (STFT) method faces a big disadvantage; it gives poor frequency resolution because it shows constant window size for all frequencies.

The wavelet transform (WT) technique has perhaps been the most persistent development in recent decades. It has drawn the attention of several researchers in various fields, such as signal processing, image processing, communication, computer science, and mathematics. Numerous works describing the advancement in wavelet theory and its applications in various fields have been published. Among others, it is one of the most attractive techniques in the field of

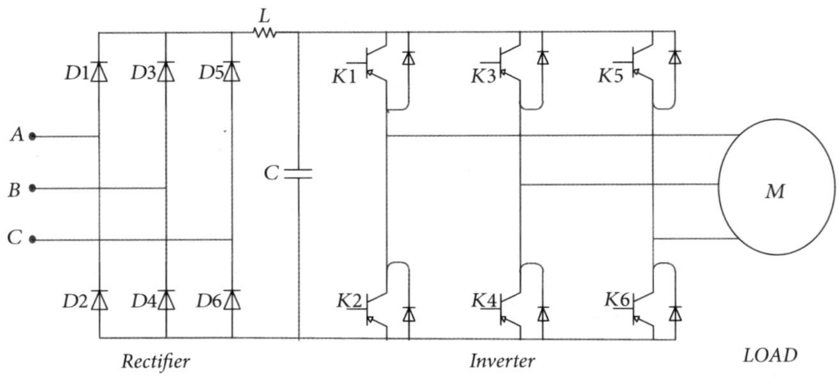

FIGURE 1: Structure of a three-phase two-level inverter.

rotary machines and static converter for fault diagnosis and it particularly matches well for the nonstationary signals [12, 13]. The main disadvantage of this technique is that the length of the range (scale) is fixed. Therefore, it is necessary to be able to keep the time-frequency representation and carry out an analysis based on a concept somewhat different from the concept of frequency: the scale concept [14]. In 1982, J. Morlet opens the way to the solution by constructing the wavelet analysis, based on the concept of scale. By manipulating the scale factor, one can zoom in and out a portion of the signal [14, 15]. Unlike the short-term Fourier transform, the wavelet transform uses the notion of time-scale involving dynamic length analysis windows. Because the continuous wavelet transform is very sensitive to noise, the discrete wavelet transform is preferred and used in our paper.

The application of artificial intelligence (AI) based techniques can be advantageous in fault diagnosis since these diagnoses have several advantages. For instance, since AI-based techniques do not require mathematical models, the development time can be significantly reduced. A literature review of recent developments in the field of AI-based diagnostic systems in power inverters has been presented [16]. In addition, some authors have studied the application of a neural network (NN) to establish a fault diagnosis system and judge the faults of power transistors [17]. These NN-based fault diagnosis methods allow an accurate solution to a particular fault problem without accurate knowledge of the faulty system. However, their main drawback is due to the fact that the exact architecture of the NN to be used is generally not known in advance.

The work proposed in this paper addresses an open-circuit fault detection of an IGBT switch of an inverter controlled by a DSPACE 1104 card based on the SVM control strategy feeding an induction motor. The analysis tools and fault diagnosis are based on the use of a combined DWT-NN approach. The DWT algorithm focuses on the investigation of the details of the stator current signal. The variation in these details for both healthy and faulty inverter cases enables us to extract useful information related to the open-circuit inverter switch faults. The NN approach is introduced in order to automate the fault diagnostic system by including a learning phase that helps in developing a very rich database storing relevant information about the open-circuit faults. To assess the effectiveness and merits of the proposed approach and validate the obtained results, experimental tests are conducted by the group diagnostic at the LDEE laboratory.

2. Voltage Source Inverter Structure

Figure 1 shows the structure of a three-phase two-level voltage source inverter feeding an induction motor.

This inverter is controlled by the SVM control strategy. For each leg of the inverter, there are two possible states:

(i) *State 1*: the higher switch K_X ($X = 1$, 2 or 3) is closed, while the lower switch K_X ($X = 1$, 2 or 3) is open. The output voltage relative to the neutral of the DC source is V_{dc}.

(ii) *State 0*: the lower switch K_X ($X = 1$, 2 or 3) is closed, while the higher switch K_X ($X = 1$, 2 or 3) is open. The output voltage relative to the neutral of the DC source is 0 v.

Unfortunately during its operation, various failures can affect the inverter especially in terms of its so-called power components (IGBT in our case study) because of their fragility. Two types of faults can be reported [1]:

(i) Short-circuit faults affecting the IGBT switches are the most serious faults. In the presence of such a fault, the current reaches limits which can cause the fusion of its chip or its connection. If the detection of this type of fault does not occur rapidly (less than 10 microseconds), then the IGBT switch which is still active on the same leg undergoes the same phenomenon and so the whole inverter leg is shorted.

(ii) Open-circuit faults affecting the IGBT switches may occur when, for any reason, the IGBT is disconnected, is damaged, or had a problem in its grid control signal. This type of fault is very difficult to perceive directly because the motor can continue to operate but with a degradation of its performance due to the occurrence of fluctuations in the mechanical parameters (speed and torque) as well as an imbalance of the currents where the currents of the other two healthy legs take high values to maintain the average torque and the

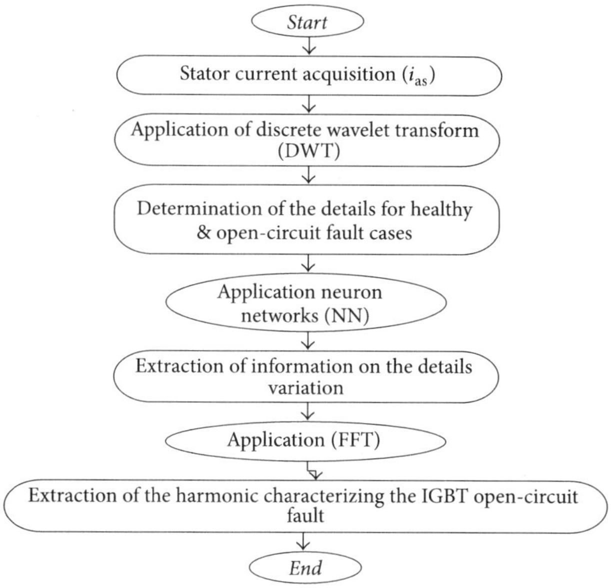

FIGURE 2: Flowchart of the proposed DWT-NN approach.

speed. The starting of the motor in the presence of this type of fault cannot always be ensured; everything will depend on the value of the torque which can be close to zero for certain rotor position.

3. Theoretical Bases of the Combined DWT-NN Approach

The flowchart presented in Figure 2 is to illustrate the main steps required when using the combined DWT-NN approach.

The following two subsections tackle in more detail both the DWT algorithm and the NN approach.

3.1. DWT Algorithm Principle. The wavelet transform is a mathematical tool which allows the decomposition of a temporal signal into a series of coefficients called approximation and detail; the approximations represent the slow variations of the signal, whereas the details represent the fastest [6, 7].

A wavelet $\psi(t)$ is an oscillating function of zero mean, with a certain degree of regularity and whose support is finite (which explains the word "wavelet," which means small wave). Mathematically, a wavelet $\psi(t)$ is a zero mean function that is expanded by a scale parameter and is translated by the time parameter b "translation in time parameter"; it is defined as follows:

$$\psi_{a,b}(t) = \frac{1}{\sqrt{a}} \Psi\left(\frac{t-b}{a}\right). \quad (1)$$

The expansions and translations of this "mother" wavelet are used to allow the projection of the signal over different time scales. These scales depend mainly on the sampling frequency F_e of the signal to be processed.

Under these conditions, a signal $x(t)$ can be represented as wavelets given by the following equation:

$$x(t) = \sum_{k=-\infty}^{\infty} c_k \varphi(t-k) + \sum_{k=-\infty}^{\infty} \sum_{j=0}^{\infty} d_{j,k} 2^{j/2} \psi\left(2^j t - k\right). \quad (2)$$

$\psi(t)$ is the derivative of the function $\varphi(t)$ with

$$c_k = \int_{-\infty}^{\infty} x(t)\, \varphi(t-k)\, dt,$$
$$d_{j,k} = \int_{-\infty}^{\infty} x(t)\, 2^{j/2} \psi\left(2^j t - k\right) dt, \quad (3)$$

where c_k are called the approximation coefficients and $d_{j,k}$ the detail coefficients.

In the case of a discrete wavelet transform (DWT) which is the tool of our paper work, the expansion and translation parameters "a" and "b" of $\psi(t)$ are limited only to discrete values leading to the following expression:

$$\psi_{m,n}(t) = \frac{1}{\sqrt{a_0^m}} \psi\left(\frac{t - n b_0 a_0^m}{a_0^m}\right), \quad (4)$$

where the parameters "m" and "n" are integers allowing the control of the dilation and the translation of the mother wavelets

Obviously, according to the last equation, different wavelets (Haar, Daubechies, Coiflet, Meyer, Morlet, etc.) generate different wavelet classes and consequently the behavior of the decomposed signal can be very different. In fact, each wavelet has particular characteristics whose choice depends on the desired application. In this paper, the Coiflet wavelets are to be applied for the detection of the IGBT open-circuit. Figure 3 depicts a typical Coiflet wavelet family graphical representation.

3.1.1. Number of Decomposition Levels Required. The DWT using the Coiflet algorithm is based on signal the decomposition using low-pass filters (LPF) and high-pass filters (HPF) followed by subsampling. Then, in each level of decomposition, the coefficients of approximation and details are computed. Figure 4 shows the implementation of the DWT for three decomposition levels as an example.

Where the LPF frequency band is defined as

$$\left[0 \ \frac{Fe}{2^n}\right], \quad n = 1, 2, \ldots, Nlevels. \quad (5)$$

And the HPF frequency band is defined as

$$\left[\frac{Fe}{2^n}, \frac{Fe}{2^{(n-1)}}\right], \quad n = 1, 2, \ldots, Nlevels, \quad (6)$$

where n indicates the number of decomposition levels and the Nlevels are the maximum number of decomposition levels.

Prior knowledge of N_{levels} of the signal to be processed is essential for a reliable and fast analysis by the Coiflet mother wavelet. The following equation gives this required parameter [11]:

$$N_{\text{levels}} = \text{int}\left(\frac{\log(F_e/f_s)}{\log(2)}\right) + 2, \quad (7)$$

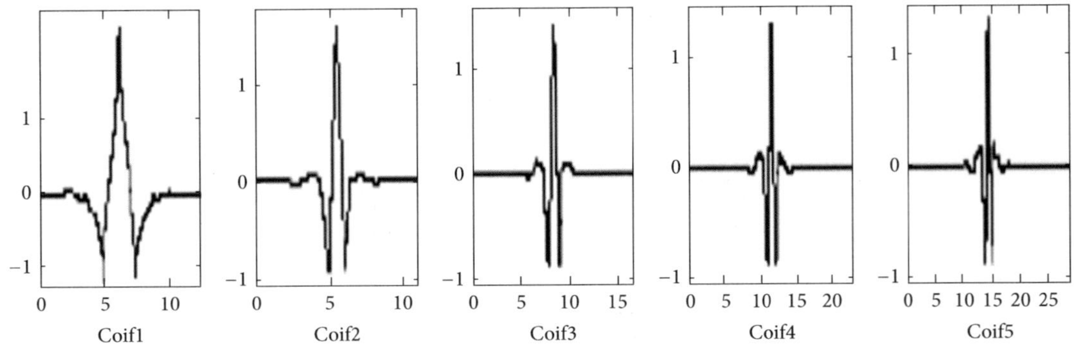

FIGURE 3: Coiflet wavelet families [16].

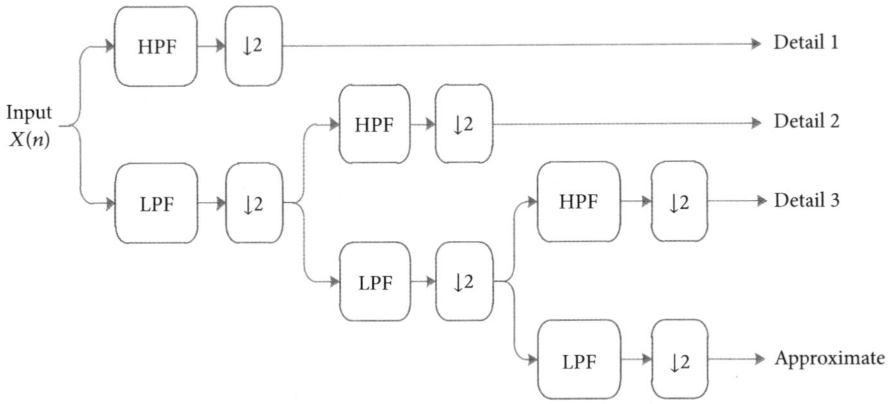

FIGURE 4: DWT implementation procedure.

where f_s is the supply frequency and F_e is the sampling frequency. Note that N_{levels} must obviously be an integer.

Knowing f_s and F_e, one can calculate the number of appropriate decompositions. For our case paper study, considering a sampling frequency of 1500 Hz and a supply frequency of 50 Hz, the number of decomposition levels required is

$$N_{\text{levels}} = \text{int}\left(\frac{\log(1500/50)}{\log(2)}\right) + 2 = \text{int}(4.9068) + 2 = 6 \text{ levels}. \quad (8)$$

Using (5) and (6) and the number of levels computed from (8), the decomposition of the motor stator current signal as a function of the sampling frequency for the six levels of decomposition can thus be obtained as illustrated in Figure 5.

3.2. Description of the Neural Network Approach. The proposed NN is a multilayer network of (6-6-1) whose adopted architecture is illustrated in Figures 6 and 7. Note that the smallest error is obtained after 38 iterations.

Each neuron is connected to all the neurons of the next layer by connections whose weights are any real numbers.

The neuron network study used in this paper is carried out through the three main steps:

(i) The construction of the network NN block using the Levenberg algorithm

(ii) The data acquisition (learning base)

(iii) The network test

3.3. Construction of the NN Block System. Figure 6 shows that our network consists of three layers:

(i) An input layer composed of six neurons, whose role is to transmit the values of the inputs that correspond to the maxima of details (max($d1$), max($d2$), max($d3$), max($d4$), max($d5$), and max($d6$)) to the next layer, called hidden layer

(ii) A hidden layer with six neurons with selected sigmoid activation functions

(iii) An output layer, which is composed of a neuron, where output of each neuron is 0 or 1

3.4. Acquisition of the Data (Learning Basis). Before building the NN block system, one must first access the learning phase. This can be in the form of a table. The latter consists of vectors (which represent the input layer of the NN), where each vector consists of 2 parameters.

A very rich database for healthy and faulty (open-circuit) cases can be developed, which has a lot of information about the open-circuit fault. During this phase, the maxima details of the healthy case are taken as the references; then the maxima of details for the faulty case are extracted and

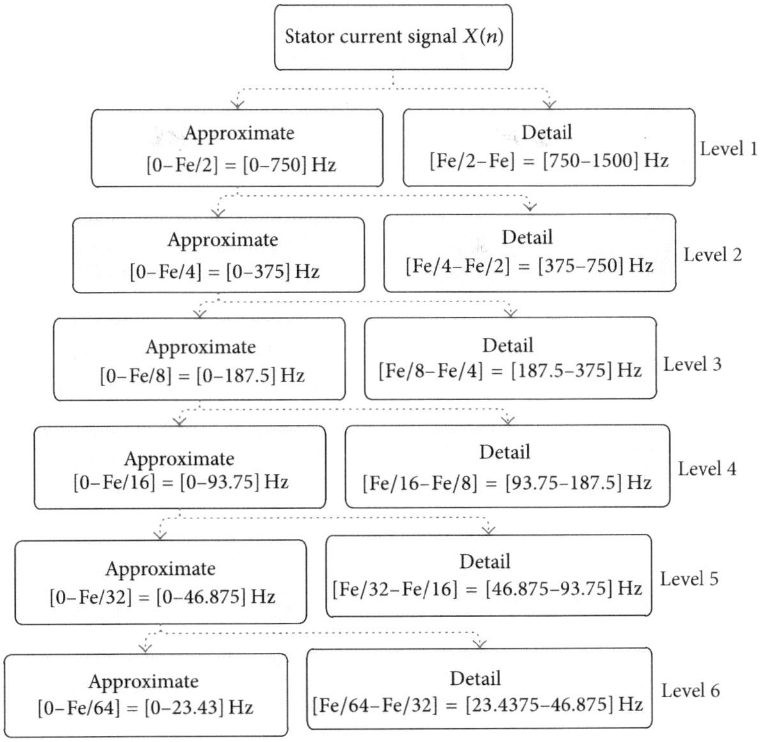

FIGURE 5: Decomposition of stator current signal $x(t)$.

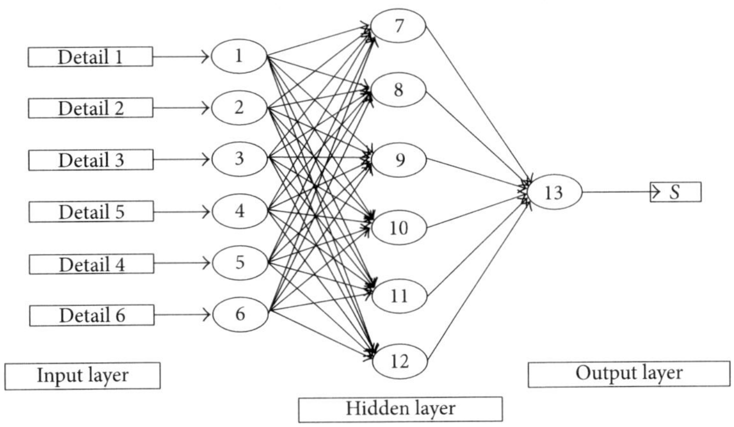

FIGURE 6: Proposed neural network architecture.

TABLE 1: Fault classification.

State	Fault type	Symbol	Code
1	Healthy state	HS	0
2	Open-circuit at K_1	OC	1

compared with the healthy case. From this comparison, it can be deduced as either state 0 (i.e., no variation in detail) or state 1 (i.e., variation in detail). Table 1 is to resume the task of this phase.

3.5. Network Test Results. An automatic learning was performed using the MATLAB software. The learning is reached once a small quadratic error of value $2.2181e^{-16}$ is obtained; see Figure 7. Note that the smallest error is obtained after 38 iterations.

4. Experimental Test-Rig

4.1. Experimental Test-Rig Description. The experimental test-rig used in this paper work includes a three-phase induction squirrel-cage motor fed by a three-phase two-level voltage source inverter. The detailed characteristics of the motor are given in the Appendix. Furthermore, the motor is mechanically coupled to a DC generator supplying resistors which allows varying the load torque. Moreover, the

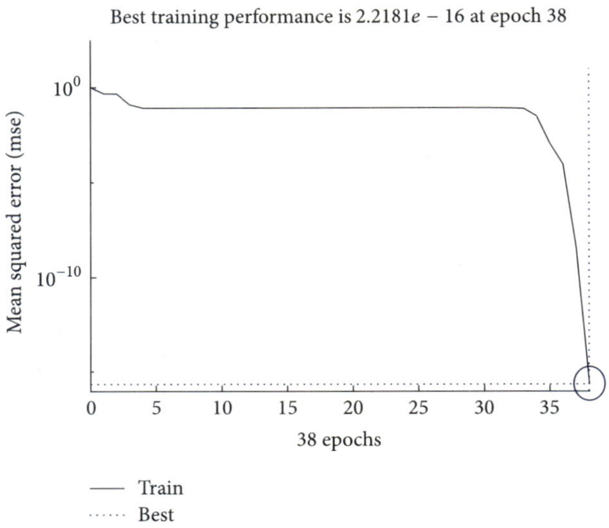

FIGURE 7: Evaluation of the quadratic error as a function of the number of learning iterations (using the method of retrogradient propagation).

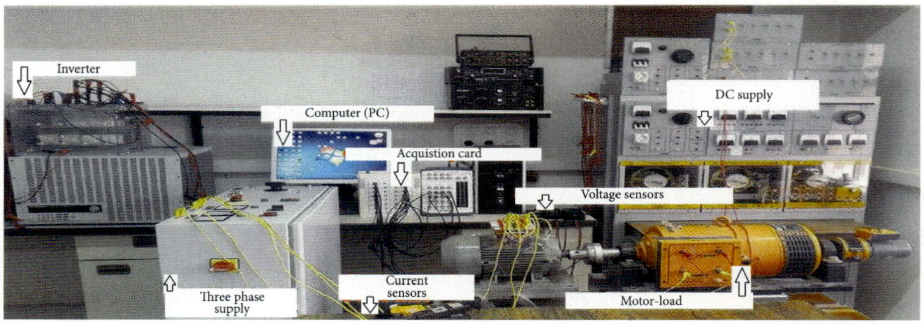

FIGURE 8: Photo of the experimental test-rig.

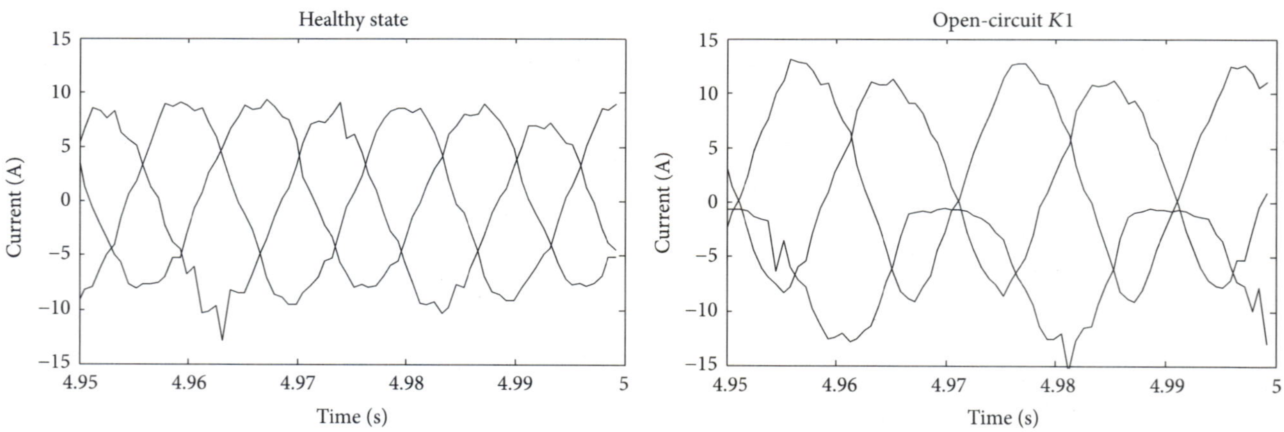

FIGURE 9: Currents waveforms of an induction motor for healthy and faulty open-circuit K_1 case.

measuring system includes three current Hall effect sensors and three voltage sensors and a DSPACE 1104 acquisition card to generate pulses for triggering the IGBTs gates. The whole set is connected to a computer for visualizing the processed sensed signal as shown in the photo of Figure 8. The acquisition time is taken as $T\text{acq} = 5$ s and the sampling frequency $Fe = 1500$ Hz.

4.2. Experimental Results Presentation and Discussion. Figure 9 depicts the phase current waveforms of the induction motor for both a healthy state and an IGBT open-circuit faulty inverter.

From the experimental results depicting the currents waveforms of the motor in Figure 9, following an IGBT open-circuit fault K_1 of the inverter leg, the phase current

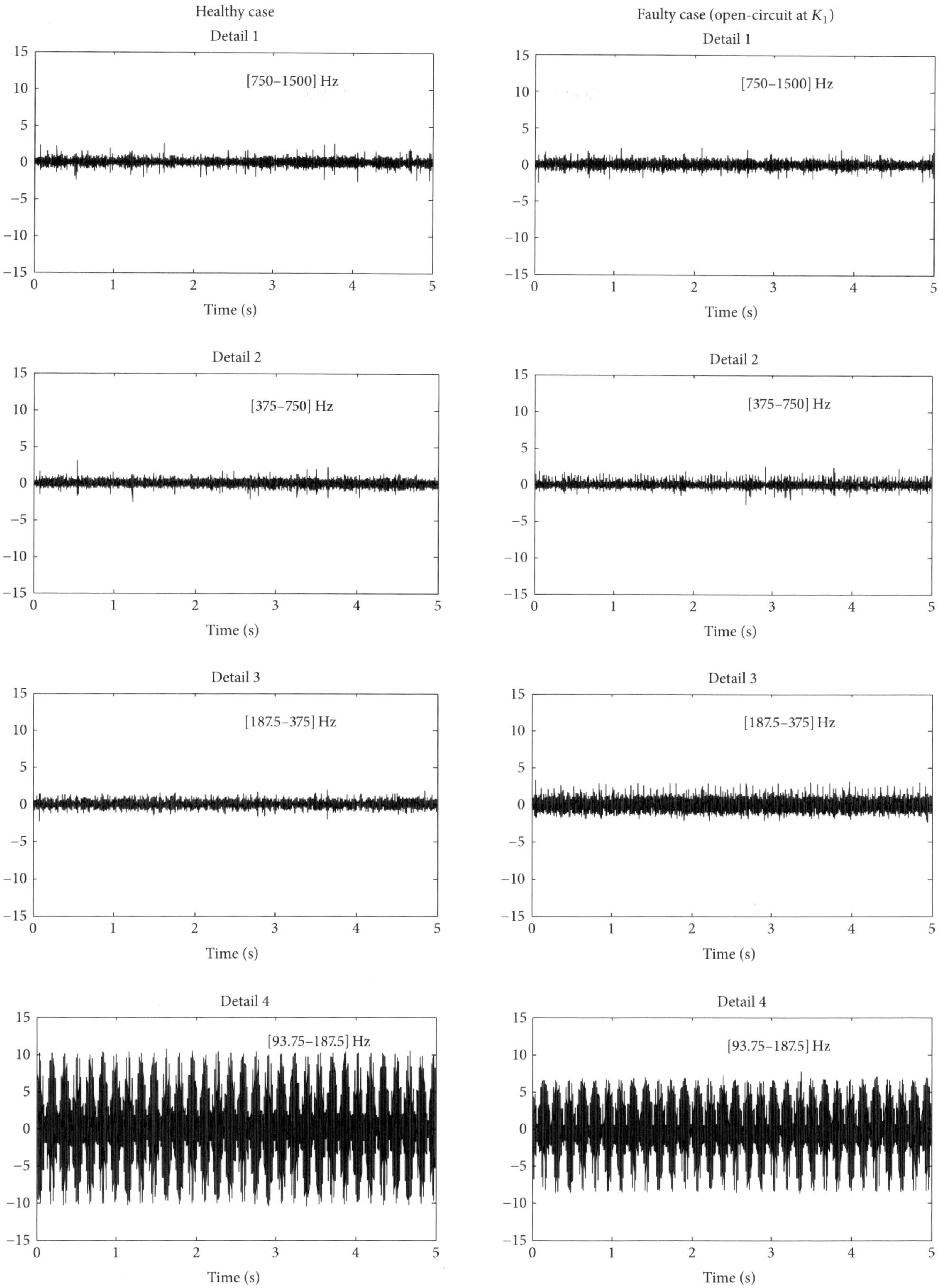

FIGURE 10: Continued.

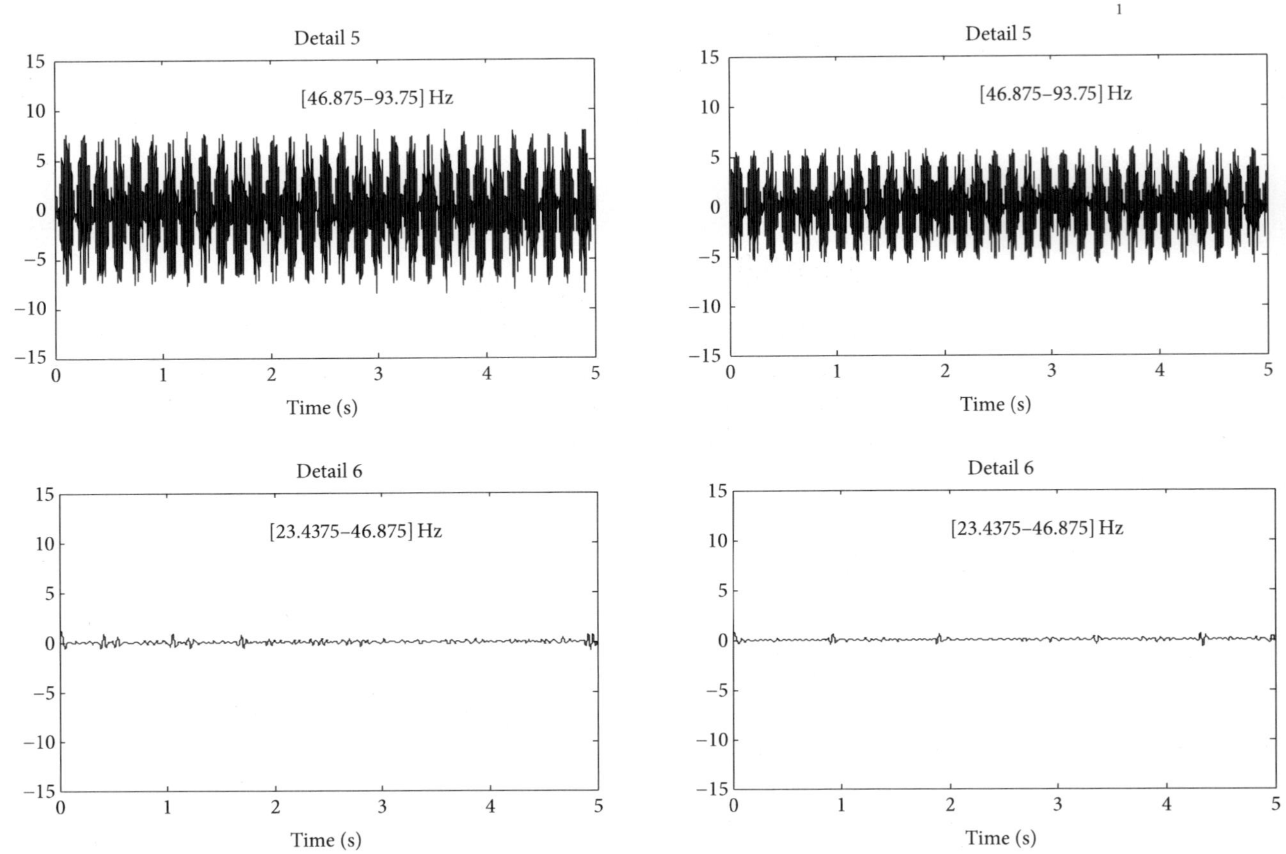

FIGURE 10: Details signal decomposition for healthy and faulty open-circuit at K_1 cases.

connected to this leg can no longer be controlled as it can only be negative or zero. The sum of the currents of the other two healthy phases is zero which may make it impossible to start the motor or to ensure for long term the continuity of service of the motor.

Figure 10 shows the various details extracted from the acquired stator current signal using the DWT technique for the healthy case and the faulty (open-circuit switch at K_1) case.

Figure 11 depicts the fault diagnostic system and Table 2 shows some examples of the diagnostic results.

Table 2 presents the input and output of the NN.

By comparing the details for the case of the healthy inverter state and that of the open-circuit switch fault state as depicted in Figure 11 or Table 2, a noticeable change in the amplitude of both details $d3$, $d4$ is easily observed, while the rest of the details remain almost the same independently of the open-circuit switch fault presence or not. The change in these two details $d3$ and $d4$ indicates the existence of certain information within our stator current signal. To extract and explain this information, a stator current spectral analysis based on the FFT is performed for both details.

Figure 12 presents the spectra of the frequency ranges $d3$: [187.5–375] and $d4$: [93.75–187.5] for the healthy and the faulty cases, respectively.

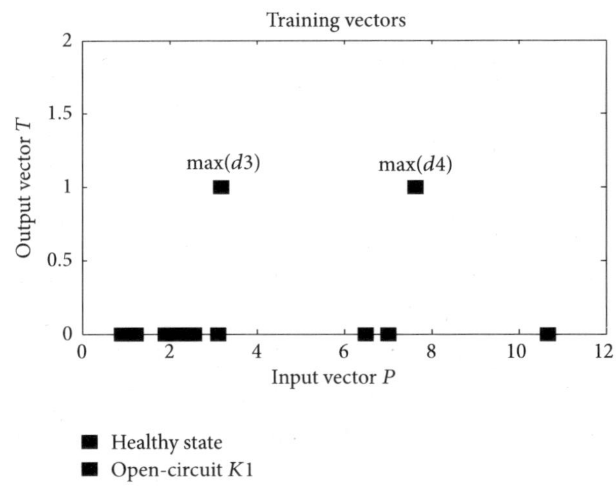

FIGURE 11: Variation of the details maxima.

Comparing both spectra in Figure 12 for healthy state and open-circuit fault state, we can easily notice the presence of additional harmonics characterizing the inverter open-circuit fault at frequencies 100 Hz in the first band [93.75–187.5] and 200 Hz and 300 Hz in the second band [187.5–375]. The amplitude and frequency of the various

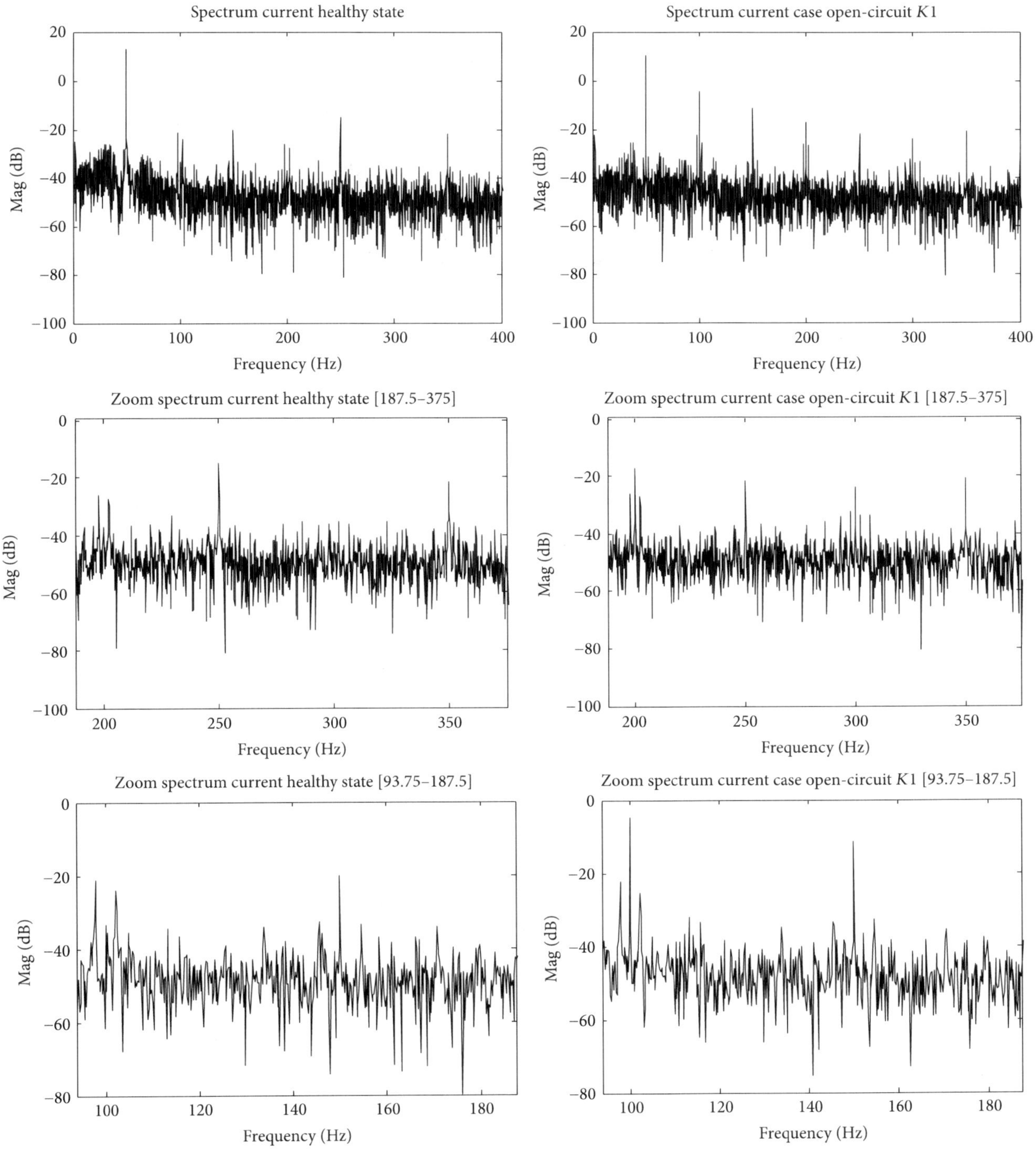

FIGURE 12: FFT of stator current signal for healthy and faulty open-circuit at K_1 cases.

TABLE 2: Details of one stator current phase for both healthy and faulty open-circuit at K_1 cases.

Max details	$\max(d_1)$	$\max(d_2)$	$\max(d_3)$	$\max(d_4)$	$\max(d_5)$	$\max(d_6)$
Input						
Healthy state	2.568	3.126	1.928	10.67	7.014	1.228
Open-circuit state K_1	2.261	2.412	3.206	7.632	6.502	0.909
Output	−0.0000	0.0000	1.0000	1.0000	0.0000	0.0000

TABLE 3: Amplitudes and frequencies of harmonics.

Harmonic (Hz)	$f_{oc} = 2f_s =$ 100 Hz	$4f_s = 200$ Hz	$6f_s = 300$ Hz
Healthy state (db)	−33.51 db	−37.25 db	0 db
Open-circuit state (db)	−4.623 db	−17.26 db	−23.77 db

obtained harmonics for the healthy and faulty cases are summarized in Table 3.

A comparative analysis between both healthy and faulty states shows with more clarity a particular frequency signature around 100 Hz for the spectrum level $d3$. Note that the open-circuit frequency $f_{oc} = 2f_s = 100$ Hz corresponds to the harmonic frequency that characterizes the open-circuit fault of the IGBT switch.

5. Conclusion

In this paper, a research area dealing with the technique of diagnosis and detection of open-circuit fault in a three-phase two-level voltage source inverter fed induction motor is investigated. The paper proposes a diagnosis approach based on the association of both the discrete wavelet transform (DWT) and the neural network (NN) for the detection of the IGBT open-circuit fault of an inverter.

The study focuses first on the extraction of the details for the cases of the healthy and the open-circuit faulty IGBT by using the DWT algorithm. The investigation of the harmonics related to the obtained details particularly $d3$ and $d4$ is then conducted by using the FFT technique. The NN enables the development of a rich database for both healthy and faulty cases resulting in the automation of the diagnostic system.

The various obtained results are validated by several experimental tests conducted in the LDEE laboratory by the group diagnostic to assess the effectiveness and merits of the combined DWT-NN proposed approach.

Appendix

Rated power: 3 KW

Supply frequency: 50 Hz

Rated voltage: 380 V

Rated current: 7 A

Rotor speed: 1440 rev/min

Number of rotor bars: 28

Number of stator slots: 36

Power factor: 0.83

Number of pair of poles: 2

Conflicts of Interest

The authors declare that there are no conflicts of interest regarding the publication of this paper.

References

[1] W. Zhang, D. Xu, P. N. Enjeti, H. Li, J. T. Hawke, and H. S. Krishnamoorthy, "Survey on fault-tolerant techniques for power electronic converters," *IEEE Transactions on Power Electronics*, vol. 29, no. 12, pp. 6319–6331, 2014.

[2] T. Orlowska-Kowalska and P. Sobanski, "Simple diagnostic technique of a single IGBT open-circuit faults for a SVM-VSI vector controlled induction motor drive," *Bulletin of the Polish Academy of Sciences—Technical Sciences*, vol. 63, no. 1, pp. 281–288, 2015.

[3] B. D. E. Cherif, A. Bendiabdellah, and M. A. Khelif, "Detection of open-circuit fault in a three-phase voltage inverter fed induction motor," *International Review of Automatic Control*, vol. 9, no. 6, pp. 374–382, 2016.

[4] B. D. E. Cherif, M. Bendjebbar, N. Benouzza, H. Boudinar, and A. Bendiabdellah, "A comparative study between two open-circuit fault detection and localization techniques in a threephase inverter fed induction motor," in *Proceedings of the 2016 8th International Conference on Modelling, Identification and Control (ICMIC)*, pp. 1–7, Algiers, November 2016.

[5] H. Keskes, A. Braham, and Z. Lachiri, "Broken rotor bar diagnosis in induction machines through stationary wavelet packet transform and multiclass wavelet SVM," *Electric Power Systems Research*, vol. 97, pp. 151–157, 2013.

[6] R. Yan, R. X. Gao, and X. Chen, "Wavelets for fault diagnosis of rotary machines: a review with applications," *Signal Processing*, vol. 96, pp. 1–15, 2013.

[7] A. F. Aimer, A. H. Boudinar, N. Benouzza, and A. Bendiabdellah, "Simulation and experimental study of induction motor broken rotor bars fault diagnosis using stator current spectrogram," in *Proceedings of the 3rd International Conference on Control, Engineering and Information Technology, CEIT 2015*, dza, May 2015.

[8] Z. Liu, Y. Cui, and W. Li, "Combined power quality disturbances recognition using wavelet packet entropies and S-transform," *Entropy*, vol. 17, no. 8, pp. 5811–5828, 2015.

[9] Q. Yang and J. Wang, "Multi-level wavelet shannon entropy-based method for single-sensor fault location," *Entropy*, vol. 17, no. 10, pp. 7101–7117, 2015.

[10] S. Kaur, K. Gaganpreet, and S. Dheerendra, "Comparative analysis of haar and coiflet wavelets using discrete wavelet transform in digital image compression," *International Journal of Computer Applications (IJERA)*, vol. 3, no. 3, pp. 669–673, 2013.

[11] J. A. Antonino-Daviu, M. Riera-Guasp, J. R. Folch, and M. P. Molina Palomares, "Validation of a new method for the diagnosis of rotor bar failures via wavelet transform in industrial induction machines," *IEEE Transactions on Industry Applications*, vol. 42, no. 4, pp. 990–996, 2006.

[12] R. R. Errabelli and P. Mutschler, "Fault-tolerant voltage source inverter for permanent magnet drives," *IEEE Transactions on Power Electronics*, vol. 27, no. 2, pp. 500–508, 2012.

[13] T. Orlowska-Kowalska and P. Sobanski, "Simple sensorless diagnosis method for open-switch faults in SVM-VSI-fed induction motor drive," in *Proceedings of the 39th Annual Conference of the IEEE Industrial Electronics Society, IECON 2013*, pp. 8210–8215, Austria, November 2013.

[14] W.-S. Im, J.-S. Kim, J.-M. Kim, D.-C. Lee, and K.-B. Lee, "Diagnosis methods for IGBT open switch fault applied to 3-phase AC/DC PWM converter," *Journal of Power Electronics*, vol. 12, no. 1, pp. 120–127, 2012.

[15] I. Jlassi, S. Khojet, and E. Khil, "A MRAS-Luenberger observer based fault tolerant control of PMSM drive," *Journal of Electrical Systems*, vol. 10, no. 1, pp. 48–62, 2014.

[16] B. Fan, Y. Yin, and C. Fu, "A method of inverter circuit fault diagnosis based on BP neural network and D-S evidence theory," in *Proceedings of the 2010 8th World Congress on Intelligent Control and Automation, WCICA 2010*, pp. 2249–2253, China, July 2010.

[17] L. hong and W. yan-qiu, "Study of Fault Diagnosis on Three-phase Sine-PWM Inverter Based on Rough Set-neural Network System," *Journal of Liaoning Istitute of Technology*, vol. 25, no. 6, pp. 351–353, 2005.

15

Position Tracking Control for Permanent Magnet Linear Motor via Continuous-Time Fast Terminal Sliding Mode Control

Wei Gao,[1] **Xiuping Chen**,[2] **Haibo Du**,[2] **and Song Bai**[3]

[1]*School of Electrical Engineering, Wuhu Vocational Institute of Technology, Wuhu, China*
[2]*School of Electrical Engineering and Automation, Hefei University of Technology, Hefei, China*
[3]*NARI-TECH Control Systems Ltd., Nanjin, China*

Correspondence should be addressed to Wei Gao; gaowei926@126.com

Academic Editor: Manuel Pineda-Sanchez

For the position tracking control problem of permanent magnet linear motor, an improved fast continuous-time nonsingular terminal sliding mode control algorithm based on terminal sliding mode control method is proposed. Specifically, first, for the second-order model of position error dynamic system, a new continuous-time fast terminal sliding surface is introduced and an improved continuous-time fast terminal sliding mode control law is proposed. Then rigorous theoretical analysis is provided to demonstrate the finite-time stability of the closed-loop system by using the Lyapunov function. Finally, numerical simulations are given to verify the effectiveness and advantages of the proposed fast nonsingular terminal sliding mode control method.

1. Introduction

Permanent magnet linear motor (PMLM) is a conversion device that directly converts electrical energy into linear motion without any intermediate switching mechanism [1, 2]. The study of PMLM has attracted many researchers' interests from motor's design, material, and control due to its many advantages such as high speed, large pushing force, and high precision. And PMLM has been successfully applied in industry, military, and other kind of motion occasions which require high-speed, low thrust, small displacement, and high-precision positioning control [3, 4]. However, the model of PMLM is a typical nonlinear multivariable and coupled system and the PMLM control performance is easily affected by nonlinear factors particularly at unknown load and friction, which vary with different operating conditions. Thus the control problem of PMLM has been an important issue in the filed of PMLM and how to improve the control performance of PMLM system has obtained certain attentions in the literature [5–7].

With the emergence of demand based on PMLM equipment, a considerable amount of nonlinear control methods has been devoted to effectively control the PMLM. For example, the authors in [8] proposed a periodic adaptive compensation control method. And for the modelling and considering both ripples and friction compensations, the improved control scheme was given in [9]. Considering the applications in real industry, the sliding mode control algorithms have been designed to solve the motor control problem due to its significant advantages. The sliding mode control algorithm is easy to use and makes the system state have a good robustness. In particular, even if the controlled systems are suffering from the uncertainty of parameters and external disturbances, SMC can theoretically determine the final tracking accuracy by constructing the reaching law and designing the sliding surface [10–12]. For example, in [13], an equivalent disturbance observer based on sliding mode control method and proportional-integral (PI) was proposed. In order to overcome the uncertainty and interference, based on radial-basis function-network (RBFN), a smart complementary sliding mode control (ICSMC) method was proposed in [14].

However, most of the designed sliding surfaces only guarantee that the system state asymptotically converges to

the equilibrium with infinite convergence time. To improve the closed-loop system's dynamic performance and guarantee that the state of the system can converge to the equilibrium within a finite-time, the terminal sliding mode control (TSMC) method is introduced in [15–18]. Due to the superiority of the terminal sliding mode control method, this method was designed as a controller in [19] for PMLM. In addition, when the state of the system is far from the equilibrium, a fast terminal sliding mode control (FTSMC) method was proposed in [20, 21] to improve the convergence rate of the system state. The (finite-time) transient convergence both at a distance from and at a close range of the equilibrium can be obtained since the merits of the TSMC control.

For the PMLM position tracking control problem and the advantages of TSMC, in this paper, a fast nonsingular TSMC law for PMLM will be designed. The contribution/novelty of this paper is that a new nonlinear control algorithm is designed, i.e., the fast terminal sliding mode control (FTSMC) algorithm. The main advantage of this algorithms is that the fast convergent rate of the closed-loop system can be guaranteed no matter the state is near or far from the equilibrium. To improve the tracking accuracy of the system state, the fast terminal sliding mode surface is obtained and the improved continuous-time nonsingular fast terminal sliding mode control law based on TSMC method is designed. Meanwhile, the rigorous stability analysis for the closed-loop system is presented. The validity and stability of the scheme are verified by employing the Lyapunov function analysis method. Simulation results are provided to show that the continuous-time fast nonsingular TSMC can improve the closed-loop system dynamic performance and robustness against uncertainties and disturbances by comparing with the traditional PID control method.

Note that the work [22] also considered the position tracking control for permanent magnet linear motor. However, the main differences of this manuscript are listed as follows. (i) The model is different: the model considered in this paper is the continuous-time model of permanent magnet linear motor while the work [22] considered the discrete-time model of permanent magnet linear motor based on Euler's discretization. (ii) The method is different: the design method of this paper is based on the continuous-time fast nonsingular terminal sliding mode control method while the work [22] employed discrete-time fast terminal sliding mode control method plus time-delayed disturbance compensation technique. (iii) The stability analysis is different: the Lyapunov function used in this paper is based on a continuous-time Lyapunov function which shows that the finite-time convergence of the closed-loop system can be guaranteed. Note that the work [22] employed a discrete-time Lyapunov function and analyzed the ultimate bound for the steady state.

The rest of the paper is organized as follows. Section 2 provides the description of system model and control objective. Section 3 presents the proposed SMC laws for PMLM. Section 4 discusses the simulation results, and Section 5 concludes this paper.

2. Description of System Model and Control Objective

2.1. Continuous-Time Model of PMLM. For a permanent magnet linear motor, the mathematical model can be described as [3]

$$\begin{aligned} \dot{x}_1(t) &= x_2(t), \\ \dot{x}_2(t) &= -\frac{k_f k_e}{Rm} x_2(t) + \frac{k_f}{Rm} u(t) - \frac{d(t)}{m}, \\ y(t) &= x_1(t), \end{aligned} \tag{1}$$

where x_1 is the linear displacement, x_2 is the linear velocity, $u(t)$ is the input voltage, R is the resistance, m is the motor mass, k_f is the force constant, k_e is the back electromotive force, and $d(t)$ can be counted as the lumped disturbances including the friction and ripple force.

2.2. Control Objective. The control objective of PMLM is to design a controller such that the reference trajectory can be tracked by the linear displacement. Generally, assume that the reference signal is $x_r(t)$, whose first-order and second-order derivatives are bounded.

For the brevity, denote

$$\begin{aligned} a &= \frac{k_f k_e}{Rm}, \\ b &= \frac{k_f}{Rm}, \\ F &= \frac{d(t)}{m}, \end{aligned} \tag{2}$$

under which (1) is rewritten as

$$\begin{aligned} \dot{x}_1(t) &= x_2(t), \\ \dot{x}_2(t) &= -a x_2(t) + b u(t) - F, \\ y(t) &= x_1(t). \end{aligned} \tag{3}$$

Define

$$\begin{aligned} e_1(t) &= x_r(t) - x_1(t), \\ e_2(t) &= \dot{x}_r(t) - x_2(t), \end{aligned} \tag{4}$$

as the tracking errors for linear displacement and speed signal. Then it can be obtained from (3) that the error dynamic equation is

$$\begin{aligned} \dot{e}_1(t) &= e_2(t), \\ \dot{e}_2(t) &= -a e_2(t) - bu + F + a\dot{x}_r + \ddot{x}_r, \\ y &= x_1. \end{aligned} \tag{5}$$

The main objective of this paper is to employ the method of sliding mode control to achieve this control objective due to its many advantages, such as simple design idea and good

robustness [10, 12]. There have been many results about the sliding control algorithms for PMLM in the literature, such as [13, 19], and the most of control laws are based on the design of continuous-time SMC laws with asymptotical convergence. The main objective of this paper is to design a continuous-time fast TSMC law for PMLM.

About the disturbance, the following case will be considered in this paper.

Assumption 1. The disturbance F is assumed to be bounded, i.e., $|F| \le d^*$ with a constant d^*.

2.3. Some Definitions and Lemmas

Definition 2 (finite-time stability [23]). Consider the system

$$\begin{aligned} \dot{x} &= f(x), \\ f(0) &= 0, \\ x(0) &= x_0, \\ x &\in R^n, \end{aligned} \tag{6}$$

where $f(\cdot) : R^n \longrightarrow R^n$ is continuous. The equilibrium $x = 0$ of system (6) is finite-time stable if it is Lyapunov stable and finite-time convergent; i.e., there exists a finite time $T(x_0)$ which is dependent on the initial condition x_0 such that $\lim_{t \longrightarrow T(x_0)} x(t) = 0$ and $x(t) = 0$ for all $t \ge T(x_0)$.

Lemma 3 (see [23]). *For system (6), suppose that there exists a positive definite and proper function $V(x) : R^n \longrightarrow R$ such that $(\partial V(x)/\partial x) f(x) + c(V(x))^{\alpha} \le 0$ for all $x \in R^n$, where $c > 0$, $\alpha \in (0, 1)$. Then, this system is globally finite-time stable.*

3. Design of Nonsingular Fast Terminal Sliding Mode Controller

Theorem 4. *For the error dynamic system (5), if the nonsingular terminal sliding mode controller is designed as*

$$\begin{aligned} u &= u_a + u_b, \\ u_a &= \frac{1}{b}\left[-ae_2 + a\dot{x}_r + \ddot{x}_r + F + \frac{1}{\beta_1\gamma_1}|e_2|^{2-\gamma_1}\operatorname{sign}(e_2) \right. \\ &\quad \left. + \frac{\beta_2}{\beta_1}|e_2|^{2-\gamma_1}|e_1|^{\gamma_1-1}\right], \\ u_b &= \frac{1}{b}\left[k_1 s + k_2 \operatorname{sign}(s)\right], \\ s &= e_1 + \beta_1 |e_2|^{\gamma_1}\operatorname{sign}(e_2) + \beta_2 |e_1|^{\gamma_1}\operatorname{sign}(e_1), \end{aligned} \tag{7}$$

where $k_1 > 0, k_2 > d^*$, $\beta_1 > 0, \beta_2 > 0$, $1 < \gamma_1 < 2$, then the error system state (e_1, e_2) will converge to zero in a finite time.

Proof.

Step 1. Choose the fast nonsingular terminal sliding mode surface.

For the error dynamic system (5), the nonsingular fast terminal sliding mode surface is chosen as

$$s = e_1 + \beta_1 |e_2|^{\gamma_1}\operatorname{sign}(e_2) + \beta_2 |e_1|^{\gamma_1}\operatorname{sign}(e_1) \tag{8}$$

with $\beta_1 > 0, \beta_2 > 0$, $1 < \gamma_1 < 2$.

If the sliding mode surface $s = 0$ can be reached in a finite time, then one obtains that

$$e_1 + \beta_1 |e_2|^{\gamma_1}\operatorname{sign}(e_2) + \beta_2 |e_1|^{\gamma_1}\operatorname{sign}(e_1) = 0, \tag{9}$$

which results into

$$\begin{aligned} e_2 &= -\frac{1}{\beta_1}\left|e_1 + \beta_2|e_1|^{\gamma_1}\operatorname{sign}(e_1)\right|^{1/\gamma_1} \\ &\quad \times \operatorname{sign}\left(e_1 + \beta_2|e_1|^{\gamma_1}\operatorname{sign}(e_1)\right). \end{aligned} \tag{10}$$

Since $\dot{e}_1 = e_2$, then

$$\begin{aligned} \dot{e}_1 &= -\frac{1}{\beta_1}\left|e_1 + \beta_2|e_1|^{\gamma_1}\operatorname{sign}(e_1)\right|^{1/\gamma_1} \\ &\quad \times \operatorname{sign}\left(e_1 + \beta_2|e_1|^{\gamma_1}\operatorname{sign}(e_1)\right). \end{aligned} \tag{11}$$

Construct the following Lyapunov function $V_1 = (1/2)e_1^2$, which leads to

$$\dot{V}_1 \le -\frac{1}{\beta_1}|e_1|^{1+1/\gamma_1} = -\frac{1}{\beta_1}2^{(1+1/\gamma_1)/2}V_1^{(1+1/\gamma_1)/2}. \tag{12}$$

It follows from Lemma 3 that the system state e_1 will converge to zero in a finite time.

Step 2. Design the nonsingular fast terminal sliding mode controller.

Choose a Lyapunov function as $V_2 = (1/2)s^2$, whose derivative along the error system (5) is

$$\begin{aligned} \dot{V}_2 &= s\dot{s} = s\left(e_2 + \beta_1\gamma_1|e_2|^{\gamma_1-1}\dot{e}_2 + \beta_2\gamma_1|e_1|^{\gamma_1-1}e_2\right) \\ &= s\left(e_2 + \beta_2\gamma_1|e_1|^{\gamma_1-1}e_2\right. \\ &\quad \left. + \beta_1\gamma_1|e_2|^{\gamma_1-1}\left[-ae_2(t) - bu + F + a\dot{x}_r + \ddot{x}_r\right]\right) \end{aligned} \tag{13}$$

□

Substituting the terminal sliding mode control law (7) into (13) leads to

$$\begin{aligned} \dot{V}_2 &= s\left[e_2 + \beta_1\gamma_1|e_2|^{\gamma_1-1}\left(-\frac{1}{\beta_1\gamma_1}|e_2|^{2-\gamma_1}\operatorname{sign}(e_2)\right.\right. \\ &\quad \left.\left. - \frac{\beta_2}{\beta_1}|e_2|^{2-\gamma_1}|e_1|^{\gamma_1-1} + F - k_1 s - k_2\operatorname{sign}(s)\right)\right] \\ &= s\left[e_2 - |e_2|\operatorname{sign}(e_2) - \beta_1\gamma_1|e_2|^{\gamma_1-1}(F + k_1 s\right. \\ &\quad \left. + k_2\operatorname{sign}(s))\right] \le -\beta_1\gamma_1 k_1|e_2|^{\gamma_1-1}s^2 - \beta_1\gamma_1(k_2 \\ &\quad - d^*)|e_2|^{\gamma_1-1}|s|. \end{aligned} \tag{14}$$

There are two possibilities for the state e_2. For the first case $e_2 \neq 0$, there is a constant $\rho > 0$ such that

$$\dot{V}_2 \leq -\rho s^2 - \rho |s| \leq -2\rho V_2 - \rho\sqrt{2V_2}. \quad (15)$$

It follows from Lemma 3 that the sliding mode state s will converge to zero in a finite time. Next, we show that $e_2 = 0$ is not an attractor in the reaching phase. Note that

$$\begin{aligned}\dot{e}_2 = &-\frac{1}{\beta_1\gamma_1}|e_2|^{2-\gamma_1}\operatorname{sign}(e_2) - \frac{\beta_2}{\beta_1}|e_2|^{2-\gamma_1}|e_1|^{\gamma_1-1} + F \\ &- k_1 s - k_2 \operatorname{sign}(s).\end{aligned} \quad (16)$$

Under the condition $e_2 = 0$ and $s \neq 0$, then it follows from (17) that

$$\dot{e}_2 = F - k_1 s - k_2 \operatorname{sign}(s) \neq 0, \quad (17)$$

which means that $e_2 = 0$ is not an attractor in the reaching phase. Thus, the finite-time stability of the error system (5) can be achieved.

Remark 5. From the work [15], we know that the traditional terminal sliding mode surface is usually chosen as

$$s = e_1 + \beta_1 |e_2|^{\gamma_1}\operatorname{sign}(e_2), \quad (18)$$

with $0 < \gamma_1 < 1$. Once the sliding mode face is reached and kept, i.e., $s = 0$, then

$$e_2 = -\frac{1}{\beta_1}|e_1|^{1/\gamma_1}\operatorname{sign}(e_1) = \dot{e}_1. \quad (19)$$

It implies that the state e_1 will converge to zero in a finite time. However, if the sliding mode surface is chosen as in this paper, i.e.,

$$s = e_1 + \beta_1 |e_2|^{\gamma_1}\operatorname{sign}(e_2) + \beta_2 |e_1|^{\gamma_1}\operatorname{sign}(e_1), \quad (20)$$

then on the sliding mode face $s = 0$, it follows from (11) that

$$\begin{aligned}\dot{e}_1 = &-\frac{1}{\beta_1}\left|e_1 + \beta_2 |e_1|^{\gamma_1}\operatorname{sign}(e_1)\right|^{1/\gamma_1} \\ &\times \operatorname{sign}\left(e_1 + \beta_2 |e_1|^{\gamma_1}\operatorname{sign}(e_1)\right).\end{aligned} \quad (21)$$

The term $\beta_2|e_1|^{\gamma_1}\operatorname{sign}(e_1)$ will guarantee that there is a faster convergent rate for the state e_1 compared with the traditional terminal sliding mode control (19). Therefore, it is called the fast terminal sliding mode control. In addition, it should be pointed out that there is no singularity problem in the proposed controller (7) since $1 < \gamma_1 < 2$.

4. Simulation Results

In this section, numerical simulations results are supported to illustrate the efficiencies of the designed fast nonsingular terminal sliding mode controller (FTSMC). All the simulation data is based on the Matlab/Simulink model. The system's parameter values of PMLM are given as that in [22], i.e., the mass $m = 5.4$ kg, the resistance $R = 16.8$ ohms, the force constant $k_f = 130$ N/A, and the back electromotive force $k_e = 123$ V/m/s.

The disturbance is composed of two parts, i.e., friction force and ripple force. Definitely, let

$$d = F_{fric} + F_{ripple}, \quad (22)$$

where F_{fric} is the friction force and F_{ripple} is ripple force. The friction force is defined as

$$F_{fric} = \left[f_c + (f_s - f_c)\, e^{-(\dot{x}/\dot{x}_s)^2} + f_v\dot{x}\right]\operatorname{sign}(\dot{x}), \quad (23)$$

where $f_c = 10$N is the Coulomb friction coefficient, $f_v = 10$N is the static friction coefficient, $f_s = 20$N is the static friction coefficient, and $\dot{x}_s = 0.1$ is the lubricant parameter. The ripple force is given as

$$F_{fipple} = A_1 \sin(\omega x) + A_2 \sin(3\omega x) + A_3 \sin(5\omega x), \quad (24)$$

with $A_1 = 8.5$, $A_2 = 4.25$, $A_3 = 2.0$, and $\omega = 314$ rad/s.

In this section, a step signal with amplitude of 200mm and a sinusoidal signal with amplitude of 5mm and the frequency of 1 rad/s, i.e., $x_r = 5\sin(t)$, are, respectively, considered as the desired displacement.

To achieve the position tracking control, the PID control algorithm and the proposed nonsingular fast terminal sliding mode controller (FTSMC) are employed. The controllers' parameters are summarized in Table 1.

Table 1: Controllers' gains.

Control schemes	Values
PID	$k_p = 300,\ k_i = 5,\ k_d = 2$
FTSMC	$k_1 = 100,\ k_2 = 500,\ \beta_1 = \beta_2 = 0.1,\ \gamma_1 = 1.1$

(1) Step response: the step signal with amplitude of 200mm is chosen as the desired displacement. Under the PID controller and proposed continuous-time nonsingular FTSMC, the response curves for the displacement of PMLM are shown in Figure 1. It can be found that the proposed fast TSMC can offer a faster convergent rate and a smaller steady-state error.

(2) Tracking a sinusoid signal: a sinusoidal signal for displacement with amplitude of 5mm and the frequency of 1rad/s is investigated. Similarly, the response curves are given in Figure 2. The tracking error is given in Figure 3. It is shown that the proposed continuous-time nonsingular FTSMC can profoundly reduce the steady-state error.

In summary, according to simulation results, it can be concluded that the closed-loop system's performance can be improved under the proposed continuous-time fast nonsingular terminal sliding mode control method.

5. Conclusions

This paper has investigated the position control problem for permanent magnet linear motors. Based on the terminal sliding mode control theory, an improved continuous-time fast nonsingular terminal sliding mode control method has

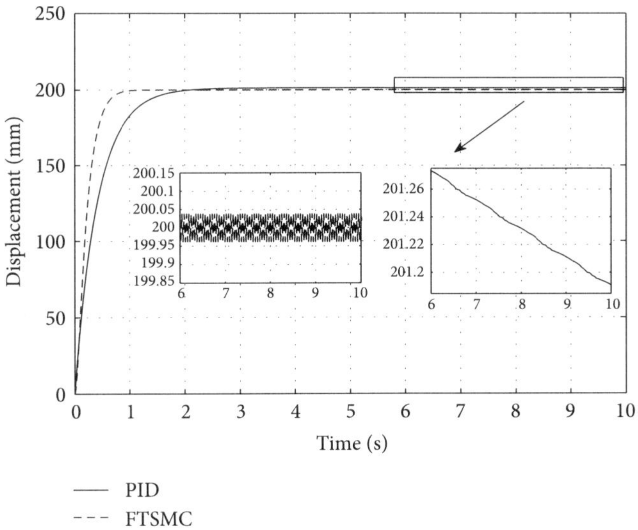

Figure 1: The response curves for displacement under step response.

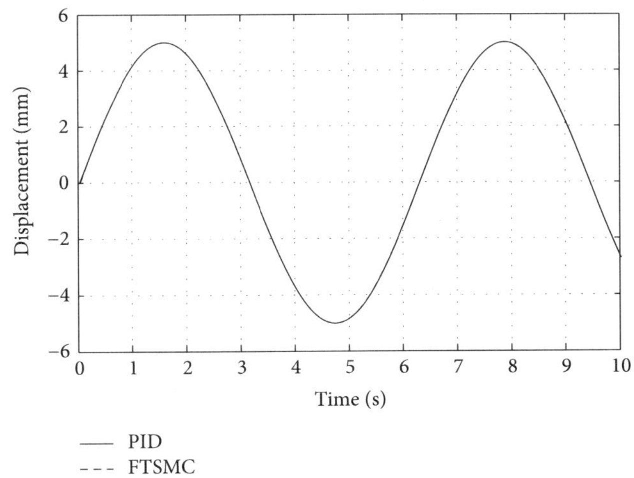

Figure 2: The response curves for displacement under step response.

been proposed. In addition, it also has been theoretically proved that the closed-loop system is finite-time stable by the Lyapunov function analysis method. Simulation results have been verified by the results of theoretical analysis and the effectiveness of the proposed control algorithm by comparing with traditional PID controller.

Conflicts of Interest

The authors declare that there are no conflicts of interest regarding the publication of this paper.

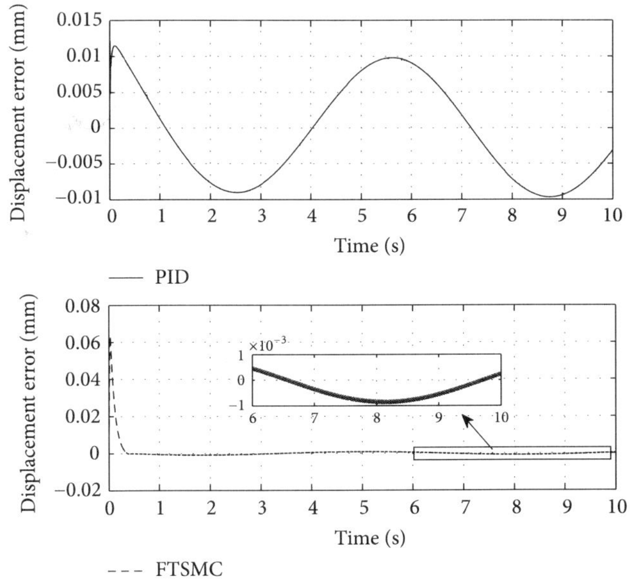

Figure 3: The response curves for displacement error for tracking a sinusoid signal.

Acknowledgments

This work is supported by Natural Science Research Project of Anhui Provincial Department of Education (KJ2016A760), National Natural Science Foundation of China (61673153), and the Fundamental Research Funds for the Central Universities of China (JZ2017HGPA0163).

References

[1] K. K. Tan, S. N. Huang, and T. H. Lee, "Robust adaptive numerical compensation for friction and force ripple in permanent-magnet linear motors," *IEEE Transactions on Magnetics*, vol. 38, no. 1, pp. 221–228, 2002.

[2] T. Hu, W. Xue, and Y. Huang, "Active disturbance rejection control for permanent magnet linear motor," in *Proceedings of the 31st Chinese Control Conference*, pp. 296–301, 2012.

[3] H.-S. Ahn, Y. Chen, and H. Dou, "State-periodic adaptive compensation of cogging and Coulomb friction in permanent-magnet linear motors," *IEEE Transactions on Magnetics*, vol. 41, no. 1 I, pp. 90–98, 2005.

[4] C.-L. Lei, J. Wu, Y.-R. Chen, and J.-M. Yang, "Auto-disturbance-rejection controller used in permanent-magnet linear motor control system," *Control Theory and Applications*, vol. 22, no. 3, pp. 423–428, 2005.

[5] K. K. Tan, T. H. Lee, H. F. Dou, S. J. Chin, and S. Zhao, "Precision motion control with disturbance observer for pulsewidth-modulated-driven permanent-magnet linear motors," *IEEE Transactions on Magnetics*, vol. 39, no. 3, pp. 1813–1818, 2003.

[6] M. Yan and Y. Shiu, "Theory and application of a combined feedback-feedforward control and disturbance observer in linear motor drive wire-EDM machines," *The International Journal of Machine Tools and Manufacture*, vol. 48, no. 3-4, pp. 388–401, 2008.

[7] O. V. Tozoni, "Self-regulating permanent magnet linear motor," *IEEE Transactions on Magnetics*, vol. 35, no. 4, pp. 2137–2145, 1999.

[8] K. Tan, T. Lee, and H. Dou, "Adaptive ripple suppression/compensation apparatus for permanent magnet linear motors," Tech. Rep. Patent US6853158, 2005.
[9] S.-L. Chen, K. K. Tan, S. Huang, and C. S. Teo, "Modeling and compensation of ripples and friction in permanent-magnet linear motor using a hysteretic relay," *IEEE/ASME Transactions on Mechatronics*, vol. 15, no. 4, pp. 586–594, 2010.
[10] V. I. Utkin, *Sliding Modes in Control Optimization*, Springer, Berlin, Germany, 1992.
[11] C. Edwards and S. Spurgeon, *Sliding Mode Control: Theory and Applications*, Taylor and Francis, London, UK, 1998.
[12] J. Zhang, M. Lyu, T. Shen, L. Liu, and Y. Bo, "Sliding Mode Control for a Class of Nonlinear Multi-agent System with Time Delay and Uncertainties," *IEEE Transactions on Industrial Electronics*, vol. 65, no. 1, pp. 865–875, 2018.
[13] F. Cupertino, D. Naso, E. Mininno, and B. Turchiano, "Sliding-mode control with double boundary layer for robust compensation of payload mass and friction in linear motors," *IEEE Transactions on Industry Applications*, vol. 45, no. 5, pp. 1688–1696, 2009.
[14] F.-J. Lin, J.-C. Hwang, P.-H. Chou, and Y.-C. Hung, "FPGA-based intelligent-complementary sliding-mode control for PMLSM servo-drive system," *IEEE Transactions on Power Electronics*, vol. 25, no. 10, pp. 2573–2587, 2010.
[15] M. Zhihong, A. P. Paplinski, and H. R. Wu, "A robust MIMO terminal sliding mode control scheme for rigid robotic manipulators," *IEEE Transactions on Automatic Control*, vol. 39, no. 12, pp. 2464–2469, 1994.
[16] X. Yu and Z. Man, "Model reference adaptive control systems with terminal sliding modes," *International Journal of Control*, vol. 64, no. 6, pp. 1165–1176, 1996.
[17] Y. Wu, X. Yu, and Z. Man, "Terminal sliding mode control design for uncertain dynamic systems," *Systems & Control Letters*, vol. 34, no. 5, pp. 281–287, 1998.
[18] X. Yu and M. Zhihong, "Multi-input uncertain linear systems with terminal sliding-mode control," *Automatica*, vol. 34, no. 3, pp. 389–392, 1998.
[19] X. Mo and Q. Lan, "Finite-time integral sliding mode control for motion control of permanent-magnet linear motors," *Mathematical Problems in Engineering*, vol. 2013, Article ID 567610, 7 pages, 2013.
[20] X. Yu and M. Zhihong, "Fast terminal sliding-mode control design for nonlinear dynamical systems," *IEEE Transactions on Circuits and Systems I: Fundamental Theory and Applications*, vol. 49, no. 2, pp. 261–264, 2002.
[21] M. Xu, C. Chen, and Z. Wu, "Study of nonsingular fast terminal sliding-mode fault-tolerant control," *IEEE Transactions on Industrial Electronics*, vol. 62, no. 6, pp. 3906–3913, 2015.
[22] H. Du, X. Chen, G. Wen, X. Yu, and J. Lu, "Discrete-time fast terminal sliding mode control for permanent magnet linear motor," *IEEE Transactions on Industrial Electronics*, vol. 65, no. 12, pp. 9916–9927, 2018.
[23] S. P. Bhat and D. S. Bernstein, "Finite-time stability of continuous autonomous systems," *SIAM Journal on Control and Optimization*, vol. 38, no. 3, pp. 751–766, 2000.

Event-Triggered H_∞ Filtering for Multiagent Systems with Markovian Switching Topologies

Jiahao Li, Tingting Zhang, Jinfeng Gao, and Ping Wu

Faculty of Mechanical Engineering and Automation, Zhejiang Sci-Tech University, Hangzhou 310018, China

Correspondence should be addressed to Jinfeng Gao; gaojf163@163.com

Academic Editor: Cui-Qin Ma

This paper is concerned with the problem of event-triggered H_∞ filtering for multiagent systems with Markovian switching topologies and network-induced delay. An event-triggered mechanism is given to ease the information transmission. Consider that the network topology is directed in this paper, which represents the communication links among agents. Due to the existence of network-induced delay, the time-delay approach is adopted, which can effectively deal with filtering error system. By constructing a Lyapunov-Krasovskii functional and employing linear matrix inequality technique, sufficient conditions are established to ensure the filtering error system to achieve asymptotically stable with H_∞ performance index. A simulation example is given to illustrate the effectiveness of the proposed method.

1. Introduction

Recently, multiagent systems (MASs) have received much interest due to their widespread applications in various fields such as formation control [1, 2], sensor network [3], synchronization [4], and flocking [5, 6]. Due to the fact that communication networks consist of lots of agents, their states are usually not fully available in network outputs. Hence, filtering or state estimation problem is to estimate the states of the agent by the available output measurement, which is very important both in theory and in practice. In the past decades, the consensus-based filtering or estimation problem for a special multiagent system, i.e., the multi-sensor networked system, has received much attention. For example, Kalman filters are the filters primarily proposed in [7, 8] for multi-sensor networked systems, while H_∞ filters are given in [9] for sensor networks with multiple missing measurements. In most existing literatures, the study of multiagent systems has been investigated in the ideal situation without considering external interference. However, in practical engineering, the existence of external disturbances cannot be ignored, which may have a great impact on agent dynamics. Furthermore, the research history of multiagent systems on H_∞ filtering or state estimation is very short; there are many problems which need to be studied extensively. Thus, it is necessary to study the H_∞ filtering problem of multiagent systems with many practical factors.

The exchange of information between agents on a communication network is usually completed on the basis of sampled packets. Many articles have presented more and more methods based on rapid development of the theory of sampling data systems. In the early results, time-triggered communication is a common way to transfer every sampled signal to the controller and update control signal periodically, which wastes a lot of network bandwidth resources and increases network load. Compared with the widely used time-triggered sampling scheme [10–13], event-triggered communication schemes [14–16] can avoid the unnecessary transmission and reduce the release times of the sensor and the burden of communication network. Event-triggered filtering/estimation for different systems has received considerable attention in the past few years [17–22]. Paper [17] addresses event-triggered state estimation problem for a class of complex networks with mixed time delays and [20] presents H_∞ filter design for a class of neural network systems with quantization. The filtering problem for discrete-time networked control system (NCS) under event-triggered scheme is proposed in [18, 19, 21]. Paper [22] designs the H_∞

fuzzy filter for a class of nonlinear networked control system based on event-triggered communication scheme.

In the real world, because of the uncertainty of the network, the communication topology may change. Hence, it is significant to design a filtering network with time-varying and switching topology to estimate or monitor a target; see [23–28]. Event-based H_∞ filtering for discrete-time and continue-time Markov jump system with network-induced delay is investigated in [23, 24], respectively. Paper [25, 26] addresses asynchronous $l_2 - l_\infty$ filtering with sensor nonlinearity and delay-dependent robust H_∞ control and filtering with parameter uncertainties, respectively. Paper [27] investigates the problem of H_∞ estimation for a class of discrete-time Markov jump systems with time-varying transition probabilities. The problem of distributed state estimation about vehicle formation with time-varying measurement topology is discussed in [28]. To the best of our knowledge, there is no article that considers H_∞ filtering problem for multiagent systems with Markovian switching topologies and network-induced delay based on event-triggered strategy. The theoretical results of such systems will be attractive and have extensive practical applications, which inspired the research presented in this paper.

In the view of the above discussion, this paper addresses the event-triggered H_∞ filtering for multiagent systems with Markovian switching topologies and network-induced delay. The main contributions are summarized as follows: (1) different from the general event-triggered sampled-data strategy, this paper takes into account the effect of network-induced delay and Markovian switching topologies. A distributed event-triggered scheme is presented to release the load of the network. (2) By employing the time-delay system method, H_∞ filtering performance is derived, the co-design method of the event-triggered condition and the filter design are also given. (3) The Laplacian matrices of switching topologies are not required to be symmetric.

The rest of this paper is organized as follows. In Section 2, we give some preliminaries and present the problem formulation. The main results for stability analysis and event-triggered H_∞ filter design are elaborated in Section 3. In Section 4, a numerical example is provided to illustrate the feasibility and effectiveness of the proposed method. Finally, conclusions are drawn in Section 5.

Notations. R^n represents the n-dimension Euclidean space and $R^{n\times m}$ denotes the set of $n \times m$ real matrices. Let I_N represent the $N \times N$ identity matrix and $\mathbf{0}$ denote zero matrix with an appropriate dimension. The superscripts -1 and T stand for the inverse and the transpose of a matrix, respectively. We use the notation $\otimes$ and $*$ to represent Kronecker product and the elements below the main diagonal of some symmetric matrix, respectively. The real matrix $P > 0$ shows that P is positive definite.

2. Problem Formulation

2.1. Graph Theory. Let $G = (V, \xi, A)$ denote a directed weighted graph of N order, where the set of nodes is denotes by $V = \{v_1, v_2, \ldots, v_n\}$ and the set of edges is denoted by $\xi \subseteq V \times V$. An edge in graph G is described by (v_j, v_i), i, $j \in 1, 2, \ldots, n$ which represents that the jth agent transmit information to the ith agent. Matrix $A = [a_{ij}]_{n\times n}$ is called weighted adjacency matrix. The set of neighbors of agent i is denoted as $N_i = \{j \mid (v_j, v_i) \in \xi\}$. If $j \in N_i$, $a_{ij} > 0$; otherwise, $a_{ij} = 0$. The diagonal matrix $D = \operatorname{diag}\{d_1, d_2, \ldots, d_N\} \in R^{N\times N}$ is called the degree matrix of graph G with diagonal elements $d_i = \Sigma_{j\in N_i} a_{ij}$. The Laplacian matrix of G is defined as $L = D - A = [l_{ij}] \in R^{N\times N}$, where $l_{ii} = \sum_{j\in N_i} a_{ij}$ and $l_{ij} = -a_{ij}$ for $i \neq j$.

2.2. System Description. We consider MASs composed of N interconnected agents in this paper, where each agent is assumed to have the following state-space description:

$$\begin{aligned} \dot{x}_i(t) &= Ax_i(t) + Bw_i(t) \\ y_i(t) &= C_2\Sigma_{j\in N_i} a_{ij}^{r(t)} \left(x_i(t) - x_j(t)\right) + Dw_i(t) \\ z_i(t) &= C_1\Sigma_{j\in N_i} a_{ij}^{r(t)} \left(x_i(t) - x_j(t)\right) \end{aligned} \tag{1}$$

for $i = 1, 2, \ldots, N$, where $x_i(t)$, $y_i(t)$, and $z_i(t)$ are the state, the measured output, and the signal to be estimated of agent i. $w_i(t)$ denotes the external disturbance which belongs to $L_2[0, \infty)$.

The switching communication topologies are described by $G_{r(t)} \in \{G_1, G_2, \ldots, G_q\}$, where $r(t)$ represents the continuous process of Markov chain. Furthermore, it takes values in a given finite set $S = \{1, 2, \ldots, q\}$. We define the transition probability as follows:

$$\begin{aligned} &prob\{r(t + \Delta t) = s \mid r(t) = r\} \\ &= \begin{cases} \pi_{rs}\Delta t + o(\Delta t), & r \neq s \\ 1 + \pi_{rr}\Delta t + o(\Delta t), & r = s \end{cases} \end{aligned} \tag{2}$$

where $\lim_{\Delta t\longrightarrow 0}(o(\Delta t)/\Delta t) = 0$, $\Delta t > 0$, and the transition probability satisfies $\pi_{rs} > 0$ when $r, s \in S$ and $\pi_{rr} = -\sum_{s=1,s\neq r}^{q} \pi_{rs}$.

For MASs (1), we design the following full-order filter to estimate signal $z_i(t)$:

$$\begin{aligned} \dot{x}_{fi}(t) &= A_{fr}x_{fi}(t) + B_{fr}y_{fi}(t) \\ z_{fi}(t) &= C_{fr}x_{fi}(t) \end{aligned} \tag{3}$$

where $x_{fi}(t)$ represents the state of the ith filtering subsystem, $z_{fi}(t)$ is the estimation signal of $z_i(t)$, and A_{fr}, B_{fr}, C_{fr} are filter parameters to be determined.

2.3. Event-Triggered Scheme. In order to reduce the communication network burden, event-triggered scheme is adopted in this paper. The sampler is assumed to be time-driven; the zero-order-hold (ZOH) is event-driven. Then, we construct the following event-triggered strategy.

$$\begin{aligned} t_{k+1}^i h = t_k^i h + \min_{l_i>1} &\left\{l_i h \mid \chi_i^T\left(t_k^i h + l_i h\right)\phi_r \chi_i\left(t_k^i h + l_i h\right)\right. \\ &\left.\geq \sigma_r y_i^T\left(t_k^i h + l_i h\right)\phi_r y_i\left(t_k^i h + l_i h\right)\right\} \end{aligned} \tag{4}$$

where $l_i \in N$, $\chi_i(t_k^i h + l_i h) \triangleq y_i(t_k^i h + l_i h) - y_i(t_k^i h)$ is the threshold error, $t_k^i h$ represents the latest kth triggering instant of the ith agent, $t_k^i h + l_i h$ denotes the currently sampled instant, and $t_{k+1}^i h$ denotes the next broadcast instant. Then $\sigma_r > 0$ represents the threshold and $\phi_r > 0$ stands for a symmetric positive definite matrix.

Remark 1. From the event-triggered strategy (4), we can automatically exclude Zeno behavior.

Proof. Zeno behavior in the event-triggered communication framework is defined as the infinite number of communication in a finite time. Due to the fact that each agent's state is periodically sampled and the sampling time sequence is $\{0, h, 2h, \ldots\}$, we can obtain that the event-triggered time sequence is a subsequence of the sampling time sequence, namely, $\{t_0^i h, t_1^i h, t_2^i h, \ldots\} \subseteq \{0, h, 2h, \ldots\}$, which means that the minimum interevent time $\min_k\{t_{k+1}^i h - t_k^i h\}$, $\forall i$, is lower bounded by the sampling period h. Since the number of agents is finite, the communication events in any finite time cannot be infinite. Therefore, the event-triggered strategy (4) can rule out Zeno behavior automatically. □

Remark 2. Different from traditional filtering problem, due to the existence of network-induced delays, taking the property of ZOH into account, we get

$$y_{fi}(t) = y_i\left(t_k^i h\right) \tag{5}$$

Substituting (5) into (3), we have

$$\begin{aligned} \dot{x}_{fi}(t) &= A_{fr} x_{fi}(t) + B_{fr} y_i\left(t_k^i h\right) \\ z_{fi}(t) &= C_{fr} x_{fi}(t), \\ & t \in \left[t_k^i h + \tau\left(t_k^i\right), t_{k+1}^i h + \tau\left(t_{k+1}^i\right)\right) \end{aligned} \tag{6}$$

Remark 3. Suppose that the network delay is bounded, i.e., $0 < \tau_m \le \tau_k^i \le \overline{\tau}$, where τ_m and $\overline{\tau}$ denote the lower and upper bound of $\tau(t)$, respectively. It is worth noting that when $\sigma_r = 0$, all the sampled signals are transmitted, the event-triggered strategy (4) reduces to a periodic time-triggered scheme.

2.4. Time-Delay Modeling. Using the same technique as in [29], we assume that there is a integer $q > 0$, which satisfies $t_{k+1}^i = t_k^i + q + 1$. Hence, we can obtain the following equality:

$$\left[t_k^i h + \tau\left(t_k^i\right), t_{k+1}^i h + \tau\left(t_{k+1}^i\right)\right) = \bigcup_{n=0}^{q} \Omega_n \tag{7}$$

with $\Omega_n = [t_k^i h + nh + \tau(t_k^i + n), t_k^i h + (n+1)h + \tau(t_k^i + n + 1))$, $t_{k+1}^i = t_k^i + n + 1$.

For convenience, we define

$$\tau(t) = t - t_k^i h - nh, \quad t \in \Omega_n \tag{8}$$

Then, we can know that

$$0 < \tau_m \le \tau(t) \le h + \overline{\tau} = \tau_M \tag{9}$$

To show the effect of the event-triggered scheme (4) in deriving system stability and stabilization criterion, we define a new variable $e_{ki}(t)$, which satisfies the following equality:

$$e_{ki}(t) = y_i\left(t_k^i h\right) - y_i\left(t_k^i h + nh\right), \quad t \in \Omega_n \tag{10}$$

Substituting (8), (10) into (6), we have

$$\begin{aligned} \dot{x}_{fi}(t) &= A_{fr} x_{fi}(t) + B_{fr}\left[y_i(t - \tau(t)) + e_{ki}(t)\right] \\ z_{fi}(t) &= C_{fr} x_{fi}(t) \end{aligned} \tag{11}$$

From (10) and (4), we can obtain

$$e_{ki}^T(t)\phi_r e_{ki}(t) \le \sigma_r y_i^T(t - \tau(t))\phi_r y_i(t - \tau(t)) \tag{12}$$

where $t \in [t_k^i h + \tau(t_k^i), t_{k+1}^i h + \tau(t_{k+1}^i))$.

2.5. Event-Triggered H_∞ Filter Problem. By defining $\xi_i^T(t) = \left[x_i^T(t) \;\; x_{fi}^T(t)\right]$, $\widehat{w}_i(t) = \left[w_i^T(t) \;\; w_i^T(t - \tau(t))\right]^T$, and $e_i(t) = z_i(t) - z_{fi}(t)$, we can obtain the following filtering error system:

$$\begin{aligned} \dot{\xi}_i(t) &= \overline{A}\xi_i(t) + \Sigma_{j\in N_i} a_{ij} \overline{B} H \xi_i(t - \tau(t)) + \overline{E} e_{ki}(t) \\ &\quad + \overline{D}\widehat{w}_i(t) \\ e_i(t) &= \overline{C}_1 \xi_i(t) + \Sigma_{j\in N_i} a_{ij} \overline{C}_2\left(\xi_i(t) - \xi_j(t)\right) \end{aligned} \tag{13}$$

where

$$\begin{aligned} \overline{A} &= \begin{bmatrix} A & 0 \\ 0 & A_{fr} \end{bmatrix}, \\ \overline{B} &= \begin{bmatrix} 0 \\ B_{fr} C_2 \end{bmatrix}, \\ H &= \left[I_n \;\; 0\right], \\ \overline{D} &= \begin{bmatrix} B & 0 \\ 0 & B_{fr} D \end{bmatrix}, \\ \overline{C}_1 &= \left[0 \;\; -C_{fr}\right], \\ \overline{C}_2 &= \left[C_1 \;\; 0\right] \\ \overline{E} &= \begin{bmatrix} 0 \\ B_{fr} \end{bmatrix} \end{aligned} \tag{14}$$

Then, we define some new augment variables to compact system:

$$\begin{aligned} \xi(t) &= \left[\xi_1^T(t) \;\; \xi_2^T(t) \;\; \ldots \;\; \xi_N^T(t)\right]^T, \\ e(t) &= \left[e_1^T(t) \;\; e_2^T(t) \;\; \ldots \;\; e_N^T(t)\right]^T, \\ \widehat{w}(t) &= \left[\widehat{w}_1^T(t) \;\; \widehat{w}_2^T(t) \;\; \ldots \;\; \widehat{w}_N^T(t)\right]^T, \\ e_k(t) &= \left[e_{k1}^T(t) \;\; e_{k2}^T(t) \;\; \ldots \;\; e_{kN}^T(t)\right]^T \end{aligned} \tag{15}$$

Therefore, the filtering error system can be rewritten as follows:

$$\begin{aligned}\dot{\xi}(t) &= \left(I_N \otimes \overline{A}\right)\xi(t) + \left(L \otimes \overline{B}H\right)\xi(t-\tau(t)) \\ &\quad + \left(I_N \otimes \overline{E}\right)e_k(t) + \left(I_N \otimes \overline{D}\right)\widehat{w}(t) \\ e(t) &= \left(I_N \otimes \overline{C}_1 + L \otimes \overline{C}_2\right)\xi(t)\end{aligned} \tag{16}$$

Definition 4. The filtering error system (16) with $w(t) = 0$ is asymptotically stable in mean square, if for any initial conditions, such that

$$\lim_{t \longrightarrow +\infty} \|\xi(t)\|^2 = 0 \tag{17}$$

Definition 5. Given a positive scalar γ, and for all nonzero $w(t) \in L_2[0,\infty)$, the filtering error system (16) is asymptotically stable in mean square with a guaranteed H_∞ performance γ if the filtering error $e(t)$ satisfies

$$E\left\{\int_0^\infty e^T(t)e(t)\,dt\right\} \le \gamma^2 E\left\{\int_0^\infty w^T(t)w(t)\,dt\right\} \tag{18}$$

Before proceeding further, we introduce the following lemmas that will be helpful for deriving the following results.

Lemma 6 (see [30]). *Given scalar $\tau > 0$, matrix $M > 0$, and vector function w, $[0,\tau] \longrightarrow R^n$ satisfy the following inequality:*

$$\begin{aligned}&\left(\int_0^\tau w(s)\,ds\right)^T M\left(\int_0^\tau w(s)\,ds\right) \\ &\quad \le \tau\left(\int_0^\tau w^T(s)Mw(s)\,ds\right).\end{aligned} \tag{19}$$

Lemma 7 (see [31]). *Suppose that $f_1(t), f_2(t), \ldots, f_N(t) : \mathbf{R}^m \longmapsto \mathbf{R}^n$ is positive values, which satisfies*

$$\min \sum_{i=1}^N \frac{1}{\alpha_i} f_i(t) = \sum_{i=1}^N f_i(t) + \max \sum_{i=1}^N \sum_{j=1, j\ne i}^N g_{i,j}(t) \tag{20}$$

subject to

$$\left\{g_{i,j} : \mathbf{R}^m \longmapsto \mathbf{R}^n,\ g_{j,i}(t) \equiv g_{i,j}(t),\ \begin{bmatrix} f_i(t) & g_{i,j} \\ g_{i,j} & f_i(t)\end{bmatrix} \ge 0\right\} \tag{21}$$

3. Main Results

3.1. Asymptotical Stability Analysis

Theorem 8. *For given scalars $\sigma_r > 0$, $\tau_M > \tau_m > 0$, the filtering error system (16) without external disturbance can achieve mean square stable under Definition 4 if there exist appropriate dimensional matrices $P_r > 0$, $Q_1 > 0$, $Q_2 > 0$, $R_1 > 0$, $R_2 > 0$, $\phi_r > 0$ and S_1 such that the following LMIs hold:*

$$\begin{bmatrix} R_2 & S_1 \\ * & R_2 \end{bmatrix} > 0 \tag{22}$$

$$\begin{bmatrix} \Pi & \tau_m \Gamma^T \widehat{H}^T R_1 & (\tau_M - \tau_m)\Gamma^T \widehat{H}^T R_2 \\ * & -R_1 & 0 \\ * & * & -R_2 \end{bmatrix} < 0 \tag{23}$$

$$\sum_{s=1}^q \pi_{rs} P_s \le P_r \tag{24}$$

where

$$\begin{aligned}
\Pi &= \begin{bmatrix} \Xi_{11} & \Xi_{12} & \Xi_{13} & 0 & \Xi_{15} \\ * & \Xi_{22} & \Xi_{23} & \Xi_{24} & 0 \\ * & * & \Xi_{33} & \Xi_{34} & 0 \\ * & * & * & \Xi_{44} & 0 \\ * & * & * & * & -\widehat{\phi}_r \end{bmatrix} \\
\Gamma &= \left[I_N \otimes \overline{A} \ \ L \otimes \overline{B} \ \ 0 \ \ 0 \ \ I_N \otimes \overline{E}\right] \\
\Xi_{11} &= \left(I_N \otimes \overline{A}\right)^T P_r + P_r\left(I_N \otimes \overline{A}\right) + P_r + \widehat{H}^T Q_1 \widehat{H} \\
&\quad + \widehat{H}^T Q_2 \widehat{H} - \widehat{H}^T R_1 \widehat{H} \\
\Xi_{12} &= P_r\left(L \otimes \overline{B}\right) \\
\Xi_{13} &= \widehat{H}^T R_1 \\
\Xi_{15} &= P_r\left(I_N \otimes \overline{E}\right) \\
\Xi_{22} &= \sigma_r (L \otimes C_2)^T \widehat{\phi}_r (L \otimes C_2) - 2R_2 + S_1 + S_1^T \\
\Xi_{23} &= -S_1^T + R_2 \\
\Xi_{24} &= -S_1 + R_2 \\
\Xi_{33} &= -Q_1 - R_1 - R_2 \\
\Xi_{34} &= S_1 \\
\Xi_{44} &= -Q_2 - R_2 \\
\widehat{H} &= I_N \otimes H, \\
\widehat{H}_2 &= I_N \otimes H_2, \\
\widehat{\phi}_r &= I_N \otimes \phi_r, \\
H_2 &= [0 \ \ I_n]
\end{aligned} \tag{25}$$

Proof. Consider that there is no external disturbance, i.e., $w(t) = 0$. We construct the following Lyapunov functional:

$$V\left(t, \xi(t), \dot{\xi}(t), r(t)\right) = \sum_{i=1}^5 V_i\left(t, \xi(t), \dot{\xi}(t), r(t)\right) \tag{26}$$

where

$$V_1\left(t, \xi(t), \dot{\xi}(t), r(t)\right) = \xi^T(t) P_r \xi(t)$$

$$V_2\left(t, \xi(t), \dot{\xi}(t)\right) = \int_{t-\tau_m}^t \xi^T(s) \widehat{H}^T Q_1 \widehat{H} \xi(s)\,ds$$

$$V_3\left(t, \xi(t), \dot{\xi}(t)\right) = \int_{t-\tau_M}^t \xi^T(s) \widehat{H}^T Q_2 \widehat{H} \xi(s)\,ds$$

$$V_4\left(t,\xi(t),\dot{\xi}(t)\right) = \tau_m \int_{-\tau_m}^{0}\int_{t+\theta}^{t} \dot{\xi}^T(s)\widehat{H}^T R_1 \widehat{H}\dot{\xi}(s)\,ds\,d\theta$$
$$V_5\left(t,\xi(t),\dot{\xi}(t)\right) = (\tau_M-\tau_m)\int_{-\tau_M}^{-\tau_m}\int_{t+\theta}^{t} \dot{\xi}^T(s)\widehat{H}^T R_2 \widehat{H}\dot{\xi}(s)\,ds\,d\theta \tag{27}$$

then its derivative along system (16) with $w(t) = 0$ is

$$\dot{V}_1 = \dot{\xi}^T(t)P_r\xi(t) + \xi^T(t)P_r\dot{\xi}(t) + \xi^T(t)\left\{\sum_{s=1}^{q}\pi_{rs}P_s\right\}\xi(t)$$
$$\dot{V}_2 = \xi^T(t)\widehat{H}^T Q_1\widehat{H}\xi(t) - \xi^T(t-\tau_m)\widehat{H}^T Q_1\widehat{H}\xi(t-\tau_m)$$
$$\dot{V}_3 = \xi^T(t)\widehat{H}^T Q_2\widehat{H}\xi(t) - \xi^T(t-\tau_M)\widehat{H}^T Q_2\widehat{H}\xi(t-\tau_M)$$
$$\dot{V}_4 = \tau_m^2\dot{\xi}^T(t)\widehat{H}^T R_1\widehat{H}\dot{\xi}(t) - \tau_m\int_{t-\tau_m}^{t}\dot{\xi}^T(s)\widehat{H}R_1\widehat{H}\dot{\xi}(s)\,ds$$
$$\dot{V}_5 = (\tau_M-\tau_m)^2\dot{\xi}^T(t)\widehat{H}^T R_2\widehat{H}\dot{\xi}(t) - (\tau_M-\tau_m)\int_{t-\tau(t)}^{t-\tau_m}\dot{\xi}^T(s)\widehat{H}R_2\widehat{H}\dot{\xi}(s)\,ds - (\tau_M-\tau_m)\int_{t-\tau_M}^{t-\tau(t)}\dot{\xi}^T(s)\widehat{H}R_2\widehat{H}\dot{\xi}(s)\,ds \tag{28}$$

Applying Lemmas 6 and 7 to deal with the integral items and combining the event-triggered scheme (12), we have

$$\dot{V}\left(t,\xi(t),\dot{\xi}(t),r(t)\right) \le \sum_{i=1}^{5}V_i\left(t,\xi(t),\dot{\xi}(t),r(t)\right) - e_k^T(t)\widehat{\phi}_r e_k(t) + \sigma_r y^T(t-\tau(t))\widehat{\phi}_r y(t-\tau(t)) = \zeta^T(t)\psi\zeta(t) \tag{29}$$

where

$$\zeta^T(t) = \left[\xi^T(t)\;\; \xi^T(t-\tau(t))\widehat{H}^T\;\; \xi^T(t-\tau_m)\widehat{H}^T\;\; \xi^T(t-\tau_M)\widehat{H}^T\;\; e_k^T(t)\right]$$
$$\psi = \Pi + \tau_m^2\Gamma^T\widehat{H}^T R_1\widehat{H}\Gamma + (\tau_M-\tau_m)^2\Gamma^T\widehat{H}^T R_2\widehat{H}\Gamma \tag{30}$$

By employing the Schur complement, we can find easily that (23) ensures $\psi < 0$. According to Definition 4, the filtering error system (16) is asymptotically stable in mean square under the event-triggered scheme (12) if (22), (23), and (24) hold. The proof is completed. □

3.2. H_∞ Performance Analysis

Theorem 9. *For given scalars $\sigma_r > 0$, $\tau_M > \tau_m > 0$, considering the existence of external disturbance, the filtering error system (16) can achieve being asymptotically stable with an H_∞ performance index γ under Definition 5 if there exist appropriate dimensional matrices $P_r > 0$, $Q_1 > 0$, $Q_2 > 0$, $R_1 > 0$, $R_2 > 0$, $\phi_r > 0$ and S_1 such that the following LMIs hold:*

$$\begin{bmatrix} R_2 & S_1 \\ * & R_2 \end{bmatrix} > 0 \tag{31}$$

$$\begin{bmatrix} \Theta & \tau_m\Gamma_1^T\widehat{H}^T R_1 & (\tau_M-\tau_m)\Gamma_1^T\widehat{H}^T R_2 & \Gamma_2^T \\ * & -R_1 & 0 & 0 \\ * & * & -R_2 & 0 \\ * & * & * & -I \end{bmatrix} < 0 \tag{32}$$

$$\sum_{s=1}^{q}\pi_{rs}P_s \le P_r \tag{33}$$

where

$$\Theta = \begin{bmatrix} \Xi_{11} & \Xi_{12} & \Xi_{13} & 0 & \Xi_{15} & \Xi_{16} \\ * & \Xi_{22} & \Xi_{23} & \Xi_{24} & 0 & \Xi_{26} \\ * & * & \Xi_{33} & \Xi_{34} & 0 & 0 \\ * & * & * & \Xi_{44} & 0 & 0 \\ * & * & * & * & -\widehat{\phi}_r & \\ * & * & * & * & * & \Xi_{66} \end{bmatrix}$$
$$\Gamma_1 = \left[I_N\otimes\overline{A}\;\; L\otimes\overline{B}\;\; 0\;\; 0\;\; I_N\otimes\overline{E}\;\; I_N\otimes\overline{D}\right]$$
$$\Gamma_2 = \left[I_N\otimes\overline{C}_1 + L\otimes\overline{C}_2\;\; 0\;\; 0\;\; 0\;\; 0\;\; 0\right]$$
$$\Xi_{16} = P_r\left(I_N\otimes\overline{D}\right)$$
$$\Xi_{26} = \sigma_r\left(L\otimes C_2\right)^T\widehat{\phi}_r\left(I_N\otimes D\right)\widehat{H}_2$$
$$\Xi_{66} = -\gamma^2 I + \sigma_r\widehat{H}_2^T\left(I_N\otimes D\right)^T\widehat{\phi}_r\left(I_N\otimes D\right)\widehat{H}_2 \tag{34}$$

and the other parameters are defined as in Theorem 8.

Proof. Consider the existence of external disturbance, i.e., $w(t) \neq 0$, and choose the same Lyapunov functional like Theorem 8; we have

$$\begin{aligned} &\dot{V}\left(t,\xi(t),\dot{\xi}(t),r(t)\right) \\ &= \dot{V}\left(t,\xi(t),\dot{\xi}(t),r(t)\right) + e^T(t)e(t) \\ &\quad - \gamma^2\widehat{w}^T(t)\widehat{w}(t) - e^T(t)e(t) + \gamma^2\widehat{w}^T(t)\widehat{w}(t) \\ &= \eta^T(t)\Omega\eta(t) - e^T(t)e(t) + \gamma^2\widehat{w}^T(t)\widehat{w}(t) \end{aligned} \tag{35}$$

where

$$\begin{aligned} \eta^T(t) &= \left[\zeta^T(t) \;\; \widehat{w}^T(t)\right] \\ \Omega &= \Theta + \tau_m^2\Gamma_1^T\widehat{H}^T R_1\widehat{H}\Gamma_1 \\ &\quad + \left(\tau_M - \tau_m\right)^2\Gamma_1^T\widehat{H}^T R_2\widehat{H}\Gamma_1 + \Gamma_2^T\Gamma_2 \end{aligned} \tag{36}$$

By using the Schur complement, we can find easily that (32) guarantees $\Omega < 0$. According to Definition 5, the filtering error system (16) is asymptotically stable with H_∞ norm bound γ if (31), (32), and (33) are satisfied. The proof is completed. □

3.3. Filter Design

Theorem 10. *For given scalars $\sigma_r > 0$, $\tau_M > \tau_m > 0$ and H_∞ performance index γ, the filtering error system (16) under the event-triggered scheme (12) is asymptotically stable in mean square for $w(t) = 0$ and satisfies the H_∞ performance constraint (18) for all nonzero $w(t) \in L_2[0,+\infty)$ under the zero initial condition if there exist real matrices $P_r > 0$, $W_r > 0$, $\phi_r > 0$ $(r \in S)$, $Q_1 > 0$, $Q_2 > 0$, $R_1 > 0$, $R_2 > 0$ and matrices $S_1, \overline{A}_{fr}, \overline{B}_{fr}, \overline{C}_{fr} (r \in S)$ with appropriate dimension such that the following LMIs hold:*

$$\begin{bmatrix} R_2 & S_1 \\ * & R_2 \end{bmatrix} > 0, \tag{37}$$

$$\begin{bmatrix} \phi_{11} & \phi_{12} & \phi_{13} & \phi_{14} \\ * & -R_1 & 0 & 0 \\ * & * & -R_2 & 0 \\ * & * & * & -I \end{bmatrix} < 0, \tag{38}$$

$$\sum_{s=1}^{q}\pi_{rs}\left(U_{1s} - W_s\right) \leq U_{1r} - W_r \tag{39}$$

where

$$\begin{aligned} \phi_{11} &= \begin{bmatrix} \widehat{\Xi}_{11} & \widehat{\Xi}_{12} & \widehat{\Xi}_{13} & 0 & \widehat{\Xi}_{15} & \widehat{\Xi}_{16} \\ * & \Xi_{22} & \Xi_{23} & \Xi_{24} & 0 & \Xi_{26} \\ * & * & \Xi_{33} & \Xi_{34} & 0 & 0 \\ * & * & * & \Xi_{44} & 0 & 0 \\ * & * & * & * & -\widehat{\phi}_r & 0 \\ * & * & * & * & * & \Xi_{66} \end{bmatrix} \\ \phi_{12} &= \left[\tau_m R_1\left(I_N \otimes A\right)\widehat{H} \;\; 0 \;\; 0 \;\; 0 \;\; 0 \;\; \tau_m R_1\left(I_N \otimes B\right)\widehat{H}\right]^T, \\ \phi_{13} &= \left[\left(\tau_M - \tau_m\right)R_2\left(I_N \otimes A\right)\widehat{H} \;\; 0 \;\; 0 \;\; 0 \;\; 0 \;\; \left(\tau_M - \tau_m\right)R_2\left(I_N \otimes B\right)\widehat{H}\right]^T, \\ \phi_{14} &= \left[\left(I_N \otimes -\overline{C}_{fr}\right)\widehat{H}_2 + \left(L \otimes C_1\right)\widehat{H} \;\; 0 \;\; 0 \;\; 0 \;\; 0 \;\; 0\right]^T, \\ \widehat{\Xi}_{11} &= I_N \otimes \begin{bmatrix} U_{1r} + A^T U_{1r} + U_{1r}A & W_r + \overline{A}_{fr} + A^T W_r \\ * & W_r + \overline{A}_{fr} + \overline{A}_{fr}^T \end{bmatrix} + \widehat{H}^T\left(Q_1 + Q_2 - R_1\right)\widehat{H}, \\ \widehat{\Xi}_{12} &= L \otimes \begin{bmatrix} \overline{B}_{fr}C_2 \\ \overline{B}_{fr}C_2 \end{bmatrix}, \\ \widehat{\Xi}_{13} &= \widehat{H}^T R_1, \\ \widehat{\Xi}_{15} &= I_N \otimes \begin{bmatrix} \overline{B}_{fr} \\ \overline{B}_{fr} \end{bmatrix}, \\ \widehat{\Xi}_{16} &= I_N \otimes \begin{bmatrix} U_{1r}B & \overline{B}_{fr}D \\ W_r B & \overline{B}_{fr}D \end{bmatrix} \end{aligned} \tag{40}$$

Moreover, if the above conditions are feasible, the parameter matrices of the filter are given by

$$A_{fr} = W_r^{-1}\overline{A}_{fr}$$
$$B_{fr} = W_r^{-1}\overline{B}_{fr} \quad (41)$$
$$C_{fr} = \overline{C}_{fr}$$

Proof. Let $P_r = I_N \otimes U_r$ for conditions (32) in Theorem 8. Let matrix U_r be partitioned as

$$U_r = \begin{bmatrix} U_{1r} & U_{2r} \\ * & U_{3r} \end{bmatrix} \quad (42)$$

where $U_{1r} > 0$, U_{2r}, and U_{3r} are nonsingular matrices, satisfying $W_r = U_{2r}U_{3r}^{-1}U_{2r}^T$. By Schur complement, we can obtain that $U_r > 0$ is equivalent to $U_{1r} - W_r > 0$. Define the following invertible matrix:

$$J_1 = \begin{bmatrix} I & 0 \\ 0 & U_{2r}U_{3r}^{-1} \end{bmatrix} \quad (43)$$

Then, pre- and postmultiply (32) by J_2 and J_2^T, respectively, where $J_2 = \text{diag}\{J_1, \underbrace{I, \ldots, I}_{8}\}$ and define new variables:

$$\overline{A}_{fr} = U_{2r}A_{fr}U_{3r}^{-1}U_{2r}^T$$
$$\overline{B}_{fr} = U_{2r}B_{fr} \quad (44)$$
$$\overline{C}_{fr} = C_{fr}U_{3r}^{-1}U_{2r}^T$$

By the linear transformation above, we can find easily that (32) is equivalent to (38). Therefore, we can obtain (41). The proof is completed. □

Remark 11. Comparing with the previous results [32], lower bounds theorem [31] is adopted in this paper, which not only achieves performance behavior identical to approaches based on the integral inequality lemma, but also decreases the number of decision variables dramatically.

4. Simulation

The parameters of MASs (1) are given as follows:

$$A = \begin{bmatrix} -1 & 1 \\ 0.8 & -4 \end{bmatrix},$$
$$B = \begin{bmatrix} 0.5 \\ 0.1 \end{bmatrix}$$
$$C1 = \begin{bmatrix} 0.1 & 0 \end{bmatrix}, \quad (45)$$
$$C2 = \begin{bmatrix} 0.2 & 0 \end{bmatrix},$$
$$D = 0.1$$

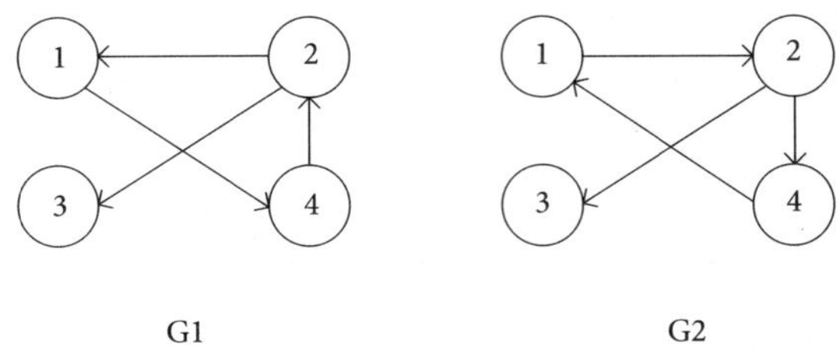

FIGURE 1: Directed communication topology graphs.

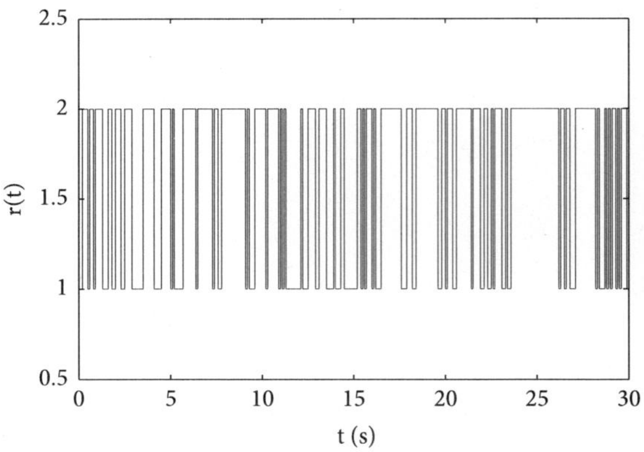

FIGURE 2: Evolution of Markov chain.

In addition, the external disturbances of MASs (1) are defined as follows:

$$w_i(t) = 0.4e^{-0.2t}\cos(0.6t). \quad (46)$$

All the possible information transmission relationships among agents are shown in Figure 1. Then, the corresponding Laplacian matrices are given as follows, respectively.

$$L_1 = \begin{bmatrix} 1 & -1 & 0 & 0 \\ 0 & 1 & 0 & -1 \\ 0 & -1 & 1 & 0 \\ -1 & 0 & 0 & 1 \end{bmatrix},$$
$$L_2 = \begin{bmatrix} 1 & 0 & 0 & -1 \\ -1 & 1 & 0 & 0 \\ 0 & -1 & 1 & 0 \\ 0 & -1 & 0 & 1 \end{bmatrix} \quad (47)$$

Figure 2 shows the switching of two modes in a Markov chain. Suppose that the probability transition matrix is defined as

$$\Pi = \begin{bmatrix} -5 & 5 \\ 2 & -2 \end{bmatrix} \quad (48)$$

Given that $h = 0.1$, $\sigma_1 = 0.1$, $\sigma_2 = 0.2$, minimum allowable delay $\tau_m = 0.1$ and the maximum allowable delay $\tau_M = 0.5$. Let the initial states $x_1(0) = [3; -1], x_2(0) = [1; -1], x_3(0) = [1; -3], x_4(0) = [-3; 3]$. By solving LMIs in

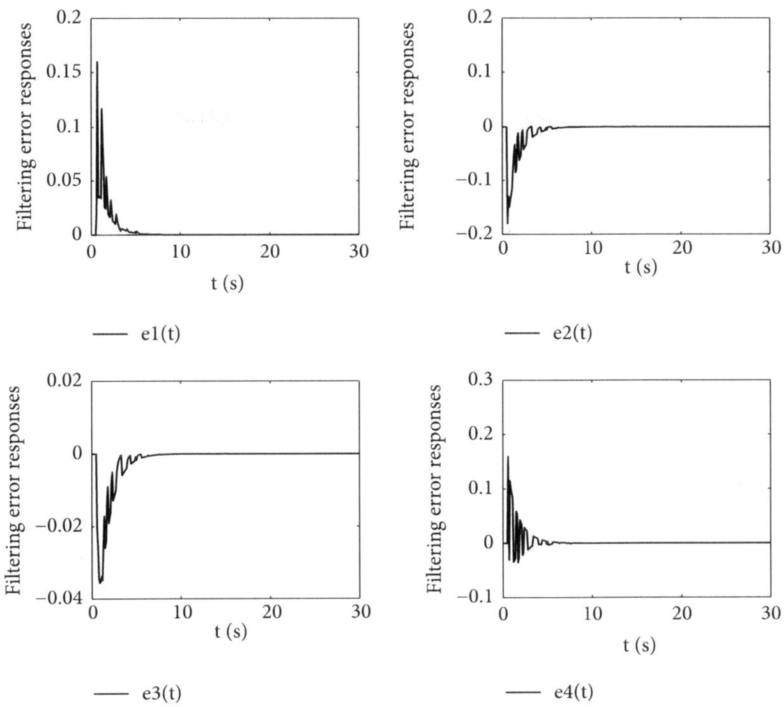

FIGURE 3: Filtering error response.

Theorem 9, we can obtain parameters of the designed filter (3),

$$\begin{aligned} A_{f1} &= \begin{bmatrix} -24.8840 & 52.462 \\ -9.6761 & 20.1822 \end{bmatrix}, \\ B_{f1} &= \begin{bmatrix} -0.6878 \\ -0.2918 \end{bmatrix} \\ C_{f1} &= \begin{bmatrix} -1.3535 & 0.0581 \end{bmatrix} \\ A_{f2} &= \begin{bmatrix} -58.5910 & 137.6004 \\ -22.9133 & 53.7343 \end{bmatrix}, \\ B_{f2} &= \begin{bmatrix} 0.8893 \\ 0.3558 \end{bmatrix} \\ C_{f2} &= \begin{bmatrix} -0.1373 & -0.0623 \end{bmatrix} \times 10^{-5}, \end{aligned} \tag{49}$$

and the event-triggered parameters,

$$\begin{aligned} \phi_1 &= 15.5816, \\ \phi_2 &= 7.9041 \end{aligned} \tag{50}$$

In Figure 3, it shows the filtering error of agent i $(i = 1, 2, 3, 4)$. In Figure 4, it depicts the state curves of $z_i(t)$ and estimation signal $z_{fi}(t)$ $(i = 1, 2, 3, 4)$. The event-triggered release instants and intervals of agent i $(i = 1, 2, 3, 4)$ are shown in Figure 5. Letting the simulation time t=30, we can get that agent 1 triggers 84 times, agent 2 triggers 36 times, agent 3 triggers 40 times, and agent 4 triggers 92 times. If $\sigma_r = 0$ (time-triggered), there are 300 times transmitted. We can find clearly that the event-triggered control strategy (12) efficiently saves the network resources.

5. Conclusion

In this paper, we study the problem of H_∞ filtering for MASs with Markovian switching topologies. Considering the effect of switching topologies and network-induced delay, we adopt a reasonable event-triggered mechanism to reduce the amount of network transmission. By employing Lyapunov stability theory and LMI technique, some sufficient conditions are obtained which can ensure filtering error system to achieve mean square stable with an H_∞ norm bound. Finally, a numerical example is provided to show that the method we proposed is feasible and effective.

Data Availability

(i) The system parameters A, B, C1, C2, D used to support the findings of this study are included within the article. (ii) The external disturbance wi(t) used to support the findings of this study titled "Event-triggered H_∞ Filtering for Discrete-Time Neural Networks with Missing Measurements" is available from the corresponding author upon request. (iii) The initial states $x1(0)$, $x2(0)$, $x3(0)$, $x4(0)$ used to support the findings of this study are included within the article. (iv) The Laplacian matrices $L1$, $L2$ and probability transition matrix Π used to support the findings of this study are included within the article. (v) The filter parameters $Af1, Af2, Bf1, Bf2, Cf1, Cf2$ and event-triggered parameters $\Phi1$, $\Phi2$ used to support the findings of this study are derived by solving LMIs, which are included in the article.

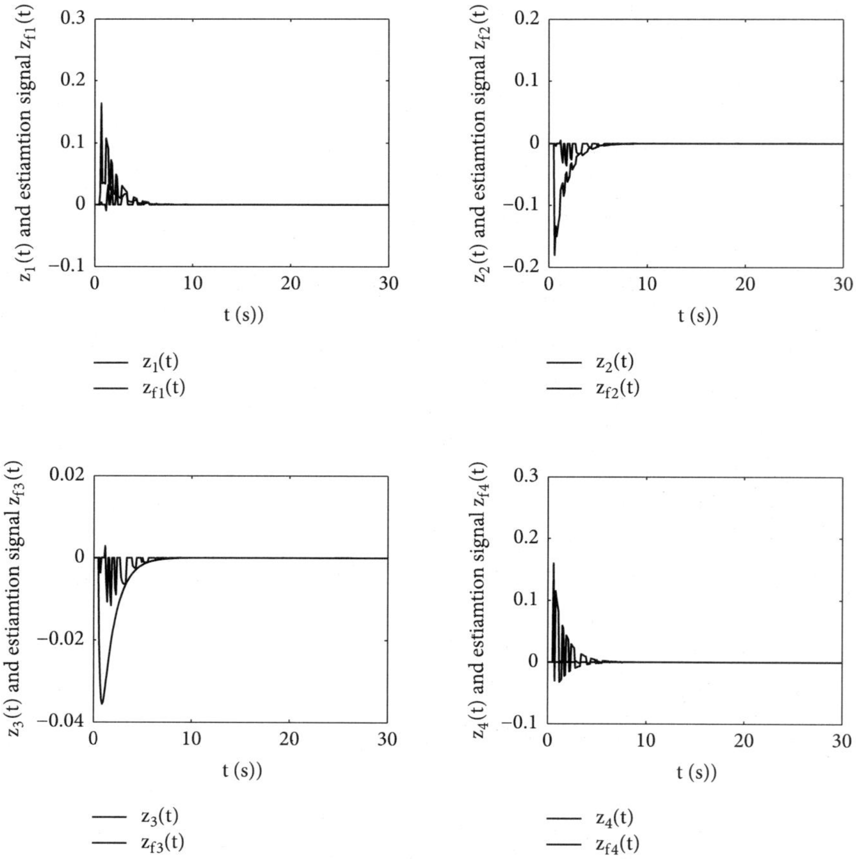

FIGURE 4: $z_i(t)$ and their estimation $z_{fi}(t)$.

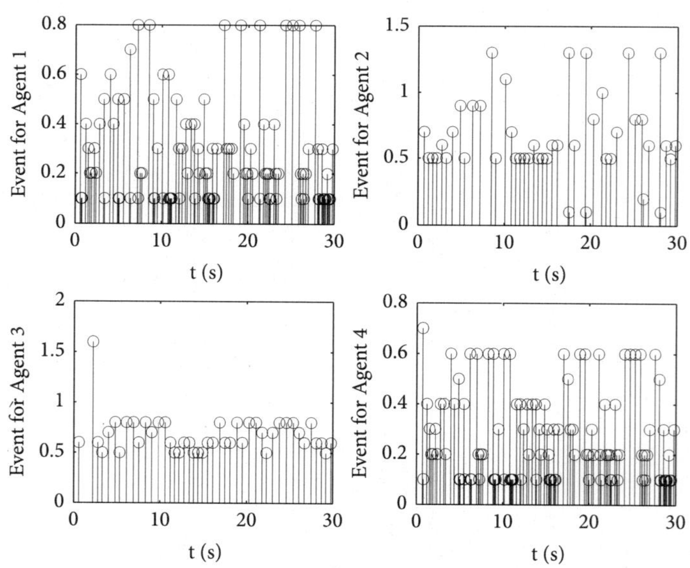

FIGURE 5: The release instants and release interval of each agent.

Conflicts of Interest

The authors declare that they have no conflicts of interest.

Acknowledgments

This work was supported by the National Natural Science Foundation of China (61374083 and 61703371).

References

[1] X. Ge and Q.-L. Han, "Distributed Formation Control of Networked Multi-Agent Systems Using a Dynamic Event-Triggered Communication Mechanism," *IEEE Transactions on Industrial Electronics*, vol. 64, no. 10, pp. 8118–8127, 2017.

[2] T. Han, Z. H. Guan, M. Chi, B. Hu, T. Li, and X. H. Zhang, "Multi-formation control of nonlinear leader-following multi-agent systems," *ISA Transactions®*, vol. 69, pp. 140–147, 2017.

[3] X. Ge and Q.-L. Han, "Distributed event-triggered H_∞ filtering over sensor networks with communication delays," *Information Sciences*, vol. 291, pp. 128–142, 2015.

[4] C. P. Bechlioulis and G. A. Rovithakis, "Decentralized robust synchronization of unknown high order nonlinear multi-agent systems with prescribed transient and steady state performance," *IEEE Transactions on Automatic Control*, vol. 62, no. 1, pp. 123–134, 2017.

[5] J. Gao, X. Xu, N. Ding, and E. Li, "Flocking motion of multi-agent system by dynamic pinning control," *IET Control Theory & Applications*, vol. 11, no. 5, pp. 714–722, 2017.

[6] L. Zhou and S. Li, "Distributed model predictive control for multi-agent flocking via neighbor screening optimization," *International Journal of Robust and Nonlinear Control*, vol. 27, no. 9, pp. 1690–1705, 2017.

[7] R. Olfati-Saber, "Distributed Kalman filtering for sensor networks," in *Proceedings of the 46th IEEE Conference on Decision and Control*, pp. 5492–5498, December 2007.

[8] R. Olfati-Saber, "Kalman-consensus filter: optimality, stability, and performance," in *Proceedings of the 48th IEEE Conference on Decision and Control Held Jointly with the 28th Chinese Control Conference (CDC/CCC '09)*, pp. 7036–7042, IEEE, Shanghai, China, December 2009.

[9] B. Shen, Z. Wang, and Y. S. Hung, "Distributed H_∞-consensus filtering in sensor networks with multiple missing measurements: the finite-horizon case," *Automatica*, vol. 46, no. 10, pp. 1682–1688, 2010.

[10] Y. Xu, H. Y. Su, Y. J. Pan, and Z. Wu, "Robust H_∞ filtering for networked stochastic systems with randomly occurring sensor nonlinearities and packet dropouts," *Signal Processing*, vol. 93, no. 7, pp. 1794–1803, 2013.

[11] X. Ge, Q. Han, D. Ding, X. Zhang, and B. Ning, "A survey on recent advances in distributed sampled-data cooperative control of multi-agent systems," *Neurocomputing*, vol. 275, pp. 1684–1701, 2018.

[12] R. Yang, P. Shi, G. P. Liu, and H. Gao, "Network-based feedback control for systems with mixed delays based on quantization and dropout compensation," *Automatica*, vol. 47, no. 12, pp. 2805–2809, 2011.

[13] Z. G. Wu, P. Shi, H. Y. Su, and J. Chu, "Network-based robust passive control for fuzzy systems with randomly occurring uncertainties," *IEEE Transactions on Fuzzy Systems*, vol. 21, no. 5, pp. 966–971, 2013.

[14] Y. Wu, R. Lu, P. Shi, H. Su, and Z. Wu, "Sampled-Data Synchronization of Complex Networks With Partial Couplings and T–S Fuzzy Nodes," *IEEE Transactions on Fuzzy Systems*, vol. 26, no. 2, pp. 782–793, 2018.

[15] Y. Wu, R. Lu, P. Shi, H. Su, and Z. Wu, "Analysis and Design of Synchronization for Heterogeneous Network," *IEEE Transactions on Cybernetics*, vol. 48, no. 4, pp. 1253–1262, 2018.

[16] Y. Wu and R. Lu, "Event-based control for network systems via integral quadratic constraints," *IEEE Transactions on Circuits and Systems I: Regular Papers*, vol. 65, no. 4, pp. 1386–1394, 2018.

[17] L. Zou, Z. Wang, H. Gao, and X. Liu, "Event-triggered state estimation for complex networks with mixed time delays via sampled data information: The continuous-time case," *IEEE Transactions on Cybernetics*, vol. 45, no. 12, pp. 2804–2815, 2015.

[18] J. Hu, Z. Wang, J. Liang, and H. Dong, "Event-triggered distributed state estimation with randomly occurring uncertainties and nonlinearities over sensor networks: a delay-fractioning approach," *Journal of The Franklin Institute*, vol. 352, no. 9, pp. 3750–3763, 2015.

[19] S. Hu and D. Yue, "Event-based H_∞ filtering for networked system with communication delay," *Signal Processing*, vol. 92, no. 9, pp. 2029–2039, 2012.

[20] J. Liu, J. Tang, and S. Fei, "Event-triggered H∞ filter design for delayed neural network with quantization," *Neural Networks*, vol. 82, pp. 39–48, 2016.

[21] S. Li, D. Sauter, and B. Xu, "Fault isolation filter for networked control system with event-triggered sampling scheme," *Sensors*, vol. 11, no. 1, pp. 557–572, 2011.

[22] H. Wang, P. Shi, and J. Zhang, "Event-triggered fuzzy filtering for a class of nonlinear networked control systems," *Signal Processing*, vol. 113, pp. 159–168, 2015.

[23] H. Wang, A. Xue, J. Wang, and R. Lu, "Event-based H∞ filtering for discrete-time markov jump systems with network-induced delay," *Journal of The Franklin Institute*, vol. 354, no. 14, pp. 6170–6189, 2017.

[24] H. Wang, D. Zhang, and R. Lu, "Event-triggered H∞ filter design for markovian jump systems with quantization," *Nonlinear Analysis: Hybrid Systems*, vol. 28, pp. 23–41, 2018.

[25] Z. G. Wu, P. Shi, H. Su, and J. Chu, "Asynchronous l_2-l_∞ filtering for discrete-time stochastic Markov jump systems with randomly occurred sensor nonlinearities," *Automatica*, vol. 50, no. 1, pp. 180–186, 2014.

[26] S. Xu, J. Lam, and X. Mao, "Delay-dependent H_∞ control and filtering for uncertain Markovian jump systems with time-varying delays," *IEEE Transactions on Circuits and Systems I: Regular Papers*, vol. 54, no. 9, pp. 2070–2077, 2007.

[27] L. X. Zhang, "H_∞ estimation for discrete-time piecewise homogeneous Markov jump linear systems," *Automatica*, vol. 45, no. 11, pp. 2570–2576, 2009.

[28] D. Viegas, P. Batista, P. Oliveira, C. Silvestre, and C. L. Chen, "Distributed state estimation for linear multi-agent systems with time-varying measurement topology," *Automatica*, vol. 54, pp. 72–79, 2015.

[29] D. Yue, E. Tian, and Q. Han, "A delay system method for designing event-triggered controllers of networked control systems," *IEEE Transactions on Automatic Control*, vol. 58, no. 2, pp. 475–481, 2013.

[30] K. Gu, "An integral inequality in the stability problem of time-delay systems," in *Proceedings of the 39th IEEE Confernce on Decision and Control*, pp. 2805–2810, Sydney, Australia, December 2000.

[31] P. G. Park, J. W. Ko, and C. Jeong, "Reciprocally convex approach to stability of systems with time-varying delays," *Automatica*, vol. 47, no. 1, pp. 235–238, 2011.

[32] H. Y. Shao, "New delay-dependent stability criteria for systems with interval delay," *Automatica*, vol. 45, no. 3, pp. 744–749, 2009.

17

Decentralized Optimization Algorithms for Variable Speed Pumps Operation based on Local Interaction Game

Shi-Qiang Wang, [1] Jian-Chun Xing, [1] Zi-Yan Jiang, [2] and Yun-Chuang Dai [2]

[1] *College of Defense Engineering, Army Engineering University of PLA, Nanjing, Jiangsu 210001, China*
[2] *Building Energy Research Center of Tsinghua University, Beijing 100084, China*

Correspondence should be addressed to Jian-Chun Xing; xjc@893.com.cn

Academic Editor: Benoit Iung

A fully distributed optimal control strategy for the parallel variable speed pumps in heating, ventilation, and air-conditioning (HVAC) systems is proposed. Compared with the traditional centralized method, the efficient control and coordination are scattered to every updated smart pump without the need for a monitoring host. Similar to the structure, mechanism, and characteristics of biological communities, a smart pump can communicate with adjacent nodes and operate collaboratively to complete pumps group operation with the least total power consumption under load demand and system constraints. And a decentralized optimization method that is decentralized estimation of distribution algorithm (DEDA) under local interaction games framework has been transplanted to the proposed structure to optimize the pumps working in parallel mode. Besides, convergence property of the two novel mechanisms is analyzed theoretically. Finally, simulation studies have been conducted based on the models of a physical pumps system, and the performance of the proposed algorithm is compared with centralized algorithms in terms of both effectiveness and stability. The results provide support for the validity of the proposed algorithms and control structure.

1. Introduction

A pumping system, which generally consists of several circulation pumps, is one of the main energy consuming parts in a chilled water based HVAC system. According to the International Energy Agency, electrical motors have consumed 46% of global electricity demand by end-use [1]. Furthermore, this rate is somewhat higher in the industrial area: up to almost 65% of generated electrical power [2]. And for the HVAC system, 60–70% of the power consumption is attributed to the water pumps and fan-based energy transmission in the water distribution system. These data has reflected the fact that there exists huge potential for energy conserving. And energy efficiency of a pumping system depends not only on the design but also, even more, on its operation. Thus an effective optimization should be proposed to improve the operating efficiency.

Currently, frequency converters are widely used with induction motor-driven pumps, enabling the pumps speed variation to control the head and flow rate of pumps under different working conditions in an energy-efficient way, that is, variable speed pumps (VSPs). However, it makes the moderating process more complicated in that not only the combination of pumps but also the frequency of each pump needs to be determined compared to the nonvariable-speed pumps.

On the one hand, the physical models of pumps operation are difficult to be designed accurately and the pumps unit commitment is always a mixed integer programming problem, which leads to conventional optimization methods difficult to be used. In this case, the majority of existing researches focus on adopting different kinds of heuristic algorithms to solve such problems. Kecebas et al. utilized the artificial neural network to optimize the geothermal district heating systems [3]. Edson Bortoni had used the dynamic programming algorithm for optimizing a complex pumping system with a set of parallel centrifugal pumps and the efficiency of the method is verified through experiment[4]. Wang et al. introduced an enhanced evolutionary algorithm for biobjective pump scheduling in water supply [5]. Arun Shankar has conducted a comprehensive review of different methods for energy efficiency enhancement initiatives in

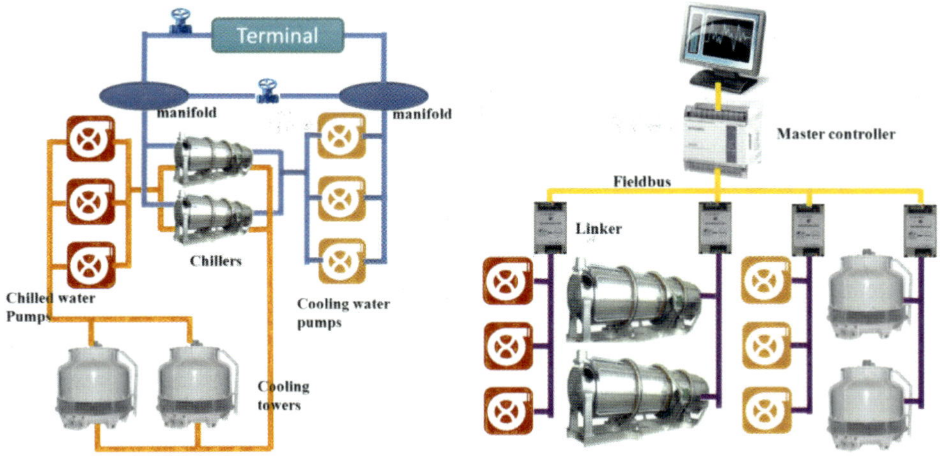

(a) Structure of the physical system (b) Structure of the centralized control system

FIGURE 1: Structure of the centralized control system of a typical HVAC system.

centrifugal pumping system in recent years, and several practical evolutionary methods were also introduced [6]. However, the theory analysis of heuristic algorithms is not easy so that its convergence characteristic has not been studied well. Therefore, a more valid solving method needs to be applied to the pumps control in some critical application.

On the other hand, in most recent studies, the keynote has been put on updating the existing optimization methods or developing more effective ones[7, 8]. However, most of the current algorithms are developed based on a centralized structure, in which all the sensors and actuators are connected to one master controller and the field bus technology is usually used to handle the communication. Figure 1 has shown a typical physical structure of an HVAC system (Figure 1(a)) and a corresponding centralized control system (Figure 1(b))[9]. The sensors and actuators of each separate equipment are gathered to the master controller level by level through the linkers (RS485/Ethernet <-> analog quantities) on the field bus.

From the engineering perspective, the centralized control structure is not consistent with the physical system, so a mapping mechanism between the two systems is needed, such as configuring and commissioning. Therefore, the development incurs high maintenance and labor costs and somehow diminishes the flexibility and universality of a construction.

After that the key procedure is to write the control model and algorithm into the monitoring center. The mathematical model for the total system is always complex and difficult to be obtained. In the algorithm writing process, a lot of work is needed such as system recognizing, software developing, and operating debugging. Moreover, traditional centralized methods require information transmission to the supervising computer during operations, which can cause severe link congestion and operational lag. Besides, since all the devices are managed by the master controller, malfunctions in one device will also bring about influence on other devices which will cause the entire control system running abnormally or even paralysis.

Recently, distributed multiagent control systems which are being focused on by many learners have provided a possible solution to this problem [10]. And Andrew Windham has analyzed the application prospects of the multiagent in HVAC control system[11]. However, the structures still require several local central nodes which are not fully distributed in our sense. And, in fact, the challenge is how to achieve the same performance as in a centralized manner where only local state information is available via each smart unit and is not considered in the above studies[12].

In this paper, a novel distributed control structure where each pump and chiller are updated to a smart unit by an embedded controller is proposed. These independent units can work independently and cooperate together to fulfill the global optimizing task.

To achieve the global optimization in a fully distributed way, a local interaction game is introduced into cognitive ratio networks in [13]. And using the log-linear algorithm[14–18], the network throughput is maximized. However, the log-linear algorithm is not applicable to the pumps optimal load distribution. Therefore, learning from the local state interaction games concept, a decentralized estimation of distribution algorithm (DEDA) is proposed in this paper. Simulation studies have been conducted based on physical models of the chilled water pumps system in a refrigeration station. The performance of DEDA is validated through comparison with centralized manner.

The rest of this paper is outlined as follows: In Section 2 the pumps system model is presented. Besides, potential game is introduced. Then a decentralized optimization model adaption is formulated in Section 3. And a decentralized algorithm is derived to solve the adapted mathematical model in this section. Section 4 focuses on performance assessment through simulations and experiments. Conclusions are drawn in Section 5.

2. System Model and Problem Formulation

2.1. Pumps Model and Its Decentralized Control System. The pump head and flow rate of a single VSP i (i=1,2,. . ., N) running at nonrated speed can be scaled from those of the pump running at the rated speed. According to the affinity laws of centrifugal pumps, the similarity working conditions of a VSP can be determined as

$$\frac{Q_i(n)}{Q_i(n_0)} = \frac{n}{n_0},$$
$$\frac{H_i(n)}{H_i(n_0)} = \left(\frac{n}{n_0}\right)^2, \quad (1)$$
$$\frac{\eta_i(n)}{\eta_i(n_0)} = 1$$

where n_0 is the rated pump speed and n is the pump speed under nonrated conditions. Q_i is the flow rate, H_i is the pump head of each branch, and η_i is the efficiency.

The speed ratio w_i for each pump i is described in (2) and w_0 is the lower limitation of the speed ratio.

$$w_i = \frac{n}{n_0} \quad (w_0 < w_i < 1) \quad (2)$$

And, the fitting formulas of pumps running at nonrated speed can be expressed as

$$H_{set} = H_i = a_i Q_i^2 + b_i w_i Q_i + c_i w_i^2, \quad i = 1, 2, \ldots, N, \quad (3)$$

$$\eta_i = j_i \left(\frac{Q_i}{w_i}\right)^2 + k_i \left(\frac{Q_i}{w_i}\right) + l_i, \quad i = 1, 2, \ldots, N, \quad (4)$$

$$W = \sum_{i=1}^{N} W_i = \sum_{i=1}^{N} \frac{\rho g H_i Q_i}{\eta_i}, \quad (5)$$

where H_{set} is the demanded pressure difference on the main pipe. W is the total energy consumption of the pumps and $W_i = \rho g H_i Q_i / \eta_i$. ρ is the density of water, g is the gravitational acceleration, N is the quantity of pumps. a_i, b_i, c_i and j_i, k_i, l_i are static physical characteristics of different branches for pump i (i=1,2,. . ., N).

If the detected pressure difference H_o is not equal to H_{set}, the control system will calculate the demanded total flow rate Q_p as shown in (6). After that the moderating process will be conducted to ensure the pressure difference by moderating w_i and make the calibration of each Q_i satisfy Q_p.

$$H_o = S Q_o, \quad (6)$$

$$H_{set} = S\left(\sum Q_i\right)^2 = S Q_p^2 \quad (7)$$

where S is the resistance of the current pipe net and H_o and Q_o are the current pressure difference and flow rate detected by the sensors.

In the largest pump systems, the pump groups are composed by several identical big pumps and one or two small pumps for energy saving. As shown in Figure 2, the decentralized control system consists of N smart updated pump by embedded microprocessors. A flow meter and a pressure meter will be connected to one of the smart units to detect the current state of the pipe net.

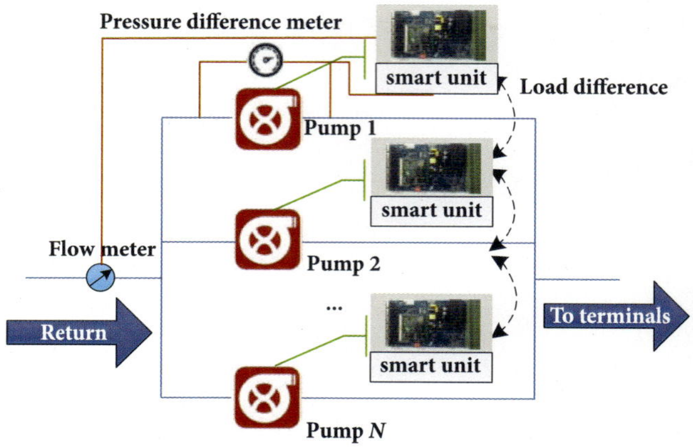

FIGURE 2: Decentralized control structure of a parallel pump system.

Each smart pump is connected hand to hand with its neighbor building up a chain-like form which corresponds to its actual engineering installation layout as shown in Figure 2. The model of each pump is stored in its smart unit as local variables and there is no need to model for the entire system as that in the centralized structure. And the control algorithm and instructions will be written into each smart pump and are identical. The load difference between supply and demand will be transmitted back and forth from one smart unit to another on the chain. And the operating state of each smart pump will be updated with the load difference between supply and demand transmitted from its neighbors. In such way, each smart device can be removed or substituted with little configuration making the system convenient to update and extend.

2.2. Problem Formulation. The operation object can be described as finding a combination of the running pumps and different pump can adopt a number of speed ratio strategies that can satisfy Q_p with the least energy consumption. Thus, it can be built up as

$$\begin{aligned} \min \quad & W \\ \text{s.t.} \quad & Q_i^{down} < Q_i < Q_i^{up}, \\ & \sum_{i=1}^{N} Q_i = Q_p \\ & i = 1, 2, \ldots, N \end{aligned} \quad (8)$$

where Q^i_{down} and Q^i_{up} denote the power outputs limit of pump i.

Implementation of the centralized model (8) requires a central processor with the total system information. The fully distributed approaches remove the requirement of monitoring host and enable each updated smart DG to complete optimal load distribution task only based on local and neighboring information. In this case, the centralized model (8) is split on each smart pump to be executed in a fully distributed and parallel manner across every processor core in each pump. Each part i (i=1, 2, . . ., N) is a suboptimization problem of (8) as decentralized optimization.

Definition 1 (decentralized optimization for pumps load distribution). The decentralized optimization problem for each node i (i=1, 2, . . ., N) is defined as

$$\min \quad W_i \tag{9}$$

$$\text{s.t.} \quad Q_i^{down} < Q_i < Q_i^{up},$$

$$\sum_{i=1}^{N} Q_i = Q_p \tag{10}$$

$$i = 1, 2, \ldots, N.$$

Indeed, (9) corresponds to part of the objective function (5). Note that each energy consumption W_i of the pump i (i=1, 2,. . ., N) is locally known by smart pump i only. The penalty term is constructed by load difference and added to the utility function as (11) for pump i (i=1, 2,. . ., N).

$$\begin{aligned} \min \quad U_i &= W_i + P_i = W_i + \beta_i \left(\sum_{i=1}^{N} Q_i - Q_p \right)^2 \\ &= W_i + \beta_i (\Delta Q)^2 \end{aligned} \tag{11}$$

where β is the penalty coefficient. The parameter β balances the tradeoff between exploration and exploitation. Smaller β implies that the smart pumps are more willing to choose a suboptimal action to explore, whereas higher β implies that they are prone to choose the best response action. Therefore, it is advisable that, at the beginning phase, the value of β is set to small number and it keeps increasing as the optimization iterates[19, 20]. ΔQ is the supply-demand difference. Then, (11) means that each pump i should select a certain speed ratio strategy from its speed space to satisfy Q_p with the least energy consumption.

Thus, the key problem lies in designing the mechanism to achieve global optimization via local interaction among nodes (the terms of agent, node, and smart pump refer to the same concept in this paper) only with direct communication links. As we all know, it is local interaction game. In this manner, optimal load distribution can be completed on the field level. In this paper, any method that solves (9)-(10) without using a central node aggregating data is called decentralized algorithm.

2.3. Potential Game. Coordination and combinatorial optimization, when viewed from a game theory perspective, can form a subset of the class of potential games.

Definition 2 (potential games, [21]). Player action sets $\{S_i\}_{i=1}^{N}$, and $\bigcup_{i=1}^{N} S_i = S$, with player objective functions; $\{U_i : S \longrightarrow R\}_{i=1}^{N}$ is a potential game if, for some potential function Φ: $S \longrightarrow R$,

$$\begin{aligned} U_i(s_i, s_{-i}) - U_i(s_i', s_{-i}) &= \phi(s_i, s_{-i}) - \phi(s_i', s_{-i}) \\ \forall s_i', s_{-i} &\in S_i \ \forall i \in \{1, 2, \ldots, N\} \end{aligned} \tag{12}$$

for every player $P_i \in P$ where P ={$P_1, \ldots, P_N$} is a group of players (or agents) and s_{-i} is the complimentary set of s_i.

And a number of learning algorithms can be found for the potential game. Most of the learning algorithms for potential games guarantee convergence to a (pure) Nash equilibrium.

Definition 3 (Nash equilibrium). A joint strategy profile, s^*, such that no individual agent has an incentive to change to a different strategy, that is,

$$\begin{aligned} U_i(s_i^*, s_{-i}^*) - U_i(s_i, s_{-i}^*) &\geq 0 \\ \forall s_i \in S_i \ \forall i &\in \{1, 2, \ldots, N\}, \end{aligned} \tag{13}$$

is a Nash equilibrium.

It is easy to see that, in potential games, any action profile maximizing the potential function is a pure Nash equilibrium. Hence, every potential game possesses at least one such equilibrium.

Consequently, each smart pump can be treated as a player and the energy consumption (11) is the utility function in potential game. And a potential game, as previously defined, requires perfect alignment between the global objective and the players' local objective functions in the following sense: if a player unilaterally changed its action, the change in its objective function would be equal to the change in the potential function. The pumps total cost function (7) is taken as the potential function in the optimization. Obviously, (9) subjected to (10) is a potential game.

However, there may also exist suboptimal pure Nash equilibrium that do not maximize the potential function[22]. Therefore, the key problem lies in how to design the mechanism to achieve the global optimization in a fully distributed way for the VSPs.

3. Decentralized Optimization Based on Probability Parallel Variable-Frequency Speed Pumps

In fact, there exist many evolutionary and swarm intelligent algorithms for coordination and combinatorial optimization in engineering[23, 24], but they might not be applicable to decentralized pumps control system due to a number of reasons. Firstly, the existing evolutionary or swarm intelligent algorithm refers to quite identical individualities. That is, every population in swarm intelligent algorithm or chromosomes in evolutionary algorithm involves a global variable, but only local information is available in the decentralized control structure of a parallel pump system. As an exploration, an example of multiple autonomous robots following simple local rules and collectively provided by [25] is quite

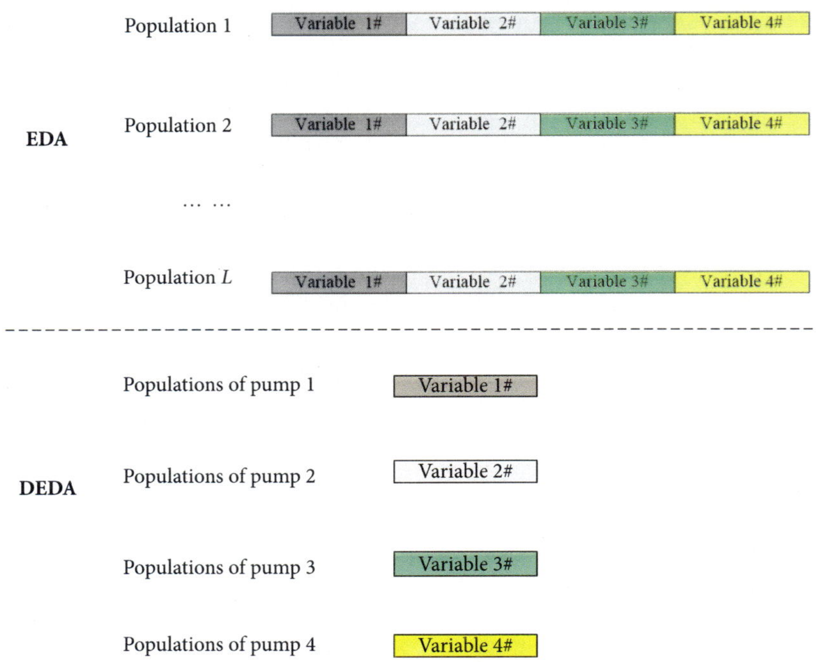

FIGURE 3: Comparison between EDA and decentralized optimization algorithm.

similar to the principles of the decentralized control system in this paper. However, the challenge is how to achieve the same performance as in a centralized manner where only local information is available.

3.1. Decentralized Estimation of Distribution Algorithm. Estimation of distribution algorithm (EDA) is an evolutionary optimization method that guides the search for the optimum by building and sampling explicit probabilistic models of promising candidate solutions[26]. Provided that a solution $\boldsymbol{x}$ has N factors$\boldsymbol{x}_i$, $i = 1, 2, \ldots, N$, the probability distribution of observing $\boldsymbol{x}$ is estimated by the product of density for each factor as follows:

$$P(\boldsymbol{x}) = \prod_{i=1}^{N} P_{\theta_i}(\boldsymbol{x}_i), \tag{14}$$

where θ_i is the vector of parameters for the probability density of factor $\boldsymbol{x}_i$.

The EDA is adapted into decentralized fashion in this study. DEDA is different from traditional evolutionary or swarm intelligent algorithms, where the populations of each smart pump only cover its own variable as shown in Figure 3.

And the detailed description of DEDA is illustrated in Figure 4. Further, from the potential game theory perspective, an adjustment mechanism can typically produce sequential or parallel actions by agents (smart pump). Observed by the characteristic of decentralized parallel pumps control structure, sequential random schedule with local interaction is adopted in the present study. And an arbitrary agent can be picked as the indicator in the pumps optimal load distribution.

3.2. Convergence Proof. The convergence proof of the DEDA is presented in Proposition 4 and illustrated in the following paragraph.

Proposition 4. *The solution $x^{de} = [x_1^{de}, \ldots, x_i^{de}, \ldots, x_N^{de}]$ for the decentralized optimization (11) generated by DEDA can converge to the optimal Nash equilibrium with a probability of 1.*

Proof. Assume the global solution space is defined as $\Omega = \Omega_1 \cap \Omega_2 \cap \ldots \cap \Omega_N$, where Ω_i, $i = 1, 2, \ldots, N$, is the space of the *ith* agent and its corresponding variable x_i. An elite-preserving operator is used in the presented algorithm. This operator favors the g elites of a population by giving them an opportunity to be directly carried over to the next generation. Thus, the new population consists of $(L - g)$ new chromosomes, with $g < L$.

Firstly, the optimality convergence of each agent i ($i = 1, 2, \ldots, N$) is discussed. The present study assumes that the search space Ω_i of problem (14) is a closed and bounded region. The objective function is taken as the fitness function $J_i(x_i)$, and the optimal solution set is expressed as

$$D_0 = \{x_i \in \Omega_i \mid |J_i(x_i) - J_i^*| < \varepsilon\} \tag{15}$$

where $J_i^* = \min\{J_i(x_i): x_i \in \Omega_i\}$.

For agent i (i=1, 2,..., N), the L chromosomes of EDA can be divided into the following properties:

(I) At least one chromosome belongs to D_0, which is written as S_0.

(II) All L chromosomes belong to D_1, which is written as S_1, where

$$D_1 = \Omega \setminus D_0. \tag{16}$$

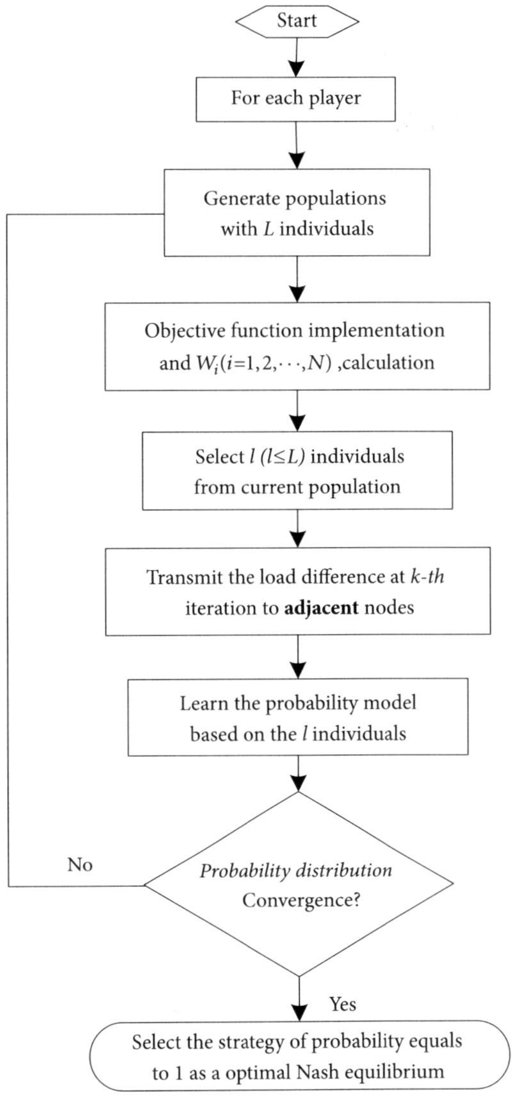

FIGURE 4: Flow chart of the DEDA in each smart pump and executed in serial mode.

And if q_{ij} $(i,j = 0,1)$ represents the probability that the $(k+1)th$ generation $x_i(k+1)$ transfers to S_j when the k-th generation $x_i(k)$ keeps the state of S_i, then the following conditions hold:

(I) $x_i(k)$ at S_0 and $q_{00}=1$.

Condition (I) is obviously valid because the elite-preserving operator is used in this paper.

(II) $x_i(k)$ at S_1 and $q_{11} \le c$, where $c \in (0,1)$.

Conditions (II) is analyzed as follows: For any $\varepsilon > 0$, if $x_{i0} \in \{x_i \in \Omega_i \mid |J_i(x_i) - J_i^*| < \varepsilon/2\}$, there exists $r > 0$ satisfying

$$\left|J_i(x_i) - J_i(x_{i0})\right| < \frac{\varepsilon}{2} \tag{17}$$

when

$$\left\|x_i - x_{i0}\right\|_\infty = \max\left|x_i - x_{i0}\right| \le r. \tag{18}$$

Define

$$Q_{x_{i0},r} = \left\{x_i \in \Omega_i \mid \left\|x_i - x_{i0}\right\|_\infty \le r\right\}, \tag{19}$$

so that

$$Q_{x_{i0},r} \subset D_0 \tag{20}$$

due to

$$\begin{aligned}\left|J_i(x_i) - J_i^*\right| &\le \left|J_i(x_i) - J_i(x_{i0})\right| + \left|J_i(x_{i0}) - J_i^*\right| \\ &< \frac{\varepsilon}{2} + \frac{\varepsilon}{2} = \varepsilon.\end{aligned} \tag{21}$$

The probability that x_i transfers to S_0 can be defined by

$$P\left\{x_{i+} \in Q_{x_{i0},r}\right\} = P\left\{\left|x_{i+} - x_{i0}\right| \le r\right\}, \tag{22}$$

where x_{i+} is the chromosome of a new generation and $i = 1, 2, \ldots, N$. Given that the evolutionary computation of EDA is based on probability, each element of the chromosome generates a series of the probability function $g(x_i)$, with $i = 1, 2, \ldots, N$.

For x_{i+},

$$x_{i+} \sim g(x_i) \tag{23}$$

and thus

$$P\left\{x_{i+} \in Q_{x_{i0},r}\right\} = \int_{x_{i0}-r}^{x_{i0}+r} g(x_i)\,dx_i. \tag{24}$$

For simplicity, $P\{x_{i+} \in Q_{xi0,r}\}$ can be denoted by

$$P_1(x_{i+}) = P\left\{x_{i+} \in Q_{x_{i0},r}\right\}, \tag{25}$$

and then

$$0 < P_1(x_{i+}) < 1. \tag{26}$$

Define

$$P_1(y_{i0}) = \min P_1(x_{i+}), \tag{27}$$

so that

$$P_1(y_{i0}) \le P_1(x_{i+}) \le q_{10}. \tag{28}$$

As

$$q_{11} + q_{10} = 1, \tag{29}$$

the following result can be achieved:

$$q_{11} = 1 - q_{10} \le c \quad (0 < c < 1), \tag{30}$$

where the constant c is

$$c = 1 - P_1(y_{i0}). \tag{31}$$

Further, for any $\varepsilon > 0$, set

$$p_k = P\left\{\left|J_i(x_i^*(k)) - J_i^*\right| \ge \varepsilon\right\} \tag{32}$$

and

$$p_k = \begin{cases} 0 & \exists T \in \{0,1,2,\ldots,k\},\ x_i^*(T) \in D_0 \\ \overline{p_k}, & x_i^*(m) \notin D_0\ \ m = 0,1,2,\ldots,k, \end{cases} \tag{33}$$

TABLE 1: Pumps parameters.

PAR	PUMP			
	1.KP1020-3 406	2.KP1020-3 406	3.KP1020-3 457	4.KP1020-3 457
a	-6.07e-6	-5.36e-6	-8.54e-6	-8.41e-6
b	-0.00822	-0.00949	-0.00048	-0.00046
c	60.41767	60.61789	74.54932	73.55836
j	-4.08e-7	-3.96e-7	-3.87e-7	-3.82e-7
k	0.000940	0.000911	0.000999	0.000985
l	0.316357	0.328377	0.249898	0.259137

and, thus, the following equation is obtained:

$$\overline{p_k} = P\{x_i^*(m) \notin D_0,\ m = 0, 1, 2, \ldots, k\} = {q_{11}}^k \le c^k, \quad (34)$$

so that

$$\sum_{k=1}^{\infty} \overline{p_k} \le \sum_{k=1}^{\infty} c^k = \frac{c}{1-c} < \infty. \quad (35)$$

Then the following theorem is introduced to complete the proof process.

Theorem 5 (Borel-Cantelli theorem). *The present study assumes that $A_1, A_2, \ldots$ is the event sequence based on probability and defines $p_k = P\{A_k\}$. If*

$$\sum_{k=1}^{\infty} p_k < \infty, \quad (36)$$

then

$$P\left\{\bigcap_{m=1}^{\infty} \bigcup_{k\ge m} A_k\right\} = 0. \quad (37)$$

If

$$\sum_{k=1}^{\infty} p_k = \infty \quad (38)$$

and A_k are independent of one another, then

$$P\left\{\bigcap_{m=1}^{\infty} \bigcup_{k\ge m} A_k\right\} = 1. \quad (39)$$

□

According to Theorem 5,

$$P\left\{\bigcap_{m=1}^{\infty} \bigcup_{k\ge m} \left[\left|J_i\left(x_i^*(k)\right) - J_i^*\right| \ge \varepsilon\right]\right\} = 0 \quad (40)$$

is true. Therefore, $x_i^t(m)$, where $t = 1, 2, \ldots, L$, is the population of the *mth* generation of EDA for agent i (i=1,2,. . ., N), which consists of L chromosomes. If the fitness function $J_i(x_i)$ is continuous on the bounded region Ω_i, the optimal solution $x_i^*(m)$ converges to the optimum for agent i with a probability of 1, and

$$x_i^*(m) = \arg\min_{1\le t\le L} J_i\left(x_i^t(m)\right). \quad (41)$$

Then, the validity of the global optimization is deduced in the subsequent paragraphs.

According to the above analysis, the solution x_i^{de} can converge to the optimum x_i^{*de} for agent i with a probability of 1. Define the probability as p_i, that is,

$$x_i^{de} \xrightarrow{p_i} x_i^{*de}, \quad (42)$$

and

$$p_i = P\left\{\bigcap_{m=1}^{\infty} \bigcup_{k\ge m} \left[\left|J\left(x_i^{de}(k)\right) - J_i^*\right| < \varepsilon\right]\right\} = 1. \quad (43)$$

According to (24), the global combined probability density can be described as

$$P = \prod_{i=1}^{N} p_i = 1. \quad (44)$$

For the coupled variables, a similar result is also obtained:

$$P = \frac{\left(\prod_{i=1}^{N} p_i\right)}{p_{ij}} = 1. \quad (45)$$

Then the present study directly obtains

$$x^{de} \xrightarrow{P} x^{*de} \quad (46)$$

where $x^{*de} = [x_1^{*de}, \ldots, x_i^{*de}, \ldots, x_N^{*de}]$ is a Nash equilibrium of problem (11). In conclusion, x^{de} converges to an optimal solution of the centralized optimization problem with a probability of 1 as the iteration step $m \longrightarrow \infty$. Then Proposition 4 can be obtained as follows.

4. Case Study

4.1. Experimental Setup. This case study is carried out to validate the performance of DOABP by applying it to the model of a physical pump system under different load demands. The physical parameters (PAR) corresponding to performance curves of the pumps are shown in Table 1, and these parameters are recognized from the practical operating data of a district freezing station. As seen from Table 1, the performances of pumps of the same type are close to each other. The pump system is composed of four pumps working

TABLE 2: Pumps discreted strategy space.

SR	PUMP			
	1.KP1020-3 406	2.KP1020-3 406	3.KP1020-3 457	4.KP1020-3 457
1	0	0	0	0
2	0.93	0.94	0.825	0.825
3	0.94	0.945	0.85	0.85
4	0.945	0.95	0.875	0.875
5	0.95	0.955	0.9	0.9
6	0.96	0.96	0.915	0.915
7	0.97	0.97	0.925	0.925
8	0.98	0.98	0.95	0.95
9	0.99	0.99	0.975	0.975
10	1	1	1	1

Note. The speed ratio of 0 means that the pump is off.

in parallel mode including two big pumps (KP1020-3 457mm in impeller diameter) and two smaller pumps (KP1020-3 406mm in impeller diameter) manufactured by Grundfos. A determined pressure difference control is applied here where the designed pressure difference and flow rate demand for the entire system are 45m and 3234 m^3/h, respectively.

4.2. Simulation Result. The strategy space is discretized, and it becomes a limited strategy space to ensure that the potential game is a finite game. And, the speed ratio (SR) of each pump, as selected action set, is shown in Table 2.

In the case study, the proposed algorithms will be triggered as follows: when the pipe net has been changed, the smart unit will detect that the pressure difference H_o is not equal to H_{set} and Q_o which can be got through the flow meter on the trunk; thus Q_i can be solved in (3).

DEDA can be downloaded to microprocessor chip in each smart pump i (i=1,2, 3,4), and then each smart pump can be active. And, each smart pump can communicate and cooperate with other smart pump nodes in serial mode. The corresponding results by the proposed DEDA are shown in Figure 5.

As shown in Figure 5, the probability distribution (PD) on each pump is dynamic during the iterative process in DEDA. And the optimal solution of each pump fluctuates and finally reaches a new equilibrium state under mutual negotiation and adjustment. By about 60 iteration steps, the pumps optimal load distribution task can be achieved. It can be seen that the probability of the optimal output flow rate tends to approach 1. Conversely, the probability of the other solution space approaches 0.

The iterative process of each pump i(i=1,2,3,4) cost is illustrated in Figure 6. Further, the iterative process of total energy cost is shown in Figure 7.

Comparing Figure 6 and Figure 7, we find that although not all the costs of the four pumps decrease, the total cost decreases during the iterative process. This can validate the effectiveness of DEDA. For a comparison, the iterative process of EDA in centralized method is also illustrated in Figure 8. As shown in Figures 5 and 8, the DEDA can achieve the actual optima which are equivalent to that obtained with the centralized algorithm.

Further, a more detailed calculation result is shown in Table 3, where EDA and DEDA are adopted, respectively.

5. Conclusion

respectivelyMost of the optimizing methods for parallel pump systems in HVAC system are based on a centralized structure under which all the management tasks have to be settled with one master controller. In this structure, the computation efficiency and stability are relatively low in that all the data and inspection have to be processed by one controller and that one malfunctioning device will influence the whole system. Besides, the control strategy is a case-by-case one which cannot be transplanted easily to other systems.

In this study, a decentralized control strategy has been applied to the parallel pump systems in which the traditional master controllers have been substituted by a list of independent smart units embedded in each pump.

(i) A new fully distributed mathematical model for pumps optimal distribution and decentralized flat architecture are presented. With the rapid development of the electronic industry, smart hardware has been widely utilized in different fields. According to the vision of the decentralized method, traditional units can be upgraded to smart agents by incorporating a microcontroller chip inside. The decentralized algorithm can be written into the chip. The smart unit communicates with neighboring nodes such that they can work collaboratively to achieve the optimal load distribution task. In this case, the complicated onsite modeling, configuration, commissioning, and other secondary developing work can be simplified to the wiring of communication connection among different nodes because the smart unit can self-organize and plug-and-play.

(ii) Decentralized optimization is used to solve pumps optimal load distribution under potential game framework. And, to achieve effectiveness of global goals, DEDA using local interaction is introduced. The convergence property of the novel method is also analyzed theoretically. Simulation

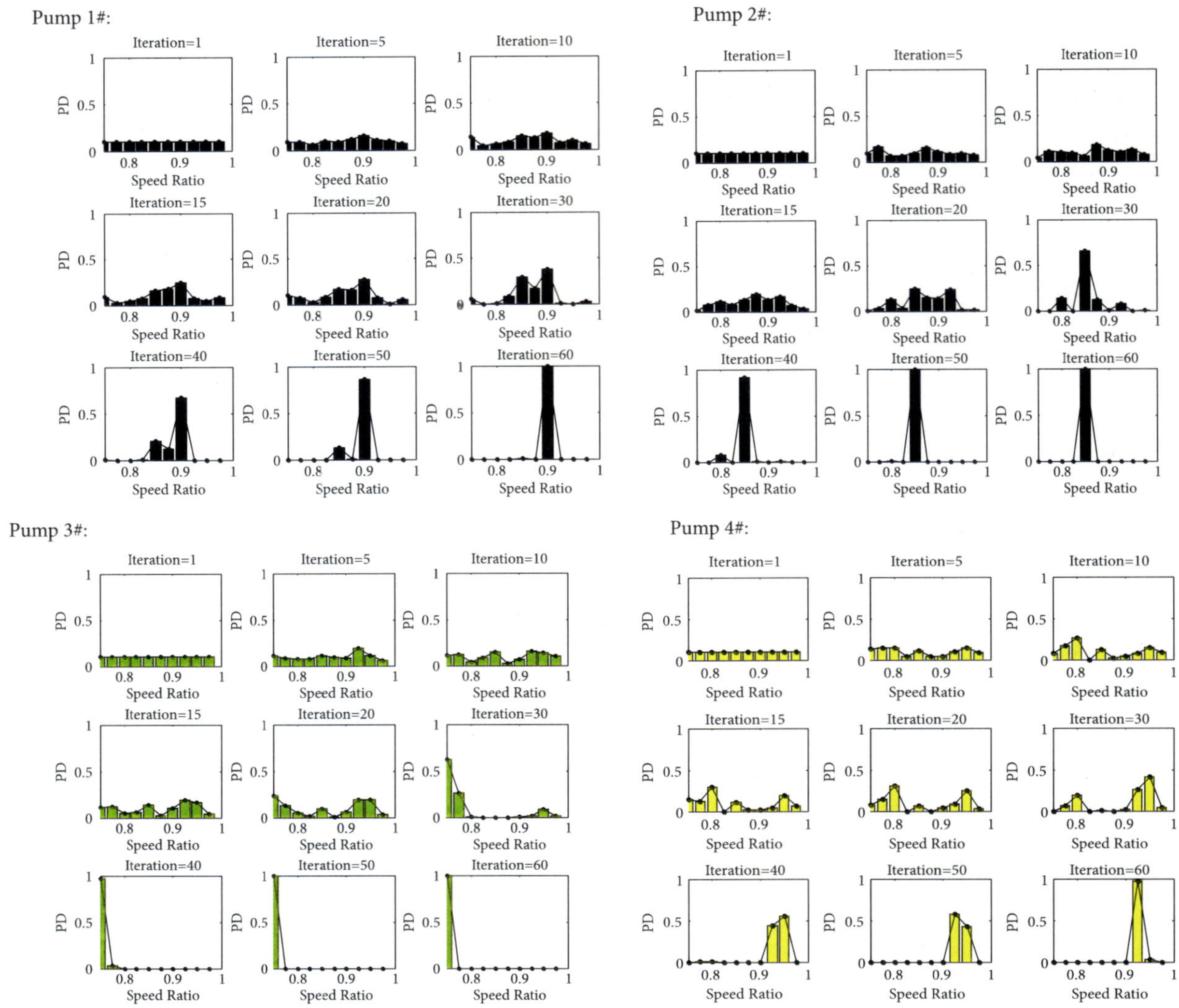

FIGURE 5: Iterative process in decentralized optimization.

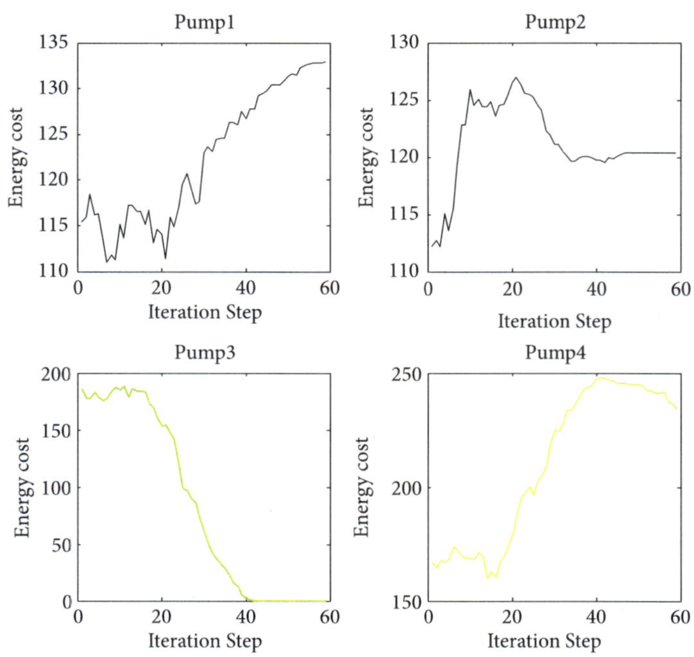

FIGURE 6: Iterative process of cost on each pump in decentralized optimization.

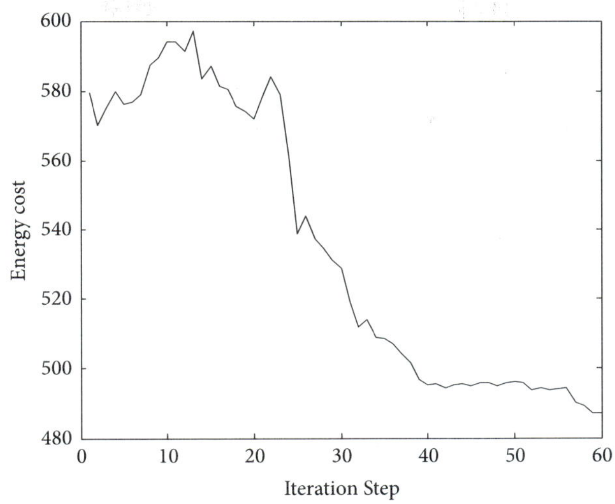

FIGURE 7: Iterative process of total cost in decentralized optimization.

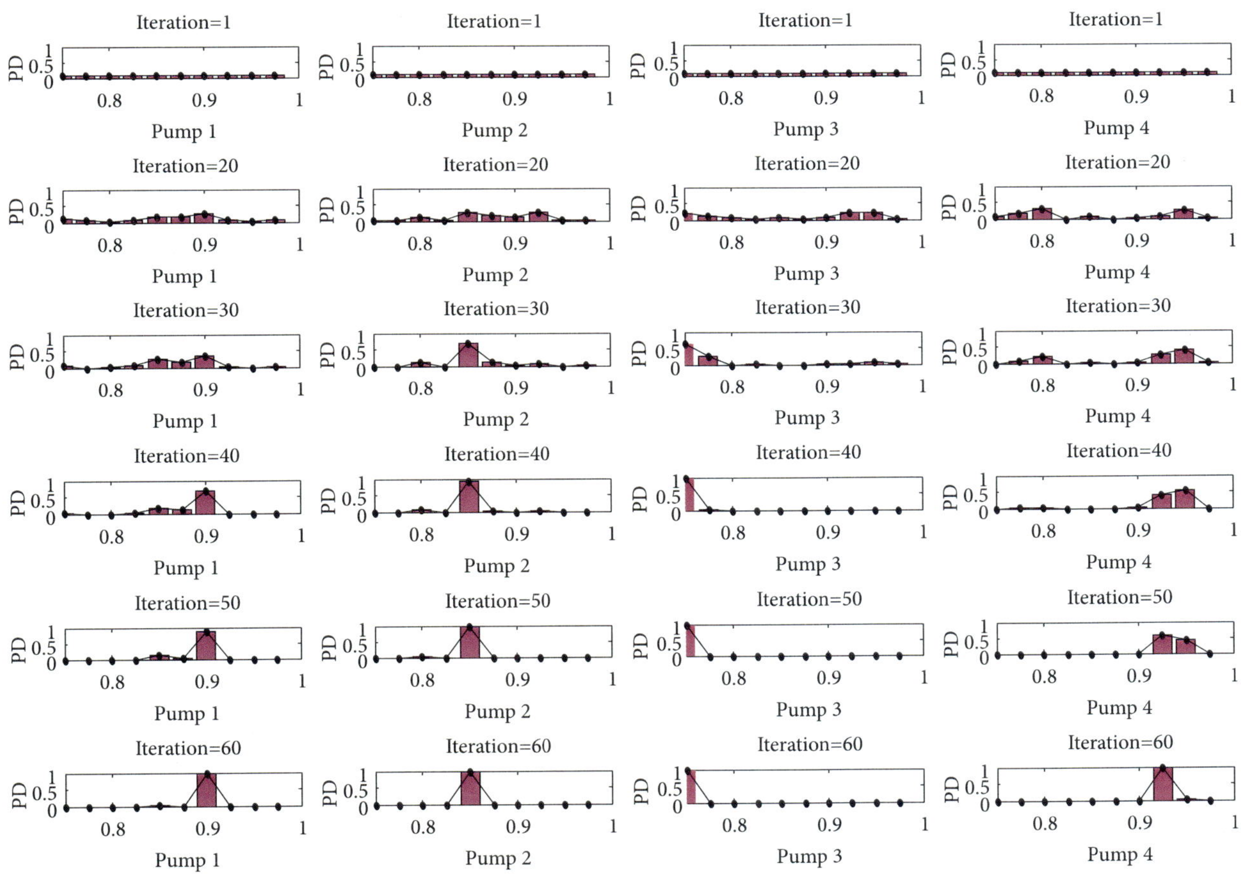

FIGURE 8: Iterative process in centralized optimization.

results presented in the preceding paragraph indicate that the proposed method can achieve optimal load distribution.

This scenario is extremely motivating for further investigation in the field of distributed networked control in other large-scale interconnected nonlinear systems. And, there are still several potential issues of the presented method that needed to be emphasized here. Firstly, the proposed mechanism is executed in sequence. Therefore, the proposed distributed algorithms should in principle be applicable to chain-like networks. Each smart pump is connected hand to hand with its neighbor, but the meshed network should also be considered. Secondly, future work will consider the hardware test and implementation on practical engineering.

TABLE 3: Comparison between centralized and decentralized results.

#Working condition	DEDA			EDA		
Demand: 1496.9 m^3/h	$n(Hz)$	$Q\ (m^3/h)$	η	$n(Hz)$	$Q\ (m^3/h)$	η
Pump 1	47.00	700.17	0.79	47.00	700.17	0.79
Pump 2	0.00	0.00	0.00	0.00	0.00	0.00
Pump 3	41.25	796.71	0.85	41.25	796.71	0.85
Pump 4	0.00	0.00	0.00	0.00	0.00	0.00
Power/kW	227.4804			227.4804		
Difference	0.0169			0.0169		
Demand: 2583.4m^3/h	$n(Hz)$	$Q\ (m^3/h)$	η	$n(Hz)$	$Q\ (m^3/h)$	η
Pump 1	48.50	886.98	0.83	48.50	886.98	0.83
Pump 2	0.00	0.00	0.00	0.00	0.00	0.00
Pump 3	0.00	0.00	0.00	0.00	0.00	0.00
Pump 4	48.75	1695.44	0.82	48.75	1695.44	0.82
Power/kW	392.2045			392.2045		
Difference	0.9772			0.9772		
Demand: 2952.9m^3/h	$n(Hz)$	$Q\ (m^3/h)$	η	$n(Hz)$	$Q\ (m^3/h)$	η
Pump 1	47.25	733.01	0.80	48.00	827.24	0.82
Pump 2	49.00	926.41	0.84	47.50	745.89	0.80
Pump 3	0.00	0.00	0	40.00	0.00	0
Pump 4	45.00	1292.65	0.89	45.75	1379.76	0.88
Power/kW	435.5667			439.1761		
Difference	0.8373			0.0181		
Demand: 3234.7m^3/h	$n(Hz)$	$Q\ (m^3/h)$	η	$n(Hz)$	$Q\ (m^3/h)$	η
Pump 1	50.00	1054.53	0.85	48.50	886.98	0.83
Pump 2	47.75	777.35	0.81	47.70	777.34	0.81
Pump 3	45.75	1402.09	0.87	0.00	0.00	0.00
Pump 4	40.00	0.00	0.00	47.50	1569.19	0.84
Power/kW	475.8191			485.6048		
Difference	0.7344			1.1869		
Demand: 3858.3m^3/h	$n(Hz)$	$Q\ (m^3/h)$	η	$n(Hz)$	$Q\ (m^3/h)$	η
Pump 1	47.00	700.17	0.79	48.50	886.98	0.83
Pump 2	49.00	926.41	0.84	50.00	1037.53	0.85
Pump 3	41.25	796.71	0.85	41.25	796.71	0.85
Pump 4	46.25	1435.66	0.87	43.75	1136.63	0.89
Power/kW	572.8274			561.5738		
Difference	0.6517			0.4537		
Demand: 4901.2 m^3/h	$n(Hz)$	$Q\ (m^3/h)$	η	$n(Hz)$	$Q\ (m^3/h)$	η
Pump 1	49.00	944.62	0.84	48.00	827.24	0.82
Pump 2	49.00	926.41	0.84	50.00	1037.51	0.85
Pump 3	50.00	1831.72	0.78	46.25	1456.97	0.86
Pump 4	46.25	1435.66	0.87	50.00	1815.93	0.79
Power/kW	778.3465			777.4647		
Difference	0.4889			1.2384		

Conflicts of Interest

The authors declare that there are no conflicts of interest regarding the publication of this paper.

Acknowledgments

This work is supported by National Key Research and Development Program of China No. 2017YFC0704100 (entitled New Generation Intelligent Building Platform Techniques). And the authors are grateful for the support provided by the Tsinghua University-UTC Research Center for Integrated Building Energy, Safety, and Control Systems.

References

[1] P. Waide and C. U. Brunner, *Energy-efficiency policy opportunities for electric motor-driven systems*, Iea Energy Papers, 2011.

[2] P. Olszewski, "Genetic optimization and experimental verification of complex parallel pumping station with centrifugal pumps," *Applied Energy*, vol. 178, pp. 527–539, 2016.

[3] A. Keçebaş and I. Yabanova, "Thermal monitoring and optimization of geothermal district heating systems using artificial neural network: A case study," *Energy and Buildings*, vol. 50, pp. 339–346, 2012.

[4] E. da Costa Bortoni, R. A. Almeida, and A. N. C. Viana, "Optimization of parallel variable-speed-driven centrifugal pumps operation," *Energy Efficiency*, vol. 1, no. 3, pp. 167–173, 2008.

[5] J.-Y. Wang, T.-P. Chang, and J.-S. Chen, "An enhanced genetic algorithm for bi-objective pump scheduling in water supply," *Expert Systems with Applications*, vol. 36, no. 7, pp. 10249–10258, 2009.

[6] V. K. Arun Shankar, S. Umashankar, S. Paramasivam, and N. Hanigovszki, "A comprehensive review on energy efficiency enhancement initiatives in centrifugal pumping system," *Applied Energy*, vol. 181, pp. 495–513, 2016.

[7] Y.-F. Wang and Q. Chen, "A direct optimal control strategy of variable speed pumps in heat exchanger networks and experimental validations," *Energy*, vol. 85, pp. 609–619, 2015.

[8] Z. Tianyi, Z. Jili, and M. Liangdong, "On-line optimization control method based on extreme value analysis for parallel variable-frequency hydraulic pumps in central air-conditioning systems," *Building and Environment*, vol. 47, no. 1, pp. 330–338, 2012.

[9] H. Yu, T. Zhao, and J. Zhang, "Development of a distributed artificial fish swarm algorithm to optimize pumps working in parallel mode," *Science and Technology for the Built Environment*, pp. 1–11, 2017.

[10] A. Viehweider and S. Chakraborty, "Multi-Agent based Distributed Optimization Architecture for Energy-Management Systems with Physically Coupled Dynamic Subsystems," *IFAC-PapersOnLine*, vol. 48, no. 30, pp. 251–256, 2015.

[11] A. Windham and S. Treado, "A review of multi-agent systems concepts and research related to building HVAC control," *Science and Technology for the Built Environment*, vol. 22, no. 1, pp. 50–66, 2015.

[12] Y. C. Dai, Z. Y. Jiang, Q. Shen et al., "A decentralized algorithm for optimal distribution in HVAC systems," *Building & Environment*, vol. 95, no. 2016, pp. 21–31, 2016.

[13] Y. Xu, J. Wang, Q. Wu, A. Anpalagan, and Y.-D. Yao, "Opportunistic spectrum access in cognitive radio networks: global optimization using local interaction games," *IEEE Journal of Selected Topics in Signal Processing*, vol. 6, no. 2, pp. 180–194, 2012.

[14] J. R. Marden, G. Arslan, and J. S. Shamma, "Regret based dynamics: Convergence in weakly acyclic games," in *Proceedings of the the 6th international joint conference*, pp. 194–201, Honolulu, Hawaii, May 2007.

[15] J. R. Marden, G. Arslan, and J. S. Shamma, "Joint strategy fictitious play with inertia for potential games," *Institute of Electrical and Electronics Engineers Transactions on Automatic Control*, vol. 54, no. 2, pp. 208–220, 2009.

[16] H. P. Young, *Individual Strategy and Social Structure*, Princeton University Press, Princeton, NJ, USA, 1998.

[17] C. H. Sun and H. B. Duan, "An evolutionary potential game theoretic approach for the K-cover problem in multi-UAV sensor networks," *Scientia sinica Technologica*, vol. 46, no. 10, pp. 1016–1023, 2016 (Chinese).

[18] Q. Zhang and H. Mühlenbein, "On the convergence of a class of estimation of distribution algorithms," *IEEE Transactions on Evolutionary Computation*, vol. 8, no. 2, pp. 127–136, 2004.

[19] Y. Song, C. Zhang, and Y. Fang, "Joint channel and power allocation in wireless mesh networks: A game theoretical perspective," *IEEE Journal on Selected Areas in Communications*, vol. 26, no. 7, pp. 1149–1159, 2008.

[20] M. V. Dolgopolik, "Smooth exact penalty functions: a general approach," *Optimization Letters*, vol. 10, no. 3, pp. 635–648, 2016.

[21] D. Monderer and L. S. Shapley, "Potential games," *Games and Economic Behavior*, vol. 14, no. 1, pp. 124–143, 1996.

[22] J. R. Marden, G. Arslan, and J. S. Shamma, "Cooperative control and potential games," *IEEE Transactions on Systems, Man, and Cybernetics, Part B: Cybernetics*, vol. 39, no. 6, pp. 1393–1407, 2009.

[23] G. Chen, L. Liu, P. Song, and Y. Du, "Chaotic improved PSO-based multi-objective optimization for minimization of power losses and *L* index in power systems," *Energy Conversion and Management*, vol. 86, pp. 548–560, 2014.

[24] N. Radhakrishnan, S. Srinivasan, R. Su, and K. Poolla, "Learning-Based Hierarchical Distributed HVAC Scheduling With Operational Constraints," *IEEE Transactions on Control Systems Technology*, 2017.

[25] J. Werfel, K. Petersen, and R. Nagpal, "Designing collective behavior in a termite-inspired robot construction team," *Science*, vol. 343, no. 6172, pp. 754–758, 2014.

[26] P.-Y. Yin and H.-L. Wu, "Cyber-EDA: estimation of distribution algorithms with adaptive memory programming," *Mathematical Problems in Engineering*, vol. 2013, Article ID 132697, 11 pages, 2013.

18

Decoupling Control Strategy of BLIM considering Rotor Mass Eccentricity

Wenshao Bu[1] **Biao Li,**[2] **and Chunxiao Lu**[2]

[1]*Electrical Engineering College, Henan University of Science and Technology, Luoyang, 471023, China*
[2]*Information Engineering College, Henan University of Science and Technology, Luoyang, 471023, China*

Correspondence should be addressed to Wenshao Bu; wsbu@163.com

Academic Editor: Carlos-Andrés García

A BLIM, i.e., bearingless induction motor, is a multivariable, nonlinear, and strong coupling object; to achieve its high performance magnetic suspension operation control and overcome the influence of the rotor mass eccentricity, a decoupling control strategy considering the rotor mass eccentricity is proposed. Firstly, the mathematical model of the torque system based on the rotor flux orientation and the mathematical model of the magnetic suspension system based on the air gap flux orientation are presented; on this basis, the inverse system decoupling control method of the BLIM is researched. Then, according to the frequency characteristics of the unbalanced displacement, an unbalance vibration compensator is designed, which can generate a compensation force to suppress or eliminate the unbalanced displacement. Simulation experimental results have shown that the decoupling control among the rotor flux-linkage, motor speed, and two radial displacement components can be achieved; in the steady state, the unbalanced displacement can be basically eliminated, while, during the mutation process of motor speed, the unbalanced displacement can be suppressed effectively.

1. Introduction

The motor supported by the mechanical bearings cannot meet the need of high-speed and long-time operation [1, 2]. Then the motor supported by the magnetic bearing is widely developed [3, 4], but it still has some disadvantages, such as the more power consumption for magnetic suspension and the overspeed difficulty [5]. The bearingless motor is a new type of AC motor, which is proposed based on the structure similarity between the magnetic bearing and the common AC motor stator [3, 4]. In the common bearingless motor, there are two sets of stator windings, i.e., torque windings (pole-pair number p_1, angle frequency ω_1 of current) and suspended windings (pole-pair number p_2, angle frequency ω_2 of current). When the two sets of stator windings meet the qualifications of "$p_1 = p_2 \pm 1$" and "$\omega_1 = \omega_2$," a stable and controllable magnetic suspension force can be generated. Figure 1 shows the schematic diagram of the generation principle of the controllable magnetic suspension force [6]. The bearingless motor has a series of advantages, such as no friction, no mechanical noise, and higher critical speed; then it is more suitable for high-speed operation, and it has become a research hotspot [5–12].

The bearingless induction motor (BLIM) is a multivariable, nonlinear, and strong coupling object. To realize its high performance control, it is necessary to achieve the dynamic decoupling between the controlled variables. As for a nonlinear system, the inverse system method is an effective one that can be used for the liberalization decoupling [12, 13]. In addition, to improve the control performance of the torque system, the rotor flux orientation is usually adopted. About the inverse system decoupling control strategy of the BLIM based on the rotor flux orientation, there has been some research progress [13–15]. But for the magnetic suspension system, the air gap flux-linkage of the torque system is required in the model of magnetic suspension force. In the coordinate system orientated by the rotor flux-linkage of the torque system, the air gap flux-linkage components of the torque system are time-varying, which will inevitably lead to the calculation complexity of the suspension control current.

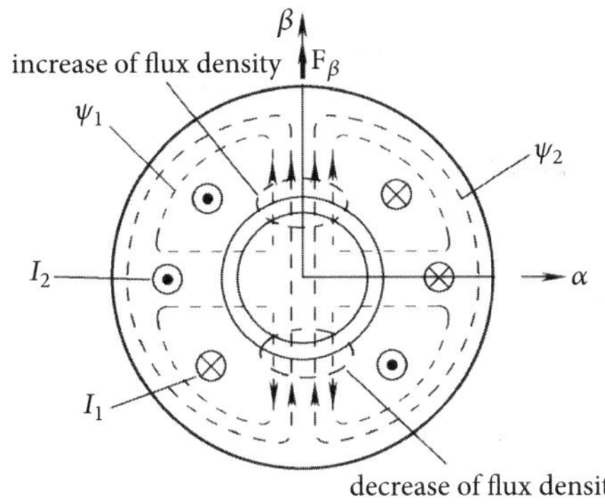

FIGURE 1: Schematic diagram of controllable magnetic suspension force generation.

In this paper, the rotor flux orientation is still applied to the torque system; but in the magnetic suspension system, the magnetic suspension force will be controlled based on the synthetic air gap flux-linkage of the torque system; the inverse system decoupling will be applied to the overall BLIM system; the required air gap flux-linkage of the torque system will be obtained by an independent air gap flux-linkage observer.

In addition, due to the uneven material, machining accuracy, assembly error, and other reasons, the mass eccentricity of rotor is unavoidable, and, during the rotor rotation, an unbalanced excitation force acting on the rotor is generated. The unbalanced excitation force has the same angle frequency with the rotational speed, which will lead to the unbalance vibration [16–19]. In view of the uncertainties of the mass eccentric direction and the mass eccentricity, it is difficult to establish an accurate model of unbalanced excitation force, and then it is difficult to solve the unbalance vibration problem by an analytic inverse system method. But the unbalanced excitation force will generate an unbalanced displacement. In this paper, the radial displacement would be divided into two parts, i.e., the random displacement and the unbalanced displacement; the inverse system decoupling control strategy is adopted to control the random displacement, and a vibration compensation controller is adopted to suppress or eliminate the unbalanced displacement. Simulation experimental results have shown that the decoupling control between the motor speed, rotor flux-linkage, and two radial displacement components can be realized; the unbalanced displacement can be effectively suppressed.

2. Mathematical Model of BLIM

2.1. Model of Torque System

Definition. d-q is the rotor flux orientation coordinate system of the torque system. Then in the d-q coordinate system, the motion equation and the rotor flux-linkage equation can be expressed as follows:

$$\dot{\omega}_r = \frac{p_1 L_{m1}}{J L_r}\psi_r i_{s1q} - \frac{1}{J}T_L \quad (1)$$

$$\dot{\psi}_r = -\frac{1}{T_r}\psi_r + \frac{L_{m1}}{T_r}i_{s1d} \quad (2)$$

$$\theta_1 = \int (p_1\omega_r + \omega_s) = \int \left(p_1\omega_r + \frac{L_{m1}}{T_{r1}\psi_{r1}}i_{s1q}\right), \quad (3)$$

where ω_r is the mechanical angular velocity of rotor; ω_1 is the electrical angular velocity of the d-q coordinate system; ψ_r and θ_1 are the amplitude and space position angle of the rotor flux-linkage vector; i_{s1d} and i_{s1q} are the current components of torque windings along the d- and q-coordinate axes; L_{m1} is the equivalent two-phase mutual inductance of the torque system; T_r is the rotor time constant; $T_r = L_r/R_r$; L_r is the rotor's self-inductance and R_r is the rotor resistance; J is the inertia moment; p_1 is the pole-pair number of torque windings; T_L is the load torque.

2.2. Model of Magnetic Suspension System. The controllable magnetic suspension force acting on the rotor is the interaction result between the air gap flux-linkage of the torque system and that of the magnetic suspension system. In the decoupling calculation of the magnetic suspension system, the air gap flux-linkage information of the torque system is required. Defining α-β as the two-phase stationary orthogonal coordinate system, its coordinate origin locates at the center of stator. Then the air gap flux-linkage components of the torque system in the α and β direction, i.e., $\psi_{1\alpha}$ and $\psi_{1\beta}$, can be expressed as follows:

$$\begin{aligned}\psi_{1\alpha} &= \int \left(u_{s1\alpha} - R_{s1}i_{s1\alpha}\right)dt - L_{s1l}i_{s1\alpha}\\ \psi_{1\beta} &= \int \left(u_{s1\beta} - R_{s1}i_{s1\beta}\right)dt - L_{s1l}i_{s1\beta},\end{aligned} \quad (4)$$

where $u_{s1\alpha}$ and $u_{s1\beta}$ are the voltage components of torque windings along the α- and β-coordinate axes; $i_{s1\alpha}$ and $i_{s1\beta}$ are the current components of torque windings along the α- and β-coordinate axes; R_{s1} is the stator resistance of torque windings; L_{s1l} is the leakage inductance of torque windings.

Definition. M-T is the air gap flux orientation coordinate system of the torque system. Then the amplitude ψ_{1m} and space position angle θ_2 of the air gap flux-linkage of the torque system can be expressed as follows:

$$\psi_{1m} = \sqrt{\psi_{1\alpha}{}^2 + \psi_{1\beta}{}^2} \quad (5)$$

$$\theta_2 = \arccos\left(\frac{\psi_{1\alpha}}{\psi_{1m}}\right). \quad (6)$$

According to the air gap flux-linkage amplitude of the torque system, the controllable magnetic suspension force components in the static α-β coordinate system can be controlled by the suspension current components i_{s2m} and i_{s2t}; the specific models can be expressed as follows [18]:

$$\begin{aligned}F_\alpha &= K_m\psi_{1m}\cdot i_{s2m}\\ F_\beta &= -K_m\psi_{1m}\cdot i_{s2t},\end{aligned} \quad (7)$$

where F_α and F_β are the controllable magnetic suspension force components along the α- and β-coordinate axes; K_m

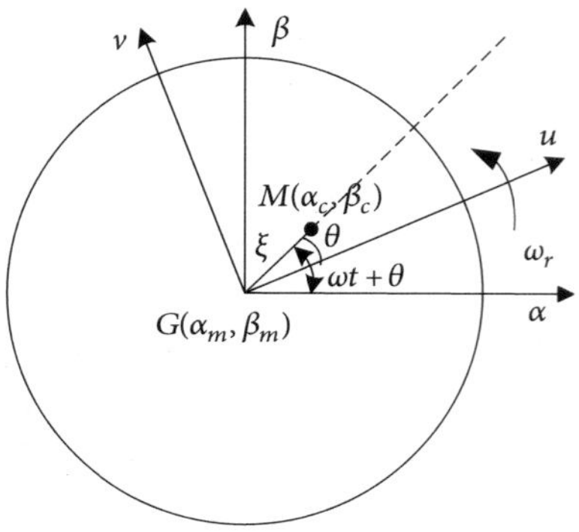

FIGURE 2: Diagram of rotor mass eccentricity.

is the magnetic suspension force coefficient determined by the BLIM structure; i_{s2m} and i_{s2t} are the control current components of magnetic suspension force along the M- and T-coordinate axes.

When the rotor deviates from the stator axis, the unilateral electromagnetic pull will be generated, and the specific expressions are as follows:

$$\begin{aligned} f_{s\alpha} &= k_s \alpha, \\ f_{s\beta} &= k_s \beta, \end{aligned} \tag{8}$$

where α and β are radial displacement components; k_s is the radial displacement stiffness coefficient; $f_{s\alpha}$ and $f_{s\beta}$ are the unilateral electromagnetic pull components along the α- and β-coordinate axes, respectively.

Figure 2 presents the rotor mass eccentricity diagram. In Figure 2, u-v is the rotor synchronous rotating coordinate system; M is the centroid of rotor; $G(\alpha_m, \beta_m)$ is the axis of rotor; ξ is the eccentric distance of rotor mass, θ is the initial mass eccentricity angle in the u-v coordinate system.

During the rotation of the mass eccentric rotor, an unbalanced excitation force F_a is generated in the mass eccentricity direction, whose angular frequency is the same with the angular velocity of rotor. The unbalanced excitation force components along the α- and β-coordinate axes can be expressed as follows [20]:

$$\begin{aligned} F_{a\alpha} &= m\xi\omega^2 \cos(\omega_r t + \theta) \\ F_{a\beta} &= m\xi\omega^2 \sin(\omega_r t + \theta), \end{aligned} \tag{9}$$

where m is the rotor mass.

The rotor's radial suspension motion equation can be expressed as follows:

$$\begin{aligned} m\ddot{\alpha} &= F_\alpha + F_{a\alpha} + f_{s\alpha} \\ m\ddot{\beta} &= F_\beta + F_{a\beta} + f_{s\beta}. \end{aligned} \tag{10}$$

3. Control System of BLIM

The decoupling control based on the inverse system method requires the analytical model of the BLIM system. But it is difficult to measure the eccentric distance ξ and the initial eccentricity angle θ of rotor mass. But the research results show that the unbalanced excitation force will generate periodic unbalanced displacement, and it has no effect on the random displacement [12, 13]. The unbalanced displacement components along the α- and β-coordinate axes are sine and cosine signals, respectively, whose frequencies are the same with the angular velocity of rotor, and they can be expressed as follows:

$$\begin{aligned} \alpha_m &= A\cos(\omega_r t + \varphi) \\ \beta_m &= A\sin(\omega_r t + \varphi), \end{aligned} \tag{11}$$

where "A" is the amplitude of unbalanced displacement, and it is proportional to the square of motor speed; φ is the initial phase angle of unbalanced displacement.

In this paper, the radial displacement is divided into a random displacement component and an unbalanced displacement component. Then by the vibration compensation controller, a vibration compensation force is generated, which is used to suppress or eliminate the unbalanced displacement. As for the random displacement, an inverse system decoupling control strategy is adopted.

3.1. Inverse System Decoupling Control. Select the system state variable x, input variable u, and output variables y as follows:

$$x = (x_1, x_2, x_3, x_4, x_5, x_6)^T = (\psi_{r1}, \omega, \alpha, \beta, \dot{\alpha}, \dot{\beta})^T \tag{12}$$

$$u = (u_1, u_2, u_3, u_4)^T = (i_{s1d}, i_{s1q}, i_{s2m}, i_{s2t})^T \tag{13}$$

$$y = (y_1, y_2, y_3, y_4)^T = (\psi_{r1}, \omega, \alpha, \beta)^T. \tag{14}$$

Substituting (12), (13), and (14) into (1), (2), and (10), then the state equations of the BLIM system can be derived as follows:

$$\begin{aligned} \dot{x}_1 &= -\frac{x_1}{T_{r1}} + \frac{L_{m1}u_1}{T_{r1}} \\ \dot{x}_2 &= \frac{p_1 L_{m1} x_1 u_2}{(JL_{r1})} - \frac{T_L}{J} \\ \dot{x}_3 &= x_5 \\ \dot{x}_4 &= x_6 \\ \dot{x}_5 &= \frac{K_m \psi_{1m} u_3}{m} + \frac{k_s x_3}{m} \\ \dot{x}_6 &= -\frac{K_m \psi_{1m} u_4}{m} + \frac{k_s x_4}{m}. \end{aligned} \tag{15}$$

where the rotor flux orientation is adopted for the torque system; based on the air gap flux-linkage of the torque system, the magnetic suspension force and the radial suspension motion of the rotor are controlled by the control current components of the magnetic suspension system, and the required air gap flux-linkage information is obtained in real time from (4)~(6).

In order to analyze the reversibility of the BLIM system, according to the Interactor algorithm, each output variable y_i (i=1, 2, 3, 4) is asked derivative to time, until the input variables are included in the derivative expression. The specific processes can be expressed as follows:

$$\dot{y}_1 = \dot{x}_1 = -\frac{x_1}{T_{r1}} + \frac{L_{m1}u_1}{T_{r1}} \tag{16}$$

$$\dot{y}_2 = \dot{x}_2 = \frac{p_1 L_{m1} x_1 u_2}{(JL_{r1})} - \frac{T_L}{J} \tag{17}$$

$$\dot{y}_3 = \dot{x}_3 = x_5$$
$$\ddot{y}_3 = \dot{x}_5 = \frac{K_m \psi_{1m} u_3}{m} + \frac{k_s x_3}{m} \tag{18}$$

$$\dot{y}_4 = \dot{x}_4 = x_6$$
$$\ddot{y}_4 = \dot{x}_6 = -\frac{K_m \psi_{1m} u_4}{m} + \frac{k_s x_4}{m}. \tag{19}$$

Setting

$$Y(u) = (\dot{y}_1, \dot{y}_2, \ddot{y}_3, \ddot{y}_4)^T \tag{20}$$

then the Jacobi matrix of the BLIM system can be derived as follows:

$$A(x,u) = \frac{\partial Y}{\partial u^T} = \begin{pmatrix} \frac{L_{m1}}{T_{r1}} & 0 & 0 & 0 \\ 0 & \frac{p_1 L_{m1} x_1}{JL_{r1}} & 0 & 0 \\ 0 & 0 & \frac{K_m \psi_{1m}}{m} & 0 \\ 0 & 0 & 0 & -\frac{K_m \psi_{1m}}{m} \end{pmatrix}. \tag{21}$$

In the normal operation of the BLIM system, $x_1 = \psi_r \neq 0$, $\psi_{1m} \neq 0$. Then $\text{rank}[A(x,u)] = 4$, and $A(x,u)$ is nonsingular. The relative order $\alpha = (\alpha_1, \alpha_2, \alpha_3, \alpha_4) = (1,1,2,2)$, while $\sum \alpha_i = 6 = n$ (i=1, 2, 3, 4). And then the BLIM system is reversible.

Select the input of inverse system as follows:

$$v = (\dot{y}_1, \dot{y}_2, \ddot{y}_3, \ddot{y}_4)^T = (v_1, v_2, v_3, v_4)^T. \tag{22}$$

Substituting (22) into (16)~(19), then the inverse system model of the BLIM can be derived as follows:

$$u_1 = \frac{T_{r1}}{L_{m1}} v_1 + \frac{1}{L_{m1}} x_1,$$
$$u_2 = \frac{L_{r1}}{p_1 L_{m1} x_1} (J v_2 + T_L) \tag{23}$$

$$u_3 = \frac{(m v_3 - k_s x_3)}{K_m \psi_{1m}},$$
$$u_4 = -\frac{(m v_4 - k_s x_4)}{K_m \psi_{1m}}. \tag{24}$$

From (23) and (24), it can be seen that when the magnetic suspension system is controlled based on the air gap flux-linkage of the torque system, it only needs to calculate the air gap flux-linkage of the torque system in real time; in addition, there is no obvious coupling between the inverse model of the torque system and that of the magnetic suspension system; this facilitates the "independent inverse system decoupling" between the torque system and the suspension system.

By connecting the inverse system in series before the original BLIM system, the torque system is dynamically decoupled into two first-order linear integral subsystems, including a rotor flux-linkage subsystem and a motor speed subsystem; meanwhile the magnetic suspension system is dynamically decoupled into two second-order linear integral subsystems, including the α random displacement component subsystem and the β random displacement component subsystem. Then by adding an appropriate regulator for each subsystem, the dynamic decoupling control among the rotor flux-linkage, motor speed, and α and β random displacement components can be realized.

3.2. Compensation Control of Unbalance Vibration. The radial displacement components caused by the mass eccentricity along the α- and β-coordinate axes are sine and cosine signals, whose frequencies are the same with the angular speed of rotor. From these characteristics, the extraction of unbalanced displacement and the compensation of unbalanced excitation force can be realized. The processes can be summed up as follows: Firstly, define u-v as the synchronous coordinate system rotating with the rotor. From the rotor synchronous rotation transformation, the radial displacement components measured in the α-β coordinate system are transformed to those in the u-v coordinate system. Then in the u-v coordinate system, the DC variable signals of unbalanced displacement, i.e., u_m and v_m, are extracted by relevant low pass filters. Secondly, in the u-v coordinate system, the given signals of unbalanced displacement components are set to zero; then through the closed-loop controls of u_m and v_m, the given signals of vibration compensation force components in the u-v coordinate system, i.e., F_{cu}^* and F_{cv}^*, are obtained. Thirdly, through the reverse rotating transformation for F_{cu}^* and F_{cv}^*, the given signals of vibration compensation force components in the α-β coordinate system, i.e., $F_{c\alpha}^*$ and $F_{c\beta}^*$, are obtained.

Figure 3 shows the schematic diagram of unbalance vibration compensation controller, where α and β are the measured radial displacements; u_m and v_m are the unbalanced displacement components in the u-v coordinate system; F_{cu}^* and F_{cv}^* are the vibration compensation force DC signals in the u-v coordinate system; F_{ca}^* and $F_{c\beta}^*$ are the vibration compensation force AC signals in the α-β coordinate system.

Because the output variables of the BLIM inverse system are current signals, the "force/current" transformation is needed to convert the vibration compensation force to the vibration compensation current. After the vibration compensation current is superposed on the control current of random displacement, the integrated control for the random and unbalanced displacements can be realized.

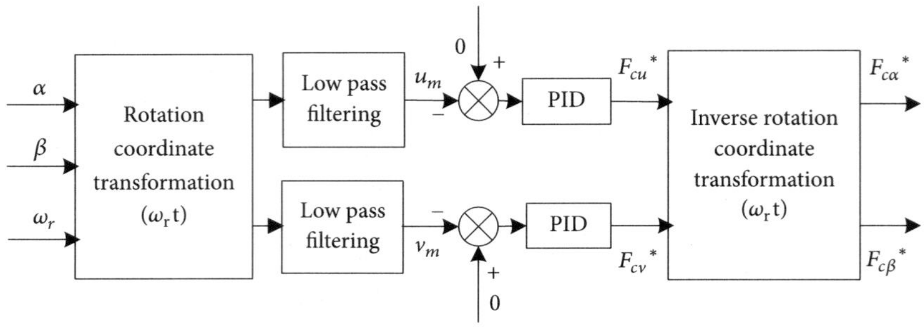

FIGURE 3: Unbalance vibration compensation controller.

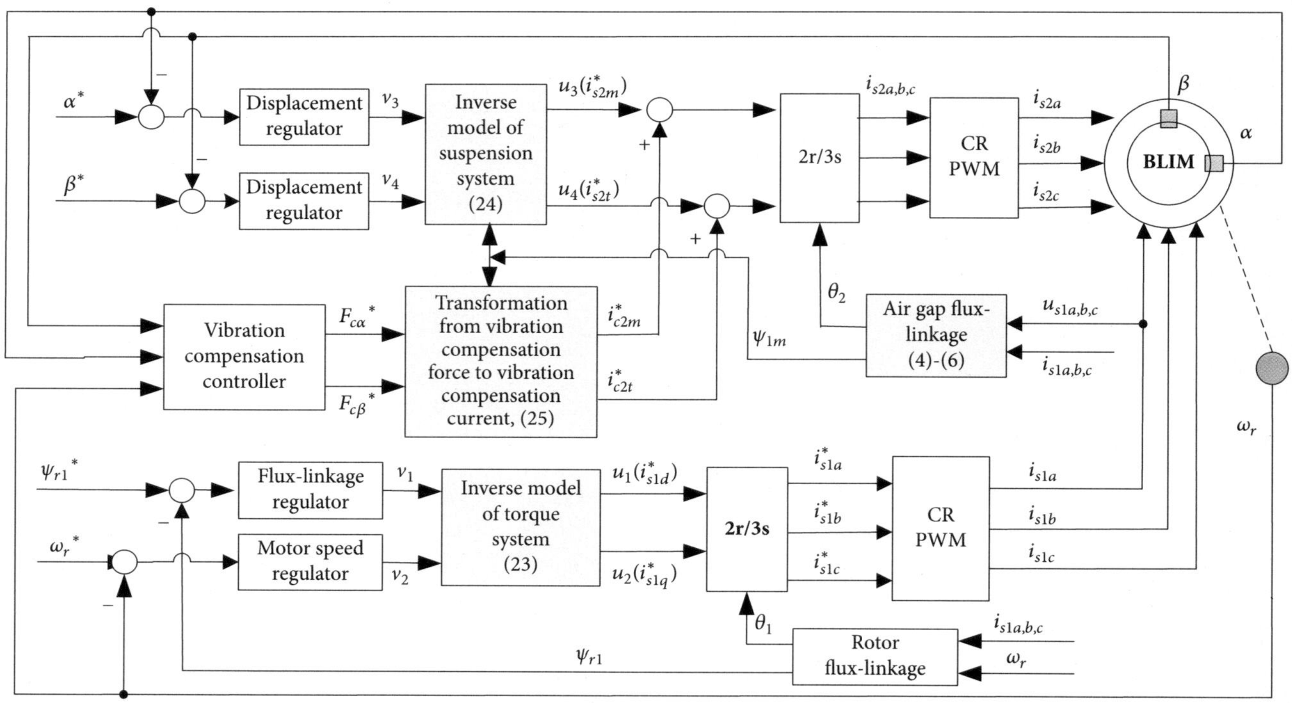

FIGURE 4: Unbalance vibration control system of bearingless induction motor.

Replacing F_α and F_β in (7) with $F_{c\alpha}{}^*$ and $F_{c\beta}{}^*$, then the analytical formula of vibration compensation current can be derived as follows:

$$i^*_{c2m} = \frac{F^*_{c\alpha}}{K_m \Psi_{1m}},$$
$$i^*_{c2t} = -\frac{F^*_{c\beta}}{K_m \Psi_{1m}}. \qquad (25)$$

In (25), $i^*{}_{c2m}$ and $i^*{}_{c2t}$ are the given signals of the vibration compensation current components along the m- and t-coordinate axes.

Figure 4 is the schematic diagram of the unbalance vibration compensation control system.

4. Simulation Verification of BLIM Control System

4.1. Decoupling Control Performance of BLIM System. According to Figure 4, a four-pole BLIM with two-pole suspension windings is taken as the object; system simulation is carried out by Matlab/Simulink. In order to verify the decoupling performance between the motor speed, rotor flux-linkage, and two displacement components, the influence of rotor unbalance vibration is temporarily ignored; the given values of related variables are suddenly changed at different times: the given rotor flux-linkage is reduced from 0.95 Wb to 0.5 Wb at 0.5 s; the given motor speed is increased to 3000 r/min from 1500 r/min at 0.75 s; the given α displacement is increased to 0.03 mm from

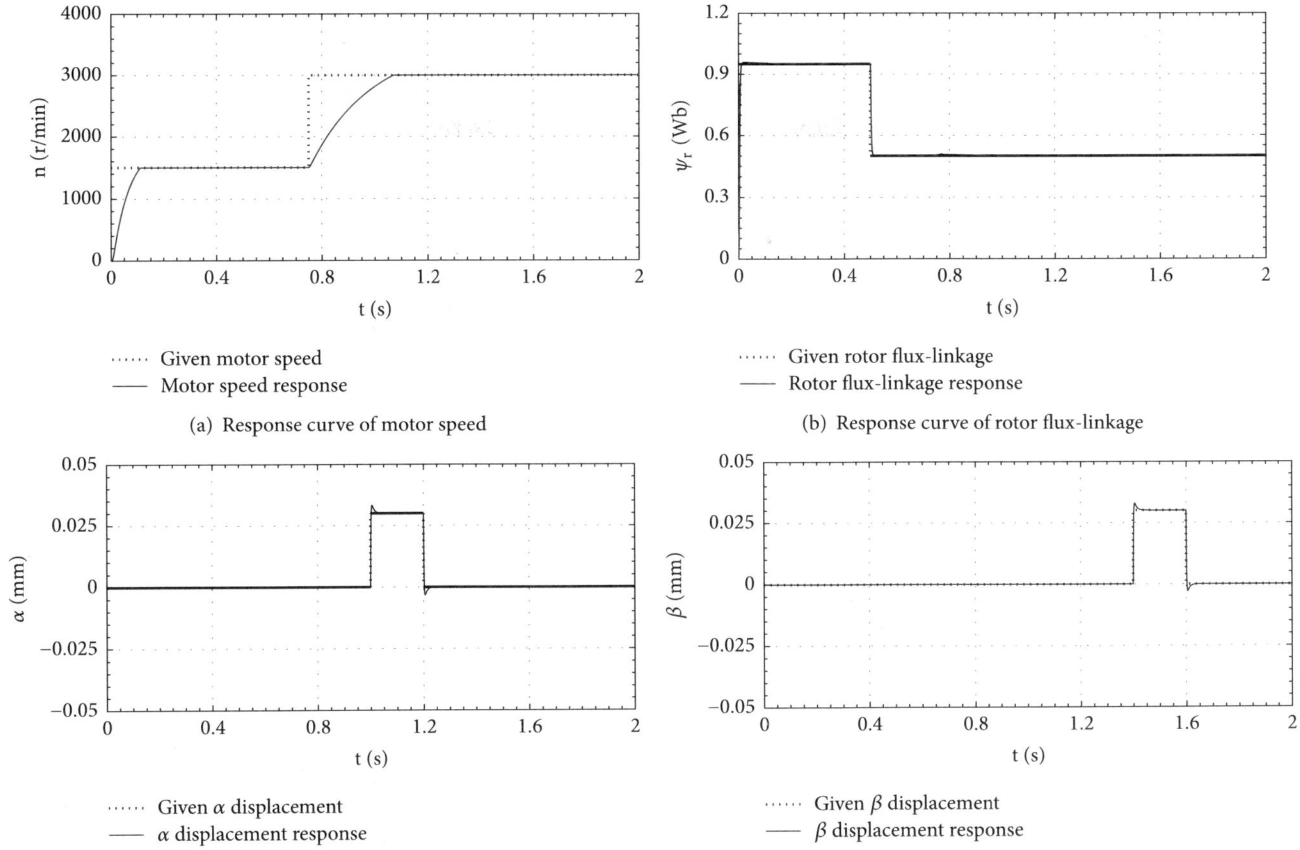

(a) Response curve of motor speed

(b) Response curve of rotor flux-linkage

(c) Radial displacement response in α direction

(d) Radial displacement response in β direction

FIGURE 5: Response curves of decoupling control system.

0.0 mm at 1.0 s and recovered to 0.0 mm at 1.2 s; the given β displacement is increased to 0.03 mm from 0.0 mm at 1.4 s and recovered to 0.0 mm at 1.6 s; a 10 N.m load torque is suddenly added at 1.5 s.

Figure 5 shows the response curves of the BLIM control system. From Figure 5, it can be seen that when any one of the motor speed, rotor flux-linkage, and α and β radial displacement components is suddenly changed, the other controlled variables basically remain unchanged or vary very little. The simulation results have shown that a good decoupling control between relevant controlled variables is achieved. In addition, the control system has some advantages, such as a faster response speed and a stronger resistance to load.

4.2. Compensation Effect of Unbalance Vibration. Set the following: rotor mass m=8 kg and eccentric distance of rotor mass ξ=0.5 mm, initial displacement α_0=-0.12 mm and β_0= -0.16 mm, given displacement $\alpha^* = \beta^* = 0.0$ mm, given rotor flux-linkage 0.95 Wb, and given motor speed 1500 r/min.

Figure 6 shows the displacement response curves before vibration compensation. From Figure 6,

(1) without vibration compensation, under the action of the displacement closed-loop control, the rotor unbalance vibration occurs; the radial displacement components along the α- and β-coordinate axes present periodic fluctuations

(2) after the BLIM control system enters its steady state, the rotor axis trajectory is a circle with an amplitude of about 6 μm. The rotor's suspension control precision is greatly influenced

Figure 7 presents the displacement response curves after vibration compensation. From Figure 7,

(1) in the initial starting stage, because the vibration compensation force has not been completely established, there exist some fluctuations with a smaller amplitude in the α- and β-displacement curves; however, with the establishment of vibration compensation force, the unbalanced displacement is quickly suppressed

(2) after the BLIM control system enters its steady state, under the combined action of the random displacement controller and the vibration compensation controller, the unbalanced displacement components along the α- and β-coordinate axes gradually decrease to zero; the rotor axis trajectory almost shrinks to the stator center; the suspension control precision of rotor has been greatly improved

Taking the horizontal α direction as an example, Figure 8 shows the comparison curves between the unbalanced excitation force and the vibration compensation force. From Figure 8,

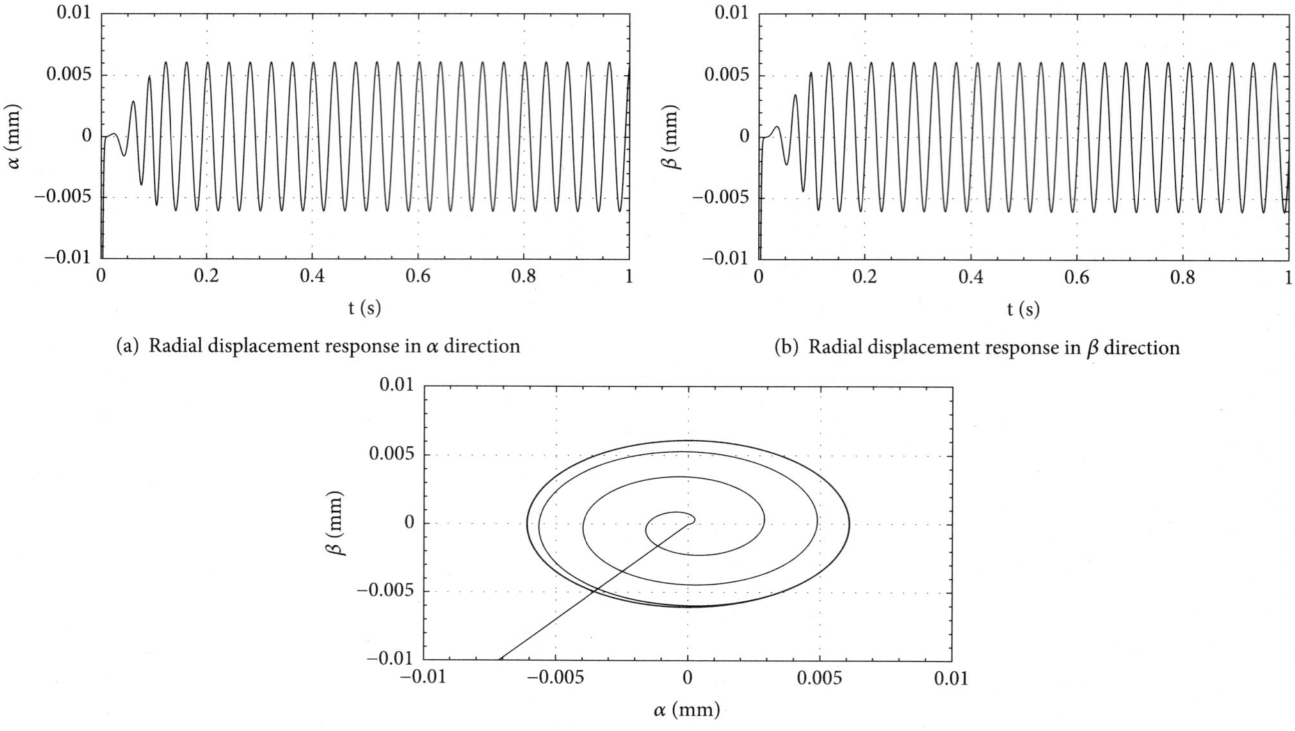

(a) Radial displacement response in α direction

(b) Radial displacement response in β direction

(c) Rotor axis trajectory

FIGURE 6: Displacement curve before vibration compensation.

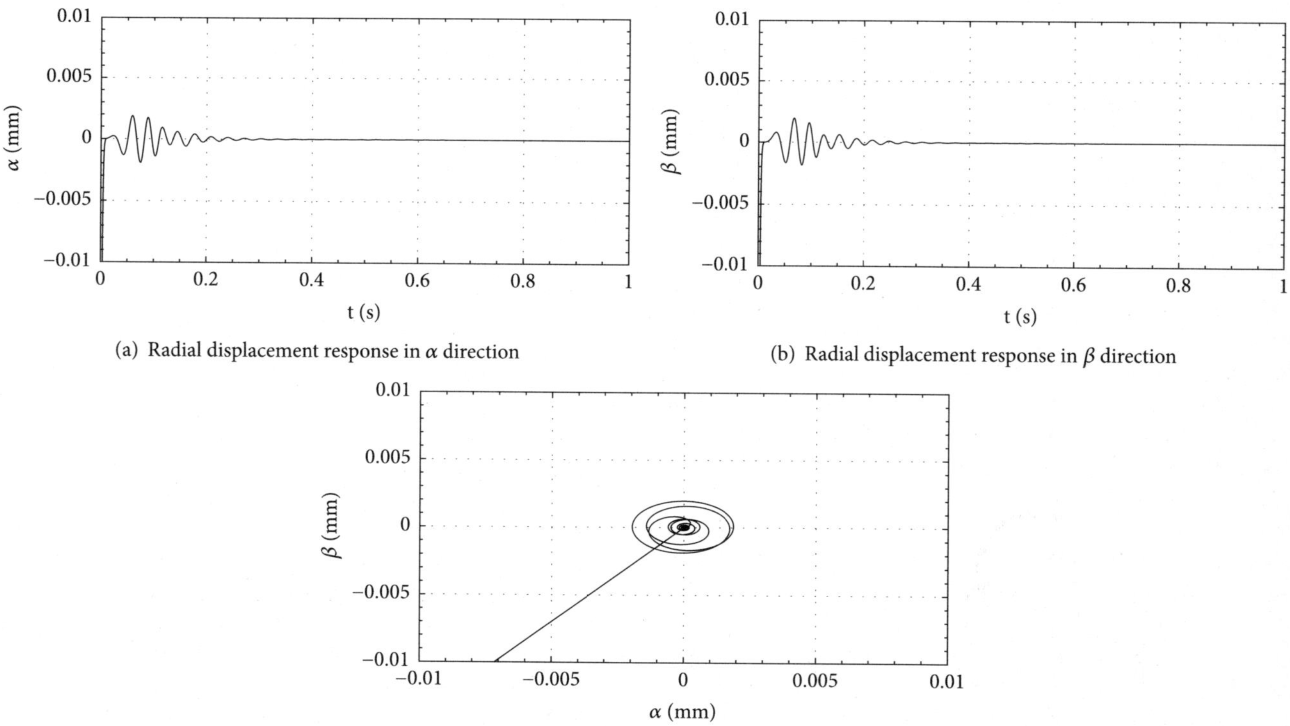

(a) Radial displacement response in α direction

(b) Radial displacement response in β direction

(c) Rotor axis trajectory

FIGURE 7: Displacement curves after vibration compensation.

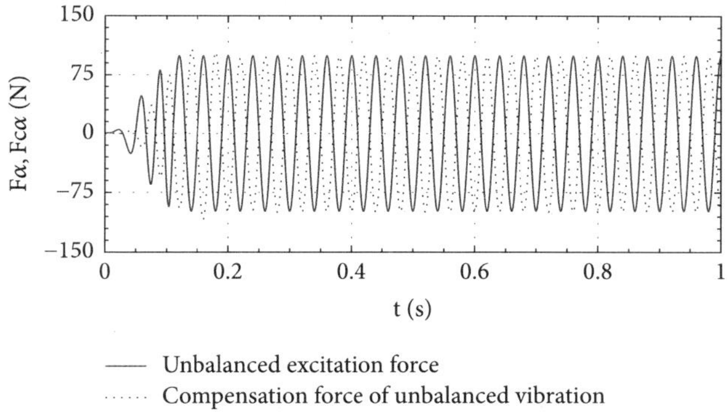

FIGURE 8: Comparison curves between the unbalanced excitation force and the vibration compensation force.

(1) in the initial starting stage, the amplitude of vibration compensation force increases with the unbalanced excitation force

(2) after the BLIM control system enters its steady state, the vibration compensation force is the same as that of the unbalanced excitation force in amplitude, and their direction is always opposite. And then, the vibration compensation force can be used to overcome the influence of the unbalanced excitation force

The comparison curves between the unbalanced excitation force and the vibration compensation force in the vertical β direction are similar to that in the horizontal α direction and will not be introduced.

5. Conclusions

To solve the strong coupling problem of the BLIM system and to solve the unbalance vibration problem caused by the rotor eccentric mass, a decoupling control strategy of the BLIM system considering the rotor mass eccentricity is studied. In order to achieve a better speed control performance, the rotor flux orientation is adopted for the torque system; to simplify the computational complexity of the magnetic suspension system, the required air gap flux-linkage of the torque system is obtained through independent and real-time calculation, and the rotor suspension motion control is achieved based on the air gap flux-linkage of the torque system. In order to improve the overall dynamic decoupling performance of the BLIM system, the whole inverse system decoupling is adopted. In order to overcome the unbalance vibration problem caused by the rotor mass eccentricity, the compensation controller of unbalance vibration is designed, and the vibration compensation force is used to suppress or eliminate the unbalanced excitation force that acts on the rotor. According to the simulation experimental results, there are the following conclusions:

(1) Adopting the presented control strategy, the dynamic decoupling control between the rotor flux-linkage, motor speed, and two radial displacement components can be achieved

(2) During the change of motor speed, the unbalanced displacement components of rotor can be suppressed effectively; meanwhile in the steady state of the BLIM control system, the unbalanced displacement components can be eliminated, and the magnetic suspension control accuracy of rotor can be greatly improved

Conflicts of Interest

The authors declare that they have no conflicts of interest.

Acknowledgments

This work is supported by the National Natural Science Foundation of China (51277053).

References

[1] A. Chiba, T. Fukao, O. Ichikawa et al., *Magnetic bearings and bearingless drives*, Elsevier Newnes Press, Boston, MA, USA, 2005.

[2] X.-g. Wu and D.-y. Zheng, “Contrastive Study on Torque Distribution of Distributed Drive Electric Vehicle under Different Control Methods,” *Journal of Control Science and Engineering*, vol. 2017, Article ID 2494712, 12 pages, 2017.

[3] X. Sun, L. Chen, H. Jiang, Z. Yang, J. Chen, and W. Zhang, “High-performance control for a bearingless permanent-magnet synchronous motor using neural network inverse scheme plus internal model controllers,” *IEEE Transactions on Industrial Electronics*, vol. 63, no. 6, pp. 3479–3488, 2016.

[4] W. S. Bu, S. H. Huang, and S. M. Wan, “General analytical models of inductance matrices of four-pole bearingless motors with two-pole controlling windings,” *IEEE Transactions on Magnetics*, vol. 45, no. 9, pp. 3316–3321, 2009.

[5] H. Sugimoto, S. Tanaka, and A. Chiba, “A vibration reduction method of one-axis actively position regulated single-drive bearingless motor with repulsive passive magnetic bearings,” *IEEE Transactions on Industry Applications*, vol. 52, no. 1, pp. 181–188, 2016.

[6] W. Bu, C. Lu, C. Zu, H. Zhang, and J. Xiao, “Inverse system decoupling control strategy of BLIM based on stator flux orientation,” *International Journal of Applied Electromagnetics and Mechanics*, vol. 48, no. 4, pp. 469–480, 2015.

[7] H. Wang, J. Bao, B. Xue, and J. Liu, “Control of suspending force in novel permanent-magnet-biased bearingless switched reluctance motor,” *IEEE Transactions on Industrial Electronics*, vol. 62, no. 7, pp. 4298–4306, 2015.

[8] S. H. Kim, J. W. Shin, and K. Ishiyama, “Magnetic bearings and synchronous magnetic axial coupling for the enhancement of the driving performance of magnetic wireless pumps,” *IEEE Transactions on Magnetics*, vol. 50, no. 1, pp. 1–4, 2014.

[9] J. Huang, B. Li, H. Jiang, and M. Kang, “Analysis and control of multiphase permanent-magnet bearingless motor with a single set of half-coiled winding,” *IEEE Transactions on Industrial Electronics*, vol. 61, no. 7, pp. 3137–3145, 2014.

[10] X. Sun, L. Chen, Z. Yang, and H. Zhu, “Speed-sensorless vector control of a bearingless induction motor with artificial neural network inverse speed observer,” *IEEE/ASME Transactions on Mechatronics*, vol. 18, no. 4, pp. 1357–1366, 2013.

[11] W.-s. Bu, C.-l. Zu, and C.-x. Lu, “Inverse System Analysis and

Modeling of Bearingless Induction Motor and Its Combined Control Strategy," *Mathematical Problems in Engineering*, vol. 2014, Article ID 698171, 8 pages, 2014.

[12] X. Cao and Z. Deng, "A full-period generating mode for bearingless switched reluctance generators," *IEEE Transactions on Applied Superconductivity*, vol. 20, no. 3, pp. 1072–1076, 2010.

[13] X. Sun, B. Su, L. Chen, Z. Yang, J. Chen, and W. Zhang, "Nonlinear flux linkage modeling of a bearingless permanent magnet synchronous motor based on AW-LSSVM regression algorithm," *International Journal of Applied Electromagnetics and Mechanics*, vol. 51, no. 2, pp. 151–159, 2016.

[14] Z. Q. Wang and X. X. Liu, "Nonlinear internal model control for bearingless induction motor based on neural network inversion," *Acta Automatica Sinica*, vol. 39, no. 4, pp. 433–439, 2013.

[15] W.-S. Bu, C.-L. Zu, and C.-X. Lu, "Decoupling control strategy of bearingless induction motor under the conditions of considering current dynamic characteristics," *Kongzhi Lilun Yu Yingyong/Control Theory and Applications*, vol. 31, no. 11, pp. 1561–1567, 2014.

[16] A. Bouaziz, S. Bouaziz, T. Hentati, J. Y. Cholley, and M. Haddar, "Vibrations monitoring of high speed spindle with active magnetic bearings in presence of defects," *International Journal of Applied Electromagnetics and Mechanics*, vol. 49, no. 2, pp. 207–221, 2015.

[17] T. Zhang and H.-Q. Zhu, "Rotor's mass unbalance compensation control in bearingless permanent magnet-type synchronous motors," *Proceedings of the Chinese Society of Electrical Engineering*, vol. 27, no. 15, pp. 33–37, 2007.

[18] H.-Q. Zhu and X.-H. Hao, "Compensation control of rotor unbalance vibration on bearingless permanent magnet slice motors," *Journal of System Simulation*, vol. 22, no. 2, pp. 453–457, 2010.

[19] W. Bu, F. He, C. Lu, Z. Li, and J. Xiao, "Unbalanced vibration control strategy of bearingless induction motor based on inverse system decoupling," *International Journal of Applied Electromagnetics and Mechanics*, vol. 51, no. 4, pp. 455–469, 2016.

[20] H. Gao and LX. Xu, "Real-time vibration compensation for active magnetic bearing systems based on LMS algorithm," *Journal of Vibration Engineering*, vol. 22, no. 6, pp. 583–587, 2009.

19

Fuzzy Control of Cold Storage Refrigeration System with Dynamic Coupling Compensation

Xiliang Ma[1,2] and Ruiqing Mao[1]

[1]*Xuzhou Institute of Technology, Xuzhou, Jiangsu 221111, China*
[2]*School of Mechanical and Electrical Engineering, China University of Mining and Technology, Xuzhou, Jiangsu 221116, China*

Correspondence should be addressed to Xiliang Ma; xlma818@sina.com

Academic Editor: Enrique Onieva

Cold storage refrigeration systems possess the characteristics of multiple input and output and strong coupling, which brings challenges to the optimize control. To reduce the adverse effects of the coupling and improve the overall control performance of cold storage refrigeration systems, a control strategy with dynamic coupling compensation was studied. First, dynamic model of a cold storage refrigeration system was established based on the requirements of the control system. At the same time, the coupling between the components was studied. Second, to reduce the adverse effects of the coupling, a fuzzy controller with dynamic coupling compensation was designed. As for the fuzzy controller, a self-tuning fuzzy controller was served as the primary controller, and an adaptive neural network was adopted to compensate the dynamic coupling. Finally, the proposed control strategy was employed to the cold storage refrigeration system, and simulations were carried out in the condition of start-up, variable load, and variable degree of superheat, respectively. The simulation results verify the effectiveness of the fuzzy control method with dynamic coupling compensation.

1. Introduction

Automatic control of cold storage refrigeration system is an important means to save energy and improve the quality of refrigeration [1, 2]. The automatic control can guarantee the stability of the storehouse temperature, avoid the unnecessary low temperature, and keep the temperature, pressure, liquid level, and other state parameters, which ensure the safety and efficiency of the system, in the required range. However, cold storage refrigeration systems possess the characteristics of multi-input and multi-output and strong coupling, which brings challenges to the dynamic modeling and the controller design [3, 4].

In the traditional control process of cold storage refrigeration system, the compressor, condenser, evaporator, expansion valve, and other main components are controlled separately [5]. Although those methods realize the automatic control of cold storage refrigeration system, it is difficult to achieve higher precision adjustment requirements because of ignoring the coupling. On the other hand, it is difficult to apply to the system with large change of loads and working conditions [6]. With the improvement of the refrigeration effect requirements of cold storage, researchers have being considered the refrigeration system as a whole and studied the combined control methods for the main components (compressor, condenser, evaporator, and expansion valve) [7]. Theoretically, the combined control methods have better control effect. However, the nonlinear characteristics of the cold storage refrigeration system lead to the difficulty of dynamic modeling, the time-varying characteristics lead to the difficulty of control parameter online adjusting, and the coupling characteristics increase the complexity of the controller design [3]. Due to the difficulty of dynamic modeling, researchers employed control methods that do not need the accurate model of the system to get a better solution [8, 9], such as fuzzy control and neural network control. Based on the coupling characteristics of cold storage refrigeration systems, scholars proposed decoupling controllers, but it is hard to achieve the desired control effect due to the strong coupling.

Studies on control strategy with dynamic coupling compensation of cold storage refrigeration system were

TABLE 1: The input and output variables and the state variables of the dynamic model.

	Symbol	Name
Input variables	f_Y (Hz)	Frequency of the compressor
	A_E	Opening of the electronic expansion valve
Output variables	T_{ri} (°C)	Temperature of the storage room
	T_{zo} (°C)	Outlet superheat of the evaporator
State variables	h_L	Outlet enthalpy of the condenser
	h_z (J/kg)	Outlet ratio of enthalpy of the evaporator
	T_{ri} (°C)	Temperature of the storage room
	T_L (°C)	Temperature of the tube wall of the condenser
	T_z (°C)	Temperature of the tube wall of the evaporator

conducted. First, the dynamic model of cold storage refrigeration system was established, and the coupling was studied. Second, a fuzzy controller with dynamic coupling compensation was designed. The control strategy consists of a fuzzy controller, which served as the primary controller, and an adaptive neural network, which was adopted to compensate the dynamic coupling. Finally, the control strategy was applied to the control of a cold storage refrigeration system.

2. Dynamic Coupling of Cold Storage Refrigeration System

The working process of refrigeration systems is a nonlinear and strong coupling dynamic process with fusion heat transfer and mass transfer flow. The dynamic model is the basis of the simulation and the control system design. In this paper, with the consideration of the requirements of refrigeration control system, a nonlinear dynamic model is established. Models of the main components such as frequency conversion compressor, electronic expansion valve, condenser, and evaporator are established under the premise of idealized assumptions [10]. Assume that the fan speed of the condenser and the evaporator are constant and the control process can be realized by controlling the frequency of the inverter compressor and the opening of the electronic expansion valve. In addition, the temperature of the cold storage room is one of the most important controlled parameters, and the control of the evaporator outlet superheat is an important way to improve the stability and efficiency of the refrigeration system.

In the overall dynamic model, the input and output variables and the state variables are listed in Table 1.

Based on the energy equation of refrigerant in the condenser [11], the heat transfer energy equation of the condenser tube wall [12], the energy equation of refrigerant in the evaporator [11], the heat energy equation of the evaporator tube wall [11], and the dynamic model of the storage room [10], the dynamic model of the cold storage refrigeration system can be established. Ordering that $\mathbf{x} = [h_L, T_L, h_z, T_z, T_{ri}]^T$ and $\mathbf{u} = [f_y, A_E]^T$, the dynamic model of the form of state space equation can be expressed as follows:

$$\dot{\mathbf{x}} = \mathbf{f}(\mathbf{x}) + \mathbf{g}(\mathbf{x})\,\mathbf{u}, \qquad (1)$$
$$\mathbf{y} = \mathbf{h}(\mathbf{x}),$$

where

$$\mathbf{f}(\mathbf{x}) = \begin{bmatrix} \frac{\alpha_{Li}A_{Li}}{m_L}(T_L - T_z) \\ \frac{\alpha_{Lo}A_{Lo}}{C_L L_L}(T_{ro} - T_L) - \frac{\alpha_{Li}A_{Li}}{C_L L_L}(T_L - T_z) \\ \frac{\alpha_{zi}A_{zi}}{m_z}(T_{zB} - T_{zL}) \\ \frac{\alpha_{zo}A_{zo}}{C_z L_z}(T_{ri} - T_{zB}) - \frac{\alpha_{zi}A_{zi}}{C_z L_z}(T_{zB} - T_{zL}) \\ \frac{\alpha_r A_r}{C_r + \rho_k C_k V_r}(T_{ro} - T_{ri}) - \frac{\alpha_z A_{zi}}{C_r + \rho_k C_k V_r}(T_{ri} - T_{zL}) + Q_r \end{bmatrix},$$

$$\mathbf{g}(\mathbf{x}) = \begin{bmatrix} g_1(\mathbf{x}) & 0 & 0 & 0 & 0 \\ 0 & 0 & g_2(\mathbf{x}) & 0 & 0 \end{bmatrix},$$

$$g_1(\mathbf{x}) = \frac{f_\lambda V_y (1-s)(h_z + V_c f_{\mathrm{Nth}} - h_L)}{p V_c m_L}, \qquad (2)$$

$$g_2(\mathbf{x}) = \frac{f_{mv}(h_L - h_z)}{m_z},$$

$$\mathbf{h}(\mathbf{x}) = \begin{bmatrix} T_{ri} \\ T_y - T_{zL} \end{bmatrix},$$

$$f_\lambda = 0.98 - 0.085\left[\left(\frac{p_i}{p_o}\right)^{1/k} - 1\right],$$

$$f_{\mathrm{Nth}} = p_o \frac{k_E}{1-k}\left[1 - \left(\frac{p_i}{p_o}\right)^{(k_E-1)/k_E}\right],$$

$$f_{mv} = C_E\sqrt{2\rho_E (p_i - p_o)}.$$

The included variables are listed in Table 2.

In the refrigeration system, inverter compressors, electronic expansion valve, evaporator, and condenser and other components are not independent; the coupling can be described as in Figure 1. The two components of condenser and evaporator are connected with the indoor and external environment. The outdoor ambient temperature T_{ro} influences the performance of the condenser, while the temperature of the storage room T_{ri} and the refrigeration system are mutually restricted. The coupling between the expansion valve and the environment is indirectly connected through the condenser and the evaporator.

TABLE 2: Variables and parameters in (2).

Symbol	Name
α_{Li}	Inner heat transfer coefficient of the condenser tube
m_L (kg)	Weight of refrigerant in the condensing tube
α_{Lo}	Air side heat transfer coefficient of the condenser tube
C_L (J /(kg°C))	Heat capacity of unit length of the condensing tube wall
T_{ro} (°C)	Temperature of the outdoor ambient
α_{zi}	Inner heat transfer coefficient of the evaporator tube
m_z (kg)	Weight of the refrigerant in the evaporator tube
T_{zL} (mm^2)	Evaporation temperature of the refrigerant
A_{zo} (mm^2)	Heat transfer area of air side the evaporator tube
L_z (mm)	Length of the evaporator pipe
A_r (mm^2)	Heat transfer area of the wall in the storage room
ρ_k (kg/mm^3)	Density of the air
V_r (mm^3)	Area of the storage room
Q_r	Heat load of the objects in the storage room
V_c	Ratio of gaseous refrigerant in suction of the compressor
p	Magnetic pole logarithm of the motor
p_o (Pa)	Outlet pressure of the electronic expansion valve
C_E	Flow coefficient of the electronic expansion valve
A_{Li} (mm^2)	Heat transfer area of the inner condenser tube
T_z (°C)	Condensation temperature of the refrigerant
A_{Lo} (mm^2)	Heat transfer area of the air side condenser tube
L_L (mm)	Length of the condensing pipe
T_L (°C)	Temperature of the condensation tube wall
A_{zi} (mm^2)	Heat transfer area of the evaporator tube
T_{zB} (mm^2)	Temperature of the evaporator tube wall
α_{zo}	Air side heat transfer coefficient of the evaporator tube
C_z (J/(kg°C))	Heat capacity of unit length of the evaporator tube wall
α_r	Comprehensive heat transfer coefficient of the wall in the storage room
C_r (J/(kg°C))	Heat capacity of the objects in the storage room
C_k	Constant pressure specific heat of the air
α_z	Comprehensive heat transfer coefficient of the evaporator
V_y	Swept volume of the compressor
s (r/s)	Speed difference of the motor
p_i (Pa)	Inlet pressure of the electronic expansion valve
k_E	Multidegree index
ρ_E (kg/mm^3)	Density of inlet refrigerant of the electronic expansion valve

3. Fuzzy Control with Dynamic Coupling Compensation

Due to the complexity of the transfer process of quality and the energy, it is difficult to establish a precise mathematical model of refrigeration system. Furthermore, cold storage refrigeration system is a typical MIMO, strong coupling, time-varying, and nonlinear complex system; it is difficult to achieve the desired control effect through a SISO control method. Fuzzy control and neural network control are not dependent on the precise mathematical model of a control object, which are effective methods to deal with the uncertainty, nonlinearity, and strong coupling in a control system [13]. In this paper, a multivariable fuzzy neural network controller with dynamic coupling compensation is proposed. The controller, which takes the advantages of fuzzy control and neural network control, is designed from the whole refrigeration system. The schematic diagram of the controller is shown in Figure 2.

As can be seen from Figure 2, the proposed controller is composed of a self-tuning fuzzy controller and an adaptive neural network coupling compensation controller. In the controller, the self-tuning fuzzy control determines the dynamic characteristics of the system, while the adaptive neural network control estimates the coupling of the multivariable.

The output deviation e is transformed to the formal control variables f_Y and A_E by the self-tuning fuzzy controller. The formal control variables are the inputs of the adaptive neural network coupling compensation controller. The

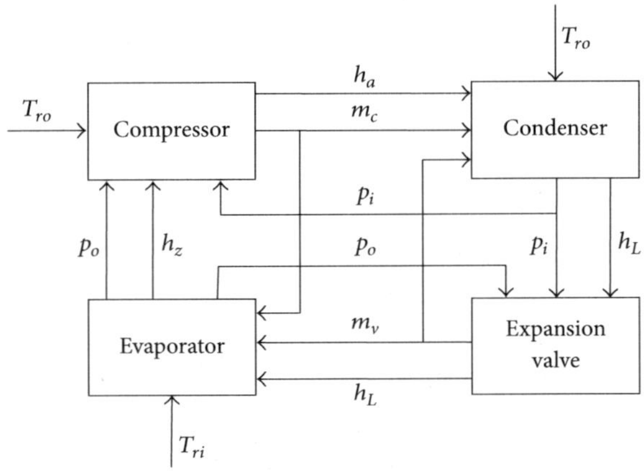

FIGURE 1: The coupling.

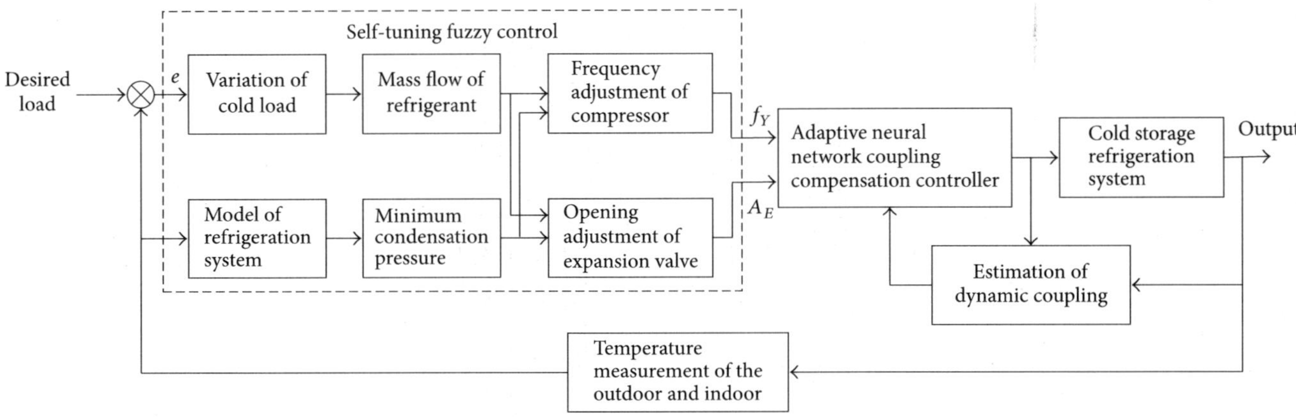

FIGURE 2: Fuzzy neural network controller with dynamic coupling compensation.

coupling effects are compensated as measurable disturbance. Through the dynamic coupling characteristic estimation and the neural network weights correction, the control object becomes a system with less coupling degree, and the output of the adaptive neural network coupling compensation controller is the input of the controlled object.

3.1. Self-Tuning Fuzzy Control. The intelligent weight function fuzzy algorithm is adopted to realize the self-tuning of the control rules. By using this method, it is not necessary to determine the membership function and fuzzy control rules of the fuzzy variables, and the interaction between other variables is not considered in the designed fuzzy controller. In a single loop of the self-tuning fuzzy controller, the selection of the fuzzy membership function is omitted, and the fuzzy quantity of the sampled discrete values $e(k)$ and $\Delta e(k)$ is multiplied by the quantization factors, shown as follows [14]:

$$E = K_E e(k), \quad E_C = K_C \Delta e(k), \tag{3}$$

where K_E and K_C are the quantitative gains of the error and the error variation, respectively.

To ensure that the outputs of the weight function are meaningful when E and E_C are zero at the same time, the weighted factors of the error and the error variation are determined as follows:

$$K_E = \frac{E}{|E| + |E_C| + \delta}, \quad K_C = \frac{E_C}{|E| + |E_C| + \delta}, \tag{4}$$

where δ is a small positive number.

The rule of fuzzy control of intelligent weight function is

$$\Delta U = K_E E + K_C E_C. \tag{5}$$

The output of the self-tuning fuzzy controller can be expressed as follows:

$$u_{0i}(k) = K_U \Delta U + K_I \sum \Delta U, \tag{6}$$

where K_U is the proportional gain and K_I is the integral gain.

3.2. Adaptive Neural Network Coupling Compensation Control. Based on the output of the self-tuning fuzzy controller (there are two output variables), ordering that the weighted coefficients of the neural networks are ω_{ij} $(i = 1,2;\ j = 1,2)$ and the output of the coupling compensation controller is

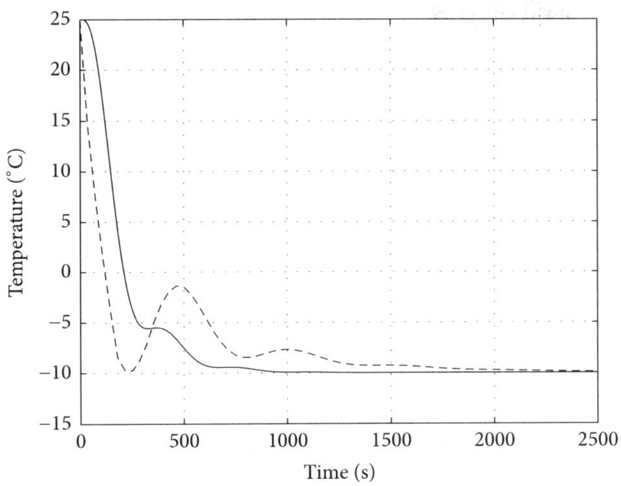

(a) The temperature of the storage room

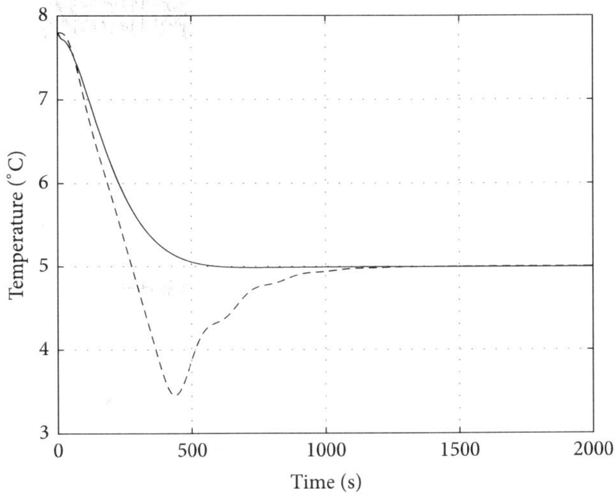

(b) The degree of superheat

FIGURE 3: Response curves after the system start-up.

$\mathbf{u}_q = [u_{q1}, u_{q2}]$, and the following relationship can be established [14]:

$$u_{qj}(k) = \sum_{i=1}^{2} \omega_{ij}(k)\, u_{0i} \quad (j = 1, 2). \tag{7}$$

The neural network and the controlled object are regarded as a whole, and the minimum value of the quadratic sum of the difference between the expected output and the actual one is used as the evaluation function of the system, shown as follows:

$$E(k) = \frac{1}{2}\sum_{i=1}^{2} [r_i(k) - y_i(k)]^2 \quad (i = 1, 2), \tag{8}$$

where $r_i(k)$ is the desired value and $y_i(k)$ is the actual value.

In (8), the value of $E(k)$ can be minimized by online adjustment and self-learning of $\omega_{ij}(k)$. In the paper, the online adjustment method of $E(k)$ is referred to as the method in [15]. The self-learning of $\omega_{ij}(k)$ is completed by adopting the Gradient descent method, and the iteration value at time $k+1$ can be expressed as follows:

$$\omega_{ij}(k+1) = \omega_{ij}(k) - \eta \frac{\partial E(k)}{\partial \omega_{ij}(k)}, \tag{9}$$

where η is the learning rate.

Therefore,

$$\frac{\partial E(k)}{\partial \omega_{ij}(k)} = -\sum_{i=1}^{2} [r_i(k) - y_i(k)] \cdot \frac{\partial y_i(k)}{\partial u_{qj}(k)} \cdot u_{0i}, \tag{10}$$

where $\partial y_i(k)/\partial u_{qj}(k)$ is the rate of the affection of the jth input $u_{qj}(k)$ on the controlled object in relation to the output. $\partial y_i(k)/\partial u_{qj}(k)$ can represent the dynamic coupling of the jth input to the ith output, and it can be calculated by the following equation:

$$\frac{\partial y_i(k)}{\partial u_{qj}(k)} \approx \operatorname{sign}\left[\frac{y_i(k+1) - y_i(k)}{u_{qi}(k) - u_{qi}(k-1)}\right]. \tag{11}$$

4. Numerical Simulation

To verify the effectiveness of the proposed control method for cold storage refrigeration system, numerical simulations were carried out. A small cold storage refrigeration system, which consists of R22 refrigerant, single-stage steam piston type compression refrigerator, electronic expansion valve, air-cooled evaporator, and air-cooled condenser, is taken as an example. In the study, $k_E = 1.18$, the sizes of the cold storage room are $L \times W \times H$: 4500 mm × 3600 mm × 2700 mm. During the process of simulation, the initial temperature of the storage room and the outdoor is 25°C and 30°C, respectively, and the desired temperature of the storage room and the degree of superheat are −10°C and 5°C, respectively. After the system start-up, the response curves of the temperature of the storage room and the degree of superheat are obtained, as shown in Figure 3.

As can be observed from Figure 3, although the temperature of the storage room and the degree of superheat can reach the predetermined steady-state values with or without the dynamic coupling compensated, there are no overshoot and fluctuation if the dynamic coupling is compensated. In another case, the temperature of the storage room is dropped from −10°C to −15°C, while the degree of superheat is still 5°C, and the response curves are shown in Figure 4.

As can be observed from Figure 4, the curve of the temperature of the storage room is smooth without overshoot after the coupling compensated. Because of the rapid increase of the cooling load, the speed of the variable frequency

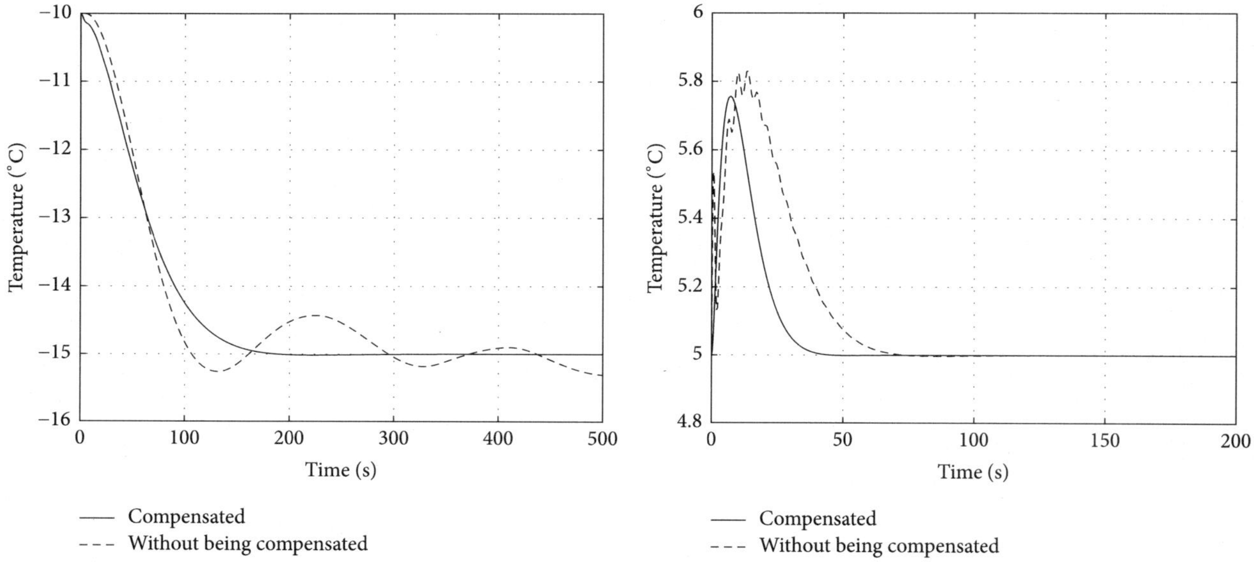

(a) The temperature of the storage room

(b) The degree of superheat

FIGURE 4: Response curves when the load changed.

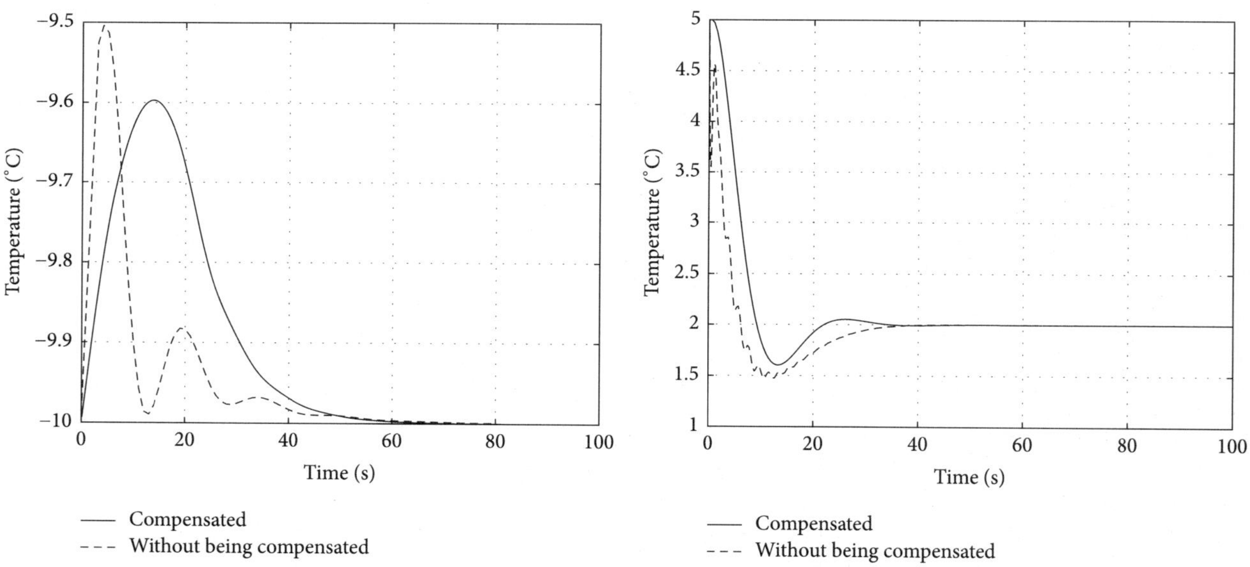

(a) The temperature of the storage room

(b) The degree of superheat

FIGURE 5: Response curves when the degree of superheat changed.

compressor causes the expansion valve to be open too late to adjust, which leads to a certain fluctuation of the superheat. The comparison of the two results in Figure 4(b) shows that the proposed controller has a better tracking performance. In another case, the degree of superheat is dropped from 5°C to 2°C while the temperature of the storage room is still −10°C, and the response curves are shown in Figure 5.

As can be observed from Figure 5, when the degree of superheat changed, the change of the expansion valve causes the fluctuation of the temperature of the storage room, and the changing process is smoother if the coupling is compensated. Overshoot appears in the adjustment process of the degree of superheat, but the fluctuation is small and the control effect is satisfied.

5. Conclusions

In the paper, studies on control strategy with dynamic coupling compensation of cold storage refrigeration system were conducted. First, the dynamic model of cold storage refrigeration system was established, and the coupling was studied. Second, a fuzzy controller with dynamic coupling compensation was designed, the self-tuning fuzzy controller was the primary controller, and the adaptive neural network was used to compensate the dynamic coupling. Finally, the designed control strategy was applied to the control of a cold storage refrigeration system, and simulations were carried out in the condition of start-up, variable load, and variable degree of superheat, respectively.

As can be observed from the results of the simulation, in the condition of start-up, there are no overshoot and fluctuation if the dynamic coupling is compensated. In the condition of the variable load and the variable degree of superheat, the curve of the temperature of the storage room is smooth without overshoot after the coupling compensated, and the comparison of the results shows that the proposed controller has a better tracking performance. The simulation results indicate that the fuzzy controller with dynamic coupling compensation has good control performance and can be used in the control process of cold storage refrigeration system.

Conflicts of Interest

The authors declare that they have no conflicts of interest.

References

[1] N. D. Mukhopadhyay and E. S. Chowdhury, "Performance analysis of solar assisted cascade refrigeration system of cold storage system," *International Journal of Advanced Research in Electrical Electronics & Instrumentation Engineering*, vol. 2, no. 4, pp. 1248–1254, 2013.

[2] A. Polzot, P. D'Agaro, P. Gullo, and G. Cortella, "Modelling commercial refrigeration systems coupled with water storage to improve energy efficiency and perform heat recovery," *International Journal of Refrigeration*, vol. 69, pp. 313–323, 2016.

[3] Y. P. Pan, *Research on Multivariable Adaptive Fuzzy Control for Cold Storage Refrigeration Syste*, Guangdong University of Technology, Guangzhou, China, 2007.

[4] J. Tian, R. Zhu, and Q. Feng, "Study on multi-variable fuzzy neural control for air cooled refrigeration system," *Journal of Xi'an Jiaotong University*, vol. 40, no. 5, pp. 514–517, 2006.

[5] T. Q. Chu and M. J. Lv, "Research on parameter-adaptive fuzzy controller for cold storage refrigeration system based on PLC," *Refrigeration and Air Conditioning*, vol. 8, no. 4, pp. 67–70, 2008.

[6] Q. S. Shao, X. W. Shi, T. X. Li et al., "Performance and control scheme of compressor and condenser in VRV air conditioning system," *Compressor Technology*, vol. 1, pp. 1–4, 2002.

[7] X. D. Wu, M. S. Jin, D. G. Fan et al., "Electronic expansion valve-compressor synchronous control," *Fluid Machinery*, vol. 33, no. 5, pp. 78–80, 2005.

[8] S. Li, T. Liang, X. Wang, and L. Kong, "Design for air conditioning evaporator superheat based on direct adaptive fuzzy controller," in *Proceedings of the 2011 IEEE International Conference on Automation and Logistics, ICAL 2011*, pp. 248–253, China, August 2011.

[9] C. Gao, "Study on the energy-saving of cold storage based on fuzzy neural-network," *Journal of Anhui Agricultural Sciences*, vol. 39, no. 18, pp. 11213-11214, 2011.

[10] W. Chen, X. Z. Cai, and X. X. Zhou, "Modeling and control research of the triple-evaporator air-conditioner with inverter," *Journal of System Simulation*, vol. 16, no. 10, pp. 2123–2127, 2004.

[11] J. Z. Xie, *MIMO Model And Decoupling Control of Refrigeration System*, Xi'an Jiaotong University, Xi'an, China, 2001.

[12] J. Nyers and G. Stoyan, "A dynamical model adequate for controlling the evaporator of a heat pump," *International Journal of Refrigeration*, vol. 17, no. 2, pp. 101–108, 1994.

[13] M. Li, J. Y. Lin, and Y. G. Ma, "Adaptive neural non-model decouple control for MIMO system," *Computer Simulation*, vol. 6, no. 3, pp. 346–350, 2003.

[14] H. Li, "Design of multivariable fuzzy-neural network decoupling controller," *Kongzhi yu Juece/Control and Decision*, vol. 21, no. 5, pp. 593–596, 2006.

[15] Z. J. Zhou and Z. J. Han, "A New multivariable fuzzy self-tuning control system," *Acta Automatica Sinica*, vol. 25, no. 2, pp. 215–219, 1999.

20

Nonfragile H_∞ Filter Design for Nonlinear Continuous-Time System with Interval Time-Varying Delay

Zhongda Lu,[1,2] **Guoliang Zhang**,[1] **Yi Sun**,[3] **Jie Sun**,[3] **Fangming Jin**,[3] **and Fengxia Xu**[1]

[1]*College of Mechatronic Engineering, Qiqihar University, Qiqihar 161006, China*
[2]*College of Computer Science and Technology, Harbin University of Science and Technology, Harbin 150080, China*
[3]*State Grid Heilongjiang Electric Power Company Qiqihar Power Supply Company, Qiqihar 161006, China*

Correspondence should be addressed to Zhongda Lu; luzhongda@163.com

Academic Editor: Jongrae Kim

This paper investigates nonfragile H_∞ filter design for a class of continuous-time delayed Takagi-Sugeno (T-S) fuzzy systems with interval time-varying delays. Filter parameters occur multiplicative gain variations according to the filter's implementation, to handle this variations, a nonfragile H_∞ filter is presented and a novel filtering error system is established. The nonfragile H_∞ filter guarantees the filtering error system to be asymptotically stable and satisfies given H_∞ performance index. By constructing a novel Lyapunov-Krasovskii function and using the linear matrix inequality (LMI), delay-dependent conditions are exploited to derive sufficient conditions for nonfragile designing H_∞ filter. Using new matrix decoupling method to reduce the computational complexity, the filter parameters can be obtained by solving a set of linear matrix inequalities (LMIs). Finally, numerical examples are given to show the effectiveness of the proposed method.

1. Introduction

As we all known, in practical control systems, nonlinearity and time delay phenomena are often encountered in various industry and control system, such as networked control system and mechanical drive control system. The control of nonlinear systems has been explored and studied by many scholars in related fields. T-S fuzzy model is a powerful tool to deal with nonlinearity; much effort has been devoted on the networked control system for T-S fuzzy system or time-delayed (see [1–4]). The actuator and sensor faults estimation based on T-S fuzzy model with unmeasurable premise variables were investigated in [5]. The problem of exponential stabilization for sampled-data T-S fuzzy control systems with packet dropouts was investigated in [6]; a switched system approach is proposed to model the data-missing phenomenon. There always exist many kinds of noise interference in the process of transmission among real industrial control system's signal, causing the error between the obtained signals and the desired signals; in order to obtain the accurate data information about the control signal and eliminate the influence of disturbances on the system, it is essential to be filtering. At present, there are the Kalman filtering, fault detection filter, L_2 filtering, H_∞ filtering, and so on. Compared with other filtering methods, H_∞ filtering does not need the exactly known statistics of the external disturbance [7, 8] and H_∞ filtering has excellent robustness against unmodeled dynamics. In recent years, H_∞ filtering system based on the Takagi-Sugeno (T-S) model has attracted much attention from the control community [9, 10], many studies have addressed H_∞ filtering for T-S fuzzy systems with time-varying delay, and the proposed H_∞ filtering technology has been applied to many actual communications system. Authors in literature [11–13] investigated the problem of H_∞ filter design for continuous-time via Takagi-Sugeno fuzzy model approach. In literature [14, 15], the problem of H_∞ filtering for a class of discrete fuzzy system has been reported. Based on discrete inequality technique and the Lyapunov-Krasovskii functional approach, sufficient conditions for the existence of admissible filters are established in terms of linear matrix inequalities. In literature [16], the event-triggered H_∞ filtering for networked control systems with quantization and network-induced delays was investigated; it improved the usage of network resource.

However, in practical system, it is difficult for an exactly implemented filter to meet the actual requirements because inaccuracies or uncertainties, which include collection error and component aging, may occur during filter implementation. It often degrades the performance of the control system and even instability; the filter has a higher sensitivity to the parameter uncertainty [17]. Thus, we need to design nonfragile H_∞ filter considering the parameter variation and uncertainty. Some achievements have been reported in journal about nonfragile H_∞ filtering for T-S fuzzy systems with time-varying delay. In literature [18], design an H_∞ filter with the gain variations such that the filtering error system was quadratically D stable and guarantees a prescribed H_∞ performance level. Literature [19] is concerned with the problem of nonfragile H_∞ filtering for discrete-time nonlinear systems and considered additive interval uncertainty. In literature [20], the designed nonfragile H_∞ filter was in standard form and the filter was designed, which have two types of multiplicative gain variations; these models were in standard form. In literature [21], the problem of nonfragile H_∞ filter design for linear continuous-time systems was studied; it proposed a notion of structured vertex separator. In literature [22], this paper studied the nonfragile H_∞ filtering problem for a class of discrete-time T-S fuzzy systems with both randomly occurring gain variations and channel fading. In literature [23], the problem of nonfragile H_∞ filter design for linear continuous-time systems has been studied. The filter has been designed; it included additive gain variations. In literature [24], it studied the nonfragile filtering design for a kind of fuzzy stochastic system with time-varying delay and parameter uncertainties. Sufficient conditions for stochastic input-to-state stability (SISS) of the fuzzy stochastic systems were obtained. Papers proposed the filter design methods with occurring additive gain variations according to the filter's implementation.

Motivated by the aforementioned discussion, in this paper, a nonfragile H_∞ filter design method is proposed to enhance the nonfragility of the filter. By considering the multiplicative gain variations and interval time-varying delays according to the filter implementation, a novel filtering error system is established. Different from some existing works, Jensen's inequality is used to tackle the integral items of the derivative of Lyapunov-Krasovskii; a more relaxed H_∞ performance stability criterion is derived. By constructing a novel Lyapunov-Krasovskii function and using the linear matrix inequality (LMI), delay-dependent conditions are exploited to derive sufficient conditions for nonfragile designing H_∞ filter. Our objective is to design nonfragile H_∞ filter which guarantees the filtering error system to be asymptotically stable and satisfies given H_∞ performance index. The filter parameters can be obtained by solving a set of linear matrix inequalities (LMIs).

The rest of this paper is organized as follows. The problem formulation is stated in Section 2; nonfragile filter scheme and filtering error system are employed to enhance system's stabilization. Stability analysis and fuzzy filter design are obtained in Section 3; by constructing a Lyapunov-Krasovskii functional, a new stability criterion is proposed to prove being less conservative than the existing ones. An applicable H_∞ filter is designed in Section 4, which guarantees stability and a desire performance of the filtering error system. In order to show the effectiveness of the proposed method, simulation results are presented in Section 5.

2. Problem Formulation

Consider a nonlinear system with time-varying delay which could be approximated by a class of T-S fuzzy systems with time-varying delays. The T-S fuzzy model with plant rules can be described by the following.

Plant Rule i. If $\theta_1(t)$ is M_{i1}... and $\theta_r(t)$ is M_{ip}, then

$$\begin{aligned}\dot{x}(t) &= A_i x(t) + A_{\tau i} x(t-\tau(t)) + B_i \omega(t)\\ y(t) &= C_i x(t) + C_{\tau i} x(t-\tau(t)) + D_i \omega(t)\\ z(t) &= E_i x(t)\end{aligned} \tag{1}$$

$x(t) = \varphi(t)$, where $\varphi(t)$ is the continuous initial vector function defined on $[-\tau_2, 0]$, M_{ij} are the fuzzy sets, $i = 1,2,3,\ldots,r$, and $j = 1,2,3,\ldots,p$ is the number of IF-THEN rules. $\theta_1(t), \theta_2(t)\ldots\theta_r(t)$ are the premise variables, $x(t) \in R^n$ is the state vector, $y(t) \in R^m$ is the measured output, $z(t) \in R^p$ is the signal vector to be estimated, $\omega(t) \in R^q$ is the disturbance signal vector which belongs to $\omega(t) \in l_2[0,\infty)$, and A_i, $A_{\tau i}$, B_i, $B_{\tau i}$, C_i, $C_{\tau i}$, D_i, $D_{\tau i}$, E_i are known constant matrices with appropriate dimensions. $\tau(t)$ is interval time-varying delay that satisfies the following inequality: $\tau_1 \le \tau(t) < \tau_2$, $\dot{\tau}(t) \le d$ where τ_1, τ_2, and d are constant scalars.

By employing the commonly used center-average defuzzifier, product interference, and singleton fuzzifier, the overall fuzzy model is inferred as follows:

$$\begin{aligned}\dot{x}(t) &= \sum_{i=1}^{r} h_i(\theta(t))\left[A_i x(t) + A_{\tau i} x(t-\tau(t)) + B_i \omega(t)\right]\\ y(t) &= \sum_{i=1}^{r} h_i(\theta(t))\left[C_i x(t) + C_{\tau i} x(t-\tau(t)) + D_i \omega(t)\right]\\ z(t) &= \sum_{i=1}^{r} h_i(\theta(t))\left[E_i x(t)\right]\end{aligned} \tag{2}$$

where $\theta = [\theta_1, \ldots, \theta_r]^T$, $h_i(\theta(t)) = h_i(\theta(t))/\sum_i^r h_i(\theta(t))$, $h_i(\theta(t)) = \prod_{j=1}^{p} M_{ij}(\theta_j(t))$, $M_{ij}(\bullet)$ represents the grade of membership for $\theta_j(t)$ in M_{ij}, $0 \le h_i(\theta(t)) \le 1$ (i = 1, 2, ..., r), and $\sum_{i=1}^{r} h_i(\theta(t)) > 0$. It can be seen that $h_i(\theta(t)) \ge 0$ ($i = 1,2,\ldots,r$), $\sum_{i=1}^{r} h_i(\theta(t)) = 1$.

Consider the nonfragile fuzzy filter with multiplicative gain uncertainties; we design the following fuzzy H_∞ filter:

$$\begin{aligned} \dot{x}_{fi}(t) &= \left(A_{fi} + \Delta A_{fi}(t)\right) x_f(t) \\ &\quad + \left(B_{fi} + \Delta B_{fi}(t)\right) y(t) \\ z_{fi}(t) &= \left(C_{fi} + \Delta C_{fi}(t)\right) x_f(t) \end{aligned} \tag{3}$$

Consider the following H_∞ filter form which is analogous to the fuzzy control form through parallel distributed compensation.

Plant Rule i. If $\theta_1(t)$ is M_{i1}... and $\theta_p(t)$ is M_{ip}, then

$$\begin{aligned} \dot{x}_f(t) &= \sum_{i=1}^{r} h_i(\theta(t)) \\ &\quad \cdot \left[\left(A_{fi} + \Delta A_{fi}(t)\right) x_f(t) + \left(B_{fi} + \Delta B_{fi}(t)\right) y(t)\right] \\ z_f(t) &= \sum_{i=1}^{r} h_i(\theta(t)) \left(C_{fi} + \Delta C_{fi}(t)\right) x_f(t) \end{aligned} \tag{4}$$

$x_f(0) = x_{f0}$, where x_{f0} is the continuous initial vector function, $x_f(t) \in R^n$ is the filter state vector, the estimated signal vector $z(t)$ is $z_f(t)$, and $A_{fi}, B_{fi}, C_{fi}, i = 1, 2, 3, \ldots, r$ are the filter parameters. $\Delta A_{fi}(t), \Delta B_{fi}(t), \Delta C_{fi}(t)$ represent the gain variations.

The multiplicative gain uncertainties are defined as

$$\begin{aligned} \Delta A_{fi}(t) &= A_{fi} M_{1i} K_A(t) N_{1i} \\ \Delta B_{fi}(t) &= B_{fi} M_{2i} K_B(t) N_{2i} \\ \Delta C_{fi}(t) &= C_{fi} M_{3i} K_C(t) N_{3i} \end{aligned} \tag{5}$$

where $M_{1i}, N_{1i}, M_{2i}, N_{2i}, M_{3i}, N_{3i}$ are constant matrices with appropriate dimensions and $K_A(t), K_B(t), K_C(t)$ are uncertain matrices bounded, such that

$$\begin{aligned} K_A^T(t) K_A(t) &< I, \\ K_B^T(t) K_B(t) &< I, \\ K_C^T(t) K_C(t) &< I. \end{aligned} \tag{6}$$

By combining (2) with (4), we can obtain the following filtering error system:

$$\begin{aligned} \dot{\xi}(t) &= \overline{A}(t)\xi(t) + \overline{A}_\tau(t)\xi(t - \tau(t)) + \overline{B}(t)\omega(t) \\ e(t) &= z(t) - z_f(t) = \overline{E}(t)\xi(t) \end{aligned} \tag{7}$$

where

$$\xi(t) = \left[x^T(t) \;\; x_f^T(t)\right]^T$$

$$\overline{A}(t) = \widehat{A}(t) + \Delta\widehat{A}(t) = \begin{bmatrix} A(t) & 0 \\ B_f(t)C(t) & A_f(t) \end{bmatrix} + \begin{bmatrix} 0 & 0 \\ \Delta B_f(t)C(t) & \Delta A_f(t) \end{bmatrix}$$

$$\overline{A}_\tau(t) = \widehat{A}_\tau(t) + \Delta\widehat{A}_\tau(t) = \begin{bmatrix} A_\tau(t) & 0 \\ B_f(t)C_\tau(t) & 0 \end{bmatrix} + \begin{bmatrix} 0 & 0 \\ \Delta B_f(t)C_\tau(t) & 0 \end{bmatrix}$$

$$\overline{B}(t) = \widehat{B}(t) + \Delta\widehat{B}(t) = \begin{bmatrix} B(t) \\ B_f(t)D(t) \end{bmatrix} + \begin{bmatrix} 0 \\ \Delta B_f(t)D(t) \end{bmatrix}$$

$$\overline{E}(t) = \widehat{E}(t) + \Delta\widehat{E}(t) = \left[E(t) \;\; -C_f(t)\right] + \left[0 \;\; -\Delta C_f(t)\right]$$

$$\widehat{A}(t) = \sum_{i=1}^{r}\sum_{j=1}^{r} h_i h_j \begin{bmatrix} A_j & 0 \\ B_{fi}C_j & A_{fi} \end{bmatrix} = \begin{bmatrix} A(t) & 0 \\ B_f(t)C(t) & A_f(t) \end{bmatrix}$$

$$\widehat{A}_\tau(t) = \sum_{i=1}^{r}\sum_{j=1}^{r} h_i h_j \begin{bmatrix} A_{\tau j} & 0 \\ B_{fi}C_{\tau j} & 0 \end{bmatrix} = \begin{bmatrix} A_\tau(t) & 0 \\ B_f(t)C_\tau(t) & 0 \end{bmatrix}$$

$$\widehat{B}(t) = \sum_{i=1}^{r}\sum_{j=1}^{r} h_i h_j \begin{bmatrix} B_j \\ B_{fi}D_j \end{bmatrix} = \begin{bmatrix} B(t) \\ B_f(t)D(t) \end{bmatrix}$$

$$\widehat{E}(t) = \sum_{i=1}^{r}\sum_{j=1}^{r} h_i h_j \left[E_j \;\; -C_{fi}\right] = \left[E(t) \;\; -C_f(t)\right]$$

$$\Delta\widehat{A}(t) = \sum_{i=1}^{r}\sum_{j=1}^{r} h_i h_j \begin{bmatrix} 0 & 0 \\ \Delta B_{fi}C_j & \Delta A_{fi} \end{bmatrix} = \begin{bmatrix} 0 & 0 \\ \Delta B_f(t)C(t) & \Delta A_f(t) \end{bmatrix}$$

$$\Delta\widehat{A}_\tau(t) = \sum_{i=1}^{r}\sum_{j=1}^{r} h_i h_j \begin{bmatrix} 0 & 0 \\ \Delta B_{fi}C_{\tau j} & 0 \end{bmatrix} = \begin{bmatrix} 0 & 0 \\ \Delta B_f(t)C_\tau(t) & 0 \end{bmatrix}$$

$$\Delta\widehat{B}(t) = \sum_{i=1}^{r}\sum_{j=1}^{r} h_i h_j \begin{bmatrix} 0 \\ \Delta B_{fi} D_j \end{bmatrix} = \begin{bmatrix} 0 \\ \Delta B_f(t) D(t) \end{bmatrix}$$

$$\Delta\widehat{E}(t) = \sum_{i=1}^{r}\sum_{j=1}^{r} h_i h_j [0 \ \ -\Delta C_{fi}] = [0 \ \ -\Delta C_f(t)] \tag{8}$$

In this paper, our purpose is to design the fuzzy H_∞ filter in the form of (3), meanwhile, satisfying the following requirements.

(1) The filtering error system (7) with $\omega(t) = 0$ is said to be asymptotic stability for any initial condition

(2) For a given positive scalar $\gamma > 0$, the filtering error system (7) is said to be asymptotically stable with guaranteed H_∞ performance γ, if it is asymptotic stability and the filtering error $e(t)$ satisfies

$$\int_0^L \|e(t)\|^2 dt \le \gamma^2 \int_0^L \|\omega(t)\|^2 dt \tag{9}$$

for all $L > 0$ and nonzero $\omega(t) \in L_2[0, \infty)$ subject to the zero initial condition.

3. Stability and H_∞ Filtering Performance Analysis

The purpose of this paper is to design nonfragile H_∞ filter such that the filtering error system (7) is asymptotically stable with H_∞ performance index. A sufficient condition is presented in the following theorem to guarantee the existence of the filter in form of (3).

Lemma 1. *Let X, Y, and K be real matrices with appropriate dimensions and $K_i^T(t)K_i(t) < I$. Then, for any scalar $\delta > 0$,*

$$XK(t)Y + Y^T K(t) X^T \le \varepsilon XX^T + \varepsilon^{-1} Y^T Y \tag{10}$$

Lemma 2. *For any vectors $x, y \in R^n$ and any scalar $\varepsilon > 0$, matrices D, F, E are real matrices of appropriate dimensions with $F^T(t)F(t) < I$; then the following inequalities hold:*

$$2xDF(t)Ey \le \varepsilon x^T DD^T x + \varepsilon^{-1} y^T EE^T y \tag{11}$$

Lemma 3. *Let V, H, E, Q, and F be real matrices of appropriate dimensions such that $Q > 0$ and $F^T F \le I$. Then, for any scalar $\varepsilon > 0$ such that $Q^{-1} - \varepsilon HH^T > 0$, we have*

$$(V + HFE)^T Q (V + HFE) \le V^T \left(Q^{-1} - \varepsilon HH^T\right)^{-1} V + \varepsilon^{-1} E^T E \tag{12}$$

Lemma 4. *Let $P > 0$ and A_i, A_l be any real matrices of appropriate dimensions. Then*

$$A_i^T PA_l + A_l^T PA_i \le A_i^T PA_i + A_l^T PA_l \tag{13}$$

Lemma 5 (Jenson's inequality). *Suppose $\tau_1 \le \tau(t) \le \tau_2$ and $x(t) \in \Re^n$; for any positive matrix $R \in \Re^{n\times n}$, the following inequality holds:*

$$-(\tau_2 - \tau_1) \int_{t-\tau_2}^{t-\tau_1} \dot{x}^T(s) R\dot{x}(s)\, ds \le \begin{bmatrix} x(t-\tau_1) \\ x(t-\tau_2) \end{bmatrix}^T \begin{bmatrix} -R & R \\ R & -R \end{bmatrix} \begin{bmatrix} x(t-\tau_1) \\ x(t-\tau_2) \end{bmatrix} \tag{14}$$

Theorem 6. *For nonlinear systems (1) and the filtering error system (7), the given positive scalar H_∞ performance γ and the filtering error system (7) are asymptotically stable with H_∞ performance γ if there exist symmetric positive scalars $\varepsilon_2 > 0$, $\varepsilon_6 > 0$, $\varepsilon_{10} > 0$, $\varepsilon_{29} > 0$, $\varepsilon_{15} > 0$, $\varepsilon_{25} > 0$, $\varepsilon_{20} > 0$, $\varepsilon_{21} > 0$, $\varepsilon_{22} > 0$, $\varepsilon_{11} > 0$, $\varepsilon_{12} > 0$ and symmetric positive definite matrices $\widehat{P} > 0$, $\widehat{Q}_1 > 0$, $\widehat{Q}_2 > 0$, $\widehat{Q}_3 > 0$, $\widehat{R}_1 > 0$, $\widehat{R}_2 > 0$ such that we have the following inequality.*

$$\begin{bmatrix} I & \widetilde{M}_3 \\ * & \varepsilon_{20}^{-1} \end{bmatrix} > 0 \tag{15}$$

$$\begin{bmatrix} \widehat{R}_1^{-1} & \widetilde{M}_2 \\ * & \varepsilon_{1k}^{-1} \end{bmatrix} > 0, \quad (k = 5, 9) \tag{16}$$

$$\begin{bmatrix} \widehat{R}_2^{-1} & \widetilde{M}_2 \\ * & \varepsilon_{2k}^{-1} \end{bmatrix} > 0, \quad (k = 5, 9) \tag{17}$$

$$\begin{bmatrix} \widehat{R}_1^{-1} & \alpha_{11}\widetilde{M}_1 & \alpha_{12}\widetilde{M}_2 \\ * & I & 0 \\ * & * & I \end{bmatrix} > 0 \tag{18}$$

$$\begin{bmatrix} \widehat{R}_2^{-1} & \alpha_{21}\widetilde{M}_1 & \alpha_{22}\widetilde{M}_2 \\ * & I & 0 \\ * & * & I \end{bmatrix} > 0 \tag{19}$$

$$\Phi = \begin{bmatrix} \Phi_{11} & \Phi_{12} & \widehat{R}_1 & 0 & \Phi_{15} \\ * & \Phi_{22} & 0 & 0 & 0 \\ * & * & \Phi_{33} & 0 & 0 \\ * & * & * & \Phi_{44} & 0 \\ * & * & * & * & \Phi_{55} \end{bmatrix} < 0 \tag{20}$$

$$\Phi_{11} = \widehat{P}\widehat{A}(t) + \widehat{A}^T(t)\widehat{P} + \widehat{Q}_1 + \widehat{Q}_2 + \widehat{Q}_3 - \widehat{R}_1 + \widehat{E}^T(t)\left[I - \varepsilon_{20}\widetilde{M}_3\widetilde{M}_3^T\right]^{-1}\widehat{E}(t) + \alpha_4^2\widehat{P}\widetilde{M}_2\widetilde{M}_2^T\widehat{P} + \alpha_3^2\widehat{P}\widetilde{M}_1\widetilde{M}_1^T\widehat{P} + \alpha_5^2\widetilde{N}_1\widetilde{N}_1 + \alpha_6^2\widetilde{N}_2\widetilde{N}_2$$

$$+\beta_1^2\widehat{A}^T(\mathrm{t})\left[\widehat{R}_1^{-1}-\varepsilon_{11}\widetilde{M}_1\widetilde{M}_1^T-\varepsilon_{12}\widetilde{M}_2\widetilde{M}_2^T\right]^{-1}\widehat{A}(\mathrm{t})$$

$$+\beta_2^2\widehat{A}^T(\mathrm{t})\left[\widehat{R}_2^{-1}-\varepsilon_{21}\widetilde{M}_1\widetilde{M}_1^T-\varepsilon_{22}\widetilde{M}_2\widetilde{M}_2^T\right]^{-1}\widehat{A}(\mathrm{t})$$

$$+\varepsilon_{20}^{-1}\widetilde{N}_3^T\widetilde{N}_3$$

$$\Phi_{12}=\widehat{P}\widehat{A}_\tau(\mathrm{t}),$$

$$\Phi_{15}=\widehat{P}\widehat{B}(\mathrm{t}),$$

$$\Phi_{33}=-\widehat{Q}_2-\left(\widehat{R}_1+\widehat{R}_2\right),$$

$$\Phi_{44}=-\widehat{Q}_3-\widehat{R}_2,$$

$$\Phi_{22}$$

$$=-(1-d)\widehat{Q}_1+\alpha_2^2\widetilde{N}_{2\tau}^T\widetilde{N}_{2\tau}$$

$$+\beta_1^2\widehat{A}_\tau^T(\mathrm{t})\left[\widehat{R}_1^{-1}-\varepsilon_{15}\widetilde{M}_2\widetilde{M}_2^T\right]^{-1}\widehat{A}_\tau(\mathrm{t})$$

$$+\beta_2^2\widehat{A}_\tau^T(\mathrm{t})\left[\widehat{R}_2^{-1}-\varepsilon_{25}\widetilde{M}_2\widetilde{M}_2^T\right]^{-1}\widehat{A}_\tau(\mathrm{t}),$$

$$\Phi_{55}$$

$$=\alpha_1^2\widetilde{N}_{2D}^T\widetilde{N}_{2D}-\gamma^2I$$

$$+\beta_1^2\widehat{B}_\tau^T(\mathrm{t})\left[\widehat{R}_1^{-1}-\varepsilon_{19}\widetilde{M}_2\widetilde{M}_2^T\right]^{-1}\widehat{B}_\tau(\mathrm{t})$$

$$+\beta_2^2\widehat{B}_\tau^T(\mathrm{t})\left[\widehat{R}_2^{-1}-\varepsilon_{29}\widetilde{M}_2\widetilde{M}_2^T\right]^{-1}\widehat{B}_\tau(\mathrm{t})$$

$$\widetilde{M}_1=\begin{bmatrix}0\\A_f(\mathrm{t})M_1\end{bmatrix},$$

$$\widetilde{M}_2=\begin{bmatrix}0\\B_f(\mathrm{t})M_2\end{bmatrix},$$

$$\widetilde{M}_3=M_3,$$

$$\widetilde{N}_1=\begin{bmatrix}0 & N_1\end{bmatrix},$$

$$\widetilde{N}_2=\begin{bmatrix}N_2C(\mathrm{t}) & 0\end{bmatrix},$$

$$\widetilde{N}_{2\tau}=\begin{bmatrix}N_2C_\tau(\mathrm{t}) & 0\end{bmatrix},$$

$$\widetilde{N}_{2\tau}=\begin{bmatrix}N_2C_\tau(\mathrm{t}) & 0\end{bmatrix},$$

$$\widetilde{N}_{2D}=N_2D(\mathrm{t}),$$

$$\widetilde{N}_3=\begin{bmatrix}0 & -N_3C_f(\mathrm{t})\end{bmatrix},$$

$$\beta_1=\sqrt{3\tau_1^2},$$

$$\beta_2=\sqrt{3\left(\tau_2^2-\tau_1^2\right)},$$

$$\alpha_1=\sqrt{\beta_1^2\varepsilon_{19}^{-1}+\beta_2^2\varepsilon_{29}^{-1}+\varepsilon_{10}^{-1}},$$

$$\alpha_2=\sqrt{\beta_1^2\varepsilon_{15}^{-1}+\beta_2^2\varepsilon_{25}^{-1}+\varepsilon_6^{-1}},$$

$$\alpha_3=\sqrt{\varepsilon_{21}},$$

$$\alpha_4=\sqrt{\varepsilon_2+\varepsilon_6+\varepsilon_{10}},$$

$$\alpha_5=\sqrt{\varepsilon_{21}^{-1}+\beta_1^2\varepsilon_{11}^{-1}+\beta_2^2\varepsilon_{22}^{-1}},$$

$$\alpha_6=\sqrt{\beta_1^2\varepsilon_{12}^{-1}+\beta_2^2\varepsilon_{22}^{-1}} \tag{21}$$

Proof. We construct a novel Lyapunov-Krasovskii function as follows:

$$\begin{aligned}V(\xi(\mathrm{t}))&=\xi^T(\mathrm{t})\widehat{P}\xi(\mathrm{t})+\int_{t-\tau(\mathrm{t})}^{t}\xi^T(s)\widehat{Q}_1\xi(s)\,ds\\&+\int_{t-\tau_1}^{t}\xi^T(s)\widehat{Q}_2\xi(s)\,ds\\&+\int_{t-\tau_2}^{t}\xi^T(s)\widehat{Q}_3\xi(s)\,ds\\&+\tau_1\int_{-\tau_1}^{0}\int_{t+\theta}^{t}\dot{\xi}^T(s)\widehat{R}_1\dot{\xi}(s)\,ds\,d\theta\\&+\left(\tau_2-\tau_1\right)\int_{-\tau_2}^{-\tau_1}\int_{t+\theta}^{t}\dot{\xi}^T(s)\widehat{R}_2\dot{\xi}(s)\,ds\,d\theta\end{aligned} \tag{22}$$

$$\begin{aligned}\dot{V}&=2\xi^T(\mathrm{t})\widehat{P}\overline{A}(\mathrm{t})\xi(\mathrm{t})\\&+2\xi^T(\mathrm{t})\widehat{P}\overline{A}_\tau(\mathrm{t})\xi(\mathrm{t}-\tau(\mathrm{t}))\\&+2\xi^T(\mathrm{t})\widehat{P}\overline{B}(\mathrm{t})\omega(\mathrm{t})\\&+\xi^T(\mathrm{t})\left(\widehat{Q}_1+\widehat{Q}_2+\widehat{Q}_3\right)\xi(\mathrm{t})\\&-(1-\dot{\tau}(\mathrm{t}))\xi^T(\mathrm{t}-\tau(\mathrm{t}))\widehat{Q}_1\xi(\mathrm{t}-\tau(\mathrm{t}))\\&-\xi^T\left(\mathrm{t}-\tau_1\right)\widehat{Q}_2\xi\left(\mathrm{t}-\tau_1\right)\\&-\xi^T\left(\mathrm{t}-\tau_2\right)\widehat{Q}_3\xi\left(\mathrm{t}-\tau_2\right)\\&+\tau_1^2\dot{\xi}^T(\mathrm{t})\widehat{R}_1\dot{\xi}(\mathrm{t})\\&+\left(\tau_2-\tau_1\right)^2\dot{\xi}^T(\mathrm{t})\widehat{R}_2\dot{\xi}(\mathrm{t})\\&-\tau_1\int_{t-\tau_1}^{t}\dot{\xi}^T(s)\widehat{R}_1\dot{\xi}(s)\,ds\\&-\left(\tau_2-\tau_1\right)\int_{t-\tau_1}^{t}\dot{\xi}^T(s)\widehat{R}_2\dot{\xi}(s)\,ds\end{aligned} \tag{23}$$

$$\begin{aligned}\dot{V}&\le 2\xi^T(\mathrm{t})\widehat{P}\widehat{A}(\mathrm{t})\xi(\mathrm{t})+2\xi^T(\mathrm{t})\widehat{P}\Delta\widehat{A}(\mathrm{t})\xi(\mathrm{t})\\&+2\xi^T(\mathrm{t})\widehat{P}(\mathrm{t})\widehat{A}_\tau(\mathrm{t})\xi(\mathrm{t}-\tau(\mathrm{t}))\\&+\xi^T(\mathrm{t})\left(\widehat{Q}_1+\widehat{Q}_2+\widehat{Q}_3\right)\xi(\mathrm{t})\\&+2\xi^T(\mathrm{t})\widehat{P}\Delta\widehat{B}(\mathrm{t})\omega(\mathrm{t})\end{aligned}$$

$$
\begin{aligned}
&+ 2\xi^T(t)\widehat{P}(t)\Delta\widehat{A}_\tau(t)\xi(t-\tau(t)) \\
&- (1-d)\xi^T(t-\tau(t))\widehat{Q}_1\xi(t-\tau(t)) \\
&- \xi^T(t-\tau_1)\widehat{Q}_2\xi(t-\tau_1) \\
&+ 2\xi^T(t)\widehat{P}\widehat{B}(t)\omega(t) \\
&- \xi^T(t-\tau_2)\widehat{Q}_3\xi(t-\tau_2) \\
&+ \tau_1^2\dot{\xi}^T(t)\widehat{R}_1\dot{\xi}(t) \\
&+ (\tau_2-\tau_1)^2\dot{\xi}^T(t)\widehat{R}_2\dot{\xi}(t) \\
&- \tau_1\int_{t-\tau_1}^{t}\dot{\xi}^T(s)\widehat{R}_1\dot{\xi}(s)\,ds \\
&- (\tau_2-\tau_1)\int_{t-\tau_2}^{t-\tau_1}\dot{\xi}^T(s)\widehat{R}_2\dot{\xi}(s)\,ds
\end{aligned} \tag{24}
$$

By Lemma 2, for $\forall\varepsilon_2 > 0$, $\forall\varepsilon_{21} > 0$, we can obtain

$$
\begin{aligned}
&2\xi^T(t)\widehat{P}\Delta\widehat{A}(t)\xi(t) = 2\xi^T(t)\widehat{P}\left[\widetilde{M}_{2i}K_i(t)\widetilde{N}_{2i}\right. \\
&\quad \left.+ \widetilde{M}_{1i}K_i(t)\widetilde{N}_{1i}\right]\xi(t) \le \xi^T(t)\left[\varepsilon_2\widehat{P}\widetilde{M}_2\widetilde{M}_2^T\widehat{P}\right. \\
&\quad \left.+ \varepsilon_{21}\widehat{P}\widetilde{M}_1\widetilde{M}_1^T\widehat{P} + \varepsilon_2^{-1}\widetilde{N}_2^T\widetilde{N}_2 + \varepsilon_{21}^{-1}\widetilde{N}_1^T\widetilde{N}_1\right]\xi(t)
\end{aligned} \tag{25}
$$

Similar to (25), if there exists $\forall\varepsilon_{20} > 0$, we can obtain

$$
\begin{aligned}
e^T(t)e(t) &= \xi^T(t)\overline{E}^T(t)\overline{E}(t)\xi(t) \\
&= \xi^T(t)\left(\widehat{E}(t) + \Delta\widehat{E}(t)\right)^T\left(\widehat{E}(t) + \Delta\widehat{E}(t)\right)\xi(t) \\
&\le \xi^T(t)\widehat{E}^T(t)\left(I - \varepsilon_{20}\widetilde{M}_3\widetilde{M}_3^T\right)^{-1}\widehat{E}(t)\xi(t) \\
&\quad + \varepsilon_{20}^{-1}\xi^T(t)\widetilde{N}_3^T\widetilde{N}_3\xi(t)
\end{aligned} \tag{26}
$$

By Lemma 2, if there exists $\forall\varepsilon_6 > 0$, we can obtain

$$
\begin{aligned}
&2\xi^T(t)\widehat{P}\Delta\widehat{A}_\tau(t)\xi(t-\tau(t)) \\
&\quad = 2\xi^T(t)\widehat{P}\left[\widetilde{M}_2K_i(t)\widetilde{N}_{2\tau}\right]\xi(t-\tau(t)) \\
&\quad \le \xi^T(t)\left[\varepsilon_6\widehat{P}\widetilde{M}_2\widetilde{M}_2^T\widehat{P}\right]\xi(t) \\
&\qquad + \xi^T(t-\tau(t))\left[\varepsilon_6^{-1}\widetilde{N}_{2\tau}^T\widetilde{N}_{2\tau}\right]\xi(t-\tau(t))
\end{aligned} \tag{27}
$$

Similar to (25), if there exists $\forall\varepsilon_{10} > 0$, we can obtain

$$
\begin{aligned}
2\xi^T(t)\widehat{P}\Delta\widehat{B}(t)\omega(t) &= 2\xi^T(t)\widehat{P}\left[\widetilde{M}_2K_i(t)\widetilde{N}_{2D}\right]\omega(t) \\
&\le \xi^T(t)\left[\varepsilon_{10}\widehat{P}\widetilde{M}_2\widetilde{M}_2^T\widehat{P}\right]\xi(t) \\
&\quad + \omega^T(t)\left[\varepsilon_{10}^{-1}\widetilde{N}_{2D}^T\widetilde{N}_{2D}\right]\omega(t)
\end{aligned} \tag{28}
$$

Combining with inequalities (25), (27), and (28) gives

$$
\begin{aligned}
&2\xi^T(t)\widehat{P}\Delta\widehat{A}(t)\xi(t) + 2\xi^T(t)\widehat{P}\Delta\widehat{A}_\tau(t)\xi(t-\tau(t)) \\
&\quad + 2\xi^T(t)\widehat{P}\Delta\widehat{B}(t)\omega(t) \le \xi^T\left[\varepsilon_2\widehat{P}\widetilde{M}_2\widetilde{M}_2^T\widehat{P}\right. \\
&\quad \left.+ \varepsilon_2^{-1}\widetilde{N}_2\widetilde{N}_2^T + \varepsilon_{21}\widehat{P}\widetilde{M}_1\widetilde{M}_1^T\widehat{P} + \varepsilon_{21}^{-1}\widetilde{N}_1\widetilde{N}_1^T\right]\xi(t) \\
&\quad + \xi^T(t)\left[\varepsilon_{10}\widehat{P}\widetilde{M}_2\widetilde{M}_2^T\widehat{P}\right]\xi(t) + \xi^T(t)\left[\varepsilon_6\widehat{P}\widetilde{M}_2\widetilde{M}_2^T\widehat{P}\right] \\
&\quad \cdot\xi(t) + \xi^T(t-\tau(t))\left[\varepsilon_6^{-1}\widetilde{N}_{2\tau}^T\widetilde{N}_{2\tau}\right]\xi(t-\tau(t)) \\
&\quad + \omega^T(t)\left[\varepsilon_{10}^{-1}\widetilde{N}_2^T\widetilde{N}_2\right]\omega(t)
\end{aligned} \tag{29}
$$

Lemma 4 gives

$$
\begin{aligned}
&\tau_1^2\dot{\xi}^T(t)\widehat{R}_1\dot{\xi}(t) = \tau_1^2\left[\overline{A}(t)\xi(t) + \overline{A}_\tau(t)\xi(t-\tau(t))\right. \\
&\quad \left.+ \overline{B}(t)\omega(t)\right]^T\widehat{R}_1\left[\overline{A}(t)\xi(t) + \overline{A}_\tau(t)\xi(t-\tau(t))\right. \\
&\quad \left.+ \overline{B}(t)\omega(t)\right] \le 3\tau_1^2\xi^T(t)\overline{A}^T\widehat{R}_1\overline{A}(t)\xi(t) \\
&\quad + 3\tau_1^2\xi^T(t-\tau(t))\overline{A}_\tau^T(t-\tau(t))\widehat{R}_1\overline{A}_\tau(t-\tau(t))\xi(t \\
&\quad - \tau(t)) + 3\tau_1^2\omega^T(t)\overline{B}^T(t)\widehat{R}_1\overline{B}(t)\omega(t)
\end{aligned} \tag{30}
$$

where

$$
\begin{aligned}
\overline{A}^T(t)\widehat{R}_1\overline{A}(t) &= \left[\widehat{A}(t) + \Delta\widehat{A}(t)\right]^T\widehat{R}_1\left[\widehat{A}(t) + \Delta\widehat{A}(t)\right] \\
&= \left[\widehat{A}(t) + \widetilde{M}_2K_i(t)\widetilde{N}_2 + \widetilde{M}_1K_i(t)\widetilde{N}_1\right]^T \\
&\quad \cdot\widehat{R}_1\left[\widehat{A}(t) + \widetilde{M}_2K_i(t)\widetilde{N}_2 + \widetilde{M}_1K_i(t)\widetilde{N}_1\right]
\end{aligned} \tag{31}
$$

$$
\begin{aligned}
&\overline{A}_\tau^T(t)\widehat{R}_1\overline{A}_\tau(t) = \left[\widehat{A}_\tau(t) + \Delta\widehat{A}_\tau(t)\right]^T \\
&\quad \cdot\widehat{R}_1\left[\widehat{A}_\tau(t) + \Delta\widehat{A}_\tau(t)\right] = \left[\widehat{A}_\tau(t) + \widetilde{M}_2K_i(t)\widetilde{N}_{2\tau}\right]^T \\
&\quad \cdot\widehat{R}_1\left[\widehat{A}_\tau(t) + \widetilde{M}_2K_i(t)\widetilde{N}_{2\tau}\right]
\end{aligned} \tag{32}
$$

$$
\begin{aligned}
\overline{B}^T(t)\widehat{P}\overline{B}(t) &= \left[\widehat{B}_\tau(t) + \Delta\widehat{B}(t)\right]^T\widehat{R}_1\left[\widehat{B}(t) + \Delta\widehat{B}(t)\right] \\
&= \left[\widehat{B}(t) + \widetilde{M}_2K_i(t)\widetilde{N}_{2D}\right]^T \\
&\quad \cdot\widehat{R}_1\left[\widehat{B}_\tau(t) + \widetilde{M}_2K_i(t)\widetilde{N}_{2D}\right]
\end{aligned} \tag{33}
$$

By Lemma 3, if there exist $\varepsilon_{11} > 0$, $\varepsilon_{12} > 0$ such that $\widehat{R}_1^{-1} - \varepsilon_{11}\widetilde{M}_1\widetilde{M}_1^T - \varepsilon_{12}\widetilde{M}_2\widetilde{M}_2^T > 0$, from (31), it follows that

$$
\begin{aligned}
&\left[\widehat{A}(t) + \widetilde{M}_2K_i(t)\widetilde{N}_2 + \widetilde{M}_1K_i(t)\widetilde{N}_1\right]^T \\
&\quad \cdot\widehat{R}_1\left[\widehat{A}(t) + \widetilde{M}_2K_i(t)\widetilde{N}_2 + \widetilde{M}_1K_i(t)\widetilde{N}_1\right]
\end{aligned}
$$

$$\leq \left[\widehat{A}^T(t) + \widetilde{M}_2 K_i(t)\widetilde{N}_2\right]^T \left[\widehat{R}_1^{-1} - \varepsilon_{11}\widetilde{M}_1\widetilde{M}_1^T\right]^{-1} \cdot \left[\widehat{A}(t) + \widetilde{M}_2 K_i(t)\widetilde{N}_2\right] + \varepsilon_{11}^{-1}\widetilde{N}_1^T\widetilde{N}_1 \leq \widehat{A}^T(t) \cdot \left[\widehat{R}_1^{-1} - \varepsilon_{11}\widetilde{M}_1\widetilde{M}_1^T - \varepsilon_{12}\widetilde{M}_2\widetilde{M}_2^T\right]^{-1}\widehat{A}(t) + \varepsilon_{12}^{-1}\widetilde{N}_2^T\widetilde{N}_2 + \varepsilon_{11}^{-1}\widetilde{N}_1^T\widetilde{N}_1 \quad (34)$$

By Lemma 3, if there exist $\varepsilon_{15} > 0$, such that $\widehat{R}_1^{-1} - \varepsilon_{15}\widetilde{M}_2\widetilde{M}_2^T > 0$, from (32), it follows that

$$\left[\widehat{A}_\tau(t) + \widetilde{M}_2 K_i(t)\widetilde{N}_{2\tau}\right]^T \widehat{R}_1 \left[\widehat{A}_\tau(t) + \widetilde{M}_2 K_i(t)\widetilde{N}_{2\tau}\right] \leq \widehat{A}_\tau^T(t)\left[\widehat{R}_1^{-1} - \varepsilon_{15}\widetilde{M}_2\widetilde{M}_2^T\right]^{-1}\widehat{A}_\tau(t) + \varepsilon_{15}^{-1}\widetilde{N}_{2\tau}^T\widetilde{N}_{2\tau} \quad (35)$$

By Lemma 3, if there exist $\varepsilon_{19} > 0$ such that $\widehat{R}_1^{-1} - \varepsilon_{19}\widetilde{M}_2\widetilde{M}_2^T > 0$, from (33), we can get

$$\left[\widehat{B}(t) + \widetilde{M}_2 K_i(t)\widetilde{N}_{2D}\right]^T \widehat{R}_1 \left[\widehat{B}_\tau(t) + \widetilde{M}_2 K_i(t)\widetilde{N}_{2D}\right] \leq \widehat{B}^T(t)\left[\widehat{R}_1^{-1} - \varepsilon_{19}\widetilde{M}_2\widetilde{M}_2^T\right]^{-1}\widehat{B}(t) + \varepsilon_{19}^{-1}\widetilde{N}_{2D}^T\widetilde{N}_{2D} \quad (36)$$

Combining with formulas (31)~(36), from (30), we have

$$\tau_1^2\dot{\xi}^T(t)\widehat{R}_1\dot{\xi}(t) \leq \xi^T(t)\left\{3\tau_1^2\widehat{A}^T(t) \cdot \left[\widehat{R}_1^{-1} - \varepsilon_{11}\widetilde{M}_1\widetilde{M}_1^T - \varepsilon_{12}\widetilde{M}_2\widetilde{M}_2^T\right]^{-1}\widehat{A}(t) + \varepsilon_{12}^{-1}\widetilde{N}_2^T\widetilde{N}_2 + \varepsilon_{11}^{-1}\widetilde{N}_1^T\widetilde{N}_1\right\}\xi(t) + \xi^T(t-\tau(t)) \cdot \left[3\tau_1^2\widehat{A}^T{}_\tau(t)\left[\widehat{R}_1^{-1} - \varepsilon_{15}\widetilde{M}_2\widetilde{M}_2^T\right]\widehat{A}_\tau^T(t) + \varepsilon_{15}^{-1}\widetilde{N}_{2\tau}^T\widetilde{N}_{2\tau}\right]\xi^T(t-\tau(t)) + \omega^T(t)\left[3\tau_1^2\widehat{B}^T(t) \cdot \left[\widehat{R}_1^{-1} - \varepsilon_{19}\widetilde{M}_2\widetilde{M}_2^T\right]\widehat{B}(t) + \varepsilon_{19}^{-1}\widetilde{N}_{2D}^T\widetilde{N}_{2D}\right]\omega(t) \quad (37)$$

Similarly, for formula (30) we have

$$(\tau_2-\tau_1)^2\dot{\xi}^T(t)\widehat{R}_2\dot{\xi}(t) \leq \xi^T(t)\left[3(\tau_2-\tau_1)^2\widehat{A}^T(t) \cdot \left[\widehat{R}_2^{-1} - \varepsilon_{21}\widetilde{M}_1\widetilde{M}_1^T - \varepsilon_{22}\widetilde{M}_2\widetilde{M}_2^T\right]^{-1}\widehat{A}(t) + 3(\tau_2-\tau_1)^2\varepsilon_{22}^{-1}\widetilde{N}_2{}^T\widetilde{N}_2 + 3(\tau_2-\tau_1)^2 \cdot \varepsilon_{21}^{-1}\widetilde{N}_1{}^T\widetilde{N}_1\right]\xi(t) + \xi^T(t-\tau(t))\left[3(\tau_2-\tau_1)^2 \cdot \widehat{A}^T(t)\left[\widehat{R}_2^{-1} - \varepsilon_{25}\widetilde{M}_2\widetilde{M}_2^T\right]^{-1}\widehat{A}(t) + 3(\tau_2-\tau_1)^2 \cdot \varepsilon_{25}^{-1}\widetilde{N}_1{}^T\widetilde{N}_1\right]\xi(t-\tau(t)) + \omega^T(t)\left[3(\tau_2-\tau_1)^2 \cdot \widehat{B}^T(t)\left[\widehat{R}_2^{-1} - \varepsilon_{29}\widetilde{M}_2\widetilde{M}_2^T\right]^{-1}\widehat{B}(t) + 3(\tau_2-\tau_1)^2 \cdot \varepsilon_{29}^{-1}\widetilde{N}_{2D}^T\widetilde{N}_{2D}\right]\omega(t) \quad (38)$$

By Lemma 5, according to Jensen's inequality, we have that

$$-\tau_1\int_{t-\tau_1}^{t}\dot{\xi}^T(s)\widehat{R}_1\dot{\xi}(s)\,ds \leq \begin{bmatrix}\xi(t)\\ \xi(t-\tau_1)\end{bmatrix}^T\begin{bmatrix}-\widehat{R}_1 & \widehat{R}_1\\ * & -\widehat{R}_1\end{bmatrix}\begin{bmatrix}\xi(t)\\ \xi(t-\tau_1)\end{bmatrix} \quad (39)$$

$$-(\tau_2-\tau_1)\int_{t-\tau_2}^{t-\tau_1}\dot{\xi}^T(s)\widehat{R}_2\dot{\xi}(s)\,ds \leq \begin{bmatrix}\xi(t-\tau_1)\\ \xi(t-\tau_2)\end{bmatrix}^T\begin{bmatrix}-\widehat{R}_2 & \widehat{R}_2\\ * & -\widehat{R}_2\end{bmatrix}\begin{bmatrix}\xi(t-\tau_1)\\ \xi(t-\tau_2)\end{bmatrix} \quad (40)$$

Combining with formulas (25)~(30), from (23), we have

$$\dot{V}(\xi(t)) \leq \eta^T(t)\overline{\Phi}\eta(t) \quad (41)$$

where

$$\overline{\Phi} = \begin{bmatrix}\overline{\Phi}_{11} & \overline{\Phi}_{12} & \widehat{R}_1 & 0 & \overline{\Phi}_{15}\\ * & \overline{\Phi}_{22} & 0 & 0 & 0\\ * & * & \overline{\Phi}_{33} & 0 & 0\\ * & * & * & \overline{\Phi}_{44} & 0\\ * & * & * & * & \overline{\Phi}_{55}\end{bmatrix} < 0$$

$$\overline{\Phi}_{11} = \widehat{P}\widehat{A}(t) + \widehat{A}^T(t)\widehat{P} + \widehat{Q}_1 + \widehat{Q}_2 + \widehat{Q}_3 - \widehat{R}_1 + \alpha_4^2\widehat{P}\widetilde{M}_2\widetilde{M}_2^T\widehat{P} + \alpha_3^2\widehat{P}\widetilde{M}_1\widetilde{M}_1^T\widehat{P} + \alpha_5^2\widetilde{N}_1\widetilde{N}_1 + \alpha_6^2\widetilde{N}_2\widetilde{N}_2 + \beta_1^2\widehat{A}^T(t)\left[\widehat{R}_1^{-1} - \varepsilon_{11}\widetilde{M}_1\widetilde{M}_1^T - \varepsilon_{12}\widetilde{M}_2\widetilde{M}_2^T\right]^{-1}\widehat{A}(t) + \beta_2^2\widehat{A}^T(t)\left[\widehat{R}_2^{-1} - \varepsilon_{21}\widetilde{M}_1\widetilde{M}_1^T - \varepsilon_{22}\widetilde{M}_2\widetilde{M}_2^T\right]^{-1}\widehat{A}(t)$$

$\overline{\Phi}_{12} = \widehat{P}\widehat{A}_\tau(t)$,

$\overline{\Phi}_{15} = \widehat{P}\widehat{B}(t)$,

$\overline{\Phi}_{33} = -\widehat{Q}_2 - \left(\widehat{R}_1 + \widehat{R}_2\right)$,

$\overline{\Phi}_{44} = -\widehat{Q}_3 - \widehat{R}_2$,

$$\overline{\Phi}_{22} = -(1-d)\widehat{Q}_1 + \varepsilon_6^{-1}\widetilde{N}_{2\tau}^T\widetilde{N}_{2\tau} + \alpha_2^2\widetilde{N}_{1\tau}^T\widetilde{N}_{1\tau} + \beta_1^2\widehat{A}_\tau^T(t)\left[\widehat{R}_1^{-1} - \varepsilon_{15}\widetilde{M}_2\widetilde{M}_2^T\right]^{-1}\widehat{A}_\tau(t)$$

$$+\beta_2^2\widehat{A}_\tau^T(t)\left[\widehat{R}_2^{-1}-\varepsilon_{25}\widetilde{M}_2\widetilde{M}_2^T\right]^{-1}\widehat{A}_\tau(t)$$

$$\overline{\Phi}_{55}=\alpha_1^2\widetilde{N}_{2D}^T\widetilde{N}_{2D}+\beta_1^2\widehat{B}^T(t)\left[\widehat{R}_1^{-1}-\varepsilon_{19}\widetilde{M}_2\widetilde{M}_2^T\right]^{-1}\widehat{B}(t)+\beta_2^2\widehat{B}^T(t)\left[\widehat{R}_2^{-1}-\varepsilon_{29}\widetilde{M}_2\widetilde{M}_2^T\right]^{-1}\widehat{B}(t) \tag{42}$$

By Schur complement and formula (26), we have

$$\dot{V}(\xi(t))+e^T(t)e(t)-\gamma^2\omega^T(t)\omega(t)\le\eta^T(t)\Phi\eta(t) \tag{43}$$

where

$$\eta^T(t)=\left[\xi^T(t)\;\;\xi^T(t-\tau(t))\;\;\xi^T(t-\tau_1)\;\;\xi^T(t-\tau_2)\;\;\omega^T(t)\right] \tag{44}$$

Consequently, it follows from inequality (20), $V(\xi(t))|_{t=0}=0$ and $V(\xi(t))|_{t=L}\ge 0$, and we have $\int_0^L(\|e(t)\|^2-\gamma^2\|\omega(t)\|^2)dt+V(\xi(t))|_{t=L}-V(\xi(t))|_{t=0}\le 0$, which implies that (9) holds.

Thus, H_∞ performance is verified. In addition, when the zero disturbance input $\omega(t)=0$, by Schur complement, we can obtain that the time derivative of Lyapunov-Krasovskii $\dot{V}(\xi(t))\le 0$; that means that the filtering error system (7) with $\omega(t)=0$ is asymptotically stable. □

4. Fuzzy H_∞ Filter Design

Theorem 6 provides a sufficient condition for H_∞ filter design with time delay and satisfied the H_∞ performance. However, there exist some coupled matrix variables in the matrix inequality (15); H_∞ filter parameter can not be calculated directly. In order to decouple the variables in (15), we will use decoupling technique. Using this method, inequality (15) can be equivalently expressed in another form; hence, we can obtain H_∞ filter parameter.

Theorem 7. *For given scalars $\delta>0$, $\varepsilon_{2i}>0$, $\varepsilon_{6i}>0$, $\varepsilon_{10i}>0$, $\varepsilon_{19i}>0$, $\varepsilon_{29i}>0$, $\varepsilon_{15i}>0$, $\varepsilon_{25i}>0$, $\varepsilon_{20i}>0$, $\varepsilon_{21i}>0$, $\varepsilon_{22i}>0$, $\varepsilon_{11i}>0$, and $\varepsilon_{12i}>0$, the filtering error system (6) is asymptotically stable as well as with the H_∞ performance level γ, if there exist matrices $\widehat{P}>0$, $\widehat{Q}_1>0$, $\widehat{Q}_2>0$, $\widehat{Q}_3>0$, $\widehat{R}_1>0$, $\widehat{R}_2>0$, A_{fi}, B_{fi}, and C_{fi} such that the following linear matrix inequalities are satisfied:*

$$\begin{bmatrix}\widehat{R}_1^{-1} & \widetilde{M}_{2i}\\ * & \varepsilon_{1ki}^{-1}\end{bmatrix}>0,\quad(k=5,9) \tag{45}$$

$$\begin{bmatrix}\widehat{R}_2^{-1} & \widetilde{M}_{2i}\\ * & \varepsilon_{2ki}^{-1}\end{bmatrix}>0,\quad(k=5,9) \tag{46}$$

$$\begin{bmatrix}\widehat{R}_1^{-1} & \alpha_{11i}\widetilde{M}_{1i} & \alpha_{12i}\widetilde{M}_{2i}\\ * & I & 0\\ * & * & I\end{bmatrix}>0 \tag{47}$$

$$\begin{bmatrix}\widehat{R}_2^{-1} & \alpha_{21i}\widetilde{M}_{1i} & \alpha_{22i}\widetilde{M}_{2i}\\ * & I & 0\\ * & * & I\end{bmatrix}>0 \tag{48}$$

$$\Psi_{ij}=\begin{bmatrix}\Psi_1 & \Psi_2\\ * & \Psi_4\end{bmatrix} \tag{49}$$

where

$$\Psi_1=\begin{bmatrix}\Psi_{111} & \Psi_{112} & \Psi_{113} & \Psi_{114} & 0 & \Psi_{115} & 0 & \Psi_{118}\\ * & -I & 0 & 0 & 0 & 0 & 0 & 0\\ * & * & \Psi_{133} & 0 & 0 & 0 & 0 & 0\\ * & * & * & \Psi_{144} & \Psi_{145} & 0 & 0 & 0\\ * & * & * & * & -I & 0 & 0 & 0\\ * & * & * & * & * & \Psi_{166} & \Psi_{167} & 0\\ * & * & * & * & * & * & -I & 0\\ * & * & * & * & * & * & * & -I\end{bmatrix}$$

$$\Psi_{111}=\begin{bmatrix}\psi_{11} & \psi_{12} & 0 & 0 & \psi_{15}\\ * & -(1-d)\widehat{Q}_1 & 0 & 0 & 0\\ * & * & -\widehat{Q}_2-\left(\widehat{R}_1+\widehat{R}_2\right) & 0 & 0\\ * & * & * & -\widehat{Q}_3-\widehat{R}_2 & 0\\ * & * & * & * & -\gamma^2 I\end{bmatrix}$$

$$\psi_{11} = \begin{bmatrix} P_{11}A_j + \overline{B}_{fi}C_j & \overline{A}_{fi} \\ P_{12}^T A_j + \overline{B}_{fi}C_j & \overline{A}_{fi} \end{bmatrix} + \begin{bmatrix} P_{11}A_j + \overline{B}_{fi}C_j & \overline{A}_{fi} \\ P_{12}^T A_j + \overline{B}_{fi}C_j & \overline{A}_{fi} \end{bmatrix}^T + \widehat{Q}_1 + \widehat{Q}_2 + \widehat{Q}_3 - \widehat{R}_1,$$

$$\psi_{12} = \begin{bmatrix} P_{11}A_{\tau j} + \overline{B}_{fi}C_{\tau j} & 0 \\ P_{12}^T A_{\tau j} + \overline{B}_{fi}C_{\tau j} & 0 \end{bmatrix},$$

$$\psi_{15} = \begin{bmatrix} P_{11}B_j + \overline{B}_{fi}D_j \\ P_{12}^T B_j + \overline{B}_{fi}D_j \end{bmatrix},$$

$$\Psi_{112} = \begin{bmatrix} \alpha_{4i}\widehat{P}\widetilde{M}_{2i} & \alpha_{3i}\widehat{P}\widetilde{M}_{1i} \\ 0 & 0 \end{bmatrix},$$

$$\Psi_{113} = \begin{bmatrix} \alpha_{5i}\widetilde{N}_{1i}^T & \alpha_{6i}\widetilde{N}_{2i}^T & \alpha_{20i}\widetilde{N}_{3i}^T & \widehat{E}_j^T \\ 0 & 0 & 0 & 0 \end{bmatrix},$$

$$\Psi_{114} = \begin{bmatrix} \beta_{2i}\widehat{A}_j^T\widehat{P} \\ 0 \end{bmatrix},$$

$$\Psi_{115} = \begin{bmatrix} \beta_{2i}\widehat{A}_j^T\widehat{P} \\ 0 \end{bmatrix},$$

$$\Psi_{118} = \begin{bmatrix} 0 \\ \alpha_{7i}\widetilde{N}_{2\tau i}^T \\ 0 \\ 0 \\ 0 \end{bmatrix},$$

$$\Psi_{133} = -\mathrm{diag}\left(I, I, I, I - \alpha_{20i}^2\widetilde{M}_{3i}\widetilde{M}_{3i}^T\right),$$

$$\Psi_{144} = -2\delta\widehat{P} + \delta^2\widehat{R}_1,$$

$$\Psi_{145} = \left[\alpha_{11i}\widehat{P}\widetilde{M}_{1i} \;\; \alpha_{12i}\widehat{P}\widetilde{M}_{2i}\right],$$

$$\Psi_{166} = -2\delta\widehat{P} + \delta^2\widehat{R}_2,$$

$$\Psi_{167} = \left[\alpha_{21i}\widehat{P}\widetilde{M}_{1i} \;\; \alpha_{22i}\widehat{P}\widetilde{M}_{2i}\right],$$

$$\Psi_2 = \begin{bmatrix} \Psi_{211} & 0 & \Psi_{213} & 0 & \Psi_{215} & \Psi_{216} & 0 & \Psi_{218} & 0 \\ 0 & 0 & 0 & 0 & 0 & 0 & 0 & 0 & 0 \\ 0 & 0 & 0 & 0 & 0 & 0 & 0 & 0 & 0 \\ 0 & 0 & 0 & 0 & 0 & 0 & 0 & 0 & 0 \\ 0 & 0 & 0 & 0 & 0 & 0 & 0 & 0 & 0 \\ 0 & 0 & 0 & 0 & 0 & 0 & 0 & 0 & 0 \\ 0 & 0 & 0 & 0 & 0 & 0 & 0 & 0 & 0 \\ 0 & 0 & 0 & 0 & 0 & 0 & 0 & 0 & 0 \end{bmatrix},$$

$$\Psi_{211} = \begin{bmatrix} 0 \\ \beta_{1i}\widehat{A}_j^T\widehat{P} \\ 0 \\ 0 \\ 0 \end{bmatrix},$$

$$\Psi_{213} = \begin{bmatrix} 0 \\ \beta_{2i}\widehat{A}_{\tau j}^T\widehat{P} \\ 0 \\ 0 \\ 0 \end{bmatrix},$$

$$\Psi_{215} = \begin{bmatrix} 0 \\ 0 \\ 0 \\ 0 \\ \alpha_{1i}\widetilde{N}_{2Di}^T \end{bmatrix},$$

$$\Psi_{216} = \begin{bmatrix} 0 \\ 0 \\ 0 \\ 0 \\ \beta_{1i}\widehat{B}_j^T\widehat{P} \end{bmatrix},$$

$$\Psi_{218} = \begin{bmatrix} 0 \\ 0 \\ 0 \\ 0 \\ \beta_{2i}\widehat{B}_j^T\widehat{P} \end{bmatrix},$$

$$\Psi_4 = \begin{bmatrix} \Psi_{411} & \alpha_{15i}\widehat{P}\widetilde{M}_{2i} & 0 & 0 & 0 & 0 & 0 & 0 & 0 \\ * & -I & 0 & 0 & 0 & 0 & 0 & 0 & 0 \\ * & * & \Psi_{433} & \alpha_{25i}\widehat{P}\widetilde{M}_{2i} & 0 & 0 & 0 & 0 & 0 \\ * & * & * & -I & 0 & 0 & 0 & 0 & 0 \\ * & * & * & * & -I & 0 & 0 & 0 & 0 \\ * & * & * & * & * & \Psi_{466} & \alpha_{19i}\widehat{P}\widetilde{M}_{2i} & 0 & 0 \\ * & * & * & * & * & * & -I & 0 & 0 \\ * & * & * & * & * & * & * & \Psi_{488} & \alpha_{29i}\widehat{P}\widetilde{M}_{2i} \\ * & * & * & * & * & * & * & * & -I \end{bmatrix}$$

$$\Psi_{411} = -2\delta\widehat{P} + \delta^2\widehat{R}_1,$$

$$\Psi_{433} = -2\delta\widehat{P} + \delta^2\widehat{R}_2,$$

$$\Psi_{466} = -2\delta\widehat{P} + \delta^2\widehat{R}_1,$$

$$\Psi_{488} = -2\delta\widehat{P} + \delta^2\widehat{R}_2,$$

$$\widetilde{M}_{2i} = \begin{bmatrix} 0 \\ B_{fi}M_{2i} \end{bmatrix},$$

$$\widetilde{M}_{1i} = \begin{bmatrix} 0 \\ A_{fi}M_{1i} \end{bmatrix},$$

$$\widetilde{M}_{3i} = M_{3i},$$

$$\widetilde{N}_{1i} = \begin{bmatrix} 0 & N_{1i} \end{bmatrix},$$

$$\widetilde{N}_{2i} = \begin{bmatrix} N_{2i}C_j & 0 \end{bmatrix},$$

$$\widetilde{N}_{2\tau i} = \begin{bmatrix} N_{2i}C_{\tau j} & 0 \end{bmatrix},$$

$$\widetilde{N}_{2Di} = N_{2i}C_{Dj},$$

$$\widetilde{N}_{3i} = \begin{bmatrix} 0 & -N_{3i}C_{fi} \end{bmatrix} \tag{50}$$

We can obtain the filter parameters as follows:

$$A_{fi} = P_{12}^{-1}\overline{A}_{fi},$$
$$B_{fi} = P_{12}^{-1}\overline{B}_{fi}, \tag{51}$$
$$C_{fi} = \overline{C}_{fi}$$

Proof. By Schur complement formula, the matrix inequality conditions (20) in Theorem 6 can be described as the following matrix inequalities:

$$\Gamma = \begin{bmatrix} \Gamma_1 & \Gamma_2 \\ * & \Gamma_4 \end{bmatrix} \tag{52}$$

where

$$\Gamma_1 = \begin{bmatrix} \Gamma_{111} & \Gamma_{112} & \Gamma_{113} & \Gamma_{114} & 0 & \Gamma_{115} & 0 & \Gamma_{118} \\ * & -I & 0 & 0 & 0 & 0 & 0 & 0 \\ * & * & \Gamma_{133} & 0 & 0 & 0 & 0 & 0 \\ * & * & * & -\widehat{R}_1^{-1} & \Gamma_{145} & 0 & 0 & 0 \\ * & * & * & * & -I & 0 & 0 & 0 \\ * & * & * & * & * & -\widehat{R}_2^{-1} & \Gamma_{167} & 0 \\ * & * & * & * & * & * & -I & 0 \\ * & * & * & * & * & * & * & -I \end{bmatrix}$$

$$\Gamma_{111} = \begin{bmatrix} \widehat{P}\widehat{A}(t) + \widehat{A}^T(t)\widehat{P} + \widehat{Q}_1 + \widehat{Q}_2 + \widehat{Q}_3 & \widehat{P}\widehat{A}_\tau(t) & 0 & 0 & \widehat{P}\widehat{B}(t) \\ * & -(1-d)\widehat{Q}_1 & 0 & 0 & 0 \\ * & * & -\widehat{Q}_2 & 0 & 0 \\ * & * & * & -\widehat{Q}_3 & 0 \\ * & * & * & * & -\gamma^2 I \end{bmatrix}$$

$$\Gamma_{112} = \begin{bmatrix} \alpha_4\widehat{P}\widetilde{M}_2 & \alpha_3\widehat{P}\widetilde{M}_1 \\ 0 & 0 \end{bmatrix},$$

$$\Gamma_{114} = \begin{bmatrix} \beta_1 \widehat{A}^T(t)\,\widehat{P} \\ 0 \end{bmatrix},$$

$$\Psi_{113} = \begin{bmatrix} \alpha_5 \widetilde{N}_1^T & \alpha_6 \widetilde{N}_2^T & \alpha_{20} \widetilde{N}_3^T & \widehat{E}^T \\ 0 & 0 & 0 & 0 \end{bmatrix},$$

$$\Gamma_{115} = \begin{bmatrix} \beta_2 \widehat{A}^T(t)\,\widehat{P} \\ 0 \end{bmatrix},$$

$$\Gamma_{118} = \begin{bmatrix} 0 \\ \alpha_7 \widetilde{N}_{2\tau}^T \\ 0 \\ 0 \\ 0 \end{bmatrix},$$

$$\Gamma_{133} = -\mathrm{diag}\left(I, I, I, I - \alpha_{20}^2 \widetilde{M}_3 \widetilde{M}_3^T\right),$$

$$\Gamma_{145} = \left[\alpha_{11} \widehat{P}\widetilde{M}_1 \quad \alpha_{12} \widehat{P}\widetilde{M}_2\right],$$

$$\Gamma_{167} = \left[\alpha_{21} \widehat{P}\widetilde{M}_1 \quad \alpha_{22} \widehat{P}\widetilde{M}_2\right],$$

$$\Gamma_2 = \begin{bmatrix} \Gamma_{211} & 0 & \Gamma_{213} & 0 & \Gamma_{215} & \Gamma_{216} & 0 & \Gamma_{218} & 0 \\ 0 & 0 & 0 & 0 & 0 & 0 & 0 & 0 & 0 \\ 0 & 0 & 0 & 0 & 0 & 0 & 0 & 0 & 0 \\ 0 & 0 & 0 & 0 & 0 & 0 & 0 & 0 & 0 \\ 0 & 0 & 0 & 0 & 0 & 0 & 0 & 0 & 0 \\ 0 & 0 & 0 & 0 & 0 & 0 & 0 & 0 & 0 \\ 0 & 0 & 0 & 0 & 0 & 0 & 0 & 0 & 0 \\ 0 & 0 & 0 & 0 & 0 & 0 & 0 & 0 & 0 \end{bmatrix},$$

$$\Gamma_{211} = \begin{bmatrix} 0 \\ \beta_1 \widehat{A}^T(t)\,\widehat{P} \\ 0 \\ 0 \\ 0 \end{bmatrix},$$

$$\Gamma_{213} = \begin{bmatrix} 0 \\ \beta_2 \widehat{A}_\tau^T(t)\,\widehat{P} \\ 0 \\ 0 \\ 0 \end{bmatrix},$$

$$\Gamma_{215} = \begin{bmatrix} 0 \\ 0 \\ 0 \\ 0 \\ \alpha_1 \widetilde{N}_{2D}^T \end{bmatrix},$$

$$
\Gamma_{216} = \begin{bmatrix} 0 \\ 0 \\ 0 \\ 0 \\ \beta_1 \widehat{B}^T(\mathrm{t}) \widehat{P} \end{bmatrix},
$$

$$
\Gamma_{218} = \begin{bmatrix} 0 \\ 0 \\ 0 \\ 0 \\ \beta_2 \widehat{B}^T(\mathrm{t}) \widehat{P} \end{bmatrix},
$$

$$
\Gamma_4 = \begin{bmatrix}
\Gamma_{411} & \alpha_{15}\widehat{P}\widetilde{M}_2 & 0 & 0 & 0 & 0 & 0 & 0 & 0 \\
* & -I & 0 & 0 & 0 & 0 & 0 & 0 & 0 \\
* & * & \Gamma_{433} & \alpha_{25}\widehat{P}\widetilde{M}_2 & 0 & 0 & 0 & 0 & 0 \\
* & * & * & -I & 0 & 0 & 0 & 0 & 0 \\
* & * & * & * & -I & 0 & 0 & 0 & 0 \\
* & * & * & * & * & \Gamma_{466} & \alpha_{19}\widehat{P}\widetilde{M}_2 & 0 & 0 \\
* & * & * & * & * & * & -I & 0 & 0 \\
* & * & * & * & * & * & * & \Gamma_{488} & \alpha_{29}\widehat{P}\widetilde{M}_2 \\
* & * & * & * & * & * & * & * & -I
\end{bmatrix}
$$

$$
\begin{aligned}
\Gamma_{411} &= -2\delta\widehat{P} + \delta^2\widehat{R}_1, \\
\Gamma_{433} &= -2\delta\widehat{P} + \delta^2\widehat{R}_2, \\
\Gamma_{466} &= -2\delta\widehat{P} + \delta^2\widehat{R}_1, \\
\Gamma_{488} &= -2\delta\widehat{P} + \delta^2\widehat{R}_2.
\end{aligned} \tag{53}
$$

Let $\widehat{P} = \begin{bmatrix} P_{11} & P_{12} \\ * & P_{12} \end{bmatrix}$, $\widehat{P}\widehat{A}(\mathrm{t}) = \begin{bmatrix} P_{11}A(\mathrm{t})+\overline{B}_f(\mathrm{t})C(\mathrm{t}) & \overline{A}_f(\mathrm{t}) \\ P_{12}^T A(\mathrm{t})+\overline{B}_f(\mathrm{t})C(\mathrm{t}) & \overline{A}_f(\mathrm{t}) \end{bmatrix}$, $\widehat{P}\widehat{A}_\tau(\mathrm{t}) = \begin{bmatrix} P_{11}A_\tau(\mathrm{t})+\overline{B}_f(\mathrm{t})C_\tau(\mathrm{t}) & 0 \\ P_{12}^T A_\tau(\mathrm{t})+\overline{B}_f(\mathrm{t})C_\tau(\mathrm{t}) & 0 \end{bmatrix}$, $\widehat{P}\widehat{B}(\mathrm{t}) = \begin{bmatrix} P_{11}B(\mathrm{t})+\overline{B}_f(\mathrm{t})D(\mathrm{t}) \\ P_{12}^T B(\mathrm{t})+\overline{B}_f(\mathrm{t})D(\mathrm{t}) \end{bmatrix}$, $\beta_1 = \sqrt{3\tau_1^2}$, $\beta_2 = \sqrt{3(\tau_2^2 - \tau_1^2)}$, $\alpha_1 = \sqrt{\beta_1^2\varepsilon_{19}^{-1} + \beta_2^2\varepsilon_{29}^{-1} + \varepsilon_{10}^{-1}}$, $\alpha_2 = \sqrt{\beta_1^2\varepsilon_{15}^{-1} + \beta_2^2\varepsilon_{25}^{-1} + \varepsilon_6^{-1}}$, $\alpha_3 = \sqrt{\varepsilon_{21}}$, $\alpha_4 = \sqrt{\varepsilon_2 + \varepsilon_6 + \varepsilon_{10}}$, $\alpha_5 = \sqrt{\varepsilon_{21}^{-1} + \beta_1^2\varepsilon_{11}^{-1} + \beta_2^2\varepsilon_{22}^{-1}}$, $\alpha_6 = \sqrt{\beta_1^2\varepsilon_{12}^{-1} + \beta_2^2\varepsilon_{22}^{-1}}$, $\alpha_7 = \sqrt{\varepsilon_6}$, $\alpha_{11} = \sqrt{\varepsilon_{11}}$, $\alpha_{12} = \sqrt{\varepsilon_{12}}$, $\alpha_{15} = \sqrt{\varepsilon_{15}}$, $\alpha_{25} = \sqrt{\varepsilon_{25}}$, $\alpha_{19} = \sqrt{\varepsilon_{19}}$, $\alpha_{20} = \sqrt{\varepsilon_{20}}$, $\alpha_{21} = \sqrt{\varepsilon_{21}}$, $\alpha_{22} = \sqrt{\varepsilon_{22}}$, $\alpha_{29} = \sqrt{\varepsilon_{29}}$, $\Delta(\mathrm{t}) = \sum_{i=1}^{r} h_i(\theta(\mathrm{t}))\Delta_i(\mathrm{t})$, where $\Delta = A, B, D, A_\tau, \overline{A}_f, \overline{B}_f, \overline{C}_f$.

According to $(\delta\widehat{R} - \widehat{P})\widehat{R}^{-1}(\delta\widehat{R} - \widehat{P}) \geq 0$, we have that $-\widehat{P}\widehat{R}^{-1}\widehat{P} \leq -2\delta\widehat{P} + \delta^2\widehat{R}$ holds for any scalar $\delta > 0$. The filter parameters in (3) can be designed as (51). This completes the proof. □

5. Simulation Example

Consider the following nonlinear systems with time-varying delays:

$$
\begin{aligned}
\dot{x}(t) &= \sum_{i=1}^{2} h_i \left[A_i x(t) + A_{\tau i} x(t - \tau(t)) + B_i \omega(t)\right] \\
y(t) &= \sum_{i=1}^{2} h_i \left[C_i x(t) + C_{\tau i} x(t - \tau(t)) + D_i \omega(t)\right] \\
z(t) &= \sum_{i=1}^{2} h_i \left[E_i x(t)\right]
\end{aligned} \tag{54}
$$

where

$$
A_1 = \begin{bmatrix} -2.1 & 0.1 \\ 1 & -2 \end{bmatrix},
$$

$$
A_2 = \begin{bmatrix} -1.9 & 0 \\ -0.2 & -1.1 \end{bmatrix},
$$

$$
A_{\tau 1} = \begin{bmatrix} -1.1 & 0.1 \\ -0.8 & -0.9 \end{bmatrix},
$$

$$\begin{aligned} A_{\tau 1} &= \begin{bmatrix} -0.9 & 0.1 \\ -1.1 & -1.2 \end{bmatrix}, \\ B_1 &= \begin{bmatrix} 1 \\ -0.2 \end{bmatrix}, \\ B_1 &= \begin{bmatrix} 0.3 \\ 0.1 \end{bmatrix}, \\ D_1 &= 0.3, \\ D_2 &= -0.6, \\ C_1 &= \begin{bmatrix} 1 & 0 \end{bmatrix}, \\ C_2 &= \begin{bmatrix} 0.5 & 0.6 \end{bmatrix} \\ C_{\tau 1} &= \begin{bmatrix} -0.8 & 0.6 \end{bmatrix}, \\ C_{\tau 2} &= \begin{bmatrix} -0.2 & 1 \end{bmatrix}, \\ E_1 &= \begin{bmatrix} 1 & -0.5 \end{bmatrix}, \\ E_2 &= \begin{bmatrix} -0.2 & 0.3 \end{bmatrix} \end{aligned} \tag{55}$$

The known matrices in (5) are given by

$$\begin{aligned} M_{11} = M_{12} &= \begin{bmatrix} 0.5 & 0.4 \\ 0.3 & 0.2 \end{bmatrix}, \\ M_{21} = M_{22} &= \begin{bmatrix} 0.5 & 0.3 \end{bmatrix}, \\ M_{31} = M_{32} &= \begin{bmatrix} 0.08 \\ 0.02 \end{bmatrix}, \\ N_{11} = N_{12} &= \begin{bmatrix} 0.3 & 0.6 \\ 0.2 & 0.1 \end{bmatrix}, \\ N_{21} = N_{22} &= \begin{bmatrix} 0.1 \\ -0.2 \end{bmatrix}, \\ N_{31} = N_{32} &= \begin{bmatrix} 0.2 & 0.3 \end{bmatrix}, \\ K_l(\mathrm{t}) &= \sin(\mathrm{t}), \quad \text{where } l = A, B, C. \end{aligned} \tag{56}$$

The disturbance signal $\omega(\mathrm{t})$ is given as follows:

$$\omega(\mathrm{t}) = \begin{cases} \sin(\mathrm{t}), & 0 < \mathrm{t} < 15 \\ 0 & other. \end{cases} \tag{57}$$

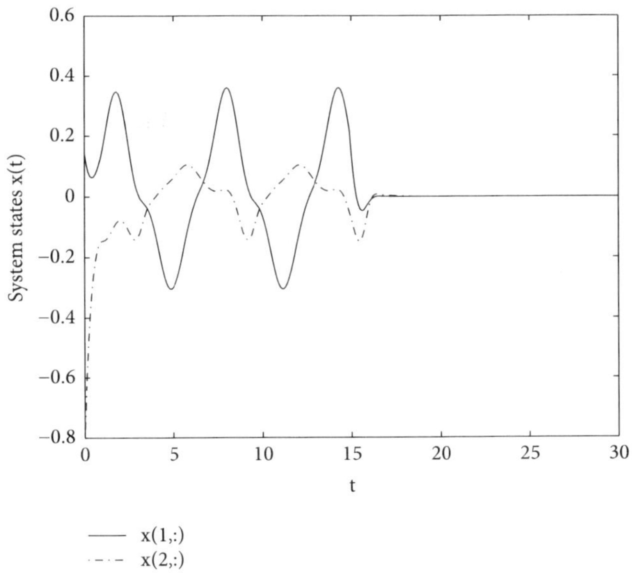

FIGURE 1: The response curve of system states x_1, x_2.

We can get the desired nonfragile filter by solving LMIS (45)-(49) in Theorem 7; the nonfragile parameter matrices are given as follows:

$$\begin{aligned} A_{f1} &= \begin{bmatrix} -3.1329 & 6.2268 \\ 0.7361 & -7.4593 \end{bmatrix}, \\ A_{f2} &= \begin{bmatrix} -2.7553 & 7.6590 \\ -0.8303 & -8.3416 \end{bmatrix}, \\ B_{f1} &= \begin{bmatrix} -0.4755 \\ -1.1658 \end{bmatrix}, \\ B_{f2} &= \begin{bmatrix} -0.2459 \\ -1.2213 \end{bmatrix}, \\ C_{f1} &= \begin{bmatrix} -0.2031 & 0.2124 \end{bmatrix}, \\ C_{f2} &= \begin{bmatrix} 0.2231 & -0.3065 \end{bmatrix} \end{aligned} \tag{58}$$

In addition, the new method provides less conservative design result; we can obtain a smaller $\gamma = 0.4996$. Figures 1 and 2 show the response of system's states $x(\mathrm{t})$ and filter's states $x_f(\mathrm{t})$, respectively. Figure 3 shows the trajectories of $z(\mathrm{t})$ and its estimates $z_f(\mathrm{t})$. The estimation error $e(\mathrm{t})$ is depicted in Figure 4.

The random disturbance signal is given as follows:

$$\omega(\mathrm{t}) = \begin{cases} \alpha(\mathrm{t}) \sin(\mathrm{t}), & 0 < \mathrm{t} < 15 \\ 0 & other, \end{cases} \tag{59}$$

where $\alpha(\mathrm{t})$ denotes Bernoulli's random event. The simulation results are obtained as shown in Figures 5–9.

The stochastic variables $\alpha(\mathrm{t})$ are Bernoulli-distributed white sequences; Figure 5 shows the event occurrence

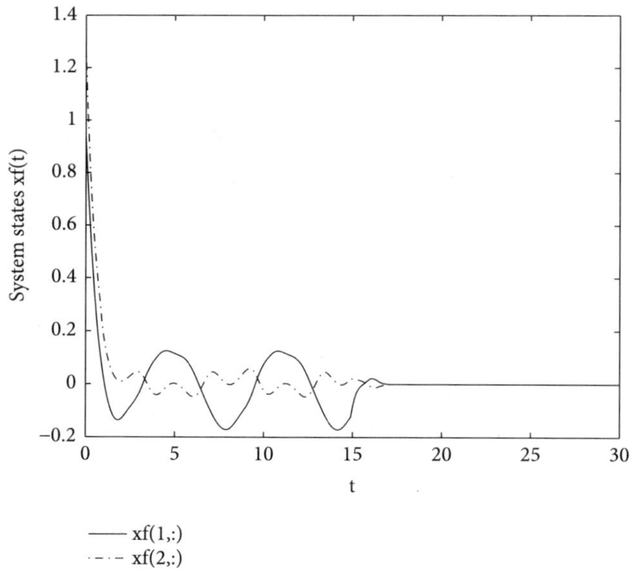

FIGURE 2: The response curve of system states x_{f1}, x_{f2}.

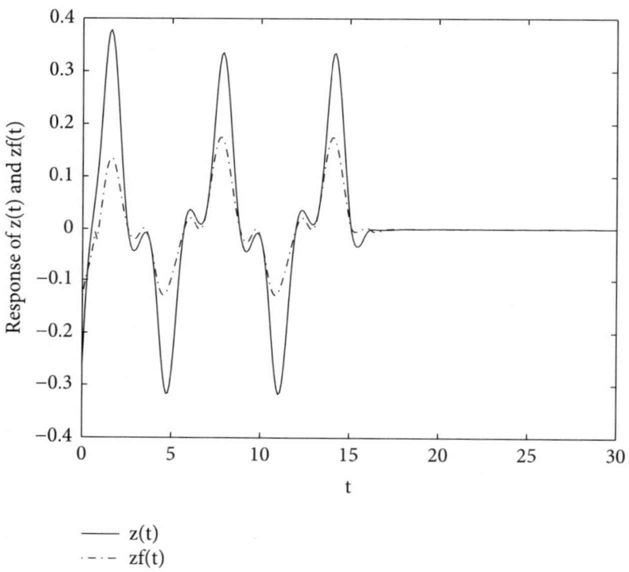

FIGURE 3: The response curve of vector $z(t), z_f(t)$.

probability. Figures 6 and 7 show the response of system's states $x_1(t), x_2(t)$ and filter's states $x_{f1}(t), x_{f2}(t)$, respectively. Figure 8 shows the trajectories of $z(t)$ and its estimates $z_f(t)$. The system error $e(t)$ is depicted in Figure 9.

The results show that the H_∞ nonfragile filter can make the system have good stability. The system generates small overshoot; this filter can efficiently reduce the influence of external disturbance and uncertainty; it can enhance control precision and dynamic qualities of the system.

6. Conclusions

This paper studied the fuzzy nonfragile H_∞ filter design problem for a class of nonlinear systems with an interval

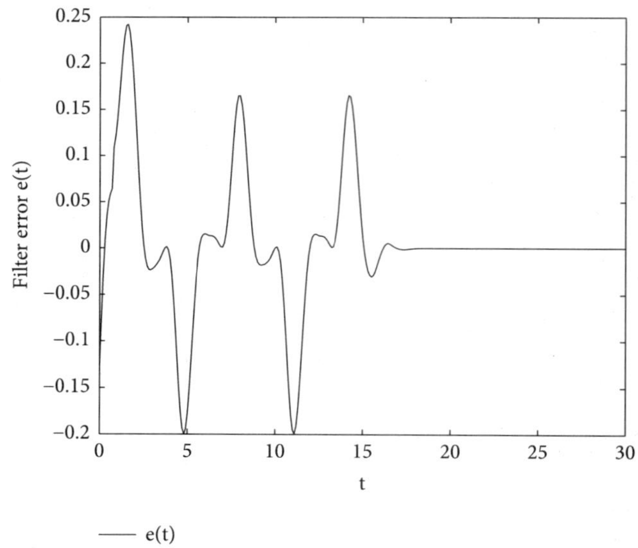

FIGURE 4: The response curve of vector $e(t)$.

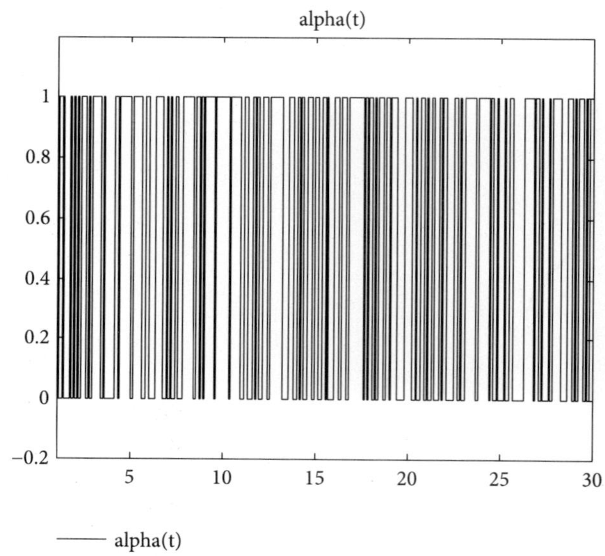

FIGURE 5: The event occurrence probability $\alpha(t)$.

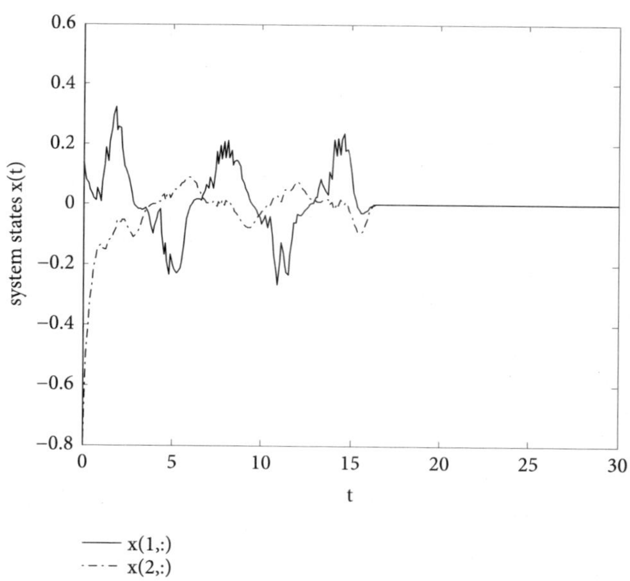

FIGURE 6: The response curve of system states x_1, x_2.

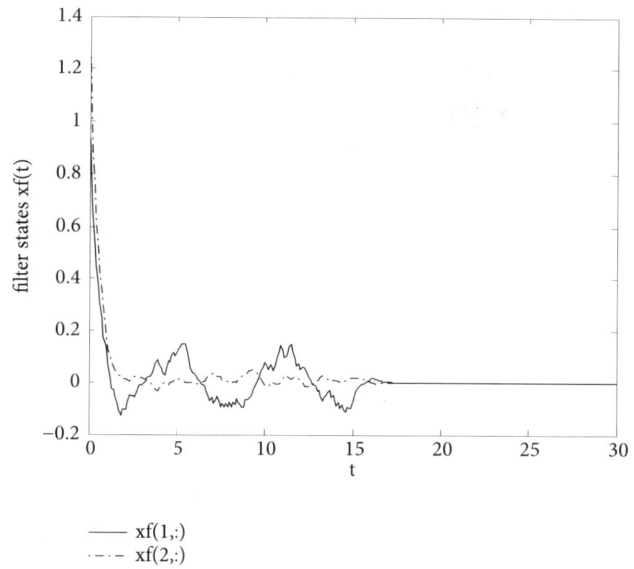

FIGURE 7: The response curve of system states x_{f1}, x_{f2}.

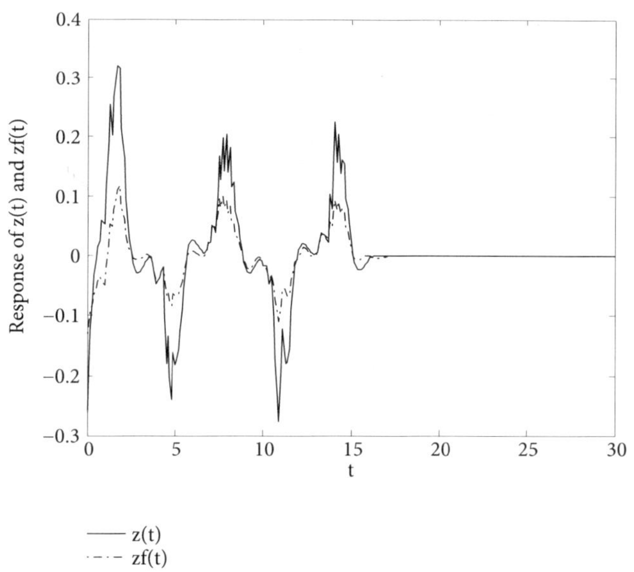

FIGURE 8: The response curve of vector $z(t)$, $z_f(t)$.

time-varying delays; meanwhile, the designing H_∞ filter with multiplicative gain variations was considered. By constructing a new Lyapunov-Krasovskii functional, we obtained a sufficient condition for designing the nonfragile H_∞ filter such that the filtering error system is asymptotically stable and satisfied the given H_∞ performance index. This nonfragile filter design method enhances the nonfragility of the filter and reduces some conservatism. A numerical example has shown the effectiveness of the proposed method. Future research includes event-triggered nonfragile H_∞ filter design for nonlinear system considering packet dropout and interval time-varying delays. Moreover, type 2 fuzzy filter design for nonlinear system with time-varying delays also can be further considered for the future investigation.

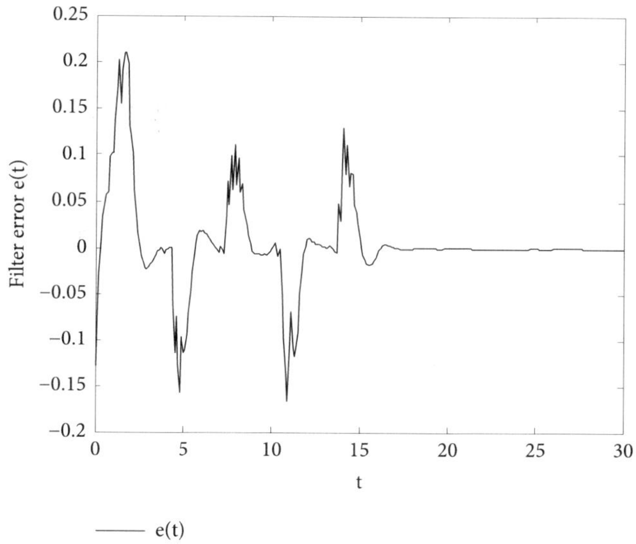

FIGURE 9: The response curve of vector $e(t)$.

Conflicts of Interest

The authors declare that there are no conflicts of interest regarding the publication of this paper.

Acknowledgments

This work was supported in part by Science and Technology project of State Grid corporation of China (SGTYHT/13-JS-175).

References

[1] T. Zhongda, L. Shujiang, W. Yanhong, and W. Xiangdong, "Scheduling method for networked control system with resource constraints based on fuzzy feedback priority and variable sampling period," *Transactions of the Institute of Measurement and Control*, vol. 40, no. 4, pp. 1136–1149, 2018.

[2] Z. Tian, S. Li, Y. Wang, and B. Gu, "Priority scheduling of networked control system based on fuzzy controller with self-tuning scale factor," *IAENG International Journal of Computer Science (IJCS)*, vol. 44, no. 3, pp. 308–315, 2017.

[3] Z. Tian, S. Li, and Y. Wang, "T-S fuzzy neural network predictive control for burning zone temperature in rotary kiln with improved hierarchical genetic algorithm," *International Journal of Modelling, Identification and Control*, vol. 25, no. 4, pp. 323–334, 2016.

[4] Z. Tian, S. Li, Y. Wang, X. Wang, and Q. Zhang, "The time-delay compensation method for networked control system based on improved fast implicit GPC," *International Journal of Control and Automation*, vol. 9, no. 1, pp. 231–240, 2016.

[5] T. Youssef, M. Chadli, H. R. Karimi, and R. Wang, "Actuator and sensor faults estimation based on proportional integral observer for TS fuzzy model," *Journal of The Franklin Institute*, vol. 354, no. 6, pp. 2524–2542, 2017.

[6] M. Wang, J. Qiu, M. Chadli, and M. Wang, "A Switched System Approach to Exponential Stabilization of Sampled-Data T–S Fuzzy Systems With Packet Dropouts," *IEEE Transactions on Cybernetics*, vol. 46, no. 12, pp. 3145–3156, 2016.

[7] J. An, G. Wen, and W. Xu, "Improved Results on Fuzzy Filter Design for T-S Fuzzy Systems," *Discrete Dynamics in Nature and Society*, vol. 21, 392915 pages, 2010.

[8] J. An, G. Wen, C. Lin, and R. Li, "New Results on a Delay-Derivative-Dependent Fuzzy H∞ Filter Design for T–S Fuzzy Systems," *IEEE Transactions on Fuzzy Systems*, vol. 19, no. 4, pp. 770–779, 2011.

[9] D. Zhang, W. Cai, L. Xie, and Q. Wang, "Non-fragile distributed filtering for T-S fuzzy systems in sensor networks," *IEEE Transactions on Fuzzy Systems*, 2014.

[10] H.-B. Zeng, J. H. Park, J.-W. Xia, and S.-P. Xiao, "Improved delay-dependent stability criteria for T-S fuzzy systems with time-varying delay," *Applied Mathematics and Computation*, vol. 235, pp. 492–501, 2014.

[11] C. Lin, Q. Wang, T. H. Lee, and B. Chen, "H_∞ filter design for nonlinear systems with time-delay through T-S fuzzy model approach," *IEEE Transactions on Fuzzy Systems*, vol. 16, no. 3, pp. 739–746, 2008.

[12] L. K. Wang, J. L. Peng, and X. D. Liu, "H∞ filtering design for continuous-time Takagi-Sugeno fuzzy model with immeasurable premise variables," *Neurocomputing*, vol. 173, pp. 2090–2096, 2016.

[13] Y. Su, Q. Zhou, and F. Gao, "A new delay-range-dependent H∞ filter design for T-S nonlinear systems," *Journal of The Franklin Institute*, vol. 351, no. 6, pp. 3305–3321, 2014.

[14] Y. Li, W. Xiao, J. Li, and L. Jiao, "H∞ filtering for discrete-time fuzzy stochastic neural networks with mixed time-delays," *Applied Mathematics and Computation*, vol. 52, no. 1-2, pp. 1–26, 2016.

[15] Y.-Q. Wu, H. Su, and Z.-G. Wu, "H∞ filtering for discrete fuzzy stochastic systems with randomly occurred sensor nonlinearities," *Signal Processing*, vol. 108, pp. 288–296, 2014.

[16] Z. Lu, G. Ran, G. Zhang, and F. Xu, "Event-Triggered H∞ Fuzzy Filtering for Networked Control Systems With Quantization and Delays," *IEEE Access*, vol. 6, pp. 20231–20241, 2018.

[17] Z. Li, Y. M. Chu, and S. Y. Xu, "Delay-dependent nonfragile robust H∞ filtering of T-S fuzzy time-delay systems," *Circuits, Systems and Signal Processing*, vol. 29, no. 3, pp. 361–375, 2010.

[18] X. Chang and G. Yang, "Non-fragile fuzzyHss filter design for nonlinear continuous-time systems with D stability constraints," *Signal Processing*, vol. 92, no. 2, pp. 575–586, 2012.

[19] D. Ding W, X. Li, Z. Shi, and X. Guo, "Nonfragile H∞ filtering for discrete-time T-S fuzzy systems," in *Proceedings of the Control Conference (CCC)*, vol. 201, pp. 3552–3557, 2012.

[20] X.-H. Chang and G.-H. Yang, "Nonfragile H∞ Filter Design for T–S Fuzzy Systems in Standard Form," *IEEE Transactions on Industrial Electronics*, vol. 61, no. 7, pp. 3448–3458, 2014.

[21] G.-H. Yang and W.-W. Che, "Non-fragile H∞ filter design with additive gain variations," in *Proceedings of the 45th IEEE Conference on Decision and Control (CDC '06)*, pp. 4775–4780, San Diego, Calif, USA, December 2006.

[22] S. Zhang, Z. Wang, D. Ding, H. Dong, F. E. Alsaadi, and T. Hayat, "Nonfragile H∞Fuzzy Filtering with Randomly Occurring Gain Variations and Channel Fadings," *IEEE Transactions on Fuzzy Systems*, vol. 24, no. 3, pp. 504–518, 2016.

[23] G. Yang and W. Che, "Non-fragile filter design for linear continuous-time systems," *Automatica*, vol. 44, no. 11, pp. 2849–2856, 2008.

[24] Z. Li, F. Gu, Y. He, and W. Hao, "Nonfragile Robust Filter Design for a Class of Fuzzy Stochastic Systems with Stochastic Input-to-State Stability," *Mathematical Problems in Engineering*, vol. 2015, Article ID 896574, 10 pages, 2015.

Permissions

All chapters in this book were first published in JCSE, by Hindawi Publishing Corporation; hereby published with permission under the Creative Commons Attribution License or equivalent. Every chapter published in this book has been scrutinized by our experts. Their significance has been extensively debated. The topics covered herein carry significant findings which will fuel the growth of the discipline. They may even be implemented as practical applications or may be referred to as a beginning point for another development.

The contributors of this book come from diverse backgrounds, making this book a truly international effort. This book will bring forth new frontiers with its revolutionizing research information and detailed analysis of the nascent developments around the world.

We would like to thank all the contributing authors for lending their expertise to make the book truly unique. They have played a crucial role in the development of this book. Without their invaluable contributions this book wouldn't have been possible. They have made vital efforts to compile up to date information on the varied aspects of this subject to make this book a valuable addition to the collection of many professionals and students.

This book was conceptualized with the vision of imparting up-to-date information and advanced data in this field. To ensure the same, a matchless editorial board was set up. Every individual on the board went through rigorous rounds of assessment to prove their worth. After which they invested a large part of their time researching and compiling the most relevant data for our readers.

The editorial board has been involved in producing this book since its inception. They have spent rigorous hours researching and exploring the diverse topics which have resulted in the successful publishing of this book. They have passed on their knowledge of decades through this book. To expedite this challenging task, the publisher supported the team at every step. A small team of assistant editors was also appointed to further simplify the editing procedure and attain best results for the readers.

Apart from the editorial board, the designing team has also invested a significant amount of their time in understanding the subject and creating the most relevant covers. They scrutinized every image to scout for the most suitable representation of the subject and create an appropriate cover for the book.

The publishing team has been an ardent support to the editorial, designing and production team. Their endless efforts to recruit the best for this project, has resulted in the accomplishment of this book. They are a veteran in the field of academics and their pool of knowledge is as vast as their experience in printing. Their expertise and guidance has proved useful at every step. Their uncompromising quality standards have made this book an exceptional effort. Their encouragement from time to time has been an inspiration for everyone.

The publisher and the editorial board hope that this book will prove to be a valuable piece of knowledge for researchers, students, practitioners and scholars across the globe.

List of Contributors

Han Peng, Chenglie Du and Zhouzhou Liu
School of Computer Science, Northwestern Polytechnical University, Xi'an, China

Lei Rao
School of Software and Microelectronics, Northwestern Polytechnical University, Xi'an, China

Zhen Zhao, Jinfeng Gao and Tingting Zhang
Faculty of Mechanical Engineering and Automation, Zhejiang Sci-Tech University, Hangzhou, Zhejiang 310018, China

Shiyun Xu, Xiaojie Chu and Long Peng
China Electric Power Research Institute, Beijing 100192, China

Lei Gao
China Electric Power Research Institute, Beijing 100192, China
Northeast Electric Power University, Jilin 132012, Jilin Province, China

Yonghui Nie
Northeast Electric Power University, Jilin 132012, Jilin Province, China

Yanchao Ma
State Grid Liaoning Power Co., Ltd., Yingkou Power Supply Company, Yingkou 115000, China

Xiongfeng Deng, Xiuxia Sun and Shuguang Liu
Equipment Management and Unmanned Aerial Vehicle Engineering College, Air Force Engineering University, Xi'an 710051, China

Ri Liu
Theory Training Department, Air Force Harbin Flight Academy, Harbin 150001, China

Sameer Qasim Hasan and Moataz Abbas Holel
Department of Mathematics, University of Mustansiriyah, Baghdad, Iraq

Godpromesse Kenné
Unitéde Recherche d'Automatique et d'Informatique Appliquée (LAIA), Département de Génié Electrique, IUT FOTSO Victor Bandjoun, Universitéde Dschang, BP 134, Bandjoun, Cameroon

RostandMarc Douanla and Armel Simo Fotso
Unitéde Recherche d'Automatique et d'Informatique Appliquée (LAIA), Département de Génié Electrique, IUT FOTSO Victor Bandjoun, Universitéde Dschang, BP 134, Bandjoun, Cameroon
Unité de Recherche de Matière Condensée, d'Electronique et de Traitement du Signal (LAMACETS), Département de Physique, Universit´e de Dschang, BP 69, Dschang, Cameroon

François Béceau Pelap
Unitéde Recherche de Mécanique et de Modélisation des Systèmes Physiques (L2MSP), Département de Physique, Université de Dschang, BP 69, Dschang, Cameroon

Guangtao Ran, Guoliang Zhang and Fengxia Xu
School of Computer and Control Engineering, Qiqihar University, Qiqihar 161006, China

Zhongda Lu
School of Computer and Control Engineering, Qiqihar University, Qiqihar 161006, China
College of Computer Science and Technology, Harbin University of Science and Technology, Harbin 150080, China

Yan Zhang
Faculty Development and Educational Technology Center, Guangdong University of Finance and Economics, Guangzhou 510320, China
School of Electric Power, South China University of Technology, Guangzhou 510640, China
Foshan Institute of Modern Services, Guangdong University of Finance and Economics, Foshan 528100, China

Yongqiang Liu
School of Electric Power, South China University of Technology, Guangzhou 510640, China

Xiaoli Chu
Center of Network, Guangdong AIB Polytechnic College 510507, China

Lin Cao
School of Business and Administration, Qilu University of Technology, Jinan 250353, China

Xiang Liu and Xiaogeng Liang
School of Automation, Northwestern Polytechnical University, Xi'an 710072, China

M. E. U. J. Araújo
Department of Informatics, Federal Institute of Paraíba, N-720, 58015-435 Jo˜ao Pessoa, PB, Brazil
Department of Automation and Computing, CT, Federal University of Rio Grande do Norte, 59078-970 Natal, RN, Brazil

J. R. T. Gadelha, W. M. Santos, A. L. Maitelli and F. M. U. Araújo
Department of Automation and Computing, CT, Federal University of Rio Grande do Norte, 59078-970 Natal, RN, Brazil

Yanxia Shen, Beibei Miao, Dinghui Wu and Kader Ali Ibrahim
Engineering Research Center of Internet of Things Technology Applications of Ministry of Education, Jiangnan University, Wuxi 214122, China

Thiago P. Chagas
Universidade Estadual de Santa Cruz (UESC), 45662-900 Ilhéus, BA, Brazil

Pierre-Alexandre Bliman
Lab. J.-L. Lions UMR CNRS, Inria, UPMC University Paris 06, Sorbonne Universit´es, 7598 Paris, France
Fundacao Getulio Vargas, Rio de Janeiro, RJ, Brazil

Karl H. Kienitz
Instituto Tecnol´ogico de Aeronáutica (ITA), 12228-900 São Josédos Campos, SP, Brazil

Bilal Djamal Eddine Cherif and Azeddine Bendiabdellah
Diagnostic Group, Laboratory LDEE, Electrical Engineering Faculty, University of Sciences and Technology of Oran, Bir El Djir, Algeria

Wei Gao
School of Electrical Engineering, Wuhu Vocational Institute of Technology,Wuhu, China

Xiuping Chen and Haibo Du
School of Electrical Engineering and Automation, Hefei University of Technology, Hefei, China

Song Bai
NARI-TECH Control Systems Ltd., Nanjin, China

Jiahao Li, Tingting Zhang, Jinfeng Gao and Ping Wu
Faculty of Mechanical Engineering and Automation, Zhejiang Sci-Tech University, Hangzhou 310018, China

Shi-Qiang Wang and Jian-Chun Xing
College of Defense Engineering, Army Engineering University of PLA, Nanjing, Jiangsu 210001, China

Zi-Yan Jiang and Yun-Chuang Dai
Building Energy Research Center of Tsinghua University, Beijing 100084, China

Wenshao Bu
Electrical Engineering College, Henan University of Science and Technology, Luoyang, 471023, China

Biao Li and Chunxiao Lu
Information Engineering College, Henan University of Science and Technology, Luoyang, 471023, China

Ruiqing Mao
Xuzhou Institute of Technology, Xuzhou, Jiangsu 221111, China

Xiliang Ma
Xuzhou Institute of Technology, Xuzhou, Jiangsu 221111, China
School of Mechanical and Electrical Engineering, China University of Mining and Technology, Xuzhou, Jiangsu 221116, China

Guoliang Zhang and Fengxia Xu
College of Mechatronic Engineering, Qiqihar University, Qiqihar 161006, China

Zhongda Lu
College of Mechatronic Engineering, Qiqihar University, Qiqihar 161006, China
College of Computer Science and Technology, Harbin University of Science and Technology, Harbin 150080, China

Yi Sun, Jie Sun and Fangming Jin
State Grid Heilongjiang Electric Power Company Qiqihar Power Supply Company, Qiqihar 161006, China

Index